AF588323

Graduate Texts in Mathematics

Volume 303

Graduate Texts in Mathematics bridge the gap between passive study and creative understanding, offering graduate-level introductions to advanced topics in mathematics. The volumes are carefully written as teaching aids and highlight characteristic features of the theory. Although these books are frequently used as textbooks in graduate courses, they are also suitable for individual study.

Claus Scheiderer

A Course in Real Algebraic Geometry

Positivity and Sums of Squares

Claus Scheiderer
Fachbereich Mathematik und Statistik
Konstanz University
Konstanz, Germany

ISSN 0072-5285 ISSN 2197-5612 (electronic)
Graduate Texts in Mathematics
ISBN 978-3-031-69212-3 ISBN 978-3-031-69213-0 (eBook)
https://doi.org/10.1007/978-3-031-69213-0

Mathematics Subject Classification (2020): 14Pxx, 06F25, 12D15, 13J25, 13J30, 90C22

This work was supported by Deutsche Forschungsgemeinschaft (DFG SCHE 281/10-2).

This Springer imprint is published by the registered company Springer Nature Switzerland AG
The registered company address is: Gewerbestrasse 11, 6330 Cham, Switzerland

Für ff

Preface

This textbook originates from a course for graduate students that I taught at Konstanz University approximately five or six times over the past twenty years. While the first part of the course remained largely the same in each iteration, the contents of the second half were updated each time to reflect recent developments in research. The book includes more than 350 exercises of varying difficulty, the majority of which have been "tested" in class.

I am deeply grateful to my students, past and present, for numerous suggestions, corrections and improvements that were incorporated over the years. In particular, I would like to mention Mario Bauer, Tim Kobert, Mario Kummer, Aaron Kunert, Thorsten Mayer, Tim Netzer, Daniel Plaumann, Rainer Sinn and Julian Vill. Some of them have their own students by now. I also profited from substantial help and advice from Gennadiy Averkov, Vicki Powers, Mihai Putinar, Konrad Schmüdgen, Marcus Tressl and Thomas Unger. Moreover, the students in my present class, notably Tobias Retzlaff, found every typo in the draft version of the manuscript. (But, as the reader will observe, I secretly introduced a few new ones.) My heartfelt thanks go out to each and every one of them.

Finally, I wish to express my gratitude to Remi Lodh from Springer Nature, for his endorsement of this project and his friendly and professional assistance.

Konstanz, June 2024 *Claus Scheiderer*

Contents

Introduction

Algebraic geometry is the study of the (complex) solutions of systems of polynomial equations

$$f_i(x_1, \dots, x_n) = 0, \quad i = 1, \dots, m,$$

both from a geometric perspective and from an algebraic point of view. In real algebraic geometry one is interested in such systems with real coefficients, and in their real solutions. Their study is harder in general, because the field of real numbers lacks the closedness property of the complex numbers. As a consequence, real algebraic geometry requires new algebraic tools. And in addition to equations $f(x) = 0$ and non-equations $f(x) \neq 0$, one is immediately led to consider inequalities $f(x) \geq 0$ or $f(x) > 0$ as well.

To the traditional technical machinery of commutative algebra, a distinct apparatus of real algebra has to be added: Ordered fields, preorderings and quadratic modules, concepts from convex geometry, to name just a few. The Tarski–Seidenberg projection theorem, together with its model-theoretic formulation, plays a significantly more important role in real algebraic geometry than the analogous Lefschetz principle does in usual algebraic geometry. Important applications require specific technical tools of their own. For example, linear matrix inequalities and spectrahedra are concepts of a genuinely real algebraic nature, and are fundamental notions in semidefinite programming.

The algebraic foundations of real algebraic geometry were established early, and the exact timing can be pinpointed. In connection with the solution of Hilbert's 17th problem, Artin and Schreier introduced ordered fields and real closed fields in 1927. In particular, they proved that every ordered field has a real closure that is unique in a strong sense.

Fundamental contributions, that would later be attributed to the field of real algebraic geometry, occurred even before Artin and Schreier. Consider, for example, the foundational work on real root counting for polynomials, by Sturm, Sylvester, Hermite and others. Another significant milestone was Hilbert's investigation into the representation of non-negative real polynomials as sums of squares. As is well known, Hilbert resolved the question completely in 1888. The methods that he used extended far beyond his time and, in fact, continue to be influential more than one hundred years later. In the context of rational functions however, Hilbert faced greater difficulties. Unable to decide the question in general, he ultimately included it as number seventeen in his famous 1900 list of unsolved mathematical problems. Artin's solution in 1927 would become one of the most important catalysts for the development of real algebraic geometry.

Bearing in mind that Tarski found his decision method for semialgebraic sets only a few years later, the table was set in the 1930s for a strong advancement of real algebraic geometry. But apparently the time was not yet ripe, and these beautiful tools were to lie dormant for almost half a century. Indeed, an interest in the systematic development of real algebra and geometry began to awaken only in the 1970s and 1980s.

One of the indicators of the beginning of a rise in interest was the discovery of the real spectrum by Coste and Roy around 1979. Brumfiel's book [35] appeared at around the same time. It explored notions of partially ordered rings and related them to semialgebraic sets. The publication of *Géométrie algébrique réelle* [25] by Bochnak, Coste and Roy in 1987 was a milestone event. This book was the first comprehensive monograph ever in the area. Inspiring as it was, it had a lasting and unifying effect on the growing community. Strong impulses were added in the early 1990s through Schmüdgen's spectacular positivstellensatz, together with consequences that were quickly starting to be built upon it. It was soon realized that these results offered a great potential for applications in optimization. Around the year 2000, Lasserre and Parrilo invented the moment relaxation method in polynomial optimization. Under conditions of a very general nature, the method offers a systematic approximate solution of polynomial optimization problems in polynomial time. The essential theoretical backbone is formed by the modern *positivstellensätze* from the early 1990s. Meanwhile, this approach is one of the most important applications of real algebraic geometry, and continues to be in the focus of intense active research.

Given this development, the current selection of available textbooks in real algebraic geometry is surprisingly small. For many years, the book by Bochnak, Coste and Roy—after the original French edition from 1987, a considerably enlarged English edition [26] appeared in 1998—was the only available comprehensive source. The 2001 monograph [159] by Prestel and Delzell addresses Archimedean positivstellensätze in great detail, and the same is true of Marshall's book [138] from 2008. Theobald's recent book [208] emphasizes the applications in optimization.

The book in your hands contains material roughly for a one year graduate course. Starting with the very first concepts of real algebra, it takes the reader to areas of active current research. The first half offers a relatively broad introduction to the basics of real algebraic geometry. It also contains Hilbert's pioneering work from 1888 and Artin's solution to the 17th problem. In the second half, starting with Chapter 5, the style starts to become slightly more demanding on the reader, and then increasingly so in the remaining chapters. The main focus in this second part is on modern positivstellensätze, and on their use in polynomial optimization.

We now give a more detailed overview. Chapter 1 introduces orderings of fields and their real closures, and proves existence and uniqueness of a real closure for every ordered field. Several methods and criteria for real root counting of univariate polynomials are discussed, and are used to prove the Tarski–Seidenberg projection theorem, one of the most important general tools in real algebraic geometry. In order to gain a greater flexibility, we introduce a bit of model-theoretic language and formulate the quantifier elimination version of this theorem as well. Having achieved this, Tarski's transfer principle is our main tool for the solution of Hilbert's 17th problem. We also use Tarski to prove the Artin–Lang theorem, which characterizes real algebraic varieties whose function field can be ordered, by their locus of real points.

In Chapter 2 we begin the study of sums of squares of polynomials. Important techniques like Gram matrices and Newton polytopes are introduced, and the Fejér–Riesz theorem is proved, featuring sums of squares on the circle. Central to the chapter are Hilbert's 1888 theorems on sums of squares representations of non-negative polynomials. Hilbert's results are proved almost in their entirety, the only exception being the case of ternary quartics. Using a proof that is tricky but elementary, we only show a slightly weaker version here (four squares instead of three). But see Chapter 7 below.

Chapter 3 introduces the real spectrum as a technical tool of prime importance. By its conceptual simplicity, the real spectrum often allows the crucial point of a problem to be pinned down in a very precise way. Orderings and preorderings are generalized from fields to arbitrary rings, and a general abstract *stellensatz* is proved almost with no effort. To deduce geometric (semialgebraic) versions (the Krivine–Stengle theorem), we again use Tarski's principle. The constructible topology on the real spectrum is introduced as an important auxiliary tool, and specializations in the real spectrum are related to valuation rings and to convex subrings of ordered fields.

Chapter 4 offers an introduction to the geometry of semialgebraic sets. From the beginning we relate semialgebraic sets to the real spectrum. The proof of the finiteness theorem is just one among several examples where we hope to convince the reader of the usefulness of the real spectrum. Cylindrical algebraic decomposition and Thom's lemma are presented, however in basic versions only. Much more elaborate formulations are possible, at the cost of a greater technical effort. Semialgebraic paths are introduced as a useful and intuitive general device. The chapter ends with a discussion of the dimension of semialgebraic sets.

Starting with Chapter 5 we begin to enter more advanced areas. The central notion in the chapter is the Archimedean property of semirings or modules, the single most important result is the Archimedean positivstellensatz. We first follow the traditional approach to this theorem, and then present a series of important consequences. Most significant among them is Schmüdgen's positivstellensatz, which originally was proven in quite a different way using operator theory. At the end of the chapter we offer an optional second path to the earlier positivstellensätze. It is based on pure states for convex cones in $\mathbb{R}$-algebras and uses concepts from locally convex vector spaces. Moreover, this approach quickly leads to the Archimedean local-global principle, which is a vital tool for the next chapter.

While the results of Chapter 5 apply to polynomials that are strictly positive on the domain of interest, we allow the polynomials to have zeros in Chapter 6. First a general negative result is proved regarding sum of squares representations, that applies in all dimensions ≥ 3. Then we re-prove the Archimedean local-global principle, this time avoiding pure states and using the real spectrum. According to this theorem, and in an Archimedean situation, the only obstructions against the existence of a sum of squares representation are of a local nature. Therefore we pursue a closer study of positivity versus sums of squares in local rings (Sections 6.3 and 6.4). To some extent it is even possible here to replace the local rings by their completions. A series of applications is then presented in Section 6.5. As one of several main results, we mention the fact that non-negative polynomials are sums of squares on every compact non-singular real surface. The chapter ends with a discussion of the (non-) existence of degree bounds in weighted sums of squares representations.

Chapter 7 picks up the discussion of Hilbert's theorems from Chapter 2. Generalizing the viewpoint from projective space to arbitrary projective varieties X, we ask when it is true that every non-negative form of degree $2d$ on X is a sum of squares of forms of degree d. Restricting to quadratic forms (the case $2d = 2$) is not a serious limitation. Under this assumption, the main theorem asserts that the answer is positive if and only if X is a variety of minimal degree, in the sense of classical algebraic geometry. Remarkably, this result contains Hilbert's theorems as a particular case, and this is even true for the quantitative refinement that we prove. In particular, the full Hilbert theorem on positive ternary quartics is contained in this theorem.

The last chapter addresses the importance of sum of squares representations for semidefinite programming, and in particular, for polynomial optimization. We start with an overview of basic concepts in (finite-dimensional) convexity theory. Then linear matrix inequalities and spectrahedra are introduced, the main players in semidefinite programming. A quick introduction to conic programming in general and semidefinite programming in particular follows, before we give a detailed account of the moment relaxation method in polynomial optimization. This approach relies crucially on the positivstellensätze studied in Chapters 5 and 6. The rest of the chapter investigates the expressive power of semidefinite programming. Taking up Nemirovski's question whether every convex semialgebraic set is a linear image of a spectrahedron (a so-called spectrahedral shadow), we first present theorems by Helton and Nie that give a positive answer under very general assumptions. Then we prove that spectrahedral shadows are characterized by the existence of suitably uniform sum of squares decompositions for non-negative linear forms. Using this result we are able to give many (prominent) examples of convex sets that fail to be spectrahedral shadows.

There are two appendices. With only few exceptions, they contain no proofs and have only very few motivational comments. In Appendix A we provide a quick access to notations, definitions and basic facts from commutative algebra and basic algebraic geometry, as far as they are used in the main text. Usually, this appendix will be needed for reference purposes only. Appendix B gives background for the pure states approach in Sections 5.6–5.7. It contains the Hahn–Banach and

Krein–Milman theorems in locally convex vector spaces, together with the Eidelheit–Kakutani separation theorem for vector spaces of arbitrary dimension without topology.

Throughout the text, a large number of exercises have been provided. Each section has a few of them. Some are just straightforward illustrations of concepts from the main text. Others have more substance, and a small number may be considered somewhat demanding. Most should however be doable without serious difficulties.

For a two-semester course based on this book, the first semester might roughly cover Chapters 1 to 4. The second semester should continue with Chapter 5, at least up to and including Section 5.5. After this point, several options are possible. Sections 5.6 and 5.7 offer an *ad libitum* alternative approach to the Archimedean positivstellensatz, and may be skipped if preferred. The remaining three chapters can essentially be arranged in any order. Chapter 7 is largely independent of Chapters 3 to 6, and there are in fact some obvious arguments for inserting Chapter 7 right after Chapter 2. Our reason for postponing this material was the desire to advance the basics more quickly, and also the fact that Chapter 7 requires considerably more background in algebraic geometry than the rest of the course. Chapter 8 discusses the role of positivstellensätze and sums of squares in polynomial optimization. It is independent of Chapter 7, and also largely of Chapter 6.

Up to and including Chapter 5, the background required from the reader is very modest. Apart from basic abstract algebra and elementary point set topology, only basic language and concepts from commutative algebra and algebraic geometry are needed. It is only in Chapters 6 and 7 that a somewhat more advanced background is expected. In Chapter 6 this mainly concerns regular local rings and their completions. In Chapter 7, classical theorems in projective geometry (Bézout, Bertini and others) are needed. We believe however that these results may be used as a black box, without a serious loss for the main understanding.

Finally, we have to mention sins of omission, of which there are many. As the author of this course, I am well aware that its selection of topics is based on personal preferences. It is evident that this book cannot nearly reflect all important aspects of present research in real algebraic geometry. The *leitmotif* for the entire course have been sums of squares, together with the quest for understanding how far they deviate from capturing all positivity. In this sense, Hilbert's work from 1888 and his 17th problem are our core initial motivation.

Next to the 17th problem there is Hilbert's 16th problem, which obviously had a strong impact on the development of real algebraic geometry as well. Its first part asked for the configuration of the real locus of real algebraic curves in the plane, or of surfaces in three-space. These questions, together with far-reaching generalizations, have been studied intensely since the 1980s (at least). Obviously, the study of the geometry and topology of real projective varieties is an essential part of real algebraic geometry, but it is not represented in this course. For an introduction to these questions one may consult Mangolte's book [132]. Other areas that belong to real algebraic geometry, or have a strong interaction with the latter, but are missing here are real analytic and semianalytic geometry, real tropical geometry and toric varieties, non-commutative real algebraic geometry, or also (reduced) quadratic form theory, to mention just a few. The material in Chapters 5 and 6 has substantial applications to the study of moment problems, but unfortunately it was not possible to include them in this course, for time and space limitations. Some of these applications can be found in Schmüdgen's volume [191]. Similarly, a more systematic study of complexity questions would have been desirable in many parts of this text, but would have gone beyond the scope of this course. Fortunately, the volume by Basu, Pollack and Roy [13] addresses this matter in great detail.

General Conventions: The natural numbers are denoted $\mathbb{N} = \{1, 2, 3, \dots\}$. By $\mathbb{Z}$ we denote the ring of integers, and $\mathbb{Z}_+ = \mathbb{N} \cup \{0\} = \{0, 1, 2, \dots\}$ is the set of non-negative integers. The fields of rational, real and complex numbers are $\mathbb{Q}$, $\mathbb{R}$ and $\mathbb{C}$, respectively. For $n \in \mathbb{N}$ we put $[n] := \{1, \dots, n\}$. Given a real number a, write $\lfloor a \rfloor$ for the largest integer m with $m \le a$ and $\lceil a \rceil$ for the smallest integer n with $a \le n$. The cardinality of a finite set M is written $|M|$. The union of pairwise disjoint sets $M_1, \dots, M_r$ may be written $M_1 \uplus \cdots \uplus M_r$.

If k is a commutative ring, vectors in k^n are considered as column vectors by default. The set of matrices of size $m \times n$ over k is $\mathrm{M}_{m \times n}(k)$, or briefly $\mathrm{M}_n(k)$ if $m = n$. The matrix transpose of $A \in \mathrm{M}_{m \times n}(k)$ is denoted $A^\top \in \mathrm{M}_{n \times m}(k)$. If V is a vector space over a field k and $M \subseteq V$ is a subset, we write $\mathrm{span}(M)$ (or $\mathrm{span}(v_1, \dots, v_r)$ if $M = \{v_1, \dots, v_r\}$) for the linear subspace of V that is spanned by M. The i-th vector in the canonical basis of k^n is denoted $e_i = (\delta_{ij})_{1 \le j \le n}$. The dual linear space of V is written $V^\vee = \mathrm{Hom}_k(V, k)$. Note that many textbooks write V^* instead.

If $f \colon U \to \mathbb{R}$ is a C^2-function defined on an open set $U \subseteq \mathbb{R}^n$, the notation $\nabla f(u) \in \mathbb{R}^n$ and $D^2 f(u) \in \mathrm{M}_n(\mathbb{R})$ is used for the gradient and the Hessian of f, respectively, at a point $u \in U$. Of course, when $n = 1$, we also write $\nabla f(u) = f'(u)$ and $D^2 f(u) = f''(u)$.

See Appendix A for general notation and background in algebra, algebraic geometry and topology. We point out that all rings are tacitly assumed to be commutative and to have a unit, unless explicitly stated otherwise. We refer in particular to Section A.6, where our general conventions from algebraic geometry are summarized.

The most basic concepts from convexity, like convex set, convex cone, convex or conic hull of a set, or polyhedra and polytopes, are used throughout the book without explanation. Their meaning is also recalled at the beginning of Section 8.1. Dual convex cones and convex cone duality (in finite-dimensional real vector spaces) are slightly more advanced, but do not appear before Chapter 7, and are again explained in Section 8.1.

Chapter 1
Ordered Fields

We introduce the key players of real algebra, which are ordered fields and their real closures. Real closed fields are characterized as those fields that have the same algebraic properties as the field $\mathbb{R}$ of real numbers. After discussing several approaches to counting the real roots of real univariate polynomials, we state and prove the Tarski–Seidenberg projection theorem. This is one of the most important general results in real algebraic geometry. To gain greater flexibility in applying the theorem, we borrow from model theory and introduce the formal language of ordered fields. We then use Tarski–Seidenberg to present Artin's solution to Hilbert's 17th problem. Finally we show for algebraic $\mathbb{R}$-varieties how the existence of an ordering of the function field is reflected in the real points of the variety (Artin–Lang theorem).

1.1 Orderings of Fields

1.1.1 A *partially ordered set* is a pair $(M, \le)$ consisting of a set M and a binary relation $\le$ on M that is reflexive, anti-symmetric and transitive (i.e. that satisfies $x \le x$, $x \le y \wedge y \le x \Rightarrow x = y$ and $x \le y \wedge y \le z \Rightarrow x \le z$ for all $x, y, z \in M$). The relation $\le$ is also called a *(partial) order relation* on M. It is a *total order relation*, and $(M, \le)$ is a *totally ordered set*, if in addition $x \le y \vee y \le x$ holds for any $x, y \in M$. Given a partially ordered set $(M, \le)$, one extends the symbol $\le$ in the natural way by defining

- $a < b$ if and only if $a \le b$ and $a \ne b$,
- $a \ge b$ if and only if $b \le a$,
- $a > b$ if and only if $b < a$

for $a, b \in M$. Intervals in M are denoted by

$$\begin{aligned} [a, b] &= \{x \in M \colon a \le x \le b\}, \\ [a, b[&= \{x \in M \colon a \le x < b\}, \\ ]a, b] &= \{x \in M \colon a < x \le b\}, \\ ]a, b[&= \{x \in M \colon a < x < b\}, \end{aligned}$$

C. Scheiderer, *A Course in Real Algebraic Geometry*, Graduate Texts in Mathematics 303,
https://doi.org/10.1007/978-3-031-69213-0_1

or also by $[a, b]_\leq$ if the order relation $\leq$ is to be emphasized. We also allow a or b to be $\pm\infty$, where $\infty = +\infty$ and $-\infty$ are two extra symbols that satisfy $-\infty < x < +\infty$ for all $x \in M$. So, for example, $[a, \infty[= \{x \in M\colon x \geq a\}$ etc.

The concept of orderings of a field was introduced by Artin and Schreier in 1927:

1.1.2 Definition. Let K be a field.

(a) An *ordering* of K is a total order relation $\leq$ on the set K that satisfies

  (1) $a \leq b \Rightarrow a + c \leq b + c$,
  (2) $a \leq b, c \geq 0 \Rightarrow ac \leq bc$

  for all $a, b, c \in K$.
(b) An *ordered field* is a pair $(K, \leq)$ where K is a field and $\leq$ is an ordering of K.
(c) The field K is said to be *real* if it admits at least one ordering.

An ordering of the field K is therefore a total ordering of the set K that is compatible with addition and multiplication, in the same way as we are used to for real numbers.

1.1.3 Lemma. *Let $(K, \leq)$ be an ordered field and let $a \in K$. Then $a \geq 0$ is equivalent to $-a \leq 0$, and also to $\frac{1}{a} \geq 0$ if $a \neq 0$. We always have $a^2 \geq 0$.*

Proof. $a \geq 0$ implies $0 \geq -a$ by adding $-a$ on both sides, and similarly vice versa. Since one of $a \geq 0$ or $-a \geq 0$ holds we have $a^2 = (-a)^2 \geq 0$. If $a \neq 0$ then $a = a^2 \cdot \frac{1}{a}$ and $\frac{1}{a} = (\frac{1}{a})^2 \cdot a$, from which one sees $a > 0 \Leftrightarrow \frac{1}{a} > 0$. □

Alternatively, an ordering of a field may be described by its non-negative elements:

1.1.4 Proposition. *Let K be a field and let $\leq$ be an ordering of K. Then the set $P = P_\leq := \{a \in K\colon a \geq 0\}$ of non-negative elements satisfies*

  (a) $P + P \subseteq P$, $PP \subseteq P$,
  (b) $P \cup (-P) = K$,
  (c) $P \cap (-P) = \{0\}$,
  (c') $-1 \notin P$.

Conversely, if a subset $P \subseteq K$ satisfies (a) and (b) together with (c) or (c'), then $a \leq_P b :\Leftrightarrow b - a \in P$ (for $a, b \in K$) defines an ordering $\leq_P$ of K. Any subset P of K that satisfies (a), (b) and (c) (or (c')) is called a positive cone *of K.*

Proof. Here, of course, $P + P := \{a + b\colon a, b \in P\}$ and $PP := \{ab\colon a, b \in P\}$. First let $\leq$ be an ordering of K. Then $P = \{a \in K\colon a \geq 0\}$ clearly satisfies (a) and (c), and (b) follows from 1.1.3. Moreover $1 = 1^2 > 0$, again by 1.1.3, and so $-1 < 0$.

Conversely let $P \subseteq K$ satisfy (a)–(c), and let $\leq_P$ be defined as above. Then $\leq_P$ is a total ordering on K, and compatibility with addition (property (1)) is clear. If $a \leq_P b$ and $c \geq_P 0$ then $b - a \in P$ and $c \in P$, so $(b - a)c \in P$, and therefore $ac \leq_P bc$. So $\leq_P$ is an ordering of K. On the other hand, (c) is also a consequence of (a), (b) and (c'): If there was an element $0 \neq a \in P \cap (-P)$, we would have $-a^2 = a(-a) \in P$ by (a), but also $\frac{1}{a^2} = (\pm\frac{1}{a})^2 \in P$ since $\pm\frac{1}{a} \in P$ for one choice of sign $\pm$. Together this would give $-1 = (-a^2) \cdot \frac{1}{a^2} \in P$, contradicting (c'). □

1.1.5 Remark. By Proposition 1.1.4 there is a natural bijective correspondence between the orderings $\le$ of K and the positive cones P of K. We will occasionally be sloppy and use the term "ordering" invariantly for both $\le$ and P, while tacitly translating between both concepts depending on the situation. In particular, we also refer to the pair (K, P) as an ordered field.

1.1.6 Definition and Lemma. *Let $(K, \le)$ be an ordered field.*

(a) *The* sign *of $a \in K$ with respect to $\le$, denoted* $\operatorname{sign}_{\le}(a)$ *or* $\operatorname{sign}(a)$, *is* 1, 0 *or* -1, *depending on whether $a > 0$, $a = 0$ or $a < 0$, respectively. Note that* $\operatorname{sign}(ab) = \operatorname{sign}(a) \cdot \operatorname{sign}(b)$ *for a, $b \in K$.*
(b) *The* absolute value *of $a \in K$ with respect to $\le$ is*

$$|a|_{\le} \;=\; |a| \;:=\; \operatorname{sign}(a) \cdot a.$$

The absolute value satisfies $|a| = |-a| \ge 0$, $|ab| = |a| \cdot |b|$ and $|a + b| \le |a| + |b|$ (triangle inequality) for a, $b \in K$.

If $P \subseteq K$ is the positive cone corresponding to $\le$, we also use the alternative notation $\operatorname{sign}_P(a)$ and $|a|_P$, respectively.

Proof. The triangle inequality is an equality if $ab \ge 0$. Otherwise we have $|a + b| < \max\{|a|, |b|\} < |a| + |b|$. □

1.1.7 Notation. For any ring A (commutative and unital) we denote by

$$\Sigma A^2 \;:=\; \{a_1^2 + \cdots + a_n^2 \colon n \ge 1,\ a_i \in A\}$$

the set of all *sums of squares* (often abbreviated *sos*) in A. It has become very common to use the acronym "sos" as an adjective as well, for being a sum of squares. So we'll frequently say that a ring element $a \in A$ is *sos* (in A) if $a \in \Sigma A^2$.

1.1.8 Corollary. *Let K be a field. Any positive cone P of K satisfies $\Sigma K^2 \subseteq P$. In particular, if K is real then $-1 \notin \Sigma K^2$, and so* $\operatorname{char}(K) = 0$.

Proof. Both assertions follow directly from 1.1.3 and 1.1.4. □

1.1.9 Remarks.

1. Let $\mathbb{R}$ be the field of real numbers, let $\le$ be the usual order relation on $\mathbb{R}$. Then $(\mathbb{R}, \le)$ is an ordered field. In particular, the field of real numbers is real.

2. An algebraically closed field, like the field $\mathbb{C}$ of complex numbers, is never real, since -1 is a square. For p a prime number, the field $\mathbb{Q}_p$ of p-adic numbers is not real since -1 is a sum of (four) squares in $\mathbb{Q}_p$. (Ignore this remark if you are not familiar with $\mathbb{Q}_p$.)

3. If K_0 is a subfield of the field K and P is a positive cone of K, then $P_0 := K_0 \cap P$ is a positive cone of K_0. We say that the ordering $\leq_P$ of K *extends* the ordering $\leq_{P_0}$ of K_0, or that $(K_0, P_0) \subseteq (K, P)$ is an *extension of ordered fields*. This means that the order relation $\leq_{P_0}$ on K_0 is the restriction of the order relation $\leq_P$ on K. In particular, any subfield of a real field is itself real, for example $\mathbb{Q}$, $\mathbb{Q}(\sqrt{2})$, $\mathbb{Q}(\pi)$ etc, which are subfields of $\mathbb{R}$.

4. If two positive cones P, Q of K satisfy $P \subseteq Q$, then $P = Q$ (prove this). Therefore, if ΣK^2 happens to be an ordering of K, it is the only ordering of K. For example, this is true for $K = \mathbb{R}$ or $K = \mathbb{Q}$.

5. In general, a real field will have more than one ordering. For example, consider the field $K = \mathbb{Q}(\sqrt{2}) = \mathbb{Q}(\alpha)$ where $\alpha^2 = 2$. Then K admits two embeddings $\varphi_1, \varphi_2 \colon K \to \mathbb{R}$, characterized by $\varphi_1(\alpha) = \sqrt{2} > 0$ and $\varphi_2(\alpha) = -\sqrt{2} < 0$. Pulling back the ordering of $\mathbb{R}$ via these embeddings one gets two different orderings $\leq_1, \leq_2$ of K, satisfying $\alpha >_1 0$ and $\alpha <_2 0$.

1.1.10 For a more interesting example, let us study the orderings of the rational function field $\mathbb{R}(t)$ in one variable t. Given an ordering $\leq$ of $\mathbb{R}(t)$, we locate the element t with respect to the real line $\mathbb{R}$. So we consider the subsets

$$I := I_{\leq} := \{a \in \mathbb{R} \colon a < t\}, \quad J := J_{\leq} := \{a \in \mathbb{R} \colon a > t\}$$

of $\mathbb{R}$. Clearly $I \cap J = \varnothing$, $I \cup J = \mathbb{R}$ and $I < J$ hold (the latter meaning that $a < b$ for any $a \in I$, $b \in J$). Hence the pair (I, J) is a Dedekind cut of $\mathbb{R}$, according to the next definition:

1.1.11 Definition. Let $(M, \leq)$ be a totally ordered set. A *Dedekind cut* of $(M, \leq)$ is a pair (I, J) of subsets of M such that $I \cup J = M$ and $I < J$ (i.e. $a < b$ for any $a \in I$, $b \in J$).

Loosely speaking, a Dedekind cut of $(M, \leq)$ is a disjoint decomposition of M into a lower and an upper set.

1.1.12 Let $(M, \leq)$ be a totally ordered set. For any $\xi \in M$, both

$$\xi_- := \big(\,]-\infty, \xi[\,,\ [\xi, \infty[\,\big) \text{ and } \xi_+ := \big(\,]-\infty, \xi]\,,\ ]\xi, \infty[\,\big)$$

are Dedekind cuts of $(M, \leq)$, one "located directly below" ξ, the other "directly above" ξ. Together with

$$-\infty := (\varnothing, M), \quad +\infty := (M, \varnothing),$$

the $\xi_\pm$ ($\xi \in M$) comprise the *trivial* Dedekind cuts of $(M, \leq)$. Any Dedekind cut of $(M, \leq)$ that is not trivial is said to be *free*. In other words, a Dedekind cut (I, J) is free if and only if I, J are both non-empty, I does not have a largest element and J does not have a smallest element.

From first year calculus it is known that the field $\mathbb{R}$ of real numbers is (Cauchy) complete. This means precisely that $(\mathbb{R}, \leq)$ doesn't have any free Dedekind cut. Any ordered field other than $\mathbb{R}$ does have a free Dedekind cut (Exercise 1.1.3). For example, if α is any irrational real number, the pair

$$\big(\,]-\infty, \alpha[\, \cap \mathbb{Q},\ ]\alpha, \infty[\, \cap \mathbb{Q}\big)$$

is a free Dedekind cut of $\mathbb{Q}$.

1.1.13 Given any Dedekind cut $\xi = (I, J)$ of $\mathbb{R}$ (necessarily non-free), we are going to construct an ordering $\leq\, = \,\leq_\xi$ of $\mathbb{R}(t)$ that induces the cut ξ in the sense of 1.1.10, i.e. that satisfies $I < t < J$.

Let us first carry this out for $\xi = 0_+$. So we want to construct an ordering $\leq$ of $\mathbb{R}(t)$ such that $0 < t < \varepsilon$ holds for every real number $\varepsilon > 0$ $(*)$. Proceeding heuristically, assume that such an ordering has been found. Then $0 < t^n < \varepsilon$ holds for any $n \geq 1$ and any real number $\varepsilon > 0$. Let $f \in \mathbb{R}[t]$ be a non-zero polynomial, say $f = t^k g$ where $k \geq 0$ and $g \in \mathbb{R}[t]$, $g(0) \neq 0$. Then $\operatorname{sign}_\leq(f) = \operatorname{sign}_\leq(g)$ since $t > 0$. To determine the sign of g, write

$$g = \sum_{i=0}^{r} a_i t^i = a_0 + th$$

with $a_i \in \mathbb{R}$ and $a_0 \neq 0$, and with $h = a_1 + a_2 t + \cdots + a_r t^{r-1}$. Let $|\cdot|$ denote the absolute value with respect to the ordering $\leq$, see 1.1.6. Since $|t| < 1$ we have

$$|h| < 1 + \sum_{i=1}^{r} |a_i| =: c,$$

which implies $|th| < tc < |a_0|$ since $t < \frac{|a_0|}{c}$. Summing up, it follows that $\operatorname{sign}_\leq(f) = \operatorname{sign}_\leq(g) = \operatorname{sign}(a_0)$.

This argument shows that in finding an ordering $\leq$ with $(*)$, we have no choice about the signs of polynomials. Therefore, if $\leq$ exists, it will be unique: Given an arbitrary rational function $f = \frac{p}{q} \in \mathbb{R}(t)$ where $p, q \in \mathbb{R}[t]$ and $q \neq 0$, we see from $f = \frac{1}{q^2} \cdot pq$ that necessarily $\operatorname{sign}_\leq(f) = \operatorname{sign}_\leq(pq)$.

Conversely, the preceding construction does indeed define an ordering $\leq$ of $\mathbb{R}(t)$. The associated positive cone $P = P_{0,+}$ is

$$P_{0,+} = \{0\} \cup \Big\{t^n \cdot \frac{p}{q} : n \in \mathbb{Z},\ p, q \in \mathbb{R}[t] \text{ and } p(0)q(0) > 0\Big\},$$

and $P + P \subseteq P$ follows from $\frac{p_1}{q_1} + t^n \frac{p_2}{q_2} = \frac{p_1 q_2 + t^n p_2 q_1}{q_1 q_2}$. The other axioms for a positive cone are even more immediate.

In a similar way we may construct a positive cone $P_{0,-}$ whose associated ordering satisfies $-\varepsilon < t < 0$ for all real $\varepsilon > 0$. More generally, for arbitrary $a \in \mathbb{R}$ we construct positive cones $P_{a,\pm}$ of $\mathbb{R}(t)$ that realize the Dedekind cuts $\xi = a_\pm$. We also find positive cones $P_{\pm\infty}$ of $\mathbb{R}(t)$ that satisfy $t - a \in P_{+\infty}$ and $a - t \in P_{-\infty}$ for all $a \in \mathbb{R}$.

We remark that all these positive cones can be obtained from $P_{0,+}$ via the standard action of the group $\mathrm{PGL}_2(\mathbb{R})$ of Möbius transformations on $\mathbb{R}(t)$ (see Exercise 1.1.5).

Altogether we have shown that any Dedekind cut of $\mathbb{R}$ is realized by exactly one ordering of $\mathbb{R}(t)$:

1.1.14 Proposition. *The orderings of the field $\mathbb{R}(t)$ are in natural bijection with the Dedekind cuts of $\mathbb{R}$, which are the $a_\pm$ ($a \in \mathbb{R}$) together with $\pm\infty$.* □

1.1.15 Remark. Let (K, P) be an arbitrary ordered field. In a completely similar fashion, we may construct orderings $P_{a,\pm}$ ($a \in K$) and $P_{\pm\infty}$ of the rational function field $K(t)$, that extend the ordering P of K (compare Exercise 1.1.7). However, there will in general be orderings of $K(t)$ that cannot be obtained in this way (Exercise 1.1.8).

1.1.16 Definition. The ordered field $(K, \leq)$, or the ordering $\leq$ of K, is called *Archimedean* if the *axiom of Archimedes* holds:

$$\forall\, a,\ b \in K\ (b > 0\ \Rightarrow\ \exists\, n \in \mathbb{N}\ \ nb > a).$$

1.1.17 Examples.

1. The ordering $\leq$ of K is Archimedean if, and only if, for any $a \in K$ there is a positive integer n with $a < n$. Clearly, the unique ordering of $\mathbb{R}$ is Archimedean. Therefore any subfield of $\mathbb{R}$ is Archimedean as well, if equipped with the ordering induced from $\mathbb{R}$.

2. The field $\mathbb{R}(t)$ has only non-Archimedean orderings. For example, with respect to $P_{a,+}$ we have $\frac{1}{t-a} > n$ for every $n \in \mathbb{N}$.

3. The field $\mathbb{Q}(t)$ has both Archimedean and non-Archimedean orderings (Exercise 1.1.8).

The next result says that, up to order-preserving isomorphism, the Archimedean ordered fields are just the subfields of $\mathbb{R}$ (with the ordering induced from $\mathbb{R}$):

1.1.18 Theorem. (Hölder) *Given an Archimedean ordered field $(K, \leq)$, there exists an order-compatible homomorphism $\varphi\colon K \to \mathbb{R}$. In addition, φ is uniquely determined.*

Here we are using the following terminology:

1.1.19 Definition. If (K, P) and (L, Q) are ordered fields, a homomorphism $\varphi\colon K \to L$ is *order-compatible* with respect to P and Q, if $\varphi(P) \subseteq Q$ (and hence $P = \varphi^{-1}(Q)$) holds. We may also say that $\varphi\colon (K, P) \to (L, Q)$ is an *order embedding.*

Proof of 1.1.18. Note that $\mathbb{Q}$ is a subfield of K. Given $a \in K$, we consider the subsets[1]

$$I_a\ :=\ ]-\infty, a]_K \cap \mathbb{Q}, \quad J_a\ :=\ [a, \infty[_K \cap \mathbb{Q}$$

[1] intervals are formed with respect to the fixed ordering $\leq$ of K

of $\mathbb{Q}$. It is obvious that $I_a \le J_a$ and $I_a \cup J_a = \mathbb{Q}$. So we are *forced* to define φ by

$$\varphi(a) \;:=\; \sup I_a = \inf J_a \in \mathbb{R}.$$

Since $\le$ is an Archimedean ordering, both I_a, J_a are non-empty, and so $\varphi(a)$ is a well-defined real number. It is immediate to see that $I_a + I_b \subseteq I_{a+b}$ and $J_a + J_b \subseteq J_{a+b}$ hold for $a, b \in K$. Applying sup to both sides of the first inclusion gives $\varphi(a) + \varphi(b) \le \varphi(a+b)$, and applying inf to the second inclusion gives the opposite inequality. Therefore φ is an additive map. On the other hand, for positive elements $a, b \in K$ we have $I_a^+ \cdot I_b^+ \subseteq I_{ab}^+$ and $J_a^+ \cdot J_b^+ \subseteq J_{ab}^+$, where we put $I_c^+ := I_c \cap [0, \infty[$ and $J_c^+ := J_c \cap [0, \infty[$. Similar to before, this implies $\varphi(ab) = \varphi(a)\varphi(b)$. □

1.1.20 Corollary. *The Archimedean orderings of a field K are in natural bijection with the set* $\mathrm{Hom}(K, \mathbb{R})$ *of field embeddings $K \to \mathbb{R}$.* □

1.1.21 Corollary. *The only ring endomorphism of the field $\mathbb{R}$ is the identity map.* □

1.1.22 Corollary. *If K is a proper field extension of $\mathbb{R}$, then K does not have any Archimedean ordering.*

Proof. Otherwise there would be an embedding $\varphi\colon K \to \mathbb{R}$, by Hölder's theorem. Since $\varphi|_{\mathbb{R}} = \mathrm{id}$ by 1.1.21, the map φ cannot be injective, contradiction. □

We now introduce an important generalization of orderings.

1.1.23 Definition. A *preordering* of a field K is a subset $T \subseteq K$ that satisfies $T + T \subseteq T$, $TT \subseteq T$ and $a^2 \in T$ for any $a \in K$. The preordering T is said to be *proper* if $-1 \notin T$.

1.1.24 Remarks.

1. Any positive cone is a proper preordering.

2. The preordering T is proper if and only if $T \cap (-T) = \{0\}$. If $\mathrm{char}(K) \ne 2$ then the only improper preordering of K is $T = K$, since $-1 \in T$ implies $T = K$ by the identity

$$x = \Big(\frac{x+1}{2}\Big)^2 - \Big(\frac{x-1}{2}\Big)^2.$$

3. If $\mathcal{T} = (T_i)_{i\in I}$ is any family of preorderings of K, the intersection $\bigcap_{i\in I} T_i$ is again a preordering. The unique smallest preordering in K is ΣK^2, the set of all sums of squares in K. If $\mathcal{T}$ is upward filtering (meaning that for any $i_1, i_2 \in I$ there exists $i \in I$ with $T_{i_1} \cup T_{i_2} \subseteq T_i$), then also the union $\bigcup_{i\in I} T_i$ is a preordering (and is proper if all the T_i are proper).

4. Given any subset S of K we may consider the preordering $T = PO(S)$ generated by S. By definition, this is the intersection of all preorderings T' with $S \subseteq T'$. Explicitly, T consists of all elements in K of the form

$$\sum_{e\in\{0,1\}^r} s_e \cdot t_1^{e_1} \cdots t_r^{e_r}$$

with $r \in \mathbb{N}$, $t_1, \dots, t_r \in S \cup \{1\}$ and $s_e \in \Sigma K^2$ for all e. This is immediate to check.

Conversely we'll now prove that every (proper) preordering is an intersection of orderings. Always let K be a field.

1.1.25 Lemma. *Let T be a proper preordering of K, and let $a \in K$ with $a \notin T$. Then $T - aT = \{s - at \colon s, t \in T\}$ is again a proper preordering of K.*

Proof. Obviously $T - aT$ is a preordering. Assuming $-1 \in T - aT$ would give $-1 = s - at$ with $s, t \in T$. Here $t \neq 0$ since $-1 \notin T$, and so $a = \frac{1+s}{t} = \frac{1}{t^2}(1+s)t \in T$. This contradicts the assumption. □

1.1.26 Proposition. *Any proper preordering of K is contained in a positive cone of K.*

Proof. Let T be a proper preordering of K, and let $\mathcal{T}$ be the set of all proper preorderings T' that contain T. Zorn's lemma can be applied to $\mathcal{T}$ (Remark 1.1.24.3), and so $\mathcal{T}$ contains a maximal element P. Let us show that any such P is a positive cone of K. We need to show $P \cup (-P) = K$, so let $a \in K$. If $a \notin P$ then $P - aP$ is a proper preordering (Lemma 1.1.25) that contains P, hence $P - aP = P$ by maximality of P. Therefore $-a \in P$. □

1.1.27 Corollary. *Any maximal proper preordering of K is a positive cone of K.* □

1.1.28 Proposition. *Any proper preordering of K is an intersection of positive cones of K.*

Proof. Let T be a proper preordering, and let $a \in K$ with $a \notin T$. Then $T - aT$ is a proper preordering (1.1.25), and hence is contained in a positive cone P (1.1.26). Therefore $a \notin P$. □

If K is any real field then $-1 \notin \Sigma K^2$, as we saw directly from the definition of orderings (1.1.8). Now we can prove a converse:

1.1.29 Corollary. *For any field K, the following are equivalent:*

(i) *K is real, i.e. K has an ordering;*
(ii) *K has a proper preordering;*
(iii) *$-1 \notin \Sigma K^2$;*
(iv) *$a_1^2 + \dots + a_n^2 = 0$ with $a_1, \dots, a_n \in K$ implies $a_1 = \dots = a_n = 0$.*

Proof. The implications (i) ⇒ (ii) ⇒ (iii) and (iii) ⇔ (iv) are clear. If (iii) holds then ΣK^2 is a proper preordering of K. So K has an ordering by Proposition 1.1.26. □

1.1.30 Theorem. (Artin) *Let K be a field,* $\mathrm{char}(K) \neq 2$. *An element $a \in K$ is non-negative with respect to every ordering of K if, and only if, a is a sum of squares in K.*

The "only if" part fails when K is a non-perfect field of characteristic two. This is why the case $\mathrm{char}(K) = 2$ has been excluded in the theorem.

Proof. If K is real then ΣK^2 is a proper preordering, and so it is the intersection of all positive cones by Proposition 1.1.28. If K is not real then $-1 \in \Sigma K^2$ (Corollary 1.1.29), and so $\Sigma K^2 = K$ by Remark 1.1.24.2. □

1.1.31 Remark. This is a remarkable theorem, even though its proof was not very difficult. Artin and Schreier introduced the concept of orderings of fields in order to approach the following question:

> **Hilbert's 17th Problem** (1900): *Let $f(x_1, \dots, x_n)$ be a polynomial with real coefficients that takes non-negative values on all of $\mathbb{R}^n$. Then, can f be written as a sum of squares of rational functions in $(x_1, \dots, x_n)$?*

So Hilbert had asked if there exists an identity $fh^2 = f_1^2 + \cdots + f_r^2$ with real polynomials $f_1, \dots, f_r$ and h, where $h \neq 0$.

The case $n = 1$ being elementary, Hilbert settled the $n = 2$ case in [92]. But for larger values of n the question remained completely open. The introduction of the "abstract" concept of field orderings, combined with Artin's theorem 1.1.30, offers a radically new perspective for approaching the problem. Using the notion of orderings, and in particular using Theorem 1.1.30, Artin (1927) proved that the answer is always positive. We can see a possible strategy for such a proof: From the assumption $f(\xi) \geq 0$ for all $\xi \in \mathbb{R}^n$, try to conclude that f is non-negative with respect to every ordering of the rational function field $\mathbb{R}(x_1, \dots, x_n)$. If this can be shown, we are done, using Theorem 1.1.30. Indeed, this approach works, and we will later see the full details in Theorem 1.5.21.

Note that the proof of Theorem 1.1.30 was entirely non-constructive. In general, from just knowing that an element $a \in K$ lies in every positive cone of K, we get no information at all about how to find a sum of squares representation of a. We'll get back to this question, in the situation of Hilbert's 17th Problem.

Exercises

1.1.1 Let P, Q be positive cones of a field K that satisfy $P \subseteq Q$. Show that $P = Q$.

1.1.2 A totally ordered set $(M, \leq)$ is *Dedekind complete* if, whenever I, J are non-empty subsets of M with $I < J$, there exists $\xi \in M$ with $I \leq \xi \leq J$. Prove that M is Dedekind complete if and only if M does not have a free Dedekind cut.

1.1.3 Let $(K, \leq)$ be an ordered field that is Dedekind complete. Show that $K = \mathbb{R}$.

1.1.4 Let $(M, \leq)$ be a totally ordered set. A pair (I, J) of subsets of M is called a *generalized Dedekind cut* of M if $I \cup J = M$ and $I \leq J$ (i.e. $a \leq b$ for all $a \in I$, $b \in J$).

(a) A generalized Dedekind cut of M is either a Dedekind cut of M, or it has the form

$$\phi(x) := (]-\infty, x], [x, \infty[)$$

for some $x \in M$.

(b) The set $\widehat{M}$ of all generalized Dedekind cuts of M gets totally ordered by setting $(I, J) \leq (I', J')$ if and only if $I \subseteq I'$ and $J \supseteq J'$.

(c) The map $\phi\colon M \to \widehat{M}$ is order-compatible and injective.
(d) The totally ordered set $(\widehat{M}, \leq)$ is Dedekind complete. It is called the *Dedekind completion* of $(M, \leq)$.

1.1.5 Given an invertible matrix $A = \left(\begin{smallmatrix} a & b \\ c & d \end{smallmatrix}\right) \in \mathrm{GL}_2(\mathbb{R})$, let σ_A be the field automorphism of $\mathbb{R}(t)$ over $\mathbb{R}$ defined by

$$\sigma_A(t) = \frac{at+b}{ct+d}$$

(and $\sigma_A(f(t)) = f(\sigma_A(t))$ for every $f \in \mathbb{R}(t)$). This defines an action of the group $\mathrm{PGL}_2(\mathbb{R}) = \mathrm{GL}_2(\mathbb{R})/\mathbb{R}^*$ on $\mathbb{R}(t)$ by field automorphisms. Show that there is an induced action of $\mathrm{PGL}_2(\mathbb{R})$ on the set of positive cones of $\mathbb{R}(t)$, and that this action is transitive. The latter means that, for every positive cone P of $\mathbb{R}(t)$, there exists $A \in \mathrm{GL}_2(\mathbb{R})$ with $P = \sigma_A(P_{0,+})$.

1.1.6 Let $\mathbb{R}(t)$ be the rational function field in one variable t over $\mathbb{R}$. For every $f \in \mathbb{R}(t)$, the value $f(a)$ of f at $a \in \mathbb{R}$ is defined for all but finitely many $a \in \mathbb{R}$.

(a) Show that $P_{a,+} = \{f \in \mathbb{R}(t)\colon \exists \varepsilon > 0 \text{ s.t. } f \geq 0 \text{ on }]a, a+\varepsilon[\}$ (and similarly for the positive cones $P_{a,-}$, $P_{\pm\infty}$, see 1.1.13).
(b) Let C be a (non-rational) irreducible algebraic curve over $\mathbb{R}$, and assume $|C(\mathbb{R})| = \infty$. Mimicking (a), try to find positive cones of the function field $C(\mathbb{R})$. Can you prove your guess?

1.1.7 Let (K, P) be an ordered field, let t be a variable, and let $Q \subseteq K(t)$ be the set of all fractions $\frac{f}{g}$ of polynomials with $g \neq 0$, such that either $f = 0$, or the leading (highest) coefficients of f and g have the same sign with respect to P.

(a) Q is a positive cone of $K(t)$ that extends P.
(b) For any $a \in K$ one has $a \leq_Q t$. In particular, the ordered field $(K(t), Q)$ is non-Archimedean.

1.1.8 For each $n \geq 1$, show that the rational function field $\mathbb{Q}(x_1, \ldots, x_n)$ over $\mathbb{Q}$ has both Archimedean and non-Archimedean orderings. (Use Exercise 1.1.7 and the fact that $\mathbb{R}$ is uncountable.)

1.1.9 Let L/K be an algebraic field extension, let Q be a positive cone of L and $P = K \cap Q$ its restriction to K. Show that the identity is the only K-automorphism of L that preserves Q.

1.1.10 Let k be a field and $x = (x_1, \ldots, x_n)$ a tuple of indeterminates. If $f, p \in k[x]$ are polynomials and p is irreducible, let $v_p(f)$ be the largest integer $m \geq 0$ such that p^m divides f, with $v_p(f) = \infty$ if $f = 0$.

(a) Assume that the quotient field of $k[x]/\langle p\rangle$ is real. Given any non-zero polynomials $f_1, \ldots, f_r \in k[x]$, show that

$$v_p(f_1^2 + \cdots + f_r^2) = 2\min\{v_p(f_i)\colon i = 1, \ldots, r\}.$$

(b) If k is real, show that $\deg(f_1^2 + \cdots + f_r^2) = 2\max\{\deg(f_i)\colon i = 1, \ldots, r\}$ for $f_1, \ldots, f_r \in k[x]$.
(c) Assuming you have familiarized yourself with valuations of fields (see Appendix A.5), state and prove a generalization of (a) for valuations with real residue field. In what sense is (b) a particular case of this generalization?

For a version that is still more general than (c), see Exercise 3.5.2 later in the book.

1.2 Extension of Orderings and Real Closed Fields

1.2.1 Always let K be a field. Let L/K be a field extension, and let P be a positive cone of K. The smallest preordering of L that contains P is

$$T_L(P) := \Big\{\sum_{i=1}^{n} a_i y_i^2 : n \in \mathbb{N},\ a_i \in P,\ y_i \in L\Big\}.$$

According to 1.1.28, $T_L(P)$ is the intersection of all positive cones Q of L that extend P, i.e. that satisfy $Q \cap K = P$. We conclude:

1.2.2 Lemma. *A positive cone P of K extends to a positive cone of L if and only if $-1 \notin T_L(P)$.* □

Let us consider a few particular cases.

1.2.3 Proposition. *Let $L = K(\sqrt{a})$ where $a \in K$. A positive cone P of K extends to L if and only if $a \in P$.*

Proof. If Q is an extension of P to L, then $a = (\sqrt{a})^2 \in Q \cap K = P$. Conversely suppose that $a \in P$ and $L \neq K$, so $\sqrt{a} \notin K$. Assuming $-1 \in T_L(P)$ we get an identity

$$-1 = \sum_{i=1}^{n} a_i\,(x_i + y_i\sqrt{a})^2 = \sum_{i=1}^{n} a_i\,(x_i^2 + a y_i^2 + 2x_i y_i\sqrt{a})$$

with $a_i \in P$ and $x_i, y_i \in K$. This implies $-1 = \sum_{i=1}^{n} a_i(x_i^2 + a y_i^2) \in P$, a contradiction. □

1.2.4 Proposition. *Let L/K be a finite extension of odd degree. Then every ordering of K extends to L.*

Proof. Assuming this is false, we fix a counter-example L/K of smallest possible odd degree n. Let P be a positive cone of K that doesn't extend to L. Since $\mathrm{char}(K) = 0$ there exists $\alpha \in L$ with $L = K(\alpha)$. Let $f \in K[t]$ be the minimal polynomial of α, then $\deg(f) = n$ and L is K-isomorphic to $K[t]/\langle f\rangle$. By Lemma 1.2.2 we have $-1 \in T_L(P)$, so there is an identity $1 + \sum_{i=1}^{r} a_i y_i^2 = 0$ in L with $0 \neq a_i \in P$ and $0 \neq y_i \in L$ for $i = 1, \dots, r$. Choose polynomials $g_i \in K[t]$ with $y_i = g_i(\alpha)$ and $\deg(g_i) < n$ $(i = 1, \dots, r)$. Then

$$1 + \sum_{i=1}^{r} a_i g_i(t)^2 = f(t)h(t) \tag{1.1}$$

holds in $K[t]$ for some polynomial $h \in K[t]$.

Let $d := \max_i \deg(g_i)$. Then $d < n$, and the left-hand side of (1.1) has degree at most $2d$. The coefficient of t^{2d} is of the form $\sum_{i=1}^{r} a_i b_i^2$ with $b_i \in K$ and $b_i \neq 0$ for at least one index i, so it is > 0 with respect to P. Hence the degree of the polynomial (1.1) is equal to $2d$, and so $\deg(h)$ is odd. Moreover $\deg(h) < n$ since $d < n$. There is at least one irreducible factor h_1 of h of odd degree, and so the field $L_1 := K[t]/\langle h_1\rangle$ over K has odd degree $< n$ over K. Moreover $-1 \in T_{L_1}(P)$ by (1.1). So the ordering P does not extend to L_1 (1.2.2) which contradicts the minimal choice of L. □

1.2.5 Proposition. *Any ordering of K can be extended to the rational function field $K(x_1, \dots, x_n)$, for every $n \in \mathbb{N}$.*

Proof. By an inductive argument it suffices to extend a given positive cone P of K to $K(x)$ (one variable). See Exercise 1.1.7 where an explicit extension is constructed. Here is another (non-constructive) argument:

Assuming that P doesn't extend to $K(x)$, we would have $-1 \in T_{K(x)}(P)$, once more by 1.2.2. This would mean an identity $f_0^2 + \sum_{i=1}^r a_i f_i^2 = 0$ in $K[x]$ with $r \geq 1$ and $0 \neq a_i \in P$ $(i = 1, \dots, r)$, $0 \neq f_i \in K[x]$ $(i = 0, \dots, r)$. By an argument similar to the previous proof, the left-hand side has degree $2d$ where $d = \max\{\deg(f_i) \colon 0 \leq i \leq r\}$. In particular, it is non-zero, which is a contradiction. □

1.2.6 Definition. A field K is *real closed* if K is real, but every proper algebraic extension of K is non-real.

For example, the field $\mathbb{R}$ of real numbers is real closed, since $\mathbb{C}$ is the only proper algebraic extension of $\mathbb{R}$.

1.2.7 Proposition. *For a field K, the following properties are equivalent:*

(i) *K is real closed;*
(ii) *K has an ordering that doesn't extend to any proper algebraic extension of K;*
(iii) *K is real, doesn't have any proper odd degree extension, and $K^* = K^{*2} \cup (-K^{*2})$.*

Here we write $K^{*2} = \{a^2 \colon a \in K^*\}$.

Proof. (i) $\Rightarrow$ (ii) is obvious from the definitions. (ii) $\Rightarrow$ (iii): Let P be a positive cone of K that doesn't extend to any proper finite extension of K. By Proposition 1.2.4, K doesn't have any proper odd degree extension. Any element $a \in P$ is a square in K, since otherwise $K(\sqrt{a})$ would be a properly larger field to which P can be extended (1.2.3). So we have $P = \{a^2 \colon a \in K\}$, and (iii) follows from $P \cup (-P) = K$.

Now assume that (iii) holds, so in particular $\mathrm{char}(K) = 0$. We use a Galois-theoretic argument to prove (i). By (iii), the only non-squares in K are the elements $-a^2$ with $a \neq 0$. Therefore $K(\sqrt{-1})$ is the unique quadratic extension of K. Let us show that every proper algebraic extension of K contains a quadratic subfield, which will imply property (i). Let L/K be a proper finite extension, let E be the Galois hull of L over K and G the Galois group of E over K. Moreover let S be a Sylow 2-subgroup of G. If F denotes the subfield of elements of L that are fixed by S, the degree $[F : K] = [G : S]$ is odd. Therefore $F = K$ by hypothesis (iii), and so G is a 2-group. Hence any proper subgroup of G is contained in a subgroup of index 2, which implies that L contains a quadratic extension of K, as was claimed. □

From (i) $\Rightarrow$ (iii) we see in particular:

1.2.8 Corollary. *If K is a real closed field then $P := \{a^2 : a \in K\}$ is the unique positive cone of K.* □

It is customary to denote a general real closed field by the letter R, the symbol $\mathbb{R}$ being reserved for the field of (classical) real numbers. If R is a real closed field, we will write $\le$ for the unique ordering of R, and often denote the unique positive cone of R by $R_+ = \{a^2 : a \in R\}$.

1.2.9 Proposition. (Fundamental Theorem of Algebra) *If R is any real closed field, the field $R(\sqrt{-1})$ is algebraically closed.*

Proof. Let us write $i = \sqrt{-1}$ as usual. Every finite extension of $R(i)$ has 2-power degree, as we saw in the proof of 1.2.7, (iii) $\Rightarrow$ (i). Therefore it suffices to prove that $R(i)$ doesn't have a quadratic extension. This is elementary: If $w = u + iv$ with $u, v \in R$ and $v \neq 0$, then $w = z^2$ for $z := x + \frac{iv}{2x}$ and $x := \sqrt{\frac{1}{2}(u + \sqrt{u^2 + v^2})} \in R$, where one has to take the *positive* value for the inner square root. □

1.2.10 Remark. Proposition 1.2.9 admits a strong converse: *If K is any field with algebraic closure $\overline{K}$, and if $1 < [\overline{K} : K] < \infty$, then K is real closed (and therefore $\overline{K} = K(\sqrt{-1})$).* See [109] Theorem 1.6.1 for the proof.

1.2.11 Corollary. *Let R be a real closed field. The monic irreducible polynomials in $R[t]$ are precisely the following:*

(1) *$t - a$, for $a \in R$;*
(2) *$t^2 + at + b$, for $a, b \in R$ with $a^2 < 4b$.*

Hence any monic polynomial in $R[t]$ is a product of finitely many factors of type (1) or (2).

Proof. Let $f \in R[t]$ be monic and irreducible. Then $\deg(f) \le 2$ by 1.2.9. This implies the assertion since t^2+at+b is irreducible over R if, and only if, the discriminant $a^2 - 4b$ is not a square (i.e., is negative). □

1.2.12 Remarks.

1. If $f(t) \in R[t]$ is a monic irreducible polynomial of degree 2, then $f(t) > 0$ for every $t \in R$.

2. Let $(K, \le)$ be an ordered field, let $L = K(i)$ with $i = \sqrt{-1}$, and let $z \mapsto \overline{z}$ denote the non-trivial automorphism of L/K. Writing $|z|^2 := z\overline{z}$ for $z \in L$, the *Cauchy–Schwarz inequality*

$$\left|\sum_{j=1}^{n} x_j \overline{y}_j\right|^2 \le \left(\sum_{j=1}^{n} |x_j|^2\right) \cdot \left(\sum_{j=1}^{n} |y_j|^2\right)$$

holds for any $x, y \in L^n$. Moreover, equality holds if and only if x and y are linearly dependent over L. The proof is the same as over $\mathbb{R}$ or $\mathbb{C}$, respectively.

3. If the field R is real closed, if $C = R(\sqrt{-1})$ and $z \in C^n$, let

$$|z| \;:=\; \sqrt{|z_1|^2 + \cdots + |z_n|^2}$$

(non-negative square root). The familiar triangle inequality $|z + z'| \le |z| + |z'|$ holds for all z, $z' \in C^n$, as a consequence of Cauchy–Schwarz.

4. Every positive cone P of a field K defines a topology $\mathcal{T}_P$ on K, by declaring the open intervals $]a,b[_P$ (for a, $b \in K$) to be a basis of open sets in K. This is the *order topology* of (K,P). All field operations (addition, multiplication, division) are continuous with respect to $\mathcal{T}_P$, which means that $(K,\mathcal{T}_P)$ is a topological field (Exercise 1.2.2). For any $n \in \mathbb{N}$ the order topology $\mathcal{T}_P$ gives the product topology on K^n, which has the open balls

$$B_r(x) \;=\; \{y \in K^n \colon |y - x|^2 <_P r^2\}$$

($x \in K^n$, $r \in K$) as a basis of open sets. We will use the order topology without any further comment, mostly in the case where K is real closed.

Exercises

1.2.1 Complete the proof of the fundamental theorem of algebra (Proposition 1.2.9).

1.2.2 Let (K,P) be an ordered field, and let $\mathcal{T}_P$ be the order topology as defined in Remark 1.2.12.4. Clearly $\mathcal{T}_P$ is a Hausdorff topology.

(a) Show that $(K,\mathcal{T}_P)$ is a topological field, i.e. that addition and multiplication (as maps $K \times K \to K$) and inversion (as a map $K^* \to K^*$) are continuous.
(b) If $K \ne \mathbb{R}$, show that the topological space $(K,\mathcal{T}_P)$ is totally disconnected.
(c) If $K \ne \mathbb{R}$, show that the topological space $(K,\mathcal{T}_P)$ is not locally compact.
(d) For the field $K = \mathbb{R}(t)$, discuss how the topology $\mathcal{T}_P$ depends on the ordering P.

1.2.3 Let $P = \{f \in \mathbb{R}(x,y) \colon \exists\varepsilon > 0$ with $f(t,e^t) \ge 0$ for $0 < t < \varepsilon\}$. Prove that P is a positive cone in the field $\mathbb{R}(x,y)$. (You will have to use the fact that the exponential function is transcendental, i.e. does not satisfy a polynomial identity.)

1.2.4 Let $(K,P) \subseteq (L,Q)$ be a finite extension of ordered fields.

(a) The extension is *relatively Archimedean*, meaning that for every $b \in L$ there exists $a \in K$ with $b \le_Q a$.
(b) In general, K need not be dense in L in the order topology of Q. Show this using $L = \mathbb{R}(t)$, $K = \mathbb{R}(t^2)$ and the positive cone Q of L with $0 <_Q nt <_Q 1$ for all $n \in \mathbb{N}$.

1.2.5 Let k be a field with $\mathrm{char}(k) = 0$, let $k[\![t]\!]$ be the ring of formal power series over k in one variable t, let $\mathfrak{m}$ be its maximal ideal and $k(\!(t)\!) = \mathrm{qf}(k[\![t]\!])$ its quotient field.

(a) For every $f \in \mathfrak{m}$ and any $n \in \mathbb{N}$, there exists $g \in \mathfrak{m}$ with $(1+g)^n = 1 + f$. (*Hint*: Binomial series.)
(b) Every $f \in k(\!(t)\!)^*$ can be written $f = c \cdot t^n \cdot (1+g)^2$ with uniquely determined $c \in k^*$, $n \in \mathbb{Z}$ and $g \in \mathfrak{m}$.
(c) Prove that every ordering of k can be extended to $k(\!(t)\!)$ in precisely two different ways.

1.3 Real Zeros of Univariate Polynomials

Given a polynomial f with real coefficients, we would like to count, or at least estimate, the number of real zeros of f, without actually having to calculate them. We will see several methods to achieve this goal and more refined ones. They mostly date back to the 19th century or are even older, and they can be regarded to be among the earliest manifestations of what we now call "real algebra".

For the sequel it will be important that the results presented here hold over arbitrary real closed fields R, and not just over $\mathbb{R}$. Some of the basic results below are well-known from first year calculus in the case $R = \mathbb{R}$, but we treat them again, using only strictly algebraic arguments. We stress that the traditional tools of calculus are unavailable over a general real closed field R, because R lacks the required completeness properties.

Throughout, R denotes a real closed field. We start with an easy a priori bound for the size of the roots, in terms of the coefficients:

1.3.1 Lemma. *Let $f = t^n + a_1 t^{n-1} + \cdots + a_{n-1}t + a_n$ be a monic polynomial with coefficients in $R(\sqrt{-1})$. Then*

$$|\alpha| \le \max\{1, |a_1| + \cdots + |a_n|\}$$

holds for every root α of f in $R(\sqrt{-1})$.

Proof. See 1.2.12.3 for the definition of absolute values of elements in $R(\sqrt{-1})$. Let $\alpha \in R(\sqrt{-1})$ with $f(\alpha) = 0$, we may assume $|\alpha| \ge 1$. Then

$$\alpha = -\Big(a_1 + \frac{a_2}{\alpha} + \cdots + \frac{a_n}{\alpha^{n-1}}\Big),$$

so the claim follows from the triangle inequality. □

Next we prove some results for rational functions. When $R = \mathbb{R}$, these are well-known from elementary calculus, and hold more generally for suitably differentiable functions.

1.3.2 Let us recall some basic notions. Let k be a field and let $f = f(t) \in k(t)$ be a non-zero rational function. Given $a \in k$, we can (uniquely) write $f = (t-a)^n \cdot \frac{g}{h}$ with $n \in \mathbb{Z}$ and $g, h \in k[t]$ such that $g(a)h(a) \ne 0$. The *vanishing order* of f at $a \in k$ is $\operatorname{ord}_a(f) = n$. If $n \ge 1$ then a is called a zero of f, and n is its order or multiplicity. If $n < 0$ then a is a pole of f, of pole order $|n|$. A given rational function $f \ne 0$ has only finitely many zeros or poles. If $a \in k$ is not a pole of f, the value $f(a) \in k$ at $t = a$ is well-defined.

Now let the field $k = R$ be real closed. Any non-zero rational function $f \in R(t)$ has a product decomposition

$$f(t) = \pm g(t) \cdot \prod_{i=1}^{r} (t - a_i)^{e_i} \tag{1.2}$$

with $r \geq 0$, $a_i \in R$ and $e_i \in \mathbb{Z}$ $(i = 1, \dots, r)$, where $g(t) \in R(t)$ is such that g has no poles in R and $g(\xi) > 0$ holds for every $\xi \in R$. This is clear from 1.2.11, see also 1.2.12.1. Thus we get:

1.3.3 Proposition. (Intermediate value theorem) *Let $f \in R(t)$, and let $a < b$ in R be such that f has no pole in $[a, b]$ and $f(a)f(b) < 0$. Then the number of zeros of f in $[a, b]$, counted with multiplicities, is odd. In particular, f has at least one zero in $[a, b]$.*

Proof. In (1.2) we can ignore $g(t)$, as well as any factor $t - a_i$ with $a_i \notin [a, b]$. □

1.3.4 Let $0 \neq f \in R(t)$ and $a \in R$. There is $\varepsilon > 0$ in R such that $|b - a| \geq \varepsilon$ holds for any zero or pole $b \neq a$ of f. By 1.3.3, f has constant signs s_- on $]a - \varepsilon, a[$ and s_+ on $]a, a + \varepsilon[$, where s_-, $s_+ \in \{\pm 1\}$. We'll say that *f changes sign at a (from s_- to s_+, if we want to specify)* if $s_- \neq s_+$. If $s_- = s_+$ then *f doesn't change sign at a.*

1.3.5 Lemma. *A non-zero rational function $f \in R(t)$ changes sign at $a \in R$ if and only if* $\mathrm{ord}_a(f)$ *is odd.*

Proof. Immediate from $f = (t - a)^n \cdot g$ where $n = \mathrm{ord}_a(f)$ and $g(a) \neq 0$. □

1.3.6 Derivatives of rational functions are defined in a purely formal way. If $f = \sum_{i \geq 0} a_i t^i$ is a polynomial with $a_i \in R$, then $f' = \sum_{i \geq 1} i a_i t^{i-1}$. More generally, if $f = \frac{p}{q}$ is a rational function with polynomials p, $q \in R[t]$ and $q \neq 0$, then $f' = \frac{p'q - pq'}{q^2}$, which is again a rational function. Instead of f' we also use the symbolic notation $\frac{df}{dt}$. As usual, the higher derivatives f'', f''', ... are defined by iteration of the first derivative: $f^{(k)} = \frac{d^k f}{dt^k} := \frac{d}{dt}(f^{(k-1)})$ for $k \geq 1$.

1.3.7 Lemma. *Let $0 \neq f \in R(t)$. If $a \in R$ is a root (resp. a pole) of f, the "logarithmic derivative" f'/f of f changes sign from minus to plus (resp. from plus to minus).*

Proof. Let $a \in R$ with $n = \mathrm{ord}_a(f) \neq 0$. Writing $f = (t - a)^n \cdot g$ with $g(a) \neq 0$, we get

$$\frac{f'(t)}{f(t)} = \frac{n}{t - a} + \frac{g'(t)}{g(t)}$$

from which one reads off the assertion. □

1.3.8 Proposition. *Let $f \in R(t)$, and let $a < b$ in R with $f(a) = f(b) = 0$. If f has neither zeros nor poles in the open interval $]a, b[$, the number of zeros of f' in $]a, b[$ is odd, counted with multiplicities.*

Proof. Choose $a < a_1 < b_1 < b$ in R such that f' doesn't vanish anywhere in $]a, a_1] \cup [b_1, b[$. From Lemma 1.3.7 we see that $\frac{f'}{f}$ is positive at $a + \varepsilon$ and negative at $b - \varepsilon$, for sufficiently small $\varepsilon > 0$. Hence $\frac{f'}{f}$ is positive at a_1 and negative at b_1. Therefore, by the intermediate value theorem 1.3.3, the number of zeros of $\frac{f'}{f}$ in $[a_1, b_1]$ is odd, counted with multiplicities. This is also the number of zeros of f' in $]a, b[$. □

1.3.9 Corollary. (Rolle's theorem) *Let $f \in R(t)$. If $a < b$ in R are such that $f(a) = f(b)$, and if f has no poles in $]a, b[$, there exists $a < \xi < b$ with $f'(\xi) = 0$.*

Proof. Replacing f by $f - f(a)$ we can assume $f(a) = f(b) = 0$. We may also assume $f \neq 0$, and may replace b with the smallest zero c of f that is larger than a. Now we are in the situation of Proposition 1.3.8. □

We get the following consequences, well-known from calculus:

1.3.10 Corollary. *Let $f \in R(t)$, and let $a < b$ in R be such that f has no pole in $[a, b]$.*

(a) (Mean value theorem) *There exists $a < \xi < b$ with $f'(\xi) = \frac{f(b)-f(a)}{b-a}$.*
(b) *If $f'(x) \geq 0$ for all $x \in]a, b[$ then f is increasing on $[a, b]$.*

Proof. To prove (a), apply Rolle's theorem to

$$g(t) = f(t) - \frac{t-a}{b-a}\big(f(b) - f(a)\big).$$

Indeed we have $g(a) = g(b) = f(a)$, and any $a < \xi < b$ with $g'(\xi) = 0$ satisfies the claim. Statement (b) is a consequence of (a). □

We are now going to discuss the main topic of this section, which is counting the number of real roots of a polynomial, either globally or in a given interval. To be clear, if R is a real closed field and $f \in R[t]$, by a *real root* of f we mean an element $\alpha \in R$ with $f(\alpha) = 0$ (as opposed to the *non-real roots* of f, which are the $\alpha \in R(\sqrt{-1}) \setminus R$ with $f(\alpha) = 0$). The non-zero polynomial $f \in R[t]$ is *real-rooted* if all its roots are real. We start by giving an upper bound to the number of real roots.

1.3.11 Definition. Given a finite sequence $c = (c_1, \dots, c_n)$ in R (with $n \geq 1$), let

$$\mathrm{Var}(c) = \mathrm{Var}(c_1, \dots, c_n)$$

denote the number of sign changes in c after deleting all zeros. So this is the number of pairs (i, j) with $1 \leq i < j \leq n$ for which $c_i c_j < 0$ and $c_k = 0$ for all $i < k < j$.

1.3.12 Theorem. (Descartes' rule of signs) *Let $f = \sum_{i=0}^n a_i t^i$ be a non-zero polynomial in $R[t]$. The number of strictly positive real roots of f, counted with multiplicity, is at most* $\mathrm{Var}(a_0, \dots, a_n)$.

Proof. For $n \leq 1$ this is clear. So let $n = \deg(f) > 1$, and assume that the theorem has been proved for all smaller degrees. Clearly we can assume $a_0 = f(0) \neq 0$. All root countings are done with multiplicities. Write $N_+(f)$ for the number of strictly positive roots of f, and let $r \geq 1$ be minimal with $a_r \neq 0$. Since $\mathrm{ord}_\xi(f') = \mathrm{ord}_\xi(f) - 1$ for any root ξ of f, Rolle's theorem gives $N_+(f') \geq N_+(f) - 1$. Since $N_+(f') \leq \mathrm{Var}(a_r, \dots, a_n) \leq \mathrm{Var}(a_0, \dots, a_n)$ holds by the inductive hypothesis, we are done if $N_+(f) \leq N_+(f')$.

This means that we can assume that $N_+(f) = 1 + N_+(f')$. If ξ denotes the smallest positive zero of f, this assumption implies that $f'(\eta) \neq 0$ for $0 < \eta < \xi$. For small enough $\eta > 0$, the sign of $f'(\eta)$ is the sign of a_r, so we conclude $\operatorname{sign} f'(\eta) = \operatorname{sign}(a_r)$ for all $0 < \eta < \xi$. Hence f is monotone on $[0, \xi]$ by 1.3.10(b). If $a_r > 0$ then f is increasing on $[0, \xi]$, and so $a_0 = f(0) < f(\xi) = 0$. If $a_r < 0$ then f is decreasing on $[0, \xi]$, and so $a_0 = f(0) > f(\xi) = 0$. In either case, therefore, $a_0 a_r < 0$, which means $\operatorname{Var}(a_0, \dots, a_n) = 1 + \operatorname{Var}(a_r, \dots, a_n)$. So $\operatorname{Var}(a_0, \dots, a_n) \geq 1 + N_+(f') = N_+(f)$, which completes the inductive step. □

1.3.13 Remarks.

1. The argument in the previous proof can be refined to show that the difference $\operatorname{Var}(a_0, \dots, a_n) - N_+(f)$ is always even. In particular, this implies that if $\operatorname{Var}(a_0, \dots, a_n)$ is odd then f has at least one positive root. See [109] Corollary 1.10.3 for the details.

2. In general, the upper bound in 1.3.12 cannot be improved, as shown by any polynomial with only real and positive roots. On the other hand, the actual number of positive roots may well be smaller than the number of sign changes. For example, $f = t^2 - t + 1$ doesn't have a real root.

1.3.14 Corollary. *Let $f \in R[t]$ be a non-zero polynomial with exactly m monomials. Then f has at most $2m - 2$ non-zero roots in R, counted with multiplicity.*

Again this bound is reached by $f(t^2)$, if f is a polynomial with only real and positive roots.

Proof. Both $f(t)$ and $f(-t)$ have at most $m-1$ positive roots, by Descartes' theorem. So f has at most $2m - 2$ non-zero roots. □

For real-rooted polynomials f, we can even read off the exact numbers of positive and of negative roots, directly from the sequence of coefficients:

1.3.15 Corollary. *Let $f = \sum_{i=0}^n a_i t^i \in R[t]$ be a non-zero real-rooted polynomial. Then the number of strictly positive (resp. strictly negative) roots of f is equal to $\operatorname{Var}(a_0, a_1, \dots, a_n)$ (resp. $\operatorname{Var}(a_0, -a_1, \dots, (-1)^n a_n)$), again counting with multiplicities.*

Proof. We may assume $\deg(f) = n \geq 1$ and $f(0) \neq 0$. Let p (resp. p') be the number of strictly positive (resp. strictly negative) roots of f, and put

$$\begin{aligned} W(x) &:= \operatorname{Var}(a_0 + x,\ a_1 + x,\ \dots,\ a_n + x), \\ W'(x) &:= \operatorname{Var}(a_0 + x,\ -(a_1 + x),\ \dots,\ (-1)^n(a_n + x)) \end{aligned}$$

for $x \in R$. According to Descartes we have $p \leq W(0)$, and similarly $p' \leq W'(0)$ (replace x by $-x$). Choose $x > 0$ in R so small that $x < |a_i|$ for every index i with $a_i \neq 0$. Then $W(x) + W'(x) = n$ since $a_i + x \neq 0$ for each index i, and since at each position exactly one of the two sequences changes sign. On the other hand, $\operatorname{Var}(a, 0, \dots, 0, b) \leq \operatorname{Var}(a + c, c, \dots, c, b + c)$ whenever a, b, $c \in R$ satisfy $0 < c <$

$\min\{|a|, |b|\}$, which shows that $W(x) \geq W(0)$ and $W'(x) \geq W'(0)$. Since $p + p' = n$ holds by hypothesis, the identity $W(x) + W'(x) = n$ and the inequalities

$$p \leq W(0) \leq W(x), \quad p' \leq W'(0) \leq W'(x)$$

combine to give $p = W(0)$ and $p' = W'(0)$. □

1.3.16 Remarks.

1. Theorem 1.3.12 and its corollaries can be seen as particular cases of a theorem of Budan–Fourier. For $f \in R[t]$ of degree n and $a < b$ in R, this theorem asserts that the number of real roots of f in $]a, b]$ is at most $V(a) - V(b)$ where $V(x) = \mathrm{Var}(f(x), f'(x), \ldots, f^{(n)}(x))$ $(x \in R)$. Again, roots are counted with multiplicities here. A proof can be found in [109] Section 1.10 or in [13] Theorem 2.35.

2. Descartes' theorem is just the tip of an iceberg. The theory of *fewnomials*, initiated by Khovanskii in the 1980s, pursues the idea that a "simple" equation, or system of equations, with real coefficients should have a "simple" set of real solutions. Slightly more concretely, the "complexity" of the real solution set of a system $f_i(x_1, \ldots, x_n) = 0$ $(i = 1, \ldots, m)$ of real polynomial equations can be bound from above, for given m and n, in terms of the total number of monomials occurring in these equations. Complexity could mean, for example, the sum of the Betti numbers. In classical algebraic geometry, where complex solutions are considered, there exists no analogue of this phenomenon. See Khovanskii's original monograph [106], or [200], for much more information.

We are now going to see two methods that allow us to find the exact number of real roots of an arbitrary polynomial $f \in R[t]$, without having to calculate any of them. In contrast to the preceding discussion, we now *disregard* all multiplicities.

1.3.17 Definition. Let $f \in R[t]$ be a non-constant polynomial. The *Sturm sequence* for f is the sequence $(f_0, f_1, \ldots, f_r)$ of polynomials in $R[t]$, recursively defined as follows: $f_0 = f$, $f_1 = f'$ and

$$\begin{aligned} f_0 &= q_1 f_1 - f_2\,, \\ f_1 &= q_2 f_2 - f_3\,, \\ &\cdots \\ f_{r-2} &= q_{r-1} f_{r-1} - f_r\,, \\ f_{r-1} &= q_r f_r \end{aligned} \tag{1.3}$$

where $q_1, \ldots, q_r \in R[t]$ are such that $f_1, \ldots, f_r$ are non-zero and satisfy $\deg(f_i) < \deg(f_{i-1})$ $(i = 1, \ldots, r)$. Note that this determines $r \in \mathbb{N}$ uniquely, as well as the f_i and q_i. For $x \in R$ we put

$$v_f(x) := \mathrm{Var}(f_0(x), f_1(x), \ldots, f_r(x)).$$

If we change the minus signs on the right of (1.3) to plus signs, we get the usual form of the Euclidean algorithm for the pair (f, f'). Recall that the Euclidean algorithm calculates the greatest common divisor of the two polynomials. The minus signs are crucial for Sturm's algorithm to give correct results (see below), but they are inessential for Euclidean division. In particular, we see that $f_r = \gcd(f, f')$ (up to a non-zero scalar factor).

1.3.18 Theorem. (Sturm algorithm) *Given $a < b$ in R with $f(a)f(b) \neq 0$, the number of distinct roots of f in the interval $[a, b]$ is equal to $v_f(a) - v_f(b)$.*

Proof. Consider $v_f(x)$ as a function of $x \in R$. On the intervals between the finitely many real roots of $f_0 f_1 \cdots f_r$, this function is constant. How does $v_f(x)$ change when we pass through a root of $f_0 f_1 \cdots f_r$?

First assume that f and f' are relatively prime, so f has simple roots. Then the sequence $(f_0, f_1, \dots, f_r)$ satisfies

(1) $\gcd(f_0, f_1) = 1$,
(2) at any real root of f_0, the product $f_0 f_1$ changes sign from minus to plus,
(3) for $1 \leq i < r$ and every $c \in R$ with $f_i(c) = 0$, we have $f_{i-1}(c) f_{i+1}(c) < 0$.

Indeed, (2) is Lemma 1.3.7, and (3) follows from $f_{i-1} = q_i f_i - f_{i+1}$ and $\gcd(f_{i-1}, f_i) = 1$.

Let $c \in R$. By $x = c_\pm$ we denote values of x for which $\operatorname{sign}(x - c) = \pm 1$ and $|x - c|$ is sufficiently small. Let $0 \leq i < r$ with $f_i(c) = 0$. Then $f_{i+1}(c) \neq 0$ by (3), and we put $\varepsilon := \operatorname{sign} f_{i+1}(c)$. We are only going to use (1)–(3):

- If $i = 0$ then (2) implies the following scheme of signs:

	$x = c_-$	$x = c$	$x = c_+$
$f_0(x)$	$-\varepsilon$	0	ε
$f_1(x)$	ε	ε	ε
contribution to $v_f(x)$	1	0	0

- If $1 \leq i < r$ then (3) implies:[2]

	$x = c_-$	$x = c$	$x = c_+$
$f_{i-1}(x)$	$-\varepsilon$	$-\varepsilon$	$-\varepsilon$
$f_i(x)$	?	0	?
$f_{i+1}(x)$	ε	ε	ε
contribution to $v_f(x)$	1	1	1

We see, for every real zero c of f, that the function $v_f(x)$ drops by 1 at c, going from $x = c_-$ to $x = c_+$. Outside the real zeros of f the function $v_f(x)$ is locally constant everywhere. Therefore, if $a < b$ in R satisfy $f(a)f(b) \neq 0$ then $v_f(a) - v_f(b)$ is the number of distinct roots of f in the interval $[a, b]$.

Now let f be arbitrary, possibly with multiple roots, and put $g_i := f_i/f_r$ for $i = 0, \dots, r$. The sequence $(g_0, g_1, \dots, g_r)$ of polynomials satisfies the analogues of properties (1)–(3) above, although it need not be the Sturm sequence of g_0.

[2] We cannot decide the value of ? in this table, but it won't matter in what follows.

Moreover, if $x \in R$ satisfies $f_r(x) \neq 0$, then the sequences $(g_0(x), \dots, g_r(x))$ and $(f_0(x), \dots, f_r(x))$ have the same number of sign changes, since the second sequence is $f_r(x)$ times the first. Moreover, $f_0 = f$ and g_0 have the same roots. So the previous argument implies the assertion of the theorem in general. □

Given the Sturm sequence $(f_0, f_1, \dots, f_r)$ of f, let $d_i := \deg(f_i)$, and let c_i be the leading (highest) coefficient of f_i, for $i = 0, \dots, r$. Then for $x \ll 0$ we have

$$v_f(x) = \operatorname{Var}((-1)^{d_0}c_0, (-1)^{d_1}c_1, \dots, (-1)^{d_r}c_r) =: v_f(-\infty),$$

while

$$v_f(x) = \operatorname{Var}(c_0, c_1, \dots, c_r) =: v_f(+\infty)$$

for $x \gg 0$. Hence we get:

1.3.19 Corollary. *The total number of distinct real roots of non-constant $f \in R[t]$ is $v_f(-\infty) - v_f(+\infty)$.* □

1.3.20 Sturm's method may be refined further. Let polynomials $f, g \in R[t]$ be given where f is non-constant and $g \neq 0$. We would like to count the distinct real zeros ξ of f that satisfy $g(\xi) > 0$. To this end we define the *generalized Sturm sequence* for the pair (f, g), as follows: Let $f_0 = f$, $f_1 = f'g$, and let $f_2, \dots, f_r$ be defined as in (1.3) by $f_{i-1} = q_i f_i - f_{i+1}$ and $\deg(f_{i+1}) < \deg(f_i)$ $(i = 1, \dots, r-1)$, together with $f_i \neq 0$ and $f_{r-1} = q_r f_r$. For $x \in R$ put

$$v_{f,g}(x) := \operatorname{Var}(f_0(x), f_1(x), \dots, f_r(x)).$$

1.3.21 Theorem. (Generalized Sturm algorithm) *If $a < b$ in R satisfy $f(a)f(b) \neq 0$, then*

$$v_{f,g}(a) - v_{f,g}(b) = \sum_{\substack{a<c<b \\ f(c)=0}} \operatorname{sign} g(c).$$

So this is the number of (different) roots c of f with $g(c) > 0$, minus the number of roots c of f with $g(c) < 0$, both in the interval $[a, b]$. The previously stated version Theorem 1.3.18 corresponds to the case $g = 1$.

Proof. First we assume $\gcd(f, f'g) = 1$. The sequence $(f_0, \dots, f_r)$ satisfies (1), (3) and

(2′) for any $c \in R$ with $f_0(c) = 0$, the product $f_0 f_1$ changes sign from $-\operatorname{sign} g(c)$ to $\operatorname{sign} g(c)$. (Note that $g(c) \neq 0$.)

Again, this follows from Lemma 1.3.7. As before we study how $v_{f,g}(x)$ changes when we pass through a real root c of f_i. If $i \geq 1$, the former argument remains valid, showing that $v_{f,g}(x)$ does not change. Write $\varepsilon := \operatorname{sign} f_1(c)$ and note that $\varepsilon \neq 0$. If $i = 0$ we get the following modified sign scheme, using (2′):

	$x = c_-$	$x = c$	$x = c_+$
$f_0(x)$	$-\varepsilon \operatorname{sign} g(c)$	0	$\varepsilon \operatorname{sign} g(c)$
$f_1(x)$	ε	ε	ε
contribution to $v_{f,g}(x)$:			
if $g(c) < 0$	0	0	1
if $g(c) > 0$	1	0	0

Therefore $v_{f,g}(c_-) - v_{f,g}(c_+) = \operatorname{sign} g(c)$. In the general case, replace the sequence $(f_0, \dots, f_r)$ by $(f_0/f_r, \dots, f_r/f_r)$ and observe that (1), (2′) and (3) remain valid for the new sequence. Both sequences have the same number of sign changes at $x = a$ and at $x = b$, and so we get the claim in the same way as before. □

1.3.22 Example. Let us calculate the Sturm sequence for a cubic polynomial $f = t^3 + at + b$ with $a, b \in R$. Assuming $a \neq 0$ we find $f_0 = f$, $f_1 = f' = 3t^2 + a$ and $f_2 = -(\frac{2a}{3}t + b)$, $f_3 = -\frac{1}{4a^2}(4a^3 + 27b^2) = \frac{D}{4a^2}$ where $D = -(4a^3 + 27b^2)$ is the discriminant of f. When $a = 0$, the sequence gets shorter, namely it ends with $f_2 = -b$ if $b \neq 0$, and even with $f_1 = f'$ if $a = b = 0$. Although this is a baby example, it exhibits a characteristic feature of the Sturm algorithm: The iteration branches into a tree of subcases, depending on the concrete coefficients of the polynomial.

1.3.23 We now discuss a second approach to exact real root counting. It is due to Hermite and Sylvester and avoids the branching feature of the Sturm algorithm. First let K be an arbitrary field and let $f = t^n + a_1t^{n-1} + \cdots + a_n$ be a monic non-constant polynomial over K. Let $\alpha_1, \dots, \alpha_n$ be the roots of f in an algebraic closure $\overline{K}$ of K, so $f = \prod_{j=1}^n (t - \alpha_j)$. For $k = 0, 1, 2, \dots$ let

$$p_k = p_k(f) := \alpha_1^k + \cdots + \alpha_n^k,$$

the k-th *Newton sum* of f. Note that $p_k \in K$ since p_k is symmetric in the α_i. Explicitly we have $p_0 = n$, $p_1 = -a_1$, $p_2 = a_1^2 - 2a_2$, $p_3 = -a_1^3 + 3a_1a_2 - 3a_3$ etc. The Newton sums can be calculated recursively from the identity

$$p_k + p_{k-1}\, a_1 + p_{k-2}\, a_2 + \cdots + p_1\, a_{k-1} + k\, a_k = 0$$

for $k \geq 0$, where $a_j := 0$ for $j > n$ (Exercise 1.3.7). For an alternative approach to calculating the Newton sums see Exercise 1.3.8.

1.3.24 Definition. The symmetric $n \times n$ matrix

$$H(f) := (p_{j+k}(f))_{0 \leq j,k \leq n-1} = \begin{pmatrix} p_0 & p_1 & \cdots & p_{n-1} \\ p_1 & p_2 & \cdots & p_n \\ \vdots & \vdots & & \vdots \\ p_{n-1} & p_n & \cdots & p_{2n-2} \end{pmatrix}$$

(with coefficients in K) will be called the *Hermite matrix* of f.

The (j, k)-coefficient of $H(f)$ depends only on $j + k$. Matrices with this property are called *Hankel matrices*.

1.3.25 Remarks.

1. For $n \in \mathbb{N}$ and any ring A, the set (or A-module) of all symmetric $n \times n$ matrices over A is denoted $\mathrm{Sym}_n(A)$. Sylvester's inertia theorem for symmetric matrices over $\mathbb{R}$ is well known from linear algebra. The standard proof (induction on n) generalizes directly to give the following result: If K is an ordered field with positive cone P and $M \in \mathrm{Sym}_n(K)$ is any symmetric matrix, there exists an invertible matrix $S \in \mathrm{GL}_n(K)$ such that $S^\top MS = \mathrm{diag}(a_1, \ldots, a_n)$ is a diagonal matrix. The integer

$$\mathrm{sign}_P(M) := \sum_{i=1}^{n} \mathrm{sign}_P(a_i)$$

depends on M and P but not on S, and is called the *(Sylvester) signature* of M with respect to P. The matrix M is *positive definite* with respect to P, denoted $M >_P 0$, if $\mathrm{sign}_P(M) = n$, i.e. if $a_i >_P 0$ for $i = 1, \ldots, n$. If only the weak inequalities $a_i \geq_P 0$ hold for $i = 1, \ldots, n$, one says that M is *positive semidefinite*, or briefly *psd*, with respect to P, written $M \geq_P 0$.

2. If R is a real closed field and $M \in \mathrm{Sym}_n(R)$, every eigenvalue of M is real, i.e. contained in R. Indeed, the usual proof over $\mathbb{R}$ works over R as well. Corollary 1.3.15 therefore implies that the signature of M can be read off directly from the characteristic polynomial $p_M(t) = t^n + a_1 t^{n-1} + \cdots + a_n$ of M, namely

$$\mathrm{sign}(M) = \mathrm{Var}(1, a_1, a_2, \ldots, a_n) - \mathrm{Var}(1, -a_1, a_2, \ldots, (-1)^n a_n).$$

In particular, the matrix M is positive definite iff $(-1)^i a_i > 0$, and positive semidefinite iff $(-1)^i a_i \geq 0$ for $i = 1, \ldots, n$ (see also Exercise 1.3.5). In fact, these observations regarding the signature are true over any field with respect to any ordering. This will be clear once we have proved that every ordered field has a real closure (Proposition 1.4.2 below).

1.3.26 Theorem. *Let K be a field with* $\mathrm{char}(K) = 0$, *and let $f \in K[t]$ be a monic non-constant polynomial.*

(a) *The rank of $H(f)$ is the number of distinct roots of f in an algebraic closure $\overline{K}$ of K.*
(b) *If $K = R$ is real closed, the signature of $H(f)$ is the number of distinct roots of f in R.*

In particular, for $K = R$ real closed, the Sylvester signature of $H(f)$ is always non-negative.

1.3.27 Corollary. *Let $K = R$ be real closed. A polynomial $f \in R[t]$ is real-rooted if and only if its Hermite matrix $H(f)$ is positive semidefinite.*

Proof. A symmetric matrix over R is positive semidefinite if and only if its signature is equal to its rank. □

Proof of Theorem 1.3.26. (a) Let $\alpha_1, \dots, \alpha_n \in \overline{K}$ with $f = \prod_{j=1}^n (t - \alpha_j)$. The Hermite matrix can be factored as $H(f) = V^\top \cdot V$, where

$$V := \begin{pmatrix} 1 & \alpha_1 & \cdots & \alpha_1^{n-1} \\ 1 & \alpha_2 & \cdots & \alpha_2^{n-1} \\ \vdots & \vdots & & \vdots \\ 1 & \alpha_n & \cdots & \alpha_n^{n-1} \end{pmatrix}$$

is the Vandermonde matrix. This gives a diagonalization (over $\overline{K}$) of the quadratic form represented by $H(f)$: Writing $x = (x_0, \dots, x_{n-1})^\top$ for a column vector of variables, we get

$$x^\top H(f)\, x = (Vx)^\top (Vx) = \sum_{j=1}^n L_{\alpha_j}(x)^2,$$

where L_α denotes the linear form $L_\alpha(x) = \sum_{k=0}^{n-1} \alpha^k x_k$ ($\alpha \in \overline{K}$). If we label the α_j in such a way that $\alpha_1, \dots, \alpha_r$ are the distinct roots of f, the linear forms $L_{\alpha_1}, \dots, L_{\alpha_r}$ are linearly independent since the r-th principal minor of V is non-zero. This already proves $\operatorname{rk} H(f) = r$.

(b) Now let $K = R$ be real closed. Let $i = \sqrt{-1}$, and let $z \mapsto \overline{z}$ denote the non-trivial automorphism of $R(i)$ over R ("complex conjugation"). As usual, write $\operatorname{Re}(z) = \frac{1}{2}(z + \overline{z})$, $\operatorname{Im}(z) = \frac{1}{2i}(z - \overline{z})$ for $z \in R(i)$. For $\alpha \in R(i)$ let $\operatorname{Re}(L_\alpha) := \frac{1}{2}(L_\alpha + L_{\overline{\alpha}})$ and $\operatorname{Im}(L_\alpha) := \frac{1}{2i}(L_\alpha - L_{\overline{\alpha}})$. These are linear forms with coefficients in R. With this notation we have $L_\alpha = \operatorname{Re}(L_\alpha) + i \operatorname{Im}(L_\alpha)$, $L_{\overline{\alpha}} = \operatorname{Re}(L_\alpha) - i \operatorname{Im}(L_\alpha)$, and so

$$L_\alpha^2 + L_{\overline{\alpha}}^2 = 2\operatorname{Re}(L_\alpha)^2 - 2\operatorname{Im}(L_\alpha)^2.$$

This shows that every pair $\alpha_j \neq \overline{\alpha}_j$ of non-real complex conjugate roots contributes a difference of two squares of real linear forms. From this observation, assertion (b) is clear. □

1.3.28 Corollary. *The determinant of the Hermite matrix is* $\det H(f) = D(f)$, *the discriminant of* f.

Proof. $H(f) = V^\top V$, as we saw in the previous proof. So the assertion follows from the formula for the Vandermonde determinant since $D(f) = \prod_{i<j} (\alpha_j - \alpha_i)^2$. □

1.3.29 Example. For an illustration we again consider a real closed field R and a cubic polynomial $f = t^3 + at + b$ with $a, b \in R$ (compare Example 1.3.22). From abstract algebra it is well-known that the splitting pattern of f over R depends only on the sign of the discriminant $D = D(f) = -(4a^3 + 27b^2)$. Let us see how to recover this fact via the Hermite matrix. We have

$$H(f) = \begin{pmatrix} 3 & 0 & -2a \\ 0 & -2a & -3b \\ -2a & -3b & 2a^2 \end{pmatrix}$$

with characteristic polynomial $p_{H(f)} = t^3 + c_1 t^2 + c_2 t + c_3$ where $c_1 = -(2a^2 - 2a + 3)$, $c_2 = -4a^3 + 2a^2 - 6a - 9b^2$ and $c_3 = -D$. To calculate the signature of $H(f)$ we use Remark 1.3.25.2. Observe that $c_1 = -\frac{1}{2}((2a-1)^2 + 5) < 0$ and $c_2 = D + 2a(a-3) + 18b^2$. When $D > 0$, we have $a < 0$ and hence $c_2 > 0$, so in this case the signature of $H(f)$ is (with slightly sloppy notation)

$$\mathrm{Var}(1,\ c_1 < 0,\ c_2 > 0,\ -D < 0) - \mathrm{Var}(1,\ -c_1 > 0,\ c_2 > 0,\ D > 0) \;=\; 3$$

(*casus irreducibilis* of classical algebra, corresponding to three distinct real roots). For $D = 0$ and $(a, b) \neq (0, 0)$ it remains true that $c_2 > 0$, and we obtain $\operatorname{sign} H(f) = 2$ (one double real root plus one simple real root). If $D < 0$ we get

$$\operatorname{sign} H(f) \;=\; \mathrm{Var}(1,\ c_1 < 0,\ c_2,\ -D > 0) - \mathrm{Var}(1,\ -c_1 > 0,\ c_2,\ D < 0) \;=\; 1$$

independent of the sign of c_2 (one real root and a pair of non-real complex conjugate roots).

The discussion becomes easier if we calculate the signature of $H(f)$ from the principal minors of $H(f)$, as outlined in Remark 1.3.38 below. The sequence of principal minors is 3, $-6a$, D, and the criterion in 1.3.38 applies unless $a = b = 0$, giving the same result somewhat more smoothly.

1.3.30 Remark. The quadratic form represented by the Hermite matrix $H(f)$ is an instance of a trace form. Recall that if k is a ring and A is a k-algebra that is finitely generated and free[3] as a k-module, the A/k-trace of an element $a \in A$, denoted $\mathrm{tr}_{A/k}(a)$, is defined to be the trace of the k-linear map $\mu_a \colon A \to A$, $\mu_a(x) = ax$. The trace (bilinear) form of A over k is the symmetric k-bilinear form $A \times A \to k$, $(a, b) \mapsto \mathrm{tr}_{A/k}(ab)$.

Let K be a field and let $f = \prod_{j=1}^{n}(t - \alpha_j)$ be a monic polynomial in $K[t]$, with $\alpha_1, \ldots, \alpha_n$ in some algebraic closure $\overline{K}$ of K. Consider the residue ring $A := K[t]/\langle f \rangle$ as a K-algebra, and write $\overline{g} := g + \langle f \rangle \in A$, for $g \in K[t]$. Then

$$\mathrm{tr}_{A/K}(\overline{g}) \;=\; \sum_{j=1}^{n} g(\alpha_j),$$

see Exercise 1.3.13. Therefore, with respect to the natural K-basis $1, \overline{t}, \ldots, \overline{t}^{n-1}$ of A, the matrix of the trace bilinear form is just the Hermite matrix $H(f)$, since

$$p_k \;=\; \sum_{j=1}^{n} \alpha_j^k \;=\; \mathrm{tr}_{A/R}(\overline{t}^k)$$

for all $k \geq 0$.

[3] free could be replaced by projective

1.3.31 Similar as for Sturm's method, there exists a generalization of Hermite's approach that allows real zeros to be counted under side conditions. As before, let $f \in K[t]$ be monic of degree $n \geq 1$ and with roots $\alpha_1, \dots, \alpha_n$ in $\overline{K}$. For arbitrary $g \in K[t]$, define *relative Newton sums* by

$$p_k(f, g) := \sum_{j=1}^{n} \alpha_j^k \, g(\alpha_j) \quad (k \geq 0).$$

The $n \times n$ matrix

$$H(f, g) := (p_{j+k}(f, g))_{0 \leq j,k \leq n-1} = \begin{pmatrix} \tilde{p}_0 & \tilde{p}_1 & \cdots & \tilde{p}_{n-1} \\ \tilde{p}_1 & \tilde{p}_2 & \cdots & \tilde{p}_n \\ \vdots & \vdots & & \vdots \\ \tilde{p}_{n-1} & \tilde{p}_n & \cdots & \tilde{p}_{2n-2} \end{pmatrix}$$

(where $\tilde{p}_\nu := p_\nu(f, g)$) will be called the *generalized Hermite matrix* of f and g. If $g = \sum_l b_l t^l$ then

$$p_k(f, g) = \sum_j \alpha_j^k \sum_l b_l \alpha_j^l = \sum_{j,l} b_l \alpha_j^{k+l} = \sum_l b_l p_{k+l}(f) \tag{1.4}$$

for all k. So relative Newton sums are linear combinations of ordinary Newton sums. In particular $p_k(f, g) \in K$. For later use we record:

1.3.32 Lemma. *If f, $g \in K[t]$ and f is monic, the coefficients $p_k(f, g)$ of $H(f, g)$ are integer polynomials in the coefficients of f and g.*

We can make the statement more precise: If $m \geq 0$, $n \geq 1$ are fixed, then for every $k \geq 0$ there exists a universal $\mathbb{Z}$-polynomial

$$P_k(a, b) = P_k(a_1, \dots, a_n, \, b_0, \dots, b_m)$$

in the variables $a = (a_1, \dots, a_n)$ and $b = (b_0, \dots, b_m)$, such that for all polynomials $f = t^n + \sum_{i=1}^n a_i t^{n-i}$ and $g = \sum_{j=0}^m b_j t^j$ one has

$$p_k(f, g) = P_k(a, b).$$

Proof. This follows inductively from the recursive formula for the Newton sums, see Exercise 1.3.7. By (1.4), $P_k(a, b)$ is linear in the b_j and has degree $k + m$ in the a_i. □

1.3.33 Theorem. *Let K be field,* $\mathrm{char}(K) = 0$, *and let $f \in K[t]$ be a monic polynomial of degree $n \geq 1$. For any polynomial $g \in K[t]$ we have*

$$\mathrm{rk}\, H(f, g) = \left| \{\alpha \in \overline{K} \colon f(\alpha) = 0 \text{ and } g(\alpha) \neq 0\} \right|.$$

If $K = R$ is real closed then

$$\operatorname{sign} H(f,g) = \sum_{\substack{\alpha\in R\\ f(\alpha)=0}} \operatorname{sign} g(\alpha)\,.$$

Note that this generalizes Theorem 1.3.26, which corresponds to the case $g = 1$.

Proof. Write $x = (x_0, \dots, x_{n-1})^\top$ (column vector) as before, then

$$x^\top \cdot H(f,g)\cdot x = \sum_{j,k=0}^{n-1}\sum_{l=1}^{n} g(\alpha_l)\,\alpha_l^{j+k}\,x_jx_k = \sum_{l=1}^{n} g(\alpha_l)\,L_{\alpha_l}(x)^2,$$

using the notation from the proof of 1.3.26. The assertions of 1.3.33 are verified in a similar way as there. Indeed, using

$$\operatorname{Re}((a+ib)(u+iv)^2) = au^2 - 2buv - av^2$$

($a,\ b,\ u,\ v \in R$) we see for $\alpha, \beta \in R(i)$ with $\alpha \neq \overline{\alpha}$ and $\beta \neq 0$ that the quadratic form

$$\beta L_\alpha(x)^2 + \overline{\beta} L_{\overline{\alpha}}(x)^2$$

is equal to $\lambda_1(x)^2 - \lambda_2(x)^2$ with linearly independent linear forms $\lambda_1,\ \lambda_2$ over R. □

1.3.34 Remarks. Let K be a field and let $f,\ g \in K[t]$ with f monic and non-constant.

1. If $f = t^3 + at + b$ and $g = t + c$ with $a,\ b,\ c \in K$, the generalized Hermite matrix is

$$H(f,g) = c\begin{pmatrix} 3 & 0 & -2a\\ 0 & -2a & -3b\\ -2a & -3b & 2a^2\end{pmatrix} + \begin{pmatrix} 0 & -2a & -3b\\ -2a & -3b & 2a^2\\ -3b & 2a^2 & 5ab\end{pmatrix}$$

(cf. Example 1.3.29).

2. Generalizing Remark 1.3.30, the generalized Hermite matrix $H(f,g)$ corresponds to the *scaled trace form*

$$A\times A \to K, \quad (\overline{p},\overline{q}) \mapsto \operatorname{tr}_{A/K}(\overline{pqg})$$

(with scaling factor $\overline{g} \in A = K[t]/\langle f\rangle$).

1.3.35 We can even count real zeros under an arbitrary finite number of side conditions, using either Sturm's or Hermite's method. To explain this, let polynomials $f,\ g_1, \dots, g_r \in R[t]$ be given with $f \neq 0$. For any tuple $e = (e_1, \dots, e_r)$ of non-negative integers write $g^e := g_1^{e_1}\cdots g_r^{e_r}$, and let

$$N_e := \sum_{c\in R:\ f(c)=0} \operatorname{sign} g^e(c).$$

As we have seen, the numbers N_e can be found effectively from the coefficients of f and the g_i, following Sturm or Hermite. The next lemma and the following remark show how to recover, from these numbers N_e, the number of real zeros c of f with prescribed signs of $g_1(c), \dots, g_r(c)$:

1.3.36 Proposition. *Let $f, g_1, \dots, g_r \in R[t]$ with $f \neq 0$ and $r \geq 0$. Then*

$$2^{-r} \sum_{e \in \{1,2\}^r} N_e = \left| \{c \in R\colon\ f(c) = 0,\ g_1(c) > 0, \dots, g_r(c) > 0\} \right|.$$

Proof. This is a beautiful trick: We have

$$\begin{aligned}\sum_{e \in \{1,2\}^r} N_e &= \sum_{\substack{c \in R:\\ f(c)=0}} \sum_{e \in \{1,2\}^r} \operatorname{sign}(g_1(c)^{e_1} \cdots g_r(c)^{e_r}) \\ &= \sum_{\substack{c \in R:\\ f(c)=0}} \prod_{i=1}^{r} (\operatorname{sign} g_i(c) + \operatorname{sign} g_i(c)^2)\end{aligned}$$

where the first equality is the definition of N_e and the second comes from distributively expanding the products in the last sum. Since for $\varepsilon_1, \dots, \varepsilon_r \in \{0, \pm 1\}$ one has

$$\prod_{i=1}^{r} (\varepsilon_i + \varepsilon_i^2) = \begin{cases} 2^r & \text{if } \varepsilon_1 = \cdots = \varepsilon_r = 1, \\ 0 & \text{else,} \end{cases}$$

we are done. □

1.3.37 Remark. Proposition 1.3.36 allows us to determine the number

$$\left| \{c \in R\colon\ f(c) = 0,\ \operatorname{sign} g_j(c) = \varepsilon_j \text{ for } j = 1, \dots, r\} \right|$$

for any prescribed tuple of signs $\varepsilon = (\varepsilon_1, \dots, \varepsilon_r) \in \{0, 1, -1\}^r$. Indeed, if $\varepsilon_j \neq 0$ for all j, this is achieved by the lemma if we replace g_j by $\varepsilon_j g_j$ for $j = 1, \dots, r$. If (e.g.) $\varepsilon_1 = 0$, replace f by $\gcd(f, g_1)$ and delete g_1, etc.

1.3.38 Remark. As mentioned in Remark 1.3.25.2, the signature of a symmetric matrix M over R can be read off from the characteristic polynomial of M. Under suitable conditions it can also be determined from the signs of the principal minors of M. If $M = (a_{ij})_{1 \leq i,j \leq n}$ has rank $r = \operatorname{rk}(M)$, and if $d_k = \det(a_{ij})_{1 \leq i,j \leq k}$ denotes the k-th principal minor of M ($k = 1, \dots, n$), then

$$\operatorname{sign}(A) = n - 2 \operatorname{Var}(1, d_1, \dots, d_r),$$

provided that $d_r \neq 0$ and the sequence $d_1, \dots, d_r$ doesn't contain two successive zeros. This is an easy exercise in linear algebra. There is also a generalization if two successive zeros occur [69]. When M is a Hankel matrix, like the (generalized) Hermite matrices, the above rule can be modified to work in general. This is due to Frobenius, see [69] (ch. X §10) or [13] (Section 6.2.2) for details.

Exercises

Let R always be a real closed field.

1.3.1 Let (k, P) be an ordered field, let A be a symmetric $n \times n$ matrix over k. For any subset $I \neq \varnothing$ of $[n] = \{1, \dots, n\}$, let $d_I(A)$ be the $I \times I$-minor of A (i.e., the determinant of the matrix obtained from A by deleting all rows and columns whose index is outside I). For $i \in I$ let $d_i(A) = d_{\{1,\dots,i\}}(A)$ be the i-th principal minor of A.

(a) $A >_P 0 \Leftrightarrow d_i(A) > 0$ for $i = 1, \dots, n$.
(b) $A \geq_P 0 \Leftrightarrow d_I(A) \geq 0$ for every non-empty subset I of $[n]$.
(c) There are easy examples where $d_i(A) \geq 0$ for $i = 1, \dots, n$ but A is not positive semidefinite. Does there exist such an example where $d_n(A) = \det(A) > 0$?

Remark: (a), (b) are classical for $k = \mathbb{R}$, and the usual proofs work over any real closed field. The proof of "$\Leftarrow$" in (b) is slightly more tricky than the rest.

1.3.2 Let $0 \neq f \in R[t]$, and assume that f is real-rooted. Show that the same is true for the derivative $f' = df/dt$. Moreover show that every root of f' of multiplicity $m \geq 2$ is a root of f (of multiplicity $m + 1$).

1.3.3 Fill in the details for the proof of Theorem 1.3.33.

1.3.4 Let $f(t) \neq 0$ be a real-rooted polynomial in $\mathbb{R}[t]$, and let $g(t) = f(t) + s \cdot f'(t)$ where $s \neq 0$ is a real number. Show that $g(t)$ is real-rooted as well. If $k \geq 1$ is the largest multiplicity of a root of $f(t)$, prove that the largest multiplicity of a root of $g(t)$ is $k - 1$.

1.3.5 Let $(K, \leq)$ be an ordered field and let

$$f = t^n + \sum_{i=1}^{n} a_i t^{n-i} = \prod_{j=1}^{n} (t - \xi_j)$$

be a monic polynomial in $K[t]$ that splits over K (with $a_i, \xi_j \in K$). Prove that

$$\xi_1, \dots, \xi_n \geq 0 \quad \Leftrightarrow \quad (-1)^i a_i \geq 0 \;\; \text{for } i = 1, \dots, n,$$

and similarly with strict instead of non-strict inequalities. (The statement is a direct consequence of Corollary 1.3.15. Try to give a direct proof here.)

1.3.6 Let $f = \sum_{i=0}^{n} a_i t^i \in R[t]$ be of degree n with $f(0) \neq 0$. If there exists $1 < i < n$ with $a_{i-1} = a_i = 0$, show that f has a non-real root. (*Hint*: Repeat the argument in the proof of Corollary 1.3.15, omitting positions $i - 1$ and i in the sequences.)

1.3.7 Let K be a field with algebraic closure $\overline{K}$. Let $f = t^n + a_1 t^{n-1} + \cdots + a_n \in K[t]$ have the roots $\alpha_1, \dots, \alpha_n$ in $\overline{K}$, and let $p_r = p_r(f) = \sum_{i=1}^{n} \alpha_i^r$ be the r-th Newton sum ($r \geq 0$) of f.

(a) Prove *Newton's identity*

$$p_r + p_{r-1} a_1 + p_{r-2} a_2 + \cdots + p_1 a_{r-1} + r a_r = 0$$

for $r \geq 0$. Here we put $a_r = 0$ for $r > n$.
(b) Show for $r \geq 0$ that $p_r(f) = p_r(a_1, \dots, a_n)$ is a polynomial in $a_1, \dots, a_n$ with integer coefficients. If we declare $\deg(a_i) = i$ then p_r is weighted homogeneous of degree r.
(c) Prove the identity

$$\frac{f'}{f} = \frac{p_0}{t} + \frac{p_1}{t^2} + \frac{p_2}{t^3} + \cdots = \sum_{r=0}^{\infty} \frac{p_r}{t^{r+1}}$$

of formal power series in $\frac{1}{t}$. In other words, the Newton sums $p_r(f)$ are the coefficients of the Taylor expansion of $\frac{f'(t)}{f(t)}$ around $t = \infty$.

1.3.8 Let again $f = t^n + \sum_{i=1}^n a_i t^{n-i} \in K[t]$ where K is a field. For an alternative approach to the Newton sums of f, consider the $n \times n$ matrix

$$A = \begin{pmatrix} 0 & & & & -a_n \\ 1 & 0 & & & -a_{n-1} \\ & \ddots & \ddots & & \vdots \\ & & 1 & 0 & -a_2 \\ & & & 1 & -a_1 \end{pmatrix}$$

(the companion matrix of f). Show that $p_r(f) = \operatorname{tr}(A^r)$ for all $r \geq 0$. (*Hint*: Calculate the characteristic polynomial of A.)

1.3.9 Let $f = t^3 + at + b \in R[t]$ with $a, b \in R$. Find polynomial conditions on the pair (a, b) that are necessary and sufficient for f to have three real zeros that are larger than -1.

1.3.10 Calculate the Hermite matrix of $f = t^4 + at^2 + bt + c$ and its sequence of principal minors. Find a condition on $a, b, c \in R$ that is equivalent to the existence of a zero of f in R. (Use a computer algebra system.)

1.3.11 Let $\alpha \in \mathbb{C}$ with $\alpha^5 - 4\alpha^3 - 2\alpha + 6 = 0$. How many orderings does the field $\mathbb{Q}(\alpha)$ have that make α positive?

1.3.12 For $a, b \in R$ consider the polynomial $f = t^5 + at^2 + b$, and assume that the discriminant D of f is non-zero. Determine the number of real roots of f

(a) using the Sturm sequence,
(b) using Hermite's method.

The answer depends only on D. (The discriminant is $D = b(3125\, b^3 + 108\, a^5)$, this need not be proved.)

1.3.13 Let K be a field, let $f \in K[t]$ be a monic polynomial of degree n, and write $A = K[t]/\langle f \rangle$ for the residue ring. If $\alpha_1, \dots, \alpha_n$ are the roots of f in an algebraic closure $\overline{K}$ of K, show that $\operatorname{tr}_{A/K}(\overline{g}) = \sum_{j=1}^n g(\alpha_j)$ for any $g \in K[t]$, where $\overline{g} = g + \langle f \rangle \in A$.

1.3.14 Let $0 \neq f \in R[t]$, and let r be the number of non-real roots of f, counting with multiplicities. The *Hawaii conjecture* (which is due to Gauss) states that

$$\frac{d}{dt}\left(\frac{f'}{f}\right) = \frac{ff'' - f'^2}{f^2}$$

has at most r real zeros (again counting with multiplicities). Prove the Hawaii conjecture in the case when all roots of f are real. (The full conjecture has been proved by Tyaglov [209].)

1.4 Real Closure of an Ordered Field

1.4.1 Definition. Let K be a field with a positive cone $P \subseteq K$. A *real closure* of the ordered field (K, P) is a real closed algebraic field extension R of K whose order extends P, i.e. such that $P = R_+ \cap K$.

1.4.2 Proposition. *Any ordered field has a real closure.*

Proof. Let (K, P) be an ordered field and fix an algebraic closure $\overline{K}$ of K. The set $\mathcal{X}$ of ordered fields (L, Q) such that $K \subseteq L \subseteq \overline{K}$ and $Q \cap K = P$ is partially ordered by

$$(L_1, Q_1) \leq (L_2, Q_2) \quad \Leftrightarrow \quad L_1 \subseteq L_2 \text{ and } Q_1 = L_1 \cap Q_2.$$

Zorn's lemma can be applied, and so $\mathcal{X}$ contains a maximal element (L, Q). This field L is a real closure of (K, P) by 1.2.7(ii). □

1.4.3 Definition. Let K be a field (with no ordering specified). A field extension R of K is *a real closure* of K if the extension $K \subseteq R$ is algebraic and the field R is real closed.

In other words, a real closure of K is simply a real closure of (K, P) for some positive cone P of K.

We are going to show that a real closure of (K, P) is unique in a strong sense.

1.4.4 Lemma. *Let K be a field, and let R_1, R_2 be two real closed field extensions of K that both induce the same ordering on K. Then every polynomial $f \in K[t]$ has the same number of distinct roots in R_1 and in R_2.*

Proof. Using Sturm's method, we argue as follows. The function $v_f(x)$ (Definition 1.3.17) counts the number of different real roots of f in a real closed extension of K. By construction, it depends only on expressions that lie in the subfield generated by the coefficients of f, and on the induced ordering of that subfield. Since R_1, R_2 induce the same ordering on K, we get the same count in R_1 and in R_2. A similar argument works if we use Hermite's method. □

1.4.5 Lemma. *Let (K, P) be an ordered field, let R be a real closure of (K, P) and let $K \subseteq L \subseteq R$ be an intermediate field with $[L : K] < \infty$. Moreover let S be another real closed field, and let $\varphi\colon (K, P) \to (S, S_+)$ be an order embedding (1.1.19). Then φ has an order-compatible extension $\psi\colon (L, R_+ \cap L) \to (S, S_+)$:*

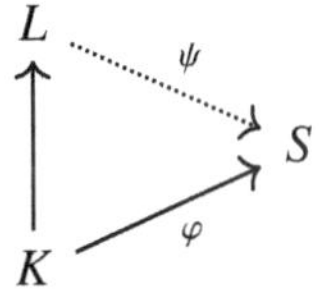

Proof. By the primitive element theorem, $L = K(\alpha)$ is a simple extension of K. The minimal polynomial f of α over K has a root in R, namely α. So f has a root in S as well, by Lemma 1.4.4. Hence there exists an extension $L \to S$ of φ. Let $\psi_1, \ldots, \psi_r$ be all such extensions, and assume that none of them is order-compatible with respect to $R_+ \cap L$. This means that there are elements $b_1, \ldots, b_r$ in L with $b_i > 0$ (in R) but $\psi_i(b_i) < 0$ (in S). Now consider the extension $L' := L(\sqrt{b_1}, \ldots, \sqrt{b_r}) \subseteq R$ of L. None of the ψ_i can be extended to L', and so φ does not extend to L'. This contradicts the first part of the proof. □

1.4.6 Theorem. *Let (K, P) be an ordered field, let R be a real closure of (K, P), and let $\varphi\colon (K, P) \to (S, S_+)$ be an order embedding into another real closed field S. Then φ has a unique extension $\psi\colon R \to S$.*

Proof. The set

$$\mathcal{X} := \Big\{(L, \psi)\colon K \subseteq L \subseteq R,\ \psi \in \mathrm{Hom}(L, S) \text{ with } \psi|_K = \varphi \text{ and } \psi(L \cap R_+) \subseteq S_+\Big\}$$

consists of all extensions of φ to intermediate fields L of R/K that are compatible with the positive cone $L \cap R_+$ of L. The set $\mathcal{X}$ is partially ordered by the extension relation, and it has a maximal element (L, ψ) by Zorn's lemma. Lemma 1.4.5 implies that $L = R$. Hence an extension $\psi\colon R \to S$ of φ exists, and ψ is automatically order-compatible since R is real closed. To prove that ψ is unique let $\alpha \in R$, let f be the minimal polynomial of α over K, and let $\alpha_1 < \cdots < \alpha_r$ (resp. $\beta_1 < \cdots < \beta_r$) be the roots of f in R (resp. in S). The number of roots in R and S is the same by Lemma 1.4.4. Then we clearly have

$$\{\psi(\alpha_1), \ldots, \psi(\alpha_r)\} = \{\beta_1, \ldots, \beta_r\}.$$

Since ψ is order-compatible, we must have $\psi(\alpha_i) = \beta_i$ for $i = 1, \ldots, r$, which shows that ψ is uniquely determined. □

1.4.7 Corollary. *Let (K, P) be an ordered field, and let R_1, R_2 be two real closures of (K, P). Then there is a unique K-homomorphism $\varphi\colon R_1 \to R_2$, and φ is an isomorphism.*

Proof. Clear from Theorem 1.4.6. □

1.4.8 Remark. The construction of a real closure of an ordered field shows many similarities to the construction of an algebraic closure. More interesting than the similarities, however, are the differences. For a general field K, the automorphism group $\mathrm{Aut}(\overline{K}/K)$ of $\overline{K}$ over K is infinite (and comes with a natural profinite topology). Therefore $\overline{K}$ is unique only up to *non-canonical* isomorphism over K. On the other hand we have seen that any real closure R of K is *rigid* over K, namely $\mathrm{Aut}(R/K) = \{\mathrm{id}\}$. This means that the real closure of an ordered field is unique up to *unique* isomorphism over K. Other than for the algebraic closure, it is therefore perfectly justified to speak of *the real closure* of an ordered field (K, P).

The difference between algebraic and real closure is also manifest in the construction of such a closure. Although we used Zorn's lemma in our existence proof 1.4.2, this can be avoided ([175], [129]). On the other hand, it can be shown that at least some weakened version of Zorn's lemma is needed to prove the existence of an algebraic closure. (There exists a model of ZF set theory in which certain fields don't have an algebraic closure [152].)

1.4.9 Remark. If K is a subfield of a real closed field R and $P = K \cap R_+$ is the induced ordering on K, then the relative algebraic closure of K in R (consisting of all elements of R that are algebraic over K) is a real closure of (K, P) (Exercise 1.4.1). For example, the subfield $R_0 \subseteq \mathbb{R}$ of real algebraic numbers is the real closure of the field $\mathbb{Q}$ (with respect to the unique ordering of $\mathbb{Q}$). Therefore R_0 is the smallest real closed field and embeds uniquely into any other real closed field.

1.4.10 Example. For a more elaborate example consider $R(t)$, the field of rational functions in one variable over the real closed field R, together with the positive cone $P = P_{0,+}$, see 1.1.13. We describe the real closure of $(R(t), P)$. Recall that a (formal) Laurent series over R is a formal series

$$f = \sum_{k=m}^{\infty} a_k t^k$$

with $m \in \mathbb{Z}$ and $a_k \in R$ for all k. These series form the field $R((t))$, which is the field of fractions of the ring $R[[t]]$ of formal power series. A (formal) *Puiseux series* over R is a formal Laurent series in $t^{1/d}$, for some integer $d \geq 1$. If $d, e \geq 1$ are integers and e is a multiple of d, one has the natural inclusion $R((t^{1/d})) \subseteq R((t^{1/e}))$ of fields. Therefore the union

$$R((t^{1/\infty})) := \bigcup_{d\geq 1} R((t^{1/d}))$$

is a field extension of R, called the field of Puiseux series over R. If we adjoin $\sqrt{-1}$ to this field we get the field of Puiseux series over $R(\sqrt{-1})$. By Puiseux' theorem, this field is algebraically closed (see A.4.10). Therefore, since $R((t^{1/\infty}))$ is a real field, it is real closed. The positive cone of $R(t)$ that is induced from the inclusion $R(t) \subseteq R((t^{1/\infty}))$ is $P_{0,+}$. Indeed, $t = (t^{1/2})^2$ is positive in $R((t^{1/\infty}))$, and from

$$1 - t = \left(\sum_{k=0}^{\infty} (-1)^k \binom{1/2}{k} t^k \right)^2$$

we see that $c - t > 0$ in $R((t^{1/\infty}))$ for every $c > 0$ in R. From the argument in 1.4.9 it follows that

$$R((t^{1/\infty}))_{\mathrm{alg}} := \{f \in R((t^{1/\infty})) \colon f \text{ is algebraic over } R(t)\}$$

is the real closure of $(R(t), P_{0,+})$.

The problem of extending an ordering to a finite field extension is equivalent to a field embedding problem:

1.4.11 Corollary. *Let (K, P) be an ordered field, let $\varphi\colon (K, P) \to S$ be an order-compatible homomorphism into a real closed field S. For every algebraic extension L/K there exists a natural bijective map*

$$\{\text{extensions } \psi\colon L \to S \text{ of } \varphi\} \longrightarrow \{\text{extensions } Q \text{ of } P \text{ to } L\},$$

given by $\psi \mapsto \psi^{-1}(S_+)$.

Proof. If $\psi\colon L \to S$ extends φ, the positive cone $\psi^{-1}(S_+)$ of L extends the positive cone P of K since $\psi^{-1}(S_+) \cap K = \varphi^{-1}(S_+) = P$. Conversely let Q be an extension of P to L. The real closure R of (L, Q) is also the real closure of (K, P). So φ has an extension $\tilde{\varphi}\colon R \to S$, by 1.4.6. The restriction $\psi := \tilde{\varphi}|_L$ of $\tilde{\varphi}$ satisfies $\psi^{-1}(S_+) = L \cap R_+ = Q$.

This shows that the map in the statement is well-defined and surjective. To prove that it is also injective, let $\psi_1, \psi_2 : L \to S$ be two homomorphisms that satisfy $\psi_i|_K = \varphi$ and $\psi_i^{-1}(S_+) = Q$ $(i = 1, 2)$. Let $\chi_i : R \to S$ be the (unique) extension of ψ_i to the real closure R of (L, Q) (Theorem 1.4.6), for $i = 1, 2$. Then $\chi_1|_K = \chi_2|_K = \varphi$, and therefore $\chi_1 = \chi_2$ by the uniqueness part of 1.4.6. Hence $\psi_1 = \psi_2$, as desired. □

1.4.12 Corollary. *Let (K, P) be an ordered field with real closure R, and let L/K be a finite extension. If α is a primitive element of L/K and f denotes the minimal polynomial of α over K, there is a natural bijective map*

$$\{\textit{roots of } f \textit{ in } R\} \longrightarrow \{\textit{extensions of } P \textit{ to } L\}.$$

Proof. Immediate from Corollary 1.4.11, since the roots of f in R correspond to the K-embeddings $L \to R$. □

1.4.13 Remark. The bijection from 1.4.12 is as follows: If β is a root of f in R and $\varphi : L \to R$ denotes the K-embedding with $\varphi(\alpha) = \beta$, then β corresponds to the positive cone $Q := \varphi^{-1}(R_+)$ of L. More explicitly,

$$\operatorname{sign}_Q g(\alpha) = \operatorname{sign}_R g(\beta)$$

holds for any polynomial $g \in K[t]$.

1.4.14 Remark. If (K, P) is an ordered field with real closure R, one may uniquely encode each element of R by data in K, as follows. An element $\alpha \in R$ is determined by its minimal polynomial f and by the position of α in the ordered list of roots of f in R. So the elements of R are in natural bijection with all pairs (f, i), where $f \in K[t]$ is monic irreducible and i is an integer satisfying $1 \le i \le m(f)$, where $m(f)$ denotes the total number of roots of f in R. The number $m(f)$ can be determined from K and P alone, by Sturm's or Hermite's method. Note that a similar encoding is impossible for the algebraic closure of K. Both facts are directly related to the rigidity property of real closures and its failure for the algebraic closure (Remark 1.4.8). Later we will see an even better encoding of the elements of R (using Thom's lemma, see Remark 4.3.21).

Exercises

1.4.1 Let R be a real closed field, let K be a subfield of R and let R' be the relative algebraic closure of K in R. Then R' is a real closure of $(K, K \cap R_+)$.

1.4.2 Let R be a real closed field, let $\xi = (I, J)$ be a Dedekind cut of R, and let

$$P_\xi := \Big\{ f \in R(t) \colon \exists\, a \in I \cup \{-\infty\}\ \exists\, b \in J \cup \{\infty\}\ \forall\, x \in\,]a, b[\ \ f(x) \ge 0 \Big\}.$$

(Here $f(x) \ge 0$ means in particular that f doesn't have a pole at x.)

(a) P_ξ is a positive cone of $R(t)$.
(b) Every positive cone of $R(t)$ has the form P_ξ for precisely one Dedekind cut ξ of R.

(c) The extension $(R, R_+) \subseteq (R(t), P_\xi)$ is relatively Archimedean (see Exercise 1.2.4) if, and only if, the Dedekind cut ξ is free.

1.4.3 Let k be a field. Prove the following amalgamation property of ordered fields (statement (b)):

(a) Let $k \subseteq L$ be a field extension, let P be a positive cone of L and Q a positive cone of $k(t)$ such that $k \cap P = k \cap Q$. Then there is a positive cone Q' of $L(t)$ for which $L \cap Q' = P$ and $k(t) \cap Q' = Q$.
(b) Let R_1, R_2 be two real closed overfields of k with $k \cap (R_1)_+ = k \cap (R_2)_+$. Then there exists a real closed field S together with k-embeddings $R_i \to S$ for $i = 1, 2$.

Hints: For (a), start by considering the case where L is the real closure of $(k, k \cap Q)$. In (b), work with a transcendence basis of R_1 over k and use transfinite induction.

1.5 The Tarski–Seidenberg Projection Theorem, and Artin's Solution of Hilbert's 17th Problem

The Tarski–Seidenberg projection theorem is a result of utmost importance in real algebraic geometry. Its impact reaches far beyond the pure statement of the theorem, and many important consequences are not immediately obvious. Our proof below rests on methods for real root counting, as developed in Section 1.3, and on the uniqueness properties of real closures, see Section 1.4. Having proved the projection theorem, we will combine it with Artin's characterization of sums of squares (Section 1.1) to give a positive answer to Hilbert's 17th problem. In the next section we'll gain greater flexibility in using Tarski–Seidenberg, after introducing some model-theoretic language.

We start by defining semialgebraic sets. Throughout let R denote a real closed field, and let A be a ring together with a fixed ring homomorphism $\varphi\colon A \to R$. For example, φ may be the inclusion of a subring of R. Let n be a natural number and write $x = (x_1, \dots, x_n)$ for a tuple of indeterminates. We will constantly use the order topology on R^n, or equivalently, the R-valued norm

$$|\xi| = \sqrt{\xi_1^2 + \cdots + \xi_n^2} \quad (\xi \in R^n)$$

(Remark 1.2.12.3). Recall that the open balls

$$B_r(\xi) = \{\eta \in R^n : |\eta - \xi|^2 < r^2\} \quad (\xi \in R^n,\ r \in R)$$

are a basis of open sets for this topology.

1.5.1 Notation. Given polynomials $f_1, \dots, f_m \in R[x] = R[x_1, \dots, x_n]$, we write

$$\mathcal{U}(f_1, \dots, f_m) = \{\xi \in R^n : f_1(\xi) > 0, \dots, f_m(\xi) > 0\},$$
$$\mathcal{S}(f_1, \dots, f_m) = \{\xi \in R^n : f_1(\xi) \ge 0, \dots, f_m(\xi) \ge 0\}$$

and

$$\mathcal{Z}(f_1, \dots, f_m) = \{\xi \in R^n : f_1(\xi) = \cdots = f_m(\xi) = 0\}.$$

Note that the sets $\mathcal{U}$ are open in R^n, the sets $\mathcal{S}$ and $\mathcal{Z}$ are closed.

1.5.2 Definition.

(a) A subset of R^n is *A-semialgebraic* if it is a finite Boolean combination (unions, intersections, complements) of sets of the form $\mathcal{U}(f)$ with $f \in A[x]$. Instead of R-semialgebraic we simply say *semialgebraic*.
(b) The subsets of R^n of the form $\mathcal{Z}(f_1, \dots, f_m)$ with $f_1, \dots, f_m \in A[x]$ are called *A-algebraic*. If $A = R$ we simply say *algebraic*.

1.5.3 Lemma. *Let $M \subseteq R^n$ be a subset.*

(a) *If M is A-algebraic there exists $f \in A[x]$ with $M = \mathcal{Z}(f)$.*
(b) *M is A-semialgebraic if, and only if, M has the form*

$$M = \bigcup_{i=1}^{m} \Big(\mathcal{Z}(f_i) \cap \mathcal{U}(g_{i1}, \dots, g_{ir_i})\Big) \tag{1.5}$$

with $m \in \mathbb{N}$, $r_i \ge 0$ and f_i, $g_{ij} \in A[x]$ $(1 \le j \le r_i,\ 1 \le i \le m)$.

Proof. (a) holds since $\mathcal{Z}(f_1, \dots, f_m) = \mathcal{Z}(f_1^2 + \cdots + f_m^2)$: Finitely many equations over R can be combined into a single one (as long as only solutions over R are considered!). As for (b), the set (1.5) is A-semialgebraic because $\mathcal{Z}(f) = R^n \setminus \mathcal{U}(f^2)$. Conversely, the system of all sets (1.5) is stable under the Boolean operations, so it coincides with the A-semialgebraic sets. □

1.5.4 Examples.

1. Given polynomials $f_1, \dots, f_r \in A[x]$ and any subset E of $\{-1, 0, 1\}^r$, the set

$$M = \{\xi \in R^n \colon (\operatorname{sign} f_1(\xi), \dots, \operatorname{sign} f_r(\xi)) \in E\}$$

is A-semialgebraic. Conversely, every A-semialgebraic set has this form for suitable $r \in \mathbb{N}$, $f_1, \dots, f_r \in A[x]$ and $E \subseteq \{-1, 0, 1\}^r$.

2. The set $\mathbb{Z}$ of integers is not semialgebraic in $\mathbb{R}$, as one checks directly from 1.5.3(b). Hence countable unions or intersections of semialgebraic sets are usually not semialgebraic. Neither is the graph $\{(t, \sin t) \colon t \in \mathbb{R}\} \subseteq \mathbb{R}^2$ of the sine function semialgebraic in $\mathbb{R}^2$, as one sees from intersecting with the first coordinate axis. The graph of the exponential function isn't semialgebraic in $\mathbb{R}^2$ either, but here a different type of argument is needed (Exercise 1.5.1).

1.5.5 Proposition. *The semialgebraic subsets of R are precisely the finite unions of intervals in R.*

Here all types of intervals (open, half-open or closed, bounded or unbounded) are allowed, in particular singletons. Note however that the intervals have to be delimited by elements in $R \cup \{\pm\infty\}$. For example, if $R = R_0$ is the (real closed) field of real algebraic numbers, the set $\{x \in R_0 \colon 0 < x < \pi\}$ is not an interval in R_0, and is not semialgebraic in R_0.

Proof. If $f \in R[t]$, the set $\mathfrak{U}(f) = \{a \in R\colon f(a) > 0\}$ is a union of finitely many (open) intervals in R, by the mean value theorem 1.3.3. On the other hand, the system of all finite unions of intervals is stable under Boolean operations. □

1.5.6 Proposition. *Let $f_1, \dots, f_m \in A[x] = A[x_1, \dots, x_n]$ and consider the polynomial map*

$$f\colon R^n \to R^m,\ f(x) = (f_1(\xi), \dots, f_m(\xi)).$$

For any A-semialgebraic set $M \subseteq R^m$, the preimage $f^{-1}(M)$ is A-semialgebraic as well. The same is true when A-semialgebraic is replaced by A-algebraic.

Proof. If $M = \mathfrak{U}(g)$ with $g \in A[y_1, \dots, y_m]$, then $f^{-1}(M) = \mathfrak{U}(h)$ for $h := g(f_1, \dots, f_m) \in A[x_1, \dots, x_n]$. This implies the general case since taking preimages commutes with Boolean operations. □

1.5.7 Remark. If $k = \overline{k}$ is an algebraically closed field and $f\colon k^n \to k^m$ is a polynomial map, the image set $f(X) \subseteq k^m$ of any algebraic subset $X \subseteq k^n$ is constructible, i.e. a finite Boolean combination of algebraic sets in k^m (Chevalley's theorem, e.g. [84] Exercise II.3.19). Over the field $\mathbb{R}$, or over any real closed field R, such a statement would fail completely, as one can see from simplest examples (think of the square map $\mathbb{R} \to \mathbb{R}$, $x \mapsto x^2$). The failure of Chevalley's theorem over $\mathbb{R}$ is one reason why, in real algebraic geometry, one cannot avoid working with semialgebraic sets, instead of only algebraic sets.

The Tarski–Seidenberg projection theorem states that images of semialgebraic sets under polynomial maps are again semialgebraic:

1.5.8 Theorem. (Projection Theorem) *Let R be a real closed field, let $A \subseteq R$ be a subring, and let*

$$\pi\colon R^m \times R^n \to R^n, \quad \pi(\xi, \eta) = \eta$$

be the projection map. Then for any A-semialgebraic subset S of $R^{m+n} = R^m \times R^n$, the image set $\pi(S) \subseteq R^n$ is A-semialgebraic as well.

1.5.9 Corollary. *Let $f_1, \dots, f_m \in A[x_1, \dots, x_n]$, let $f = (f_1, \dots, f_m)\colon R^n \to R^m$ be the associated polynomial map, and let $M \subseteq R^n$ be an A-semialgebraic set. Then the image set $f(M) \subseteq R^m$ is again A-semialgebraic.*

Proof. The graph $G = \{(\xi, \eta) \in R^n \times R^m\colon \eta = f(\xi)\}$ of f is an A-algebraic set. Let $\pi_1\colon R^m \times R^n \to R^m$ and $\pi_2\colon R^m \times R^n \to R^n$ be the two projection maps. Then $f(M) = \pi_2(G \cap \pi_1^{-1}(M))$, which is an A-semialgebraic set by Theorem 1.5.8. □

To prove the projection theorem we shall employ Hermite's method for counting real roots. First note that the Sylvester signature of a symmetric matrix depends semialgebraically on the coefficients. More precisely:

1.5.10 Lemma. *Let $n \in \mathbb{N}$ and $k \in \mathbb{Z}$. There exist finitely many $\mathbb{Z}$-polynomials $f_\mu, g_{\mu\nu} \in \mathbb{Z}[x_{ij}\colon 1 \le i \le j \le n]$ (indexed by $1 \le \nu \le s_\mu$ and $1 \le \mu \le r$, say) such that, for every real closed field R and every symmetric matrix $M \in \mathrm{Sym}_n(R)$ with coefficients $x_{ij} = x_{ji}$ $(i \le j)$, the following is true:*

$$\mathrm{sign}(M) = k \quad \Longleftrightarrow \quad \bigvee_{\mu=1}^{r} \left(f_\mu(x) = 0 \ \wedge \ \bigwedge_{\nu=1}^{s_\mu} g_{\mu\nu}(x) > 0 \right).$$

In particular, the set $\{M \in \mathrm{Sym}_n(R)\colon \ \mathrm{sign}(M) = k\}$ is $\mathbb{Z}$-semialgebraic in $\mathrm{M}_n(R) = R^{n^2}$ for every real closed field R.

Proof. Let $x = (x_{ij}\colon 1 \le i \le j \le n)$, a tuple of $\binom{n+1}{2}$ variables, and put $x_{ji} := x_{ij}$ for $i > j$. Let S be the symmetric $n \times n$ matrix whose (i, j)-coefficient is x_{ij}, and let $p = \det(tI_n - S)$ be its characteristic polynomial. Write

$$p \ = \ t^n + a_1(x)t^{n-1} + \cdots + a_n(x)$$

with polynomials $a_1(x), \ldots, a_n(x) \in \mathbb{Z}[x]$. Then for any symmetric $n \times n$ matrix $M = (\xi_{ij})$ over a real closed field R, the signature of M satisfies

$$\mathrm{sign}(M) \ = \ \mathrm{Var}(1, a_1(\xi), a_2(\xi), \ldots, a_n(\xi)) - \mathrm{Var}(1, -a_1(\xi), a_2(\xi), \ldots, (-1)^n a_n(\xi)),$$

see Remark 1.3.25.2. From this the assertion of the lemma is clear (each f_μ or $g_{\mu\nu}$ can be taken to be $\pm a_i$ for some i). □

1.5.11 Example. For $n = 2$ and $M = \left(\begin{smallmatrix} a & c \\ c & b \end{smallmatrix}\right)$ we get

$$\mathrm{sign}(M) \ = \ \begin{cases} 2 & \text{if } a > 0 \ \wedge \ ab - c^2 > 0, \\ 1 & \text{if } ab - c^2 = 0 \ \wedge \ (a > 0 \ \vee \ b > 0), \\ 0 & \text{if } ab - c^2 < 0 \ \vee \ a = b = c = 0, \\ -1 & \text{if } ab - c^2 = 0 \ \wedge \ (a < 0 \ \vee \ b < 0), \\ -2 & \text{if } a < 0 \ \wedge \ ab - c^2 < 0. \end{cases}$$

1.5.12 We start the proof of Theorem 1.5.8. Clearly it suffices to prove the case $m = 1$. So let $S \subseteq R^{n+1} = R \times R^n$ be an A-semialgebraic set, and let

$$\pi\colon R^{n+1} = R \times R^n \to R^n, \quad \pi(t, x_1, \ldots, x_n) = (x_1, \ldots, x_n)$$

be the projection. We'll show that $\pi(S)$ is an A-semialgebraic subset of R^n. By Lemma 1.5.3 we can assume that S has the form

$$S \ = \ \mathcal{Z}(f) \cap \mathcal{U}(g_1, \ldots, g_r) = \{(t, x) \in R^{n+1}\colon \ f(t, x) = 0, \ g_j(t, x) > 0 \ (1 \le j \le r)\}$$

where $x = (x_1, \ldots, x_n)$ and $f, g_1, \ldots, g_r \in A[t, x]$. We think of $f(t, x)$ and $g_j(t, x)$ as polynomials in the single variable t, parametrized by x. Using the methods from Section 1.3 we can decide, for any $\xi \in R^n$, whether there exists $t \in R$ with $(t, \xi) \in S$, i.e. whether $\xi \in \pi(S)$. We then have to show that the conditions on ξ that we obtain are A-semialgebraic.

Write $f = \sum_{i=0}^{m} a_i(x)t^i$ and $g_j = \sum_{k=0}^{m_j} b_{jk}(x)t^k$ where $a_i, b_{jk} \in A[x]$. For any polynomial $h \in A[t,x]$ and any $\xi \in R^n$ put $h_\xi(t) := h(t,\xi)$, a univariate polynomial in $R[t]$. For $-1 \le d \le m$ let

$$\Sigma_d := \{\xi \in R^n \colon \deg(f_\xi) = d\}$$

where we put $\deg(0) := -1$. Obviously, the sets Σ_d are A-semialgebraic in R^n, and $\bigcup_{d=-1}^{m} \Sigma_d = R^n$. So the proof of Theorem 1.5.8 will be finished once we have shown:

1.5.13 Lemma. *For $-1 \le d \le m$, the set $\pi(S) \cap \Sigma_d$ is A-semialgebraic.*

Proof. For $d = 0$ we have $\pi(S) \cap \Sigma_0 = \varnothing$ since a polynomial of degree 0 (i.e., a non-zero constant) has no roots. Let therefore $d \ge 1$. For $\xi \in \Sigma_d$ we have

$$f_\xi(t) = a_d(\xi)\,t^d + \cdots + a_1(\xi)\,t + a_0(\xi)$$

where $a_d(\xi) \ne 0$. By Proposition 1.3.36 we have

$$\pi(S) \cap \Sigma_d = \Big\{\xi \in \Sigma_d \colon \sum_{e\in\{1,2\}^r} \operatorname{sign} H\Big(\frac{1}{a_d(\xi)} f_\xi,\, g_\xi^e\Big) > 0\Big\},$$

where we have put $g_\xi^e := (g_1^{e_1} \cdots g_r^{e_r})_\xi$ for $e \in \{1,2\}^r$. The generalized Hermite matrices

$$H_{d,e}(\xi) := H\Big(\frac{1}{a_d(\xi)} f_\xi,\, g_\xi^e\Big)$$

(for $e \in \{1,2\}^r$) have signatures in $\{0, \pm 1, \dots, \pm d\}$. Their coefficients are $\mathbb{Z}$-polynomials in the

$$\frac{a_i(\xi)}{a_d(\xi)} \quad (i = 0, \dots, d-1)$$

(of total degree at most $\delta := 2d - 2 + \sum_j \deg(g_j)$) and in the

$$b_{jk}(\xi) \quad (k = 1, \dots, m_j,\ j = 1, \dots, r),$$

see 1.3.32. To calculate the signature, we may replace $H_{d,e}(\xi)$ by the matrix

$$\tilde{H}_{d,e}(\xi) := a_d(\xi)^{2m} \cdot H_{d,e}(\xi)$$

where $2m$ is some even integer with $2m \ge \delta$. The advantage of the latter matrix is that its coefficients are $\mathbb{Z}$-polynomials in the $a_i(\xi)$ and the $b_{jk}(\xi)$. Therefore, if we fix a tuple $e \in \{1,2\}^r$ and an integer $s \in \mathbb{Z}$, the set

$$\{\xi \in R^n \colon \operatorname{sign} \tilde{H}_{d,e}(\xi) = s\}$$

is A-semialgebraic by Lemma 1.5.10. According to Theorem 1.3.33 and Proposition 1.3.36, the set $\pi(S) \cap \Sigma_d$ is the union of the finitely many A-semialgebraic sets

$$T(\mathbf{s}) := \bigcap_{e \in \{1,2\}^r} \{\xi \in \Sigma_d \colon \operatorname{sign} \tilde{H}_{d,e}(\xi) = s_e\},$$

where the union is over all 2^r-tuples $\mathbf{s} = (s_e)_{e \in \{1,2\}^r}$ in $\{-d, \dots, d\}^{\{1,2\}^r}$ for which

$$\Sigma(\mathbf{s}) := \sum_{e \in \{1,2\}^r} s_e > 0.$$

Clearly, this implies that the set $\pi(S) \cap \Sigma_d$ is A-semialgebraic.

It remains to consider the case $d = -1$, which means $f_\xi \equiv 0$ for $\xi \in \Sigma_d$. How can one read off from the coefficients of polynomials $g_1, \dots, g_r \in R[t]$ whether there exists $\xi \in R$ with $g_1(\xi) > 0, \dots, g_r(\xi) > 0$? We use the following trick:

1.5.14 Lemma. *Let $g_1, \dots, g_r \in R[t]$, put $g := g_1 \cdots g_r$ and $h := g' \cdot (1 - g^2)$. If there is $\xi \in R$ with $g_1(\xi) > 0, \dots, g_r(\xi) > 0$, then there also exists such ξ for which in addition $h(\xi) = 0$.*

Proof. We can assume that g is not constant. Let $c_1 < \cdots < c_N$ be the real zeros of g, with $N \geq 0$. If $N = 0$ then $\deg(g)$ is even and each g_i is strictly positive on R. Since g', being of odd degree, has a zero in R, we are finished in this case. Now assume $N \geq 1$. Then g' (and hence h) has a zero in any of the intervals $]c_i, c_{i+1}[$ $(i = 1, \dots, N-1)$, by Rolle's theorem 1.3.9. For $|t|$ sufficiently large we have $1 - g(t)^2 < 0$ since g is not constant. Therefore, by the intermediate value theorem (1.3.3), $1 - g^2$ (and hence h) has a root both in $]-\infty, c_1[$ and in $]c_N, +\infty[$. By assumption, there is at least one among these $N + 1$ intervals on which all g_i are strictly positive. So the lemma is proved. □

1.5.15 In the proof of Theorem 1.5.8, the last missing step was to show that $\pi(S) \cap \Sigma_{-1}$ is an A-semialgebraic set. By Lemma 1.5.14, this set is $\pi(S') \cap \Sigma_{-1}$ where

$$S' := \{(t, \xi) \in R \times R^n \colon h(t, \xi) = 0,\ g_j(t, \xi) > 0\ (j = 1, \dots, r)\}$$

and $g := g_1 \cdots g_r$, $h := (1 - g^2)g'$. The cases $d \geq 0$ of 1.5.13 are already established, so we can apply them to S'. Then we are left again with the case $d = -1$, now however for S'. But due to the particular shape of h, this case can now be fixed directly: The point is to show that

$$\{\xi \in R^n \colon h_\xi \equiv 0,\ \exists t \in R \text{ with } g_j(t, \xi) > 0\ (j = 1, \dots, r)\}$$

is an A-semialgebraic set. Since $h_\xi \equiv 0$ implies that g_ξ is a constant, the set in question is equal to

$$\bigcap_{j=1}^{r} \{\xi \in R^n \colon (g_j)_\xi \text{ is a positive constant}\},$$

which clearly is A-semialgebraic. With this, the proof of Theorem 1.5.8 is finally complete. □

1.5.16 Remark. We used Hermite matrices to characterize the existence of a real root c of f with $g(c) > 0$. Alternatively, we could have used the generalized version of Sturm's theorem (1.3.21). However, in a parameter-dependent situation like the one at hand, Sturm's method has the disadvantage of immediately ramifying into a tree of subcases, making it a challenge to organize the discussion in a coherent manner. In contrast, Hermite's method allows a uniform treatment regardless of the parameters.

Here is an alternative version of the projection theorem:

1.5.17 Theorem. (Quantifier elimination, 1st version) *Let $x = (x_1, \dots, x_n)$, $y = (y_1, \dots, y_p)$ be tuples of variables, and let polynomials $f, g_1, \dots, g_r \in \mathbb{Z}[x, y]$ be given. There exist finitely many polynomials F_i, G_{ij} in $\mathbb{Z}[x]$ ($1 \le j \le m_i$, $1 \le i \le N$) such that the following are equivalent, for any real closed field R and any $\xi \in R^n$:*

(i) $\exists \eta \in R^p \left(f(\xi, \eta) = 0 \wedge \bigwedge_{\nu=1}^{r} g_\nu(\xi, \eta) > 0 \right)$;

(ii) $\bigvee_{i=1}^{N} \left(F_i(\xi) = 0 \wedge \bigwedge_{j=1}^{m_i} G_{ij}(\xi) > 0 \right)$.

Proof. We have actually proved this already. Indeed, let $p = 1$, which suffices by induction, and write $y = y_1$. Using the notation from the proof of 1.5.8, property (i) is equivalent to

$$\bigvee_{d=-1}^{m} \varphi_d(\xi),$$

where $f = \sum_{k=0}^{m} a_k(x)\, y^k$ with $m = \deg_y(f)$, and where $\varphi_d(\xi)$, for $d \ge 0$, denotes the condition

$$a_d(\xi) \ne 0 \ \wedge \ \bigwedge_{k=d+1}^{m} a_k(\xi) = 0 \ \wedge \ \bigvee_{\substack{s \in \{-d,\dots,d\}^{\{1,2\}^r} \\ \Sigma(s) > 0}} \ \bigwedge_{e \in \{1,2\}^r} \left(\operatorname{sign} \tilde{H}_{d,e}(\xi) = s_e \right).$$

The condition on the signature of $\tilde{H}_{d,e}(\xi)$ is equivalent, for fixed d, e and s_e, to a condition of type (ii), see Lemma 1.5.10. If $d = -1$, we take the following condition for $\varphi_d(\xi)$, according to Lemma 1.5.14:

$$\left[\bigwedge_{k=0}^{m} a_k(\xi) = 0 \right] \wedge \left[\exists y \left(h(\xi, y) = 0 \wedge \bigwedge_{\nu=1}^{r} g_\nu(\xi, y) > 0 \right) \right]$$

with $g := g_1 \cdots g_r$ and $h := (1 - g^2) \cdot \frac{\partial g}{\partial y}$. This last condition on ξ is equivalent to a quantifier-free condition (ii) on ξ, as before. □

1.5.18 Remark. Theorem 1.5.17 eliminates the existential quantifier $\exists y$ from condition (i). Note that this elimination (i.e., the proof of the theorem) was entirely constructive, regardless of its complexity: The polynomials F_i, G_{ij} were constructed in finitely many explicit steps. The complexity of this algorithm, however, is gigantic. For practical purposes this means that the algorithm is almost useless.

1.5.19 Corollary. (Transfer principle of Tarski–Seidenberg) *Let K be a field, let R_1, R_2 be two real closed field extensions of K that induce the same ordering on K. Given finitely many polynomials $f_1, \dots, f_r \in K[x] = K[x_1, \dots, x_n]$ together with signs $\varepsilon_1, \dots, \varepsilon_r \in \{-1, 0, 1\}$, we have: There exists $\xi \in R_1^n$ satisfying* $\operatorname{sign} f_\nu(\xi) = \varepsilon_\nu$ *for $\nu = 1, \dots, r$ if, and only if, there exists $\xi \in R_2^n$ satisfying the same conditions.*

Proof. Replacing a condition $f_1 = \cdots = f_m = 0$ by $f_1^2 + \cdots + f_m^2 = 0$ and a condition $f_i < 0$ by $-f_i > 0$, we may assume $\varepsilon_1 = 0$ and $\varepsilon_2 = \cdots = \varepsilon_r = +1$. We reduce the proof to Theorem 1.5.17 by first treating the coefficients of the f_i as indeterminates, before we specialize: There exist polynomials $F_1, \dots, F_r \in \mathbb{Z}[x, y]$ (with $y = (y_1, \dots, y_N)$ and N sufficiently large), together with a tuple $b = (b_1, \dots, b_N)$ in K^N, such that $f_\nu(x) = F_\nu(x, b)$ for $\nu = 1, \dots, r$. Using 1.5.17 we see, for any real closed overfield R of K, that the existence of $\xi \in R^n$ with $\operatorname{sign} f_\nu(\xi) = \varepsilon_\nu$ $(\nu = 1, \dots, r)$ is equivalent to a condition on b of type

$$\bigvee_i \Big(g_i(b) = 0 \wedge \bigwedge_j h_{ij}(b) > 0 \Big),$$

where the g_i, h_{ij} are $\mathbb{Z}$-polynomials in y. This condition depends only on the positive cone $P = K \cap R_+$ of K, since $g_i(b)$, $h_{ij}(b)$ lie in K. □

We now turn to Hilbert's 17th problem and its solution by Artin. See 1.1.31 for the statement of the problem and a first discussion. The following terminology is standard and will be used throughout the remainder of this course:

1.5.20 Definition. Let R be a real closed field. A polynomial $f \in R[x_1, \dots, x_n]$ is *positive semidefinite* (or *non-negative*, or just *psd* for short), if $f(\xi) \ge 0$ for all $\xi \in R^n$. If $f(\xi) > 0$ for all $\xi \in R^n$ we say that f is *(strictly) positive definite.*

1.5.21 Theorem. (Artin 1927) *Let R be a real closed field and suppose that $f \in R[x_1, \dots, x_n]$ is positive semidefinite. Then there exist finitely many rational functions $f_1, \dots, f_r \in R(x_1, \dots, x_n)$ such that*

$$f = f_1^2 + \cdots + f_r^2.$$

In fact we'll prove the following more precise version:

1.5.22 Theorem. *In the situation of 1.5.21, let $K \subseteq R$ be a subfield that contains the coefficients of f. Then there exist $f_1, \dots, f_r \in K(x_1, \dots, x_n)$ and $a_1, \dots, a_r \in K$ with $a_i > 0$ in R, such that*

$$f = a_1 f_1^2 + \cdots + a_r f_r^2.$$

Proof. It suffices to prove 1.5.22. Let $P := K \cap R_+$ be the ordering on K induced from R, and let

$$T = T_{K(x)}(P) = \Big\{ \sum_{i=1}^r a_i f_i^2 : r \in \mathbb{N},\ a_i \in P,\ f_i \in K(x_1, \dots, x_n) \Big\},$$

the preordering generated by P in $K(x)$, see 1.2.1. Theorem 1.5.22 claims that $f \in T$. We assume $f \notin T$ and will arrive at a contradiction. From Proposition 1.1.28 we get an ordering Q of $K(x)$ with $T \subseteq Q$ (and in particular, $K \cap Q = P$) and $f <_Q 0$. Let R_1 be the real closure of $(K(x), Q)$. Then R and R_1 are real closed field extensions of K that both induce the same ordering P on K. There exists a point $\xi \in K(x)^n \subseteq R_1^n$ with $f(\xi) <_Q 0$, namely $\xi = (x_1, \dots, x_n)$. By the Tarski–Seidenberg transfer principle (Corollary 1.5.19) there also exists a point $a \in R^n$ with $f(a) < 0$. But this contradicts the hypothesis that f was positive semidefinite. So the assumption was false, and hence f satisfies an identity as asserted. □

1.5.23 Remarks.

1. In a long and difficult paper [92], Hilbert (1893) himself had proved Theorem 1.5.21 in the case $n = 2$ (and $R = \mathbb{R}$), but couldn't decide the question for three or more variables. In his famous list of 23 unsolved mathematical problems, delivered at the 1900 International Congress of Mathematicians in Paris, he included the problem as number 17.

2. The refined version 1.5.22 responds to a question that Hilbert had raised himself. If $f \in \mathbb{R}[x]$ is psd and has coefficients in a subfield $K \subseteq \mathbb{R}$, Hilbert had asked whether f is always a sum of squares in the field $K(x)$. When asked in this form, the answer is negative since there may be an embedding $\varphi\colon K \to \mathbb{R}$ for which the polynomial f^φ in $\mathbb{R}[x]$ fails to be psd (e.g. take $K = \mathbb{Q}(\sqrt{2})$ and f the constant polynomial $\sqrt{2}$).

1.5.24 Examples.

1. The *arithmetic-geometric inequality (AGI)* states that the arithmetic mean of finitely many non-negative real numbers is at least their geometric mean:

$$\frac{1}{n}(t_1 + \cdots + t_n) \ge (t_1 \cdots t_n)^{1/n}, \tag{1.6}$$

or equivalently $(t_1 + \cdots + t_n)^n \ge n^n t_1 \cdots t_n$, for all $t_1, \dots, t_n \ge 0$. The usual proof from calculus uses concavity of the logarithm, an argument that is not available over general real closed fields. Nonetheless, the inequality holds in this generality, for example as a consequence of Tarski's transfer principle. There is a more elegant way of reasoning, though. Writing each t_i as a $2n$-th power, (AGI) can be expressed by saying that the symmetric form

$$f_n := x_1^{2n} + \cdots + x_n^{2n} - n x_1^2 \cdots x_n^2$$

is positive semidefinite. Now f_n can in fact be written as a sum of squares of forms, a fact that was discovered by Hurwitz. The proof is not obvious, see Exercise 1.5.6. Clearly, this gives an algebraic proof of AGI, valid over any real closed field R. Moreover, the argument implies for $t_1, \dots, t_n \ge 0$ that equality holds in (1.6) only if $t_1 = \cdots = t_n$ (a fact that is also well-known from calculus).

2. In 1888, Hilbert proved for every $n \geq 2$ that there exists a psd polynomial f in n variables that is not a sum of squares of polynomials. His results were in fact much more precise, we will discuss them in detail in Section 2.4. Remarkably, it was only in 1967 when the first explicit example of such a polynomial was published: Motzkin [141] showed that $f(x, y) = x^4y^2 + x^2y^4 - 3x^2y^2 + 1$ is a psd polynomial that is not a sum of squares. See Exercise 1.5.5 for the proof, and for an identity that expresses f as a sum of four squares of rational functions in $\mathbb{R}(x, y)$.

1.5.25 Remark. Artin's theorem 1.5.21 gives rise to a series of natural follow-up questions. For example, how many squares are needed? Let p_n denote the infimum of all integers $m \geq 1$ with the property that every psd polynomial $f \in R[x_1, \dots, x_n]$ is a sum of m squares of rational functions. For $n = 1$ we have $p_n = 2$ (elementary, see 1.2.11), but for $n \geq 2$ it is not even clear a priori whether p_n is finite. However Pfister [149] proved in a 1967 landmark paper that $p_n \leq 2^n$ for all $n \geq 1$. For $n = 2$ it is known that $p_2 = 4$. In fact, Cassels, Ellison and Pfister [39] proved that Motzkin's polynomial (see above) cannot be written as a sum of three squares of rational functions. For $n \geq 3$, the exact value of p_n is a wide open question, and it is only known that $n + 2 \leq p_n \leq 2^n$. The number p_n is called the *Pythagoras number* of the field $R(x_1, \dots, x_n)$.

Another question concerns degrees. Let $f \in R[x] = R[x_1, \dots, x_n]$ be a psd polynomial of degree d. By 1.5.21 there is a polynomial $p \neq 0$ in $R[x]$ such that both p and pf are sums of squares of polynomials. Is there a uniform bound $N = N(n, d)$ such that one can always find p with $\deg(p) \leq N$? The existence of *some* bound $N(n, d)$ like this is quite easy to prove (Exercise 3.3.5 later), but finding an explicit value for N is a very hard problem. For $n = 2$, the results of Hilbert's 1893 paper [92] imply such a bound, namely

$$\deg(p) \leq \left\lfloor \frac{(d-2)^2}{8} \right\rfloor.$$

The best known general bound for all $n \geq 3$ is quite discouraging: Lombardi, Perrucci and Roy [128] proved that

$$\deg(p) \leq 2^{2^{2^{d^{4^n}}}}$$

suffices!

Exercises

1.5.1 Show that the graph $\{(t, e^t) : t \in \mathbb{R}\}$ of the exponential function is not a semialgebraic subset of $\mathbb{R}^2$. (Use that the exponential function is transcendental, cf. Exercise 1.2.3.)

1.5.2 Let $(k, \leq)$ be an ordered field with real closure R, and let $f \in k[x] = k[x_1, \dots, x_n]$ be a polynomial with $f(\xi) \geq 0$ for every $\xi \in k^n$. If $\deg(f) \leq 2$, show that $f(\xi) \geq 0$ for every $\xi \in R^n$.

1.5.3 Exercise 1.5.2 does not extend to polynomials of higher degree. Consider the field $K = \mathbb{R}(x)$ with the ordering $\le$ that satisfies $0 < nx < 1$ for all $n \ge 1$. Let R be the real closure of $(K, \le)$. For the polynomial $f = t^4 - 4xt^2 + 3x^2 \in K[t]$, show that

(a) $f(a) \ge 0$ for every $a \in K$,
(b) there exists $b \in R$ with $f(b) < 0$.

In Artin's theorem 1.5.22, it is therefore not enough to assume $f(\xi) \ge 0$ for every $\xi \in K^n$.

1.5.4 Let (K, P) be an ordered field with real closure R and let $f \in K[x_1, \dots, x_n]$. The following are equivalent:

(i) $f(\xi) \ge 0$ for every $\xi \in R^n$;
(ii) $f \ge_Q 0$ for every extension Q of the ordering P to $K(x_1, \dots, x_n)$.

1.5.5 Consider the (inhomogeneous) Motzkin polynomial $f = x^4y^2 + x^2y^4 - 3x^2y^2 + 1$ in $\mathbb{R}[x, y]$ and prove:

(a) $f(a, b) \ge 0$ for all $(a, b) \in \mathbb{R}^2$.
(b) f is not a sum of squares of polynomials in $\mathbb{R}[x, y]$.
(c) Find a representation of f as a sum of four squares of rational functions in $\mathbb{R}(x, y)$.

(*Hints*: For (a) use the arithmetic-geometric inequality. For (c) multiply f by $1 + x^2$.)

1.5.6 Following Hurwitz we give an algebraic proof of the arithmetic-geometric inequality, see Example 1.5.24.1. Let k be a field of characteristic 0, let $x = (x_1, \dots, x_n)$, and let $h_n = x_1^n + \cdots + x_n^n - nx_1 \cdots x_n$. For any polynomial $f = f(x) \in k[x]$ let $\mathfrak{S}f := \frac{1}{n!} \sum_{\sigma \in S_n} f(x_{\sigma(1)}, \dots, x_{\sigma(n)})$, the symmetrization of f. Let $n \ge 2$ and consider the form

$$p_n = (x_1 - x_2) \sum_{i=1}^{n-1} (x_1^{n-i} - x_2^{n-i})\, x_3 \cdots x_{i+1}$$

in $k[x]$. Prove that $\mathfrak{S}p_n = \frac{2}{n} h_n$, and deduce that $f_n(x_1, \dots, x_n) := h_n(x_1^2, \dots, x_n^2)$ is a sum of squares of forms. (*Hint*: $p_n(x_1^2, \dots, x_n^2)$ is a sum of squares of forms.)

1.5.7 Find a psd polynomial $f \in \mathbb{R}[x_1, \dots, x_n]$ that is not a sum of squares of polynomials, but such that $f + c$ is a sum of squares for some real number c. (*Hint*: Take $f = (g + 1)^2 - 1$, where g is a psd form that is not sos.)

1.6 Model-Theoretic Formulation of Quantifier Elimination and Transfer Principle

To make use of the full strength of the Tarski–Seidenberg theorem, it is convenient to borrow from the language of logic and model theory. In model theory one studies systems of axioms (called *theories*) that are stated in *formal languages*. A fundamental result is *completeness* of first order logic: A theory is consistent (no contradiction can be deduced from its axioms) if and only if it has a model.

We will employ the model-theoretic language and formalism only for the theory of real closed fields. A rigorous introduction of the general model-theoretic framework would certainly be worthwhile. However we won't attempt to do this, both for space and time limitations. Instead we are going to introduce the necessary terminology in an *ad hoc* way, tailored towards our needs, namely the basic concepts of model theory for the first order language of ordered fields.

For a quick introduction to general model theory we recommend [160]. For a much more comprehensive account see [133], for example.

Let A be a ring (commutative and unital, as always), and let Var be a countably infinite set whose elements we call *variables*. For the first definitions we take $Var = \{x_1, x_2, \dots\}$. Later on we'll allow ourselves to be more flexible and to use arbitrary names for variables. By $A[Var]$ we denote the ring of polynomials in the variables Var with coefficients in A.

1.6.1 Definition. Let A be a ring.

(a) An *A-prime formula* is an expression of the form[4] '$f = g$' or '$f < g$', with $f, g \in A[Var] = A[x_1, x_2, \dots] = \bigcup_{n \in \mathbb{N}} A[x_1, \dots, x_n]$.
(b) The set Fml_A of all *A-formulas* is the smallest set that satisfies the following conditions (1)–(3):

  (1) Every A-prime formula is an A-formula;
  (2) if φ, ψ are A-formulas then so are '$\varphi \wedge \psi$' and '$\neg\varphi$';
  (3) if φ is an A-formula and $x \in Var$, then '$\forall\, x\, \varphi$' is again an A-formula.

1.6.2 Remark. What has been called A-formulas here are, more precisely, the formulas in the language $\mathcal{L}(+, -, \cdot, <, 0, 1)_A$ of ordered fields with constants in A. The only significance of the base ring A is to serve as a source for the constants used in formulas. To improve readability of formulas we agree on the following standard conventions:

- '$f \neq g$' stands for '$\neg(f = g)$';
- '$f \leq g$' stands for '$f < g \vee f = g$';
- '$f > g$' (resp. '$f \geq g$') stands for '$g < f$' (resp. '$g \leq f$');
- '$\varphi \vee \psi$' stands for '$\neg(\neg\varphi \wedge \neg\psi)$';
- '$\varphi \to \psi$' stands for '$\neg(\varphi \wedge \neg\psi)$';
- '$\varphi \leftrightarrow \psi$' stands for '$\varphi \to \psi \wedge \psi \to \varphi$';
- '$\exists\, x\, \varphi$' stands for '$\neg(\forall\, x\, \neg\varphi)$'.

Moreover we will usually omit quotes around formulas from now on.

1.6.3 Definition. Let φ be an A-formula. A variable $x \in Var$ may occur in φ in two possible ways, bound or free. An occurrence of x is said to be *bound* if it lies in the range of a quantifier $\exists\, x$ or $\forall\, x$. Any other occurrence is called *free*. The set of *free variables* of φ is

$$\mathrm{Fr}(\varphi) := \{x \in Var\colon x \text{ has a free occurrence in } \varphi\}.$$

[4] the quotes are not part of the formula

1.6.4 Example. To illustrate this definition, consider the following formulas:

$$\begin{aligned} \varphi_1 &: \ \exists x_1\ x_1^2 + 1 = 0; \\ \varphi_2 &: \ \forall x_2\ \exists x_3\ x_1 x_2 x_3 < 1 + x_1; \\ \varphi_3 &: \ (\exists x_1\ x_1 = x_1 + 1) \wedge x_1^2 < x_2. \end{aligned}$$

The occurrence of the variable x_1 in φ_1 is bound, in φ_2 it is free, and in φ_3 there are both a bound and a free occurrence. The set of free variables for these formulas is $\mathrm{Fr}(\varphi_1) = \varnothing$, $\mathrm{Fr}(\varphi_2) = \{x_1\}$ and $\mathrm{Fr}(\varphi_3) = \{x_1, x_2\}$.

The following rules for the set of free variables $\mathrm{Fr}(\varphi)$ are verified immediately. Alternatively, they give an inductive definition of $\mathrm{Fr}(\varphi)$:

1.6.5 Lemma. *Let φ, ψ be A-formulas.*

(a) *If $f, g \in A[Var]$ then* $\mathrm{Fr}(f = g) = \mathrm{Fr}(f < g) = \{x \in Var\colon x \text{ occurs in } f \text{ or in } g\}$.
(b) $\mathrm{Fr}(\neg\varphi) = \mathrm{Fr}(\varphi)$;
(c) $\mathrm{Fr}(\varphi \vee \psi) = \mathrm{Fr}(\varphi \wedge \psi) = \mathrm{Fr}(\varphi \to \psi) = \mathrm{Fr}(\varphi) \cup \mathrm{Fr}(\psi)$;
(d) $\mathrm{Fr}(\exists\, x\, \varphi) = \mathrm{Fr}(\forall\, x\, \varphi) = \mathrm{Fr}(\varphi) \setminus \{x\}$.

Next we explain how to replace variables in formulas by values. This only makes sense for free occurrences:

1.6.6 Definition. Let φ be an A-formula.

(a) The notation $\varphi = \varphi(x_1, \ldots, x_n)$ with $x_1, \ldots, x_n \in Var$ indicates that $\mathrm{Fr}(\varphi) \subseteq \{x_1, \ldots, x_n\}$.
(b) Let $\varphi = \varphi(x_1, \ldots, x_n)$, let B be an A-algebra. Given $b_1, \ldots, b_n \in B$, the B-formula $\varphi(b_1, \ldots, b_n)$ arises from φ by replacing every *free* occurrence of x_i in φ by b_i, for $i = 1, \ldots, n$.

Note that the B-formula $\varphi(b_1, \ldots, b_n)$ has no free variables left.

1.6.7 Definition. An (A-) formula φ with $\mathrm{Fr}(\varphi) = \varnothing$ is called an *(A-) sentence.*

So far we only explained formal manipulations on formulas. Now we assign a meaning to formulas.

1.6.8 Let R be a real closed field. Any R-sentence may be read as a statement on elements in R, by interpreting $\wedge$ and $\neg$ as 'and' and 'not', respectively (and hence $\vee$ as 'or', $\to$ as 'implies' etc.), and interpreting '$\forall\, x \psi(x)$' as 'for all x in R, $\psi(x)$ holds' (and hence '$\exists\, x\, \psi(x)$' as 'there exists x in R with $\psi(x)$'). *Therefore, any R-sentence is either true or false in R.* If φ is such an R-sentence, the notation

$$R \models \varphi$$

(read *"R is a model of φ"*, or *"φ holds in R"*) means that the statement φ is true in R.

More generally, we need to allow formulas with coefficients in arbitrary base rings A. To interpret them (and assign them a true/false value), we have to fix a ring homomorphism from A into a real closed field:

1.6.9 Definition. Let φ be an A-formula with $\mathrm{Fr}(\varphi) \subseteq \{x_1, \ldots, x_n\}$, and let $\alpha\colon A \to R$ be a homomorphism into a real closed field R. Let $\varphi^\alpha = \varphi^\alpha(x_1, \ldots, x_n)$ denote the R-formula that arises from φ by applying α to all constants. The subset

$$\mathcal{S}_{R,\alpha}(\varphi) := \{\xi = (\xi_1, \ldots, \xi_n) \in R^n : R \models \varphi^\alpha(\xi_1, \ldots, \xi_n)\}$$

of R^n is called the *relation defined by* φ (in R^n, with respect to α).

Strictly speaking we haven't defined $\mathcal{S}_{R,\alpha}(\varphi)$ uniquely, since we did not insist that $\mathrm{Fr}(\varphi) = \{x_1, \ldots, x_n\}$. This bit of imprecision is however insignificant and convenient. For the sake of lighter notation we'll often suppress α and just write $\mathcal{S}_R(\varphi)$. Frequently α will just be the inclusion $A \subseteq R$ of a subring, or even $A = R$ and $\alpha = \mathrm{id}$.

1.6.10 Example. Consider the following $\mathbb{Z}$-formula $\varphi = \varphi(x_1, x_2, x_3)$:

$$\exists y_1 \left(y_1^3 + x_1 y_1^2 + x_2 y_1 + x_3 = 0 \ \wedge \ \forall y_2 \left(y_2^3 + x_1 y_2^2 + x_2 y_2 + x_3 = 0 \ \rightarrow \ y_2 = y_1 \right) \right)$$

The relation $\mathcal{S}_{\mathbb{R}}(\varphi) \subseteq \mathbb{R}^3$ defined by φ (and formed with respect to the unique homomorphism $\mathbb{Z} \to \mathbb{R}$) is the set of $\xi \in \mathbb{R}^3$ for which the univariate polynomial $t^3 + \xi_1 t^2 + \xi_2 t + \xi_3$ has precisely one real root (possibly of higher multiplicity).

The sets $\mathcal{S}_R(\varphi)$ can be described inductively as follows (again this gives an alternative recursive definition):

1.6.11 Lemma.

(a) *If* $f, g \in R[x_1, \ldots, x_n]$ *then* $\mathcal{S}_R(f = g) = \{\xi \in R^n : f(\xi) = g(\xi)\}$ *and* $\mathcal{S}_R(f < g) = \{\xi \in R^n : f(\xi) < g(\xi)\}$.

(b) *If* $\varphi = \varphi(x_1, \ldots, x_n)$ *and* $\psi = \psi(x_1, \ldots, x_n)$ *are* R*-formulas then*

$$\begin{aligned}
\mathcal{S}_R(\varphi \vee \psi) &= \mathcal{S}_R(\varphi) \cup \mathcal{S}_R(\psi),\\
\mathcal{S}_R(\varphi \wedge \psi) &= \mathcal{S}_R(\varphi) \cap \mathcal{S}_R(\psi),\\
\mathcal{S}_R(\neg\varphi) &= R^n \setminus \mathcal{S}_R(\varphi),\\
\mathcal{S}_R(\varphi \rightarrow \psi) &= \mathcal{S}_R(\neg\varphi) \cup \mathcal{S}_R(\psi).
\end{aligned}$$

(c) *If* $\phi = \phi(x, x_1, \ldots, x_n)$ *is an* R*-formula then*

$$\begin{aligned}
\mathcal{S}_R(\exists\, x\, \phi) &= \{(a_1, \ldots, a_n) \in R^n : \exists\, a \in R\ (a, a_1, \ldots, a_n) \in \mathcal{S}_R(\phi)\},\\
\mathcal{S}_R(\forall\, x\, \phi) &= \{(a_1, \ldots, a_n) \in R^n : \forall\, a \in R\ (a, a_1, \ldots, a_n) \in \mathcal{S}_R(\phi)\}.
\end{aligned}$$

1.6.12 Proposition. *Given a homomorphism* $\alpha\colon A \to R$ *into a real closed field* R*, the* A*-semialgebraic sets in* R^n *are precisely the relations* $\mathcal{S}_{R,\alpha}(\varphi)$ *defined by* quantifier-free A*-formulas* $\varphi = \varphi(x_1, \ldots, x_n)$.

Proof. A typical A-semialgebraic set has the form

$$\bigcup_{i=1}^{m} \{x \in R^n : f_i(x) = 0,\ g_{i1}(x) > 0, \ldots, g_{ir_i}(x) > 0\}$$

with polynomials f_i, $g_{ij} \in A[x]$. This is precisely the relation $\mathcal{S}_R(\varphi)$ defined by the quantifier-free A-formula

$$\varphi : \bigvee_{i=1}^{m} \Big(f_i = 0 \wedge \bigwedge_{j=1}^{r_i} g_{ij} > 0 \Big).$$

And conversely, since the sets defined by prime A-formulas are A-semialgebraic, every quantifier-free A-formula defines an A-semialgebraic set. □

1.6.13 Definition. Let A be a ring. Two A-formulas φ, ψ are called *A-equivalent*, denoted $\varphi \equiv_A \psi$, if $\mathcal{S}_{R,\alpha}(\varphi) = \mathcal{S}_{R,\alpha}(\psi)$ holds for every homomorphism $\alpha \colon A \to R$ into a real closed field R.

The reader with a model-theoretic background will realize that our notation $\varphi \equiv \psi$ signalizes that the formula $\forall\, (\varphi \leftrightarrow \psi)$ holds in every model of $Th_{RCF,A}$, the theory of real closed fields with constants in A. Or equivalently, by completeness of first order logic, that $\forall\, (\varphi \leftrightarrow \psi)$ can be proved within $Th_{RCF,A}$.

1.6.14 Remarks.

1. The formulas $x \geq 0$ and $\exists\, y\; x = y^2$ are $\mathbb{Z}$-equivalent, since for any real closed field R and any $a \in R$ one has: $a \geq 0 \Leftrightarrow \exists\, b \in R\; a = b^2$.
2. If $\varphi \equiv_A \varphi'$ and $\psi \equiv_A \psi'$, then also $\varphi \wedge \psi \equiv_A \varphi' \wedge \psi'$, $\neg\, \varphi \equiv_A \neg \varphi'$ and $(\forall\, x\, \varphi) \equiv_A (\forall\, x\, \varphi')$ hold (for any variable x). This is obvious from Lemma 1.6.11.
3. Every quantifier-free A-formula is A-equivalent to a formula of the form

$$\bigvee_{i} \Big(f_i = 0 \wedge \bigwedge_{j} g_{ij} > 0 \Big)$$

with f_i, $g_{ij} \in A[Var]$, see Exercise 1.6.1.

1.6.15 Theorem. (Quantifier elimination, 2nd version) *Let A be a ring. Given any A-formula $\varphi = \varphi(x_1, \ldots, x_n)$, there exists a quantifier-free A-formula $\phi = \phi(x_1, \ldots, x_n)$ that is A-equivalent to φ.*

Proof. The proof is by induction on the structure of φ. If φ is quantifier-free there is nothing to show. If the theorem holds for φ and for ψ, it also holds for $\varphi \wedge \psi$ and $\neg\, \varphi$. Let now φ be the formula

$$\exists y_1\; \psi(x_1, \ldots, x_n, y_1)$$

and assume by induction that $\psi \equiv_A \psi'$ for some quantifier-free formula ψ'. By Remarks 2 and 3 in 1.6.14, we can assume that ψ has the form

$$f = 0 \wedge \bigwedge_{\nu} g_\nu > 0$$

with polynomials f, $g_\nu \in A[x_1, \ldots, x_n, y_1]$. With this we are in the situation of Theorem 1.5.17. By this theorem there exist polynomials F_i, $G_{ij} \in A[x_1, \ldots, x_n]$ such that φ is A-equivalent to the formula $\bigvee_i (F_i = 0 \wedge \bigwedge_j G_{ij} > 0)$. So we are done. □

1.6.16 Corollary. *For every A-formula $\varphi = \varphi(x_1, \ldots, x_n)$ and every homomorphism $\alpha\colon A \to R$ into a real closed field, the set $\mathcal{S}_R(\varphi) \subseteq R^n$ defined by φ is an A-semialgebraic subset of R^n.*

The emphasis here is on *every* formula: The conclusion is true not just for quantifier-free formulas.

Proof. Choose a quantifier-free formula ϕ as in Theorem 1.6.15. Then $\mathcal{S}_R(\varphi) = \mathcal{S}_R(\phi)$, and the second set is A-semialgebraic by Proposition 1.6.12. □

1.6.17 Corollary. (Transfer principle) *Let K be a field, and let R_1, R_2 be two real closed field extensions of K that induce the same ordering on K. Then for any K-sentence φ we have*

$$R_1 \models \varphi \quad \Leftrightarrow \quad R_2 \models \varphi.$$

Proof. By Theorem 1.6.15 we can choose a quantifier-free K-sentence $\phi \equiv_K \varphi$. This ϕ is a logical-Boolean combination (finitely many operations $\wedge$, $\vee$, $\neg$) of elementary sentences $f_i = g_i$ or $f_i < g_i$ with f_i, $g_i \in K$. So the claim is obvious. □

Here is an application:

1.6.18 Corollary. *Let R be a real closed field, let $A \subseteq R$ be a subring, and let $M \subseteq R^n$ be an A-semialgebraic set. Then the closure $\overline{M}$, the interior* int(M)*, the boundary ∂M and the R-convex hull* conv(M) *of M are A-semialgebraic sets as well.*

Of course, closure, interior and boundary refer to the order topology on R^n. A subset $M \subseteq R^n$ is *R-convex* if u, $v \in M$ and $0 < t < 1$ in R imply $(1-t)u + tv \in M$. The *R-convex hull* of $M \subseteq R^n$ is

$$\operatorname{conv}(M) := \left\{ \sum_{i=1}^{m} a_i \xi_i \colon m \geq 1,\ 0 \leq a_i \in R,\ \xi_i \in M\ (i = 1, \ldots, m),\ \sum_{i=1}^{m} a_i = 1 \right\},$$

and is the smallest R-convex set containing M.

Proof. Each of these sets can be described by an A-formula, see Exercise 1.6.2. So the corollary follows from 1.6.16. □

1.6.19 Remarks.

1. Let $M \subseteq R^n$ be a semialgebraic set, say $M = \bigcup_i (\mathcal{Z}(f_i) \cap \mathcal{U}(g_{i1}, \ldots, g_{ir_i}))$ with polynomials f_i, g_{ij}. A description of $\overline{M}$ will usually require new polynomials, different from the f_i. In particular, it is *not true* in general that $\overline{M}$ is obtained from M by simply relaxing strict inequalities to non-strict ones. While this is not really surprising, it will be more of a surprise to learn that the conclusion *does* hold under suitable hypotheses. More on this in Section 4.3 (Thom's lemma).

2. We could have derived Corollary 1.6.18 directly from the projection theorem 1.5.8, without using model-theoretic language. But converting a formula for the closure $\overline{M}$, say, into projections of semialgebraic sets would quickly become very complicated and awkward, and would obscure the picture. The model-theoretic language is much more elegant and transparent.

3. A word of caution. One should always be careful when checking whether a property can, or cannot, be phrased in the formal language of ordered fields. Here are a few examples:

(a) The sentence "*Every bounded increasing sequence in R has a limit in R*" is true for $R = \mathbb{R}$. It cannot be expressed in the language of ordered fields. In fact, it is usually false in general real closed fields R.

(b) The sentence "*If $M \subseteq R^n$ is a semialgebraic set that is bounded (contained in $B_r(0)$ for some $r \in R$) and closed, then any polynomial has a minimum on M*" is true for $R = \mathbb{R}$. In order to deduce it for arbitrary R via Tarski, it is not possible, in the language of ordered fields, to quantify over all semialgebraic sets in R^n. But one may fix a concrete semialgebraic description of a given M, say as in (1.5), and may fix upper bounds for the degrees of all the polynomials involved. By quantifying over all the occurring coefficients, one may then express both the assumption and the conclusion.

(c) Let $M \subseteq \mathbb{R}^m \times \mathbb{R}^n$ be a semialgebraic set. For $\eta \in \mathbb{R}^n$ let $M_\eta = \{\xi \in \mathbb{R}^m : (\xi, \eta) \in M\}$, the η-slice of M, which is a semialgebraic set in $\mathbb{R}^m$. There is no general way to express in a formal sentence whether a semialgebraic set is connected. Therefore one would not expect that the set

$$\{\eta \in \mathbb{R}^n : M_\eta \text{ is connected}\}$$

is semialgebraic. But it turns out that it always is. For this, however, there is no proof directly from Tarski's principle. Rather one has to study decompositions of semialgebraic sets into topologically simple pieces, as we will do in 4.3.

Exercises

1.6.1 Let A be a ring. Every quantifier-free A-formula is A-equivalent to a formula

$$\bigvee_i \Big(f_i = 0 \wedge \bigwedge_j g_{ij} > 0\Big)$$

with $f_i, g_{ij} \in A[Var]$.

1.6.2 Let R be a real closed field, let $\varphi = \varphi(x_1, \dots, x_n)$ be an R-formula, and let $M = \mathcal{S}_R(\varphi) \subseteq R^n$ be the relation defined by φ. Find an explicit formula $\psi(x_1, \dots, x_n)$ that defines (a) the closure $\overline{M}$, (b) the interior $\mathrm{int}(M)$, (c) the boundary ∂M, (d) the R-convex hull $\mathrm{conv}(M)$ of M. (*Remark*: For (d) you will need Carathéodory's theorem (in its version over R), see 8.1.2.)

1.6.3 Let $n \in \mathbb{N}$ and let $f \in \mathbb{Q}[x] = \mathbb{Q}[x_1, \dots, x_n]$ be a polynomial that satisfies $f(a) \geq 0$ for every $a \in \mathbb{Q}^n$. Prove, or disprove by a counter-example: For every ordered field $(K, \leq)$ and every $\xi \in K^n$ one has $f(\xi) \geq 0$.

1.6.4 Let R be a real closed field, let $m, n \geq 1$, and let $f_1, \dots, f_m, g \in R[x_1, \dots, x_n]$ be polynomials. Let $M \subseteq R^n$ be a semialgebraic set that is closed and (semialgebraically) bounded, and assume that $g(\xi) \neq 0$ for every $\xi \in M$. Then the map

$$f: M \to R^m, \quad f(\xi) = \left(\frac{f_1(\xi)}{g(\xi)}, \dots, \frac{f_m(\xi)}{g(\xi)}\right)$$

is well-defined. Use the transfer principle to prove that the image set $f(M)$ is closed and bounded in R^m. (A set in R^n is called (semialgebraically) bounded if it is contained in $[-c, c]^n$ for some positive $c \in R$.)

1.6.5 Let I be an infinite set and let 2^I be the power set of I, consisting of all subsets of I. An *ultrafilter* on I (compare also Remark 4.1.20) is a non-empty subset $\mathcal{F}$ of 2^I with $\varnothing \notin \mathcal{F}$, such that $\mathcal{F}$ is stable under finite intersections and satisfies $J \in \mathcal{F}$ or $I \setminus J \in \mathcal{F}$ for every $J \in 2^I$. For every index $i \in I$ let K_i be a field. Given a tuple $a = (a_i)_{i \in I} \in \prod_{i \in I} K_i$, let $Z(a) = \{i \in I \colon a_i = 0\}$, and put $J_{\mathcal{F}} = \left\{a \in \prod_{i \in I} K_i \colon Z(a) \in \mathcal{F}\right\}$.

(a) Show that $J_{\mathcal{F}}$ is a maximal ideal of the product ring $\prod_{i \in I} K_i$. The residue field

$$\Big(\prod_{i \in I} K_i\Big)\Big/\mathcal{F} := \Big(\prod_{i \in I} K_i\Big)\Big/J_{\mathcal{F}}$$

is called an *ultraproduct* of the fields K_i.

(b) If each field K_i is real closed, show that the field $(\prod_I K_i)/\mathcal{F}$ is real closed as well.

Assume now that R is a fixed real closed field and that $K_i = R$ for every $i \in I$. Write S for the ultrapower $R^I/\mathcal{F}$ of R. The diagonal map $R \to R^I$ induces an embedding $R \to S$ of real closed fields.

(c) Let $\phi(x_1, \dots, x_n)$ be an R-formula and let $\xi = (\xi_1, \dots, \xi_n) \in S^n$, where $\xi_\nu \in S$ is represented by the tuple $(\xi_\nu^i)_{i \in I} \in R^I$ ($\nu = 1, \dots, n$). Show that $\phi(\xi)$ holds in S if, and only if, the set $\{i \in I \colon R \models \phi(\xi_1^i, \dots, \xi_n^i)\}$ lies in $\mathcal{F}$.

1.6.6 Let R be a real closed field, let $\mathcal{F}$ be an ultrafilter on the set $\mathbb{N}$. The ultrapower $S = R^{\mathbb{N}}/\mathcal{F}$ of R is a real closed field extension of R, see Exercise 1.6.5. We assume that $\mathcal{F}$ is non-principal, meaning that $\bigcap_{J \in \mathcal{F}} J = \varnothing$. Let M^k ($k = 1, 2, \dots$) be a countable sequence of semialgebraic sets in S^n. If $M^1 \cap \dots \cap M^k \neq \varnothing$ for every $k \geq 1$, show that $\bigcap_{k \geq 1} M^k$ is non-empty.

Suggestion for proof: For every $k \geq 1$ fix a quantifier-free S-formula $\phi_k(x_1, \dots, x_n)$ that describes $M^k \subseteq S^n$. Lift all constants in ϕ_k to tuples in $R^{\mathbb{N}}$, thereby obtaining for each k a sequence M_i^k ($i \geq 1$) of semialgebraic sets in R^n. For each k, fix a tuple $a^k = (a_i^k)_{i \geq 1} \in (R^n)^{\mathbb{N}}$ that represents a point in $M^1 \cap \dots \cap M^k$. Show for each $k \geq 1$ that the set

$$I_k := \{i \in \mathbb{N} \colon i \geq k \text{ and } a_i^j \in M_i^m \text{ for each } 1 \leq m \leq j \leq k\}$$

lies in $\mathcal{F}$, and observe $I_{k+1} \subseteq I_k$ for all k. Show that, for each $i \geq 1$, a point $b_i \in R^n$ is well-defined by $b_i := a_i^j$ if $i \in I_j \setminus I_{j+1}$, $b_i := 0$ if $i \notin I_1$. The point $[b] \in S^n$ represented by b lies in M^k for every k. (Use Exercise 1.6.5(c).)

The property of S proved in this exercise is known as $\aleph_1$-*saturatedness* of S.

1.7 The Artin–Lang Theorem

We now study orderings of function fields of algebraic varieties over a real closed field R. In particular, we'll arrive at several characterizations of those irreducible varieties whose function field is real, i.e., can be ordered. We are also going to generalize the statement of Hilbert 17 to arbitrary irreducible R-varieties.

We start with a few preliminaries on the order topology on (the R-points of) algebraic R-varieties. For this we work in the more general setting of a topological base field, since nothing is special in the case when the field topology comes from an ordering,

1.7.1 For general conventions on algebraic varieties we refer to Appendix A.6. The viewpoint of schemes is nowhere necessary in this course. All varieties that we talk about can be assumed to be quasi-projective. Note that, in this book, the term "variety" does not imply irreducibility. Recall that the coordinate ring of an affine k-variety X is denoted $k[X]$. The function field of an irreducible k-variety X is $k(X)$. By $X(k)$ we denote the set of k-rational points of X. For the main results to follow, we will need the notion of non-singular points on k-varieties. It is recalled in A.6.17.

1.7.2 A *topological field* is a field k together with a Hausdorff topology $\mathcal{T}$ on k, such that the sum and product map $k \times k \to k$, as well as the inversion map $k^* \to k^*$, $x \mapsto x^{-1}$, are continuous. Examples are $\mathbb{R}$ or $\mathbb{C}$ with the standard (Euclidean) topology. Other examples (that otherwise will never play a role in this book) are the field $\mathbb{Q}_p$ of p-adic numbers and its finite extensions (for p any prime), or the field $F(\!(t)\!)$ of formal Laurent series over a finite field F. It is a well-known fact that together these fields comprise the list of all fields with a non-discrete and locally compact field topology. Note that the field $\mathbb{R}$ of real numbers is the only real field in this list (a fact that was proved in Exercise 1.2.2). For the purpose of this section it is important that every ordered field (K, P) is a non-discrete topological field with respect to its order topology $\mathcal{T}_P$, see 1.2.12.4 and Exercise 1.2.2.

1.7.3 Let $(k, \mathcal{T})$ be a topological field. For every k-variety X, the field topology on k induces a natural topology on $X(k)$. It is sometimes called the *strong topology*, to distinguish it from the k-Zariski topology. When $\mathcal{T} = \mathcal{T}_P$ for some ordering P of k, we also speak of the *order topology* induced by P. This topology is defined as follows. A subset $W \subseteq X(k)$ is open with respect to the strong topology if, and only if, for every closed k-subvariety U of some affine space $\mathbb{A}^m$ and every k-morphism $f: U \to X$, the preimage $f^{-1}(W)$ is relatively open in $U(k)$, where $U(k)$ is given the relative topology induced from $U(k) \subseteq k^m$ and the product topology on k^m. Note that when $X \subseteq \mathbb{A}^n$ is locally closed, the strong topology is simply the restriction of the product topology from $\mathbb{A}^n(k) = k^n$ to $X(k)$. For $X \subseteq \mathbb{P}^n$ it is similar. For any morphism $X \to Y$ of k-varieties, the induced map $X(k) \to Y(k)$ is continuous with respect to the strong topologies.

1.7.4 Lemma. *Let k be a topological field, let X, Y be k-varieties.*

(a) *If $f: X \to Y$ is an open immersion of k-varieties, the induced map $f: X(k) \to Y(k)$ is an open topological embedding (homeomorphism onto its open image). Likewise with "open" replaced by "closed".*
(b) *The strong topology on $(X \times Y)(k)$ is the product topology on $X(k) \times Y(k)$.*
(c) *The strong topology on $X(k)$ is Hausdorff.*

Proof. (a) is true when Y is affine, and this implies the general case by the definition of morphisms. (b) is clear when $X \subseteq \mathbb{A}^m$ and $Y \subseteq \mathbb{A}^n$ are locally closed. This implies the general case, since if $(X_i)_i$, $(Y_j)_j$ are open-affine coverings of X and Y, then $(X_i \times Y_j)_{i,j}$ is an open-affine covering of $X \times Y$. When X is quasi-projective, it suffices for (c) to show that the strong topology on $\mathbb{P}^n(k)$ is Hausdorff. For any k-hyperplane $H \subseteq \mathbb{P}^n$, the strong topology on $(\mathbb{P}^n \smallsetminus H)(k)$ is Hausdorff since $\mathbb{P}^n \smallsetminus H \cong \mathbb{A}^n$. Since for any two points $\xi \neq \eta \in \mathbb{P}^n(k)$ there is a k-hyperplane $H \subseteq \mathbb{P}^n$ with $\xi, \eta \notin H$, this proves (c) for quasi-projective X. If one wants to prove (c) for k-varieties that are not necessarily quasi-projective (those are otherwise not considered in this course), one needs to use the fact that the diagonal Δ_X is Zariski closed in $X \times X$ (see A.6.12). Therefore the diagonal of $X(k)$ is closed in $X(k) \times X(k)$ in the strong topology, by (b), and this means that $X(k)$ is Hausdorff. □

1.7.5 Remark. If $(k, \leq)$ is an ordered field and V is an affine k-variety, the sets

$$\mathfrak{U}_V(f) := \{\xi \in V(k) : f(\xi) > 0\} \quad (f \in k[V])$$

form a basis of open sets for the strong topology on $V(k)$. In fact, if $k[V]$ is generated by $x_1, \dots, x_n$ as a k-algebra, already the balls $\mathfrak{U}_V(r^2 - \sum_i (x_i - x_i(\xi))^2)$ (for $\xi \in V(k)$ and $0 \neq r \in k$) are a basis of open sets.

1.7.6 Proposition. *Let k be a real field. Every quasi-projective k-variety X contains a Zariski open affine subset U for which $X(k) = U(k)$.*

Proof. We can assume that $X \subseteq \mathbb{P}^n$ is Zariski locally closed. The subset $W := \mathbb{P}^n \smallsetminus \mathcal{V}(x_0^2 + \cdots + x_n^2)$ of $\mathbb{P}^n$ is open affine (A.6.10) and satisfies $W(k) = \mathbb{P}^n(k)$. So we may replace X by $X \cap W$ and thereby assume that $X \subseteq \mathbb{A}^m$ is locally closed. Let $\overline{X}$ be the Zariski closure of X in $\mathbb{A}^m$. There exist $f_1, \dots, f_r \in k[\overline{X}]$ with $X = \overline{X} \smallsetminus \mathcal{V}_{\overline{X}}(f_1, \dots, f_r)$. Now $U = \overline{X} \smallsetminus \mathcal{V}_{\overline{X}}(f_1^2 + \cdots + f_r^2)$ is an open affine subset of X and satisfies $U(k) = X(k)$. □

1.7.7 Remark. Proposition 1.7.6 is in fact true for any field k that is not algebraically closed. One may argue in a similar way as above, using the following fact: For every integer $n \geq 1$, there exists a non-constant homogeneous polynomial $f_n \in k[x_1, \dots, x_n]$ such that $f_n(\xi) \neq 0$ for every $0 \neq \xi \in k^n$. See Exercise 1.7.3.

Now let R be a real closed field. Whenever X is an R-variety, we consider the set $X(R)$ of R-rational points with the strong (order) topology. The next theorem is of fundamental importance. It characterizes those irreducible R-varieties whose function field can be ordered. In fact we prove a stronger version right away:

1.7.8 Theorem. *Let R be a real closed field, let X be an irreducible R-variety and $f_1, \dots, f_r \in R(X)$. The following are equivalent:*

(i) *The function field $R(X)$ has an ordering P for which $f_1 >_P 0, \dots, f_r >_P 0$;*
(ii) *the subset $U := \{\xi \in X(R)\colon f_i$ is defined at ξ and $f_i(\xi) > 0$ for $i = 1, \dots, r\}$ of $X(R)$ is Zariski dense in X;*
(iii) *U contains a non-singular R-point of X.*

Proof. Let X' be a non-empty open affine subset of X on which $f_1, \dots, f_r$ are all defined. The R-variety X' is isomorphic to a closed subvariety V of $\mathbb{A}^n$, for some n. Let $I = \langle h_1, \dots, h_m \rangle$ be the vanishing ideal of V in $R[x] = R[x_1, \dots, x_n]$. For $i = 1, \dots, r$, the restriction of f_i to X' is a regular function, so there exists a polynomial $F_i \in R[x]$ for which $F_i + I \in R[x]/I = R[V]$ corresponds to f_i under the chosen isomorphism $X' \cong V$.

(i) ⇒ (ii): Let S be a real closure of $(R(X), P)$. By hypothesis (i), the R-sentence

$$\exists\, y = (y_1, \dots, y_n) \quad \Big(\bigwedge_{j=1}^{m} h_j(y) = 0 \ \wedge \ \bigwedge_{i=1}^{r} F_i(y) > 0 \Big)$$

holds in S, since $x = (x_1, \dots, x_n)$ is such a tuple y. By the transfer principle 1.6.16, the sentence holds in R as well. Hence there exists $\xi \in X'(R)$ with $f_i(\xi) > 0$ for $i = 1, \dots, r$. In particular $U \neq \emptyset$.

In fact, U is Zariski dense in X. For this it suffices to see that $U \cap X'(R)$ is Zariski dense in X', since X' is Zariski dense in X. But this means to find, for any non-zero $g \in R[X']$, a point ξ in $U \cap X'(R)$ that satisfies $g(\xi) \neq 0$. We achieve this by simply appending $f_{r+1} := g^2$ to the list $f_1, \dots, f_r$ and repeating the previous argument for the extended list.

(ii) ⇒ (iii): Assume that U is Zariski dense in X. Since the singular locus X_{sing} of X is a proper Zariski closed subset, the set U contains a non-singular point of X.

We still need to give the proof of (iii) ⇒ (i). This proof makes use of the interaction between orderings and valuation rings of a field. Its proper context would be Section 3.6 later, where this topic will be studied systematically. For now we give an *ad hoc* proof. More background will later be given in Remark 3.6.10. Let $\xi \in X(R)$ be a non-singular point with $f_i(\xi) > 0$ for $i = 1, \dots, r$, let $\mathcal{O}_\xi = \mathcal{O}_{X,\xi}$ be the local ring of X at ξ, and let $\mathfrak{m}_\xi$ be its maximal ideal. To prove (i) it suffices to show that the preordering T of $R(X)$ generated by $f_1, \dots, f_r$ does not contain -1. Indeed, Proposition 1.1.26 will then imply the existence of a positive cone P of $R(X)$ that satisfies $f_i >_P 0$ for $i = 1, \dots, r$.

Assume to the contrary that $-1 \in T$. So there is an identity $-1 = \sum_{i=1}^m p_i g_i^2$ in $R(X)$ with $g_i \in R(X)^*$ and $p_i \in \mathcal{O}_\xi$, where the p_i satisfy $p_i(\xi) > 0$. Clearing denominators, this implies an identity

$$p_0 q_0^2 + \cdots + p_m q_m^2 = 0 \tag{1.7}$$

where p_i, $q_i \in \mathcal{O}_\xi$ are non-zero and $p_i(\xi) > 0$ for every i. Since the local ring $\mathcal{O}_\xi$ is regular there exists, by Proposition A.5.5, a valuation ring B of $R(X)$ that dominates $\mathcal{O}_\xi$, and such that the induced map $R = \mathcal{O}_\xi/\mathfrak{m}_\xi \to B/\mathfrak{m}_B$ between the residue fields is an isomorphism. Now read (1.7) as an identity in B. Since B is a valuation ring, there is some index j such that $b_i := q_i/q_j \in B$ for $i = 0, \dots, m$. For simplicity of notation assume $j = 0$. Dividing (1.7) by q_0^2 now gives an identity $p_0 + \sum_{i=1}^m p_i b_i^2 = 0$ with $b_i \in B$. Reducing this identity modulo $\mathfrak{m}_B$ yields a contradiction, since the first summand is strictly positive in $B/\mathfrak{m}_B = R$ and the others are non-negative. □

We isolate the case $r = 0$ of Theorem 1.7.8:

1.7.9 Corollary. (Artin–Lang) *Let R be a real closed field. The following are equivalent, for every irreducible R-variety X:*

(i) *The function field $R(X)$ of X is real, i.e. can be ordered;*
(ii) *$X(R)$ is Zariski dense in X;*
(iii) *X has a non-singular R-point.* □

($R = \mathbb{R}$) Irreducible $\mathbb{R}$-varieties for which these equivalent conditions hold are sometimes said to be *real* in the literature. We will not follow this usage, to avoid the danger of confusing the meaning with "variety defined over the field of real numbers".

1.7.10 Examples. Let X an irreducible variety over a real closed field R.

1. From the theorem just proved, it follows that the existence of a non-singular R-point on X is invariant under birational equivalence over R. Indeed, this property depends only on the function field $R(X)$, by Corollary 1.7.9.

2. The function field of an irreducible curve X over R is real if and only if $|X(R)| = \infty$. This follows from 1.7.9, since a subset of $X(R)$ is Zariski dense in X if and only if it is infinite. As an example of an irreducible curve X with a non-real function field but with $X(R) \neq \emptyset$, take the plane affine curve $x^4 + x^2 + y^2 = 0$. In $R(X)$ the identity $-1 = (\frac{x}{y})^2 + (\frac{x^2}{y})^2$ holds, so the field $R(X)$ is not real. Still the origin is an R-point contained in X.

3. Even if the function field $R(X)$ is real, the non-singular R-points may fail to be dense in $X(R)$ in the strong (order) topology. Examples are the plane curve X with equation $x^2 + y^2 = x^3$, which has an isolated R-point at the origin, or the surface V in affine 3-space with equation $y^2 = x^2 z$. The surface V is known as the *Whitney umbrella*. Its singular locus consists of the line $x = y = 0$ (the "stick" of the umbrella).

$x^2 + y^2 = x^3$ $\qquad\qquad$ $x^2 = y^2 z$

4. Let X be an irreducible R-variety of dimension n. Once we have developed the notion of dimension for semialgebraic sets, we will see (Remark 4.6.5.6) that the function field $R(X)$ is real if and only if the (semi-) algebraic set $X(R)$ has dimension n.

1.7.11 Corollary. *Let X be an irreducible R-variety, and let $W \subseteq X(R)$ be a subset that is open in the order topology. Then W is Zariski dense in X if, and only if, W contains a non-singular R-point of X.*

Proof. If W is Zariski dense then it contains a non-singular point, since the non-singular locus of X is open (non-empty) in the Zariski topology. Conversely let $\xi \in W$ be a non-singular point of X. Replacing X with an open affine neighborhood of ξ we may assume that X itself is affine. Since W is open in $X(R)$ there exist $f_1, \dots, f_r \in R[X]$ with $f_i(\xi) > 0$ such that $\{\eta \in X(R) \colon f_i(\eta) > 0, i = 1, \dots, r\} \subseteq W$ (see Remark 1.7.5). So W is Zariski dense by (iii) $\Rightarrow$ (ii) of Theorem 1.7.8. □

The next result generalizes the Hilbert 17 property, from affine space to arbitrary irreducible R-varieties:

1.7.12 Theorem. *Let X be an irreducible R-variety. A rational function $f \in R(X)$ is a sum of squares in $R(X)$ if, and only if, there exists a non-empty Zariski open subset U of X on which f is defined and such that $f \geq 0$ on $U(R)$.*

Proof. We may assume that X is affine and $f \in R[X]$. Let f be a sum of squares in $R(X)$, so there is a non-zero polynomial $h \in R[X]$ such that fh^2 is a sum of squares in $R[X]$. Then $f(\xi) \geq 0$ for any $\xi \in X(R)$ with $h(\xi) \neq 0$, so the assertion holds for the principal open subset $U = D_X(h)$ of X. Conversely, if there is an open subset $U \neq \varnothing$ of X with $f \geq 0$ on $U(R)$, the set $\{\xi \in X(R) \colon f(\xi) < 0\}$ is contained in $(X \setminus U)(R)$, hence it is not Zariski dense in X. Theorem 1.7.8 therefore implies that $f \geq_P 0$ for every ordering P of $R(X)$. According to Artin's characterization 1.1.30, this means that f is a sum of squares in $R(X)$. □

1.7.13 Remark. Beware that a rational function f on X may well be a sum of squares in $R(X)$ and, at the same time, take *negative* values at some points $\xi \in X(R)$. This can be seen in the examples 1.7.10.3: On the curve X, the polynomial $f = x - 1$ is the square of a rational function (namely $f = (\frac{y}{x})^2$), but $f(\xi) < 0$ for $\xi = (0, 0) \in X(R)$. Similarly, $f = z$ is a rational square on the Whitney umbrella, but is negative on the lower part of the "stick" of the umbrella. Generally it is true that $f(\xi) < 0$ is only possible at singular R-points ξ of X, see Exercise 1.7.2.

For hypersurfaces there exists a particularly simple way to decide when the function field is real:

1.7.14 Theorem. (Sign-changing criterion) *Let R be a real closed field, let $f \in R[x_1, \dots, x_n]$ be an irreducible polynomial, and let $X = \mathcal{V}(f) \subseteq \mathbb{A}^n$, the hypersurface defined by f. The following are equivalent:*

(i) *The function field $R(X)$ is real;*
(ii) *f is indefinite on R^n, i.e. there exist $\xi, \eta \in R^n$ with $f(\xi) < 0 < f(\eta)$.*

Proof. We abbreviate $x = (x_1, \dots, x_n)$.

(i) $\Rightarrow$ (ii): Assume that f is positive semidefinite on R^n. By Artin's theorem 1.5.21 there exists an identity $fh^2 = \sum_{i=1}^r g_i^2$ with non-zero polynomials g_i, $h \in R[x]$. We may assume $\gcd(g_1, \dots, g_r, h) = 1$. Then there is an index i such that f does not divide g_i. Reading the identity modulo f therefore shows that -1 is a sum of squares in $R(X)$.

(ii) $\Rightarrow$ (i): Let f be indefinite on R^n. If $n = 1$ then f is linear and therefore $R(X) = R$. So assume $n \geq 2$ and fix ξ, $\eta \in R^n$ with $f(\xi) < 0 < f(\eta)$. After a suitable affine-linear change of coordinates we may assume that $\xi = (0, \dots, 0, a)$ and $\eta = (0, \dots, 0, b)$ for some a, $b \in R$. Let $x' = (x_1, \dots, x_{n-1})$ and let Q be an ordering of $R(x')$ satisfying $0 <_Q x_i <_Q \varepsilon$ for $i = 1, \dots, n-1$ and every positive ε in R. (Such an ordering can be constructed inductively by adjoining one variable at a time and by successively extending the ordering using the $P_{0,+}$-construction, cf. 1.1.15.) Then

$$f(x', a) <_Q 0 <_Q f(x', b)$$

as elements of $R(x')$, since the constant terms in $f(x', \alpha)$ or $f(x', \beta)$ dominate all other terms in absolute value. Let R_1 denote the real closure of $(R(x'), Q)$. By the intermediate value theorem 1.3.3, the polynomial $f = f(x', x_n)$ (in the variable x_n and with coefficients in R_1) has a zero in R_1.

The polynomial f remains irreducible if we consider it as a polynomial in the variable x_n over the field $R(x')$. Indeed, otherwise we would have $f = \frac{g}{s} \cdot \frac{h}{t}$, i.e. $stf = gh$, with g, $h \in R[x]$ and $0 \neq s$, $t \in R[x']$, where both g, h involve the variable x_n. By unique factorization, and since f is irreducible in $R[x]$, f would have to divide one of g or h in $R[x]$, say g. This in turn would imply $h \in R(x')$, contradiction. According to Corollary 1.4.12, the ordering Q can be extended from $R(x')$ to the field $K := R(x')[x_n]/\langle f \rangle$. We claim that K is isomorphic to the function field $R(X)$ of the hypersurface $f = 0$ (which finishes the proof). Indeed, the natural ring homomorphism $R[x]/\langle f \rangle \to K$ is injective by Gauss's lemma. So there is an induced field embedding of $R(X) = \mathrm{qf}(R[x]/\langle f \rangle)$ into K, and it obviously is surjective. □

1.7.15 Remarks.

1. Theorem 1.7.14 has an obvious extension to hypersurfaces over an ordered base field (k, P): If $f \in k[x_1, \dots, x_n]$ is irreducible and $X = \mathcal{V}(f) \subseteq \mathbb{A}^n$, the ordering P can be extended to the function field $k(X)$ if and only if f is indefinite on R^n, where R denotes the real closure of (k, P). Note however that it is not enough in general that f is indefinite on k^n, see Exercise 1.5.3 for an example.

2. An alternative topological proof for the implication (ii) $\Rightarrow$ (i) in Theorem 1.7.14 will be available once we have discussed the concepts of dimension and connectedness for semialgebraic sets. See Exercise 4.6.5.

Exercises

Let R always be a real closed field.

1.7.1 Let k be a real field and let X be an irreducible k-variety. If X has a non-singular k-point, show that the function field $k(X)$ is real. Give an example to show that the converse does not hold in general.

1.7.2 Let X be an irreducible R-variety and let the rational function $f \in R(X)$ be a sum of squares in $R(X)$. Suppose that $\xi \in X(R)$ is a point where the rational function f is defined and satisfies $f(\xi) < 0$. Show that ξ is a singular point of X.

1.7.3 Let k be a field that is not algebraically closed. Prove for every $n \geq 1$ that there exists a non-constant homogeneous polynomial $f_n \in k[x_1, \ldots, x_n]$ with $f_n(\xi) \neq 0$ for every $0 \neq \xi \in k^n$. Conclude that any quasi-projective k-variety X contains an open affine subset U with $X(k) = U(k)$ (cf. Proposition 1.7.6).

1.7.4 Let $K = R(x, y)$, the rational function field in two variables. Construct an explicit positive cone P of K that satisfies

$$\mathrm{sign}_P(f) = \mathrm{sign}\, f(0, 0)$$

for every $f \in R[x, y]$ with $f(0, 0) \neq 0$.

1.7.5 Let K be a field with $\mathrm{char}(K) \neq 2$, and let $f \in K[t]$ be a monic irreducible polynomial. Use the sign changing criterion to prove: The field $L = K[t]/\langle f \rangle$ is real if and only if f is not a sum of squares in $K(t)$.

Remark: If $f \in K[t]$ is a sum of squares in $K(t)$, we will later prove that f is in fact a sum of squares in $K[t]$ (Proposition 6.4.2). Try not to use this fact.

1.7.6 Show that the sign changing criterion holds for projective hypersurfaces over R as well: Let $f \in R[x_0, \ldots, x_n]$ be an irreducible homogeneous polynomial and let $X \subseteq \mathbb{P}^n$ be the projective hypersurface $f = 0$. The function field $R(X)$ is real if and only if f is indefinite on R^{n+1}.

1.8 Notes

The material in this chapter is classical, and is fundamental for real algebra and geometry. Orderings of arbitrary fields were introduced by Artin and Schreier [6] in 1927, together with real closed fields and real closures of ordered fields. The authors also prove the uniqueness properties of real closures from Section 1.4. Before, orderings were essentially only considered for number fields. The results of [6] were used by Artin [5] to prove Theorem 1.1.30, and subsequently to give the solution to Hilbert's 17th problem (Theorem 1.5.21). Hilbert's original article with descriptions of all 23 problems is in [93] or [95]. An English translation can be found in [94]. Hölder's theorem 1.1.18 was proved already in 1901. Theorem 1.2.4 is due to Springer, and is a particular case of his theorem for anisotropic quadratic forms [201].

Descartes' rule of signs 1.3.12 was described in 1637. It may be the oldest mathematical result in this book that is not considered elementary today. Sturm's method 1.3.18 dates back to 1829. Today it is seen as the historically first algorithm for real root isolation. The approach 1.3.26 via the Hermite matrix, or more generally via (scaled) trace forms, was developed around 1850 and is often referred to as the

Hermite–Sylvester method. But Borchardt and Jacobi also contributed, and the priority question appears to be subtle.

The Tarski–Seidenberg projection theorem and transfer principle were found and announced by Tarski already in 1930, but not published before 1948 (Tarski [207]). See also Seidenberg 1954 [196]. The Artin–Lang theorem is based on Lang's 1953 paper [120]. Our exposition is inspired by Becker [14]. The sign-changing criterion 1.7.14 is due to Dubois and Efroymson [60].

Chapter 2
Positive Polynomials and Sums of Squares

We start the chapter by introducing the important notion of Gram matrices (Section 2.1). To some extent, this technique allows us to linearize the problem of finding sum of squares representations of a given polynomial. Next we discuss Newton polytopes, from which natural restrictions for such representations arise. The rest of the chapter is devoted to classical results on sums of squares representations. First we present the Fejér–Riesz theorem, featuring sums of squares on the circle (Section 2.3). The highlights are Hilbert's results from 1888, presented in Section 2.4. Let f be a homogeneous real polynomial in n variables that takes non-negative values on all of $\mathbb{R}^n$. The starting point for Hilbert was the question whether such f can always be written as a sum of squares of real polynomials. His results in [91] went far beyond answering this question, since he was able to decide the problem separately for every fixed pair (n, d) where $d = \deg(f)$. In Section 2.4 we present his results almost in full completeness. Most interesting is the case $(n, d) = (3, 4)$ of ternary quartic forms, for which Hilbert proved that every non-negative such form is a sum of three squares of quadratic forms. For the moment, we give an elementary proof of a slightly weaker result (four squares instead of three). In Chapter 7 we'll return to this question and prove the full result, embedded in a much more general context.

2.1 Sums of Squares of Polynomials

2.1.1 If A is a (commutative) ring, recall (1.1.7) that ΣA^2 denotes the set of all sums of squares in A. We are often going to replace the phrase "$f \in A$ is a sum of squares in A" by the shorthand "f is sos in A", a slight abuse of language that is both very customary and convenient. For $f \in \Sigma A^2$ we define the *sum of squares length* (or *sos length*) of f as

$$\ell(f) = \ell_A(f) = \inf\{r \geq 0 \colon \exists f_1, \ldots, f_r \in A \text{ with } f = f_1^2 + \cdots + f_r^2\}.$$

C. Scheiderer, *A Course in Real Algebraic Geometry*, Graduate Texts in Mathematics 303,
https://doi.org/10.1007/978-3-031-69213-0_2

If $f \in A$ and $f \notin \Sigma A^2$ we put $\ell(f) = \infty$. The *Pythagoras number* of A is

$$p(A) := \sup\{\ell_A(f) \colon f \in \Sigma A^2\}$$

2.1.2 Remark. For a well-known example, recall that $p(\mathbb{Z}) = p(\mathbb{Q}) = 4$ by Lagrange's theorem. Generally speaking, the Pythagoras number $p(A)$ is very hard to determine, even in the case of fields. As was already mentioned (Remark 1.5.25), the Pythagoras number of the rational function fields $\mathbb{R}(x_1, \dots, x_n)$ is unknown for $n \geq 3$. Even the order of magnitude of this invariant for $n \to \infty$ is not known.

2.1.3 We can generalize these definitions to symmetric matrices. A symmetric matrix $S \in \mathrm{Sym}_n(A)$ is a *(matrix) sum of squares* if there exist column vectors $u_1, \dots, u_r \in A^n$ such that $S = \sum_{i=1}^r u_i u_i^\top$. The smallest number $r \geq 0$ for which such u_i exist is the *sos length* $\ell(S) = \ell_A(S)$ of S. See Exercise 2.1.1 for other equivalent characterizations of $\ell(S)$, and see Exercise 2.1.2 for equivalent characterizations of matrix sums of squares in the case where $A = k$ is a field.

2.1.4 We start by showing that the problem of finding sum of squares representations of polynomials can be linearized. Always let k be a (coefficient) field and let $x = (x_1, \dots, x_n)$ with $n \in \mathbb{N}$. Recall some standard notation and terminology. For $\alpha = (\alpha_1, \dots, \alpha_n) \in \mathbb{Z}_+^n$ a multi-index, let $|\alpha| = \alpha_1 + \dots + \alpha_n$ and $x^\alpha = x_1^{\alpha_1} \cdots x_n^{\alpha_n}$. The *(total) degree* $\deg(f)$ of $f \in k[x]$ is defined by $\deg(0) = -1$ and

$$\deg\Big(\sum_{\alpha \in \mathbb{Z}_+^n} c_\alpha x^\alpha\Big) = \sup\{|\alpha| \colon c_\alpha \neq 0\}$$

for $c_\alpha \in K$. For $d \geq 0$ let $k[x]_{\leq d} = \{f \in k[x] \colon \deg(f) \leq d\}$, which is a linear subspace of $k[x]$ of dimension $\binom{n+d}{n}$. A polynomial $\sum_{\alpha \in \mathbb{Z}_+^n} c_\alpha x^\alpha$ is *homogeneous of degree d* if $c_\alpha = 0$ whenever $|\alpha| \neq d$. Homogeneous polynomials of degree d are often called *forms* of degree d. We let $k[x]_d = \{f \in k[x] \colon f$ is homogeneous of degree $d\}$, a linear subspace of $k[x]$ of dimension $\binom{n+d-1}{d}$. From Exercise 1.1.10(b) we recall:

2.1.5 Lemma. *If k is a real field then* $\deg(f_1^2 + \dots + f_r^2) = 2\max_i \deg(f_i)$ *for any* $f_1, \dots, f_r \in k[x]$.

Proof. Let $d = \max_i \deg(f_i)$, and let g_i be the degree d homogeneous subform of f_i $(i = 1, \dots, r)$, so $g_i = 0$ if $\deg(f_i) < d$. The homogeneous subform of $\sum_{i=1}^r f_i^2$ of degree $2d$ is $\sum_{i=1}^r g_i^2$. This form is non-zero since the field $k(x)$ is real and $g_i \neq 0$ for at least one index i. The inequality $\leq$ in the lemma is obvious, and the lemma is proved. □

2.1.6 We continue to assume that k is real. Fix $d \geq 0$ in the following and put $J_d := \{\alpha \in \mathbb{Z}_+^n \colon |\alpha| \leq d\}$. So $(x^\alpha)_{\alpha \in J_d}$ is the standard monomial basis of $k[x]_{\leq d}$. Let $X := (x^\alpha)_{\alpha \in J_d}$, the column vector of all monomials of degree $\leq d$, whose components are indexed by the set J_d. The map

$$k^{J_d} \to k[x]_{\leq d}, \quad u = (u_\alpha)_{\alpha \in J_d} \mapsto X^\top \cdot u = \sum_{\alpha \in J_d} u_\alpha x^\alpha$$

is a linear isomorphism. Let $\mathrm{Sym}_{J_d}(k)$ be the space of symmetric matrices over k whose rows and columns are indexed by J_d. The linear map

$$\gamma\colon \mathrm{Sym}_{J_d}(k) \to k[x]_{\le 2d}, \quad S = (a_{\alpha\beta})_{\alpha,\beta\in J_d} \mapsto X^\top S X = \sum_{\alpha,\beta\in J_d} a_{\alpha\beta}\, x^{\alpha+\beta}$$

is called the *Gram map*, and it is clearly surjective. For every polynomial $f \in k[x]_{\le 2d}$ write $G_f := \gamma^{-1}(f)$, an affine-linear subspace of $\mathrm{Sym}_{J_d}(k)$. The matrices in G_f are called the *Gram matrices* of f.

2.1.7 Proposition. *Let k be a real field. A polynomial $f \in k[x]$ is a sum of squares in $k[x]$ if and only if f has a Gram matrix S with $S \ge_P 0$ for every ordering P of k. In this case the sos length is*

$$\ell(f) \;=\; \min\{\ell(S)\colon S \in G_f,\ S \ge_P 0 \textit{ for every ordering } P\}.$$

In particular, if $k = R$ is real closed then $\ell(f) = \min\{\mathrm{rk}(S)\colon S \in G_f,\ S \ge 0\}$.

Proof. Choose d with $\deg(f) \le 2d$ and put $J = J_d$. If $f = \sum_{j=1}^r f_j^2$ with $f_j \in k[x]$ then $\deg(f_j) \le d$ for all j. Let $B \in \mathrm{M}_{J\times r}(k)$ be the matrix whose j-th column is the coefficient vector of f_j ($j = 1, \dots, r$). Then $X^\top B = (f_1, \dots, f_r)$, and hence

$$f \;=\; (X^\top B)(X^\top B)^\top \;=\; X^\top (BB^\top) X.$$

So $BB^\top$ is a Gram matrix of f with $\ell(BB^\top) \le r$ that is psd with respect to every ordering. The same argument works backward: If $S \in G_f$ is psd with respect to every ordering and if $r = \ell(S)$, there is $B \in \mathrm{M}_{J\times r}(k)$ with $S = BB^\top$ (Exercise 2.1.1). Therefore $f = \sum_{j=1}^r f_j^2$ where the coefficients of f_j are those in the j-th column of B ($j = 1, \dots, r$). □

2.1.8 Definition. Let k be a real field and let $f \in k[x]_{\le 2d}$. Given an sos representation $f = f_1^2 + \cdots + f_r^2$ with $f_j \in k[x]$, let B be the $J_d \times r$ matrix whose j-th column is the coefficient vector of f_j ($j = 1, \dots, r$). The symmetric $J_d \times J_d$ matrix $S = BB^\top$ is called the *Gram matrix of f associated* with the given sos representation. We put $G_f^+ := \{S \in G_f\colon S \ge_P 0$ for every ordering P of $k\}$, the set of all *totally psd Gram matrices* of f.

2.1.9 Remark. If S is the Gram matrix associated with $f = f_1^2 + \cdots + f_r^2$, observe that the rank of S coincides with the dimension of the linear span of $f_1, \dots, f_r$ (Exercise 2.1.3). When the field $k = R$ is real closed, this is also the minimum number of squares among all sos representations with Gram matrix S.

2.1.10 Remark. As we have seen, every sos representation of f gives a totally psd Gram matrix of f, and conversely, every totally psd Gram matrix of f arises from such a representation. When do two sos representations give the same Gram matrix? If

$$f \;=\; f_1^2 + \cdots + f_r^2 \tag{2.1}$$

is an sos representation of f and $U = (u_{ij})$ is any orthogonal $r \times r$ matrix over k (i.e. $UU^\top = I$), then

$$f = \Big(\sum_i u_{i1} f_i\Big)^2 + \cdots + \Big(\sum_i u_{ir} f_i\Big)^2 \tag{2.2}$$

is another representation of f, and both have the same Gram matrix. Indeed, if B is the coefficient matrix of (2.1), then (2.2) has coefficient matrix BU, and both representations have Gram matrix $BB^\top = (BU)(BU)^\top$. We turn this observation into a definition:

2.1.11 Definition. Two sum of squares representations

$$f_1^2 + \cdots + f_r^2 = g_1^2 + \cdots + g_r^2$$

of the same polynomial (with $f_i, g_i \in k[x]$) are said to be *(orthogonally) equivalent* if there exists an orthogonal $r \times r$ matrix $U = (u_{ij})$ over k such that $g_j = \sum_{i=1}^r u_{ij} f_i$ ($j = 1, \dots, r$). (Note that this definition also applies if the representations have different length, by filling up the shorter one with zeros.)

Equivalent sos representations have the same Gram matrix, as we just saw. The converse is true as well:

2.1.12 Proposition. *Let k be a real field. Two sum of squares representations $f_1^2 + \cdots + f_r^2 = g_1^2 + \cdots + g_r^2$ (with $f_i, g_i \in k[x]$) have the same Gram matrix if and only if they are orthogonally equivalent.*

2.1.13 Corollary. *The orthogonal equivalence classes of sum of squares representations of $f \in k[x]$ are in bijection with the set of totally psd Gram matrices of f.* □

To prove Proposition 2.1.12 it remains to show that two sos representations with the same Gram matrix are orthogonally equivalent. This follows from the next lemma:

2.1.14 Lemma. *Let k be a real field. Given two rectangular matrices B, C over k of the same size $n \times r$, the following are equivalent:*

(i) $BB^\top = CC^\top$;
(ii) *there exists an orthogonal $r \times r$ matrix U over k with $C = BU$.*

Proof. (ii) $\Rightarrow$ (i) is obvious (and true over any field). Conversely assume $BB^\top = CC^\top$. For $x, y \in k^r$ put $\langle x, y\rangle = \sum_{i=1}^r x_i y_i$. Let V resp. W be the linear subspace of k^r that is spanned by the rows $b_1, \dots, b_n$ of B resp. $c_1, \dots, c_n$ of C. By hypothesis we have $\langle b_i, b_j\rangle = \langle c_i, c_j\rangle$ for any $i, j = 1, \dots, n$. Whenever $\lambda_1, \dots, \lambda_n \in k$ are such that $\sum_{i=1}^n \lambda_i b_i = 0$, then

$$\Big\langle \sum_i \lambda_i c_i,\, c_j \Big\rangle = \sum_i \lambda_i \langle c_i, c_j\rangle = \sum_i \lambda_i \langle b_i, b_j\rangle = \Big\langle \sum_i \lambda_i b_i,\, b_j \Big\rangle = 0$$

for $j = 1, \dots, n$, which implies $\sum_i \lambda_i c_i = 0$ since the field k is real. So there exists a linear map $\psi\colon V \to W$ satisfying $\psi(b_i) = c_i$ $(i = 1, \dots, n)$. Moreover ψ is an isometry from V to W, which means that ψ is bijective and $\langle \psi(v), \psi(v') \rangle = \langle v, v' \rangle$ holds for all $v, v' \in V$.

Let $V^\perp = \{x \in k^r \colon \forall\, y \in V\ \langle x, y \rangle = 0\}$ be the orthogonal complement of V, and define $W^\perp$ similarly. The two subspaces $V^\perp$ and $W^\perp$ of k^r are isometric, i.e. there exists a linear isomorphism $\psi'\colon V^\perp \to W^\perp$ satisfying $\langle \psi'(y), \psi'(y') \rangle = \langle y, y' \rangle$ for $y, y' \in V^\perp$. This is clear if $k = R$ is real closed. We won't prove the case of a general (real) field k; here the desired conclusion follows directly from *Witt cancellation*, see [176] 1.5.8 or [119] I.4.2, for example.

As a consequence, the map $\phi = \psi \oplus \psi'\colon V \oplus V^\perp \to W \oplus W^\perp$ is an isometry from k^r to itself with $\phi(b_i) = c_i$ for $i = 1, \dots, n$. If U denotes the (orthogonal) matrix with $\phi(x) = Ux$ for $x \in k^r$, then $BU^\top = C$ holds by construction. □

Note that Lemma 2.1.14 becomes false whenever the field k has no ordering (Exercise 2.1.3).

2.1.15 Example. To illustrate these results, consider the case of univariate polynomials over a real closed field R. Up to equivalence, the sos representations of a polynomial $f = \sum_{i=0}^{2d} a_i t^i \in R[t]$ correspond to those psd symmetric matrices $(b_{ij})_{0 \le i,j \le d}$ over R that have fixed skew diagonal sums

$$\sum_{i=0}^{k}{}' b_{i,k-i} = a_k, \quad k = 0, \dots, 2d. \tag{2.3}$$

In particular, the polynomial f is a sum of squares if and only if there exist numbers $b_{ij} = b_{ji} \in R$ $(0 \le i, j \le d)$ such that identities (2.3) hold, and such that the symmetric matrix (b_{ij}) (of size $(d+1) \times (d+1)$) is positive semidefinite. For a concrete example let $f = t^4 + 1$. Then a direct computation shows that G_f^+ consists of all matrices

$$\begin{pmatrix} 1 & 0 & -a \\ 0 & 2a & 0 \\ -a & 0 & 1 \end{pmatrix} \tag{2.4}$$

in $\mathrm{Sym}_3(R)$ that are psd, which comes down to the condition $0 \le a \le 1$. For such a, an sos representation of f that corresponds to the matrix (2.4) is $f = (t^2 - a)^2 + 2at^2 + (1 - a^2)$.

We record a consequence and assume for simplicity that the field $k = R$ is real closed.

2.1.16 Corollary. *Let $U \subseteq R[x] = R[x_1, \dots, x_n]$ be an m-dimensional linear subspace, and let $f \in R[x]$ be a sum of squares of elements from U. Then f is a sum of m squares of elements from U.*

Proof. The Gram matrix associated with the given sos representation of f has rank $\le \dim(U)$. □

Note that this is a better (smaller) bound on the number of squares than one would get from Carathéodory's theorem (Proposition 8.1.2).

2.1.17 Corollary. *Every sum of squares f in $R[x_1, \dots, x_n]$ with $\deg(f) \le 2d$ is a sum of $\binom{n+d}{n}$ squares.* □

2.1.18 Remark. Let $f \in \mathbb{R}[x_1, \dots, x_n]$ be a polynomial. The set G_f^+ of all psd Gram matrices of f (Definition 2.1.8) is a *spectrahedron*, meaning that it is an affine-linear section of the cone of psd symmetric $N \times N$ matrices (where $N = \binom{n+d}{n}$ if $\deg(f) = 2d$). Spectrahedra will be discussed later in more detail, see Section 8.2. Therefore, deciding whether f is a sum of squares means to decide whether a certain *linear matrix inequality (LMI)* has a solution. This is the question for the existence of a tuple $(u_1, \dots, u_s)$ of real numbers such that

$$A_0 + u_1 A_1 + \cdots + u_s A_s \succeq 0,$$

where $A_0, \dots, A_s$ are symmetric $N \times N$ matrices given explicitly from f. Using techniques from convex optimization, and under mild conditions, it is effectively possible to decide solvability of an LMI, and to find a solution in case there exists one, up to any prescribed numerical precision. In particular, it is essentially possible to decide algorithmically whether f is sos, and to find a concrete (approximate) sos decomposition of f if the answer is positive. Much more on this in Chapter 8.

2.1.19 Remark. (This remark will not be used further) From a general perspective, the following basis-free approach to Gram matrices is convenient. Let R be a real closed field (for simplicity) and let A be an R-algebra. Given a finite-dimensional linear subspace V of A, let VV denote the subspace of A that is spanned by all products $p_1 p_2$ (p_1, $p_2 \in V$). Let $S_2 V \subseteq V \otimes V$ denote the space of all tensors that are invariant under the involution $p_1 \otimes p_2 \mapsto p_2 \otimes p_1$. Elements of $S_2 V$ may be identified with symmetric bilinear forms on the dual space $V^\vee$ of V. In particular, any such element can be written as a finite sum $\vartheta = \sum_i (\pm p_i \otimes p_i)$ with $p_i \in V$, and has a well-defined Sylvester signature. The Gram map $\gamma\colon S_2 V \to VV$ is the restriction of the (linear) product map $V \otimes V \to VV$, $p_1 \otimes p_2 \mapsto p_1 p_2$ to $S_2 V$. For $f \in VV$, the set

$$G_{f,V}^+ = \{\vartheta \in S_2 V\colon \gamma(\vartheta) = f,\ \vartheta \succeq 0\}$$

is in natural bijection with the equivalence classes of representations of f as sums of squares of elements of V. Under this bijection, a tensor $\sum_i p_i \otimes p_i$ in $G_{f,V}^+$ corresponds to the sos representation $f = \sum_i p_i^2$. The proof is exactly the same as for Proposition 2.1.12.

Exercises

Let R always be a real closed field.

2.1.1 Let A be a (commutative) ring, let $S \in \mathrm{Sym}_n(A)$ be a symmetric matrix over A and let $x = (x_1, \dots, x_n)^\top$ (considered as a column vector). For every $r \geq 1$, the following are equivalent:

(i) The polynomial $q_S(x) = x^\top S x$ is a sum of r squares of linear forms in $A[x_1, \dots, x_n]$;
(ii) there exist (column) vectors $u_1, \dots, u_r \in A^n$ such that $S = \sum_{i=1}^r u_i u_i^\top$;
(iii) there exists a matrix $T \in \mathrm{M}_{n\times r}(A)$ with $S = TT^\top$.

2.1.2 Let k be a field with $\mathrm{char}(k) \neq 2$, and let $S \in \mathrm{Sym}_n(k)$ with characteristic polynomial $p_S(t) = t^n + \sum_{i=1}^n a_i t^{n-i}$. Show that the following are equivalent:

(i) $S \geq_P 0$ for every ordering P of k;
(ii) S is a matrix sum of squares;
(iii) $(-1)^i a_i \in \Sigma k^2$ for $i = 1, \dots, n$.

Show $\ell(S) \leq \mathrm{rk}(S) \cdot p(k)$ if (i)–(iii) are satisfied, where $p(k)$ is the Pythagoras number of k (2.1.1). How does one have to modify (ii) and (iii), so as to make either of them equivalent to "$S >_P 0$ for every ordering P of k"?

2.1.3 Let k be a field.

(a) When the field k is real, show that $\mathrm{rk}(BB^\top) = \mathrm{rk}(B)$ holds for every (rectangular) matrix B over k.
(b) Assuming that k is non-real, show that (a) fails. Moreover find (rectangular) matrices B, C over k of the same size $n \times r$ such that $BB^\top = CC^\top$, but there is no orthogonal $r \times r$ matrix U over k with $C = BU$.

2.1.4 Let $f \in R[x_1, \dots, x_n]$ be a polynomial with $\deg(f) \leq 2d$. For the Gram spectrahedron G_f^+ of f, show that $\dim(G_f^+) \leq \frac{1}{2}N(N+1) - \binom{n+2d}{n}$ where $N = \binom{n+d}{n}$.

2.1.5 Let $x = (x_1, \dots, x_n)$, let $\Sigma_{n,\leq 2d}$ denote the cone of sums of squares in $R[x]$ of degree $\leq 2d$, and let $N = \binom{n+d}{n}$. Show that the following conditions are equivalent for $f \in R[x]$ with $\deg(f) \leq 2d$:

(i) f lies in the interior of the cone $\Sigma_{n,\leq 2d}$;
(ii) f has a positive definite Gram matrix of size $N \times N$.

2.1.6 How many representations as a sum of two squares does a sufficiently general positive polynomial $f \in R[t]$ of degree $2d$ have (up to equivalence)? Find all such representations explicitly for $f = t^6 + 1$. (*Hint*: $x^2 + y^2 = (x + iy)(x - iy)$.)

2.1.7 Let k be a field and let $f \in k[x_1, \dots, x_n]$ be a non-constant polynomial that is irreducible over the algebraic closure $\overline{k}$ of k. Prove that $\mathrm{rk}(S) \geq 3$ for every Gram matrix S of f.

2.1.8 Let $S \subseteq R^n$ be an unbounded closed semialgebraic set that is star-shaped with respect to some point $u \in S$, i.e., for every $v \in S$ the segment $[u, v] = \{(1 - t)u + tv\colon 0 \leq t \leq 1\}$ is contained in S. Then S contains a halfline centered at u: There is $0 \neq w \in R^n$ with $u + tw \in S$ for all $t \geq 0$ in R. (*Hint*: Tarski's principle may be used (how precisely?) to reduce to the case $R = \mathbb{R}$.)

2.1.9 For any polynomial $f \in R[x_1, \dots, x_n]$, the set G_f^+ of all psd Gram matrices of f is a semialgebraic R-convex set that is closed and bounded. (*Hint:* Use Exercise 2.1.8.)

2.1.10 Let A be an R-algebra, and let V be a linear subspace of A of dimension $\dim(V) = m < \infty$. Generalize Corollary 2.1.16 as follows: Every sum of squares of elements from V is a sum of m such squares.

2.1.11 Let A be an R-algebra, and let $V \subseteq A$ be a linear subspace with $\dim(V) < \infty$. We use the notation from the basis-free setup in Remark 2.1.19. Let $f \in VV$, and let $U = U_f$ be the linear subspace of V that is spanned by all $p \in V$ with $f - p^2 \in \Sigma V^2$. Prove the following:

(a) $G^+_{f,V} = \{\vartheta \in S_2V \colon \gamma(\vartheta) = f, \vartheta \succeq 0\}$ is contained in $S_2U \subseteq U \otimes U$.
(b) f has a Gram tensor that is positive definite as an element of S_2U.
(c) Conclude that the dimension of $G^+_{f,V}$ is given by $\dim(G^+_{f,V}) = \frac{1}{2}r(r+1) - s$ where $r = \dim(U)$ and $s = \dim(UU)$.

Here $\dim(G^+_{f,V})$ is defined to be the dimension of the affine-linear hull of $G^+_{f,V}$ (as in 8.1.1).

2.2 Newton Polytopes

As a second technique we now introduce Newton polytopes, and apply them to sums of squares.

2.2.1 Definition. Let k be a field and let $f = \sum_{\alpha \in \mathbb{Z}^n_+} a_\alpha x^\alpha$ be a polynomial in $k[x] = k[x_1, \dots, x_n]$. The *support* of f is the set

$$\operatorname{supp}(f) := \{\alpha \in \mathbb{Z}^n_+ \colon a_\alpha \neq 0\},$$

corresponding to those monomials that actually occur in f. The *Newton polytope* of f is the convex hull of $\operatorname{supp}(f)$ in $\mathbb{R}^n$,

$$\operatorname{New}(f) := \operatorname{conv}(\operatorname{supp}(f)).$$

2.2.2 Remark. Being the convex hull of finitely many points in $\mathbb{R}^n$, $\operatorname{New}(f)$ is indeed a polytope, i.e. a compact intersection of finitely many closed affine-linear halfspaces in $\mathbb{R}^n$. For the following discussion write $\langle u, v\rangle = \sum_{i=1}^n u_i v_i$ for u, $v \in \mathbb{R}^n$, and let

$$H_{u,c} := \{\xi \in \mathbb{R}^n \colon \langle \xi, u\rangle \leq c\}$$

for $0 \neq u \in \mathbb{R}^n$ and $c \in \mathbb{R}$. Every closed halfspace is of this form.

2.2.3 Any n-tuple $u \in \mathbb{R}^n$ defines a grading of the polynomial ring $k[x] = k[x_1, \dots, x_n]$, by declaring the variable x_i to be homogeneous of degree u_i ($i = 1, \dots, n$). So the monomial $x^\alpha = x_1^{\alpha_1} \cdots x_n^{\alpha_n}$ is homogeneous of degree $\langle \alpha, u\rangle$. If $G = \mathbb{Z}_+ u_1 + \cdots + \mathbb{Z}_+ u_n \subseteq \mathbb{R}$ denotes the semigroup generated by $u_1, \dots, u_n$ in $\mathbb{R}$, we therefore have

$$k[x] = \bigoplus_{g \in G} k[x]_g,$$

and each homogeneous component $k[x]_g$ is a k-linear subspace of $k[x]$ that is generated by monomials. We'll say that $f \in k[x]$ is *u-homogeneous* if f is homogeneous with respect to this grading.

Let $0 \neq f \in k[x]$, and write $f = \sum_{g \in G} f_g$ where $f_g \in k[x]_g$ for $g \in G$. We call

$$\deg_u(f) := \max\{g \in G \colon f_g \neq 0\} = \max\{\langle \alpha, u\rangle \colon \alpha \in \operatorname{supp}(f)\}$$

the *u-degree* of f. If $\deg_u(f) = g$ then $L_u(f) := f_g$ is the *u-leading form* of f. By definition, $L_u(f)$ is non-zero and u-homogeneous. For any two polynomials $f_1, f_2 \neq 0$ we have $\deg_u(f_1 f_2) = \deg_u(f_1) + \deg_u(f_2)$ and $L_u(f_1 f_2) = L_u(f_1)\,L_u(f_2)$. From the definitions it follows that $\mathrm{New}(f) \subseteq H_{u,c}$ if and only if $\deg_u(f) \le c$.

2.2.4 Remarks.

1. For $u = (1, \dots, 1)$, the u-grading of $k[x]$ is the standard grading, $\deg_u(f)$ is the usual total degree of f and $L_u(f)$ is the highest degree subform of f. If $u = (-1, \dots, -1)$ then $-\deg_u(f)$ is the *smallest* degree of a monomial in f, and $L_u(f)$ is the sum of all terms in f of this degree.

2. Let $k = R$ be real closed. For $0 < t \in R$ and every $q \in \mathbb{Q}$, the power t^q is a well-defined positive element in R. If $u \in \mathbb{Q}^n$ and $f \in R[x]$ is u-homogeneous with $\deg_u(f) = g \in \mathbb{Q}$, then

$$f(t^{u_1}\xi_1, \dots, t^{u_n}\xi_n) = t^g f(\xi)$$

for every $\xi \in R^n$ and $0 < t \in R$. The positive multiplicative group R_+^* of R acts on R^n via

$$R_+^* \times R^n \to R^n, \quad (t, \xi) \mapsto (t^{u_1}\xi_1, \dots, t^{u_n}\xi_n).$$

Typical pictures of the induced orbit decomposition for this action (for $n = 2$) are

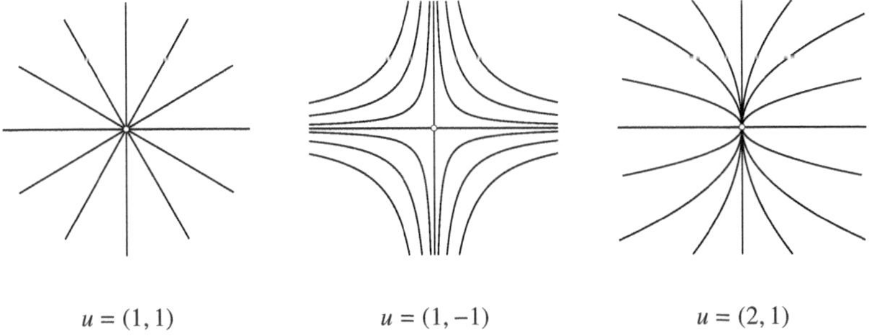

$u = (1, 1)$ $\qquad$ $u = (1, -1)$ $\qquad$ $u = (2, 1)$

The following result characterizes the Newton polytope $\mathrm{New}(f)$ in terms of the values of f along orbits:

2.2.5 Proposition. *Let R be real closed and let $0 \neq f \in R[x]$. For $u = (u_1, \dots, u_n) \in \mathbb{Q}^n$ and $c \in \mathbb{Q}$, the following are equivalent:*

(i) *$\mathrm{New}(f) \subseteq H_{u,c}$, i.e. $\deg_u(f) \le c$;*
(ii) *$t^{-c} \cdot f(t^{u_1}\xi_1, \dots, t^{u_n}\xi_n)$ remains bounded for $t \to \infty$, $t \in R$ and for every $\xi \in R^n$.*

Proof. Let $f = f_{g_1} + \cdots + f_{g_r}$ be the decomposition of f into non-zero u-homogeneous components, such that $\deg_u(f) = g_1 > \cdots > g_r$. Fix $\xi \in R^n$ and put $\eta_t := (t^{u_1}\xi_1, \dots, t^{u_n}\xi_n)$ for $t > 0$. Remark 2.2.4.2 shows that

$$f(\eta_t) = \sum_{i=1}^{r} t^{g_i} f_{g_i}(\xi) \tag{2.5}$$

for $t > 0$. Condition (i) means that $g_i \le c$ for every index i (see 2.2.3), so (i) implies that $t^{-c} f(\eta_t)$ remains bounded for $t \to \infty$. For the converse choose $\xi \in R^n$ with $f_{g_1}(\xi) \neq 0$. The highest power of t in $t^{-c} f(\eta_t)$ is $t^{g_1 - c}$. So (ii) implies $g_1 - c \le 0$, which is (i). □

We had to assume that u and c are rational since otherwise the powers don't usually make sense for $R \neq \mathbb{R}$. If $R = \mathbb{R}$ then u and c may be arbitrary.

2.2.6 Proposition. *Let* $0 \neq f, g \in R[x]$.

(a) $\mathrm{New}(fg) = \mathrm{New}(f) + \mathrm{New}(g)$.
(b) *If* f, g *are psd then* $\mathrm{New}(f+g) = \mathrm{conv}(\mathrm{New}(f) \cup \mathrm{New}(g))$.

(Part (a) is true for polynomials over an arbitrary field k.)

In (a), $A_1 + A_2 = \{x + y \colon x \in A_1, y \in A_2\}$ is the *Minkowski sum* of A_1 and A_2.

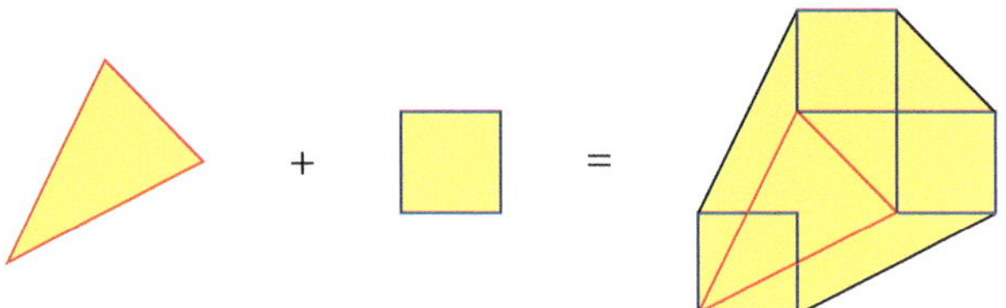

If A_1 and A_2 are both compact or both convex, the same is true for $A_1 + A_2$. In particular, $\mathrm{conv}(A_1 + A_2) = \mathrm{conv}(A_1) + \mathrm{conv}(A_2)$ holds for any subsets A_1, A_2 of $\mathbb{R}^n$.

Proof. (a) Every monomial of fg is the product of a monomial of f and a monomial of g. Hence $\mathrm{supp}(fg) \subseteq \mathrm{supp}(f) + \mathrm{supp}(g)$ and therefore $\mathrm{New}(fg) \subseteq \mathrm{New}(f) + \mathrm{New}(g)$. For the converse it suffices to show: If $0 \neq u \in \mathbb{Q}^n$ and $c \in \mathbb{Q}$ are such that $\mathrm{New}(fg) \subseteq H_{u,c}$, then also $\mathrm{New}(f) + \mathrm{New}(g) \subseteq H_{u,c}$. Let $a = \deg_u(f)$ and $b = \deg_u(g)$. By hypothesis we have $\deg_u(fg) \le c$. Therefore $a + b = \deg_u(fg) \le c$ and $\mathrm{New}(f) \subseteq H_{u,a}$, $\mathrm{New}(g) \subseteq H_{u,b}$. Therefore $\mathrm{New}(f) + \mathrm{New}(g) \subseteq H_{u,a} + H_{u,b} = H_{u,a+b} \subseteq H_{u,c}$, as desired.

(b) The inclusion "$\subseteq$" follows from $\mathrm{supp}(f+g) \subseteq \mathrm{supp}(f) \cup \mathrm{supp}(g)$. It remains to show $\mathrm{New}(f) \cup \mathrm{New}(g) \subseteq \mathrm{New}(f+g)$. For this it suffices to see that $\mathrm{New}(f+g) \subseteq H_{u,c}$ implies $\mathrm{New}(f) \cup \mathrm{New}(g) \subseteq H_{u,c}$. Let $\xi \in R^n$ and let $\eta_t = (t^{u_1}\xi_1, \dots, t^{u_n}\xi_n)$ for $t > 0$. We are going to use Proposition 2.2.5. By hypothesis, $t^{-c}(f(\eta_t) + g(\eta_t))$ is bounded for $t \to \infty$. Since f and g are psd, this implies that both $t^{-c} f(\eta_t)$ and $t^{-c} g(\eta_t)$ are bounded as well. So the conclusion follows from 2.2.5. □

2.2.7 Remark. If $K \subseteq \mathbb{R}^n$ is any convex set, the m-fold Minkowski sum $K + \cdots + K$ is the same as $mK = \{m\xi \colon \xi \in K\}$ (check this). So Proposition 2.2.6(a) implies $\mathrm{New}(f^m) = m\,\mathrm{New}(f)$ for every $m \ge 1$.

We now draw some conclusions and sketch applications. We keep assuming that R is a real closed field.

2.2.8 Corollary. *Let* $f = f_1^2 + \cdots + f_r^2$ *with polynomials* $f_i \in R[x]$. *Then* $\mathrm{New}(f_i) \subseteq \frac{1}{2}\mathrm{New}(f)$ *for* $i = 1, \dots, r$.

Proof. 2.2.6(b) implies $\text{New}(f_i^2) \subseteq \text{New}(f)$ for all i, and this means $\text{New}(f_i) \subseteq \frac{1}{2}\,\text{New}(f)$ according to 2.2.6(a). □

2.2.9 Corollary. *Let $f \in R[x]$ be a sum of squares. If N is the number of integral (lattice) points in $\frac{1}{2}\,\text{New}(f)$, every sum of squares representation of f is equivalent to a sum of at most N squares.*

Proof. If $f = \sum_i f_i^2$ then each monomial of each f_i lies in $\frac{1}{2}\,\text{New}(f)$, by 2.2.8. So the assertion follows from Corollary 2.1.16. □

Here are a few applications.

2.2.10 Examples.

1. In Exercise 1.5.5 it was shown that the Motzkin polynomial $f = x^4y^2 + x^2y^4 - 3x^2y^2 + 1$ is psd but not sos. The psd property follows directly from the arithmetic-geometric inequality

$$\frac{1}{3}\left(1 + x^4y^2 + x^2y^4\right) \geq \sqrt[3]{1 \cdot x^4y^2 \cdot x^2y^4}, \tag{2.6}$$

as remarked in Example 1.5.24.1. By a term inspection of a hypothetical sos representation, one checks that f cannot be a sum of squares. This last argument simplifies if one uses the Newton polytope: $\frac{1}{2}\text{New}(f)$ is the convex hull of $(2,1)$, $(1,2)$, $(1,1)$ and $(0,0)$, and these are the only lattice points it contains.

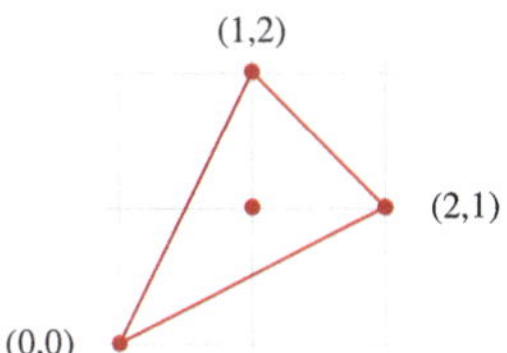

So if f were sos, there would have to be an identity

$$f = \sum_i (a_i x^2 y + b_i x y^2 + c_i x y + d_i)^2$$

with real numbers a_i, b_i, c_i, d_i. But the coefficient of x^2y^2 in the right-hand sum is $\sum_i c_i^2 \geq 0$, which shows that such an identity cannot exist.

Note that f has exactly four zeros in $\mathbb{R}^2$, namely $(\pm 1, \pm 1)$ (this follows from inequality (2.6), see Remark 1.5.24.1). For every real number $c > 0$, the polynomial $f_c := f + c$ is strictly positive on $\mathbb{R}^2$, but is still not sos since nothing changes in the argument just given.

2. The following example is due to Choi and Lam [42]: The form

$$f = w^4 + x^2y^2 + x^2z^2 + y^2z^2 - 4wxyz$$

in $\mathbb{R}[w, x, y, z]$ is psd, again by the arithmetic-geometric inequality (applied to w^4, x^2y^2, x^2z^2 and y^2z^2). To see that f is not sos, note that $\frac{1}{2}\text{New}(f)$ is the convex hull of the five points

$$2e_1,\ e_2 + e_3,\ e_2 + e_4,\ e_3 + e_4,\ \frac{1}{2}\big(e_1 + e_2 + e_3 + e_4\big)$$

in $\mathbb{R}^4$. The last point is the barycentre of the others, hence it doesn't contribute to the convex hull. The only lattice points in $\frac{1}{2}\text{New}(f)$ are the first four points. So if f were sos, we would have to have

$$f = \sum_i (a_i w^2 + b_i xy + c_i xz + d_i yz)^2.$$

The coefficient of $wxyz$ reveals that this cannot be the case.

3. In the preceding two examples, the non-negativity of f was a consequence of the arithmetic-geometric inequality (AGI). Systematizing this observation, Iliman, de Wolff *et al.* were led to the notion of *circuit polynomials* ([98], [99], [55]). Let $\alpha^{(0)}, \dots, \alpha^{(n)} \in 2\mathbb{Z}_+^n$ be $n + 1$ even tuples that are affinely independent, so the convex hull $P := \text{conv}(\alpha^{(0)}, \dots, \alpha^{(n)})$ is an n-simplex. Let $\beta \in \mathbb{Z}_+^n$ be a further tuple that lies in the interior of P. Given non-zero real numbers $a_0, \dots, a_n, b$, the support of the polynomial

$$f = \sum_{i=0}^{n} a_i x^{\alpha(i)} + bx^\beta$$

is a *circuit*, in the sense that supp(f) is a minimal affinely dependent set. Let $\lambda_0, \dots, \lambda_n \geq 0$ with $\sum_i \lambda_i = 1$ be the unique numbers for which

$$\beta = \sum_{i=0}^{n} \lambda_i \alpha^{(i)}.$$

Generalizing the preceding examples, one shows that the polynomial f is psd on $\mathbb{R}^n$ provided that $a_i > 0$ for $i = 0, \dots, n$ and

$$|b| \leq \prod_{i=0}^{n} \Big(\frac{a_i}{\lambda_i}\Big)^{\lambda_i} \tag{2.7}$$

(Exercise 2.2.6). Conversely one can show that if f is psd, either (2.7) holds, or else f is a non-negative linear combination of monomial squares [98]. The convex cone generated by non-negative circuit polynomials is known as the *SONC cone* (sums of non-negative circuit polynomials). It is not contained in the sos cone except for obvious cases, and was much studied in the last years as an alternative and complement to the sos cone in polynomial optimization.

Exercises

2.2.1 Let R be a real closed field and let $f = \sum_\alpha c_\alpha x^\alpha$ be a psd polynomial in $R[x] = R[x_1, \dots, x_n]$. For every vertex (extreme point) β of the Newton polytope $\mathrm{New}(f)$, show that $\beta \in 2\mathbb{Z}^n$ and $c_\beta > 0$.

2.2.2 Let $f \in R[x] = R[x_1, \dots, x_n]$ be a psd polynomial that is not a sum of squares. Show that f contains at least four monomials.

2.2.3 Find all zeros of the Choi–Lam form 2.2.10.2 in real projective space $\mathbb{P}^3(\mathbb{R})$.

2.2.4 The Robinson form is the symmetric ternary form

$$f = x^6 + y^6 + z^6 - x^4y^2 - x^2y^4 - x^4z^2 - x^2z^4 - y^4z^2 - y^2z^4 + 3x^2y^2z^2.$$

Show that f is psd but not sos (in $\mathbb{R}[x, y, z]$).

Hints: Projectively, f has ten real zeros that are easy to guess. Use them to prove that f is not sos. To show that f is psd, write the product $(x^2 + y^2)f$ as a sum of squares.

2.2.5 (Choi) Let $x = (x_1, x_2, x_3)$ and $y = (y_1, y_2, y_3)$, and let f be the biquadratic form

$$(x_1y_1)^2 + (x_2y_2)^2 + (x_3y_3)^2 - 2(x_1x_2y_1y_2 + x_2x_3y_2y_3 + x_3x_1y_3y_1) + 2(x_1y_2)^2 + 2(x_2y_3)^2 + 2(x_3y_1)^2$$

in $\mathbb{R}[x, y]$. Show that f is psd but not a sum of squares. (*Hint*: f is invariant under cyclic permutation of the indexes 1, 2, 3. The monomials $(x_1y_3)^2$, $(x_2y_1)^2$ and $(x_3y_2)^2$ are absent in f.)

We add the remark that every psd biform $f(x_1, \dots, x_n; y_1, y_2)$ of bidegree $(2, 2d)$ can be shown to be a sum of squares of biforms. (This fact is contained in the main result of Chapter 7, see Remark 7.2.15.) Choi's example above shows that, in this result, the limitation on the number of variables in the second group cannot be avoided.

2.2.6 With the notation of Remark 2.2.10.3, let $f = \sum_{i=0}^n a_i x^{\alpha(i)} + bx^\beta$ be a circuit polynomial with real coefficients $a_1 > 0, \dots, a_n > 0$ and b. Show that f is psd provided that

$$|b| \le \prod_{i=0}^n \Big(\frac{a_i}{\lambda_i}\Big)^{\lambda_i}.$$

2.3 The Fejér–Riesz Theorem

We now turn to concrete results on sum of squares representations. Given a polynomial f that is non-negative on the (unit) circle, we prove that f can be written as a sum of squares of polynomials on the circle. In fact, the main result (Theorem 2.3.4) will be quite a bit stronger.

Let R denote an arbitrary real closed field, and let X be the plane affine curve $x^2 + y^2 = 1$ over R. So X has coordinate ring $R[X] = R[x, y]/\langle x^2 + y^2 - 1\rangle$, and has the unit circle $X(R) = \{(s, t) \in R^2 \colon s^2 + t^2 = 1\}$ as the set of its R-points.

2.3.1 Proposition. *Every element of $R[X]$ that is psd on $X(R)$ is a sum of two squares in $R[X]$.*

We start with an elementary lemma:

2.3.2 Lemma. *Every binary form $f \in R[x_1, x_2]$ that is non-negative on R^2 is a sum of two squares of binary forms.*

Proof. By the fundamental theorem of algebra 1.2.9, f is a product of squares of linear forms and of positive definite quadratic forms. Since each of these factors is a sum of two squares, the same holds for f, by the identity

$$(f_1^2 + f_2^2)(g_1^2 + g_2^2) = (f_1 g_1 - f_2 g_2)^2 + (f_1 g_2 + f_2 g_1)^2. \qquad \square$$

Proof of Proposition 2.3.1. The coordinate ring $R[X]$ is isomorphic to the homogeneous localization

$$R[u,v]_{(u^2+v^2)} = \left\{ \frac{f(u,v)}{(u^2+v^2)^r} : r \geq 0,\ f \in R[u,v]_{2r} \right\} \tag{2.8}$$

via the homomorphism $R[X] \to R[u,v]_{(u^2+v^2)}$ given by $x \mapsto \frac{u^2-v^2}{u^2+v^2}$ and $y \mapsto \frac{2uv}{u^2+v^2}$. This can be checked directly, e.g. by writing down the inverse isomorphism. So we may replace $R[X]$ by the ring (2.8). Its elements are the rational functions on the projective line $\mathbb{P}^1_R$ with poles at most at $(1 : \pm i)$. Let $f = \frac{p(u,v)}{(u^2+v^2)^r}$ be a psd such function, where $p \in R[u,v]$ is homogeneous of degree $\deg(p) = 2r$. Clearly p is psd itself, and we may assume that $r = 2s$ is even, by multiplying the numerator and denominator by u^2+v^2 if necessary. Then, by 2.3.2, there exist forms $p_1,\ p_2 \in R[u,v]$ of degree $r = 2s$ with $p = p_1^2 + p_2^2$. Therefore

$$f = \left(\frac{p_1}{(u^2+v^2)^s} \right)^2 + \left(\frac{p_2}{(u^2+v^2)^s} \right)^2$$

is a decomposition of f in $R[u,v]_{(u^2+v^2)}$ into a sum of two squares, as desired. $\square$

2.3.3 To state and prove an even stronger version, we switch to a complex view point. Let $i = \sqrt{-1}$ and $C = R(i)$. On the ring $C[z,z^{-1}]$ of Laurent polynomials over C (in one variable z) we consider the C/R-involution $f \mapsto f^*$, characterized by $z^* = z^{-1}$ and $a^* = \overline{a}$ for $a \in C$. So $f \mapsto f^*$ is an R-linear ring endomorphism of $C[z,z^{-1}]$ that satisfies $f^{**} := (f^*)^* = f$ for every f. Explicitly, if $f = \sum_{n\in\mathbb{Z}} a_n z^n \in C[z,z^{-1}]$ then $f^* = \sum_{n\in\mathbb{Z}} \overline{a}_n z^{-n}$. Note that $\overline{f^*(\alpha)} = f(\overline{\alpha}^{-1})$ holds for every $\alpha \in C$, $\alpha \neq 0$. In particular, if $f = f^*$ then $f(\alpha) \in R$ for every $\alpha \in C$ with $|\alpha| = 1$.

2.3.4 Theorem. (Fejér–Riesz) *Let $f \in C[z,z^{-1}]$ satisfy $f = f^*$. If $f \geq 0$ on the unit circle $S = \{u \in C\colon |u| = 1\}$, there exists a polynomial $g \in C[z]$ with $f = gg^*$. In fact, there is such g satisfying $g(u) \neq 0$ for any $u \in C$ with $|u| < 1$ (assuming $f \neq 0$). Under this assumption, g is even unique up to a constant factor $\gamma \in C$ with $|\gamma| = 1$. If $n \geq 0$ is the minimal integer with $z^n f \in C[z]$, then* $\deg(g) = n$.

Proof. Let $f \in C[z,z^{-1}]$ satisfy $f = f^*$ and $f \neq 0$. Let $n \geq 0$ be minimal such that $z^n f \in C[z]$. Then f has the form $f = \sum_{j=-n}^{n} c_j z^j$, where the coefficients $c_j \in C$ satisfy $c_{-j} = \overline{c}_j$ for all j and $c_n \neq 0$. Hence

$$f = c z^{-n} \prod_{j=1}^{2n} (z - \alpha_j) \tag{2.9}$$

with $c \in C$, where $\alpha_1, \dots, \alpha_{2n}$ are the zeros of f in $C^* = C \setminus \{0\}$. If α is any zero of f then $\overline{\alpha}^{-1}$ is a zero of f as well, since $f = f^*$ by assumption. So there exists a permutation σ of the numbers $1, \dots, 2n$ with $\sigma^2 = \mathrm{id}$ such that $\alpha_j \cdot \overline{\alpha_{\sigma(j)}} = 1$ for all j. The constant $c \in C$ satisfies $\overline{c} = c \prod_j \alpha_j$. Indeed, $f^* = \overline{c} z^n \prod_j (z^{-1} - \overline{\alpha}_j) = \overline{c} z^{-n} \prod_j (1 - \overline{\alpha}_j z)$, and $f = f^*$ implies $c \prod_j (z - \alpha_j) = \overline{c} \prod_j (1 - \overline{\alpha}_j z)$. Substituting $z = 0$ gives

$$c \prod_j \alpha_j = \overline{c}. \tag{2.10}$$

Let α be a zero of f with $|\alpha| = 1$, let m be the multiplicity of $z - \alpha$ in the product decomposition (2.9). The restriction $f|_S$ of f changes sign at α if and only if m is odd, as can be seen by a local Taylor expansion (Exercise 2.3.3). Therefore, the assumption $f|_S \geq 0$ implies that every zero of f in S has even multiplicity. So we may select n elements $\beta_1, \dots, \beta_n$ out of the $2n$ elements $\alpha_1, \dots, \alpha_{2n}$ in such a way that

$$f = cz^{-n} \prod_{k=1}^{n} (z - \beta_k)(z - \overline{\beta}_k^{-1}).$$

In doing this we may assume $|\beta_k| \geq 1$ for every $k = 1, \dots, n$, which makes $\beta_1, \dots, \beta_n$ unique up to permutation. The constant c satisfies $\overline{c} = c \prod_k \frac{\beta_k}{\overline{\beta}_k}$ by identity (2.10), hence $s := c / \prod_k \overline{\beta}_k \neq 0$ lies in R and satisfies $c = s \prod_k \overline{\beta}_k$. Using the identity

$$\overline{\beta} z^{-1} \cdot (z - \beta)(z - \overline{\beta}^{-1}) = (z - \beta)(\overline{\beta} - z^{-1}) = -(z - \beta)(z - \beta)^* \quad (\beta \in C^*)$$

it follows that

$$f = (-1)^n s \cdot \prod_{k=1}^{n} (z - \beta_k)(z - \beta_k)^*.$$

Consequently, $f = tgg^*$ with $0 \neq t \in R$ and $g = \prod_{k=1}^{n} (z - \beta_k)$. Since $f \geq 0$ on S we must have $t > 0$. By construction, the polynomial g has no roots in the open unit disk. □

2.3.5 Remark. Let us relate Theorem 2.3.4 to the simpler formulation 2.3.1. An isomorphism

$$\varphi\colon\ C[x, y]/\langle x^2 + y^2 - 1 \rangle \to C[z, z^{-1}] \tag{2.11}$$

of C-algebras is given by $x \mapsto \frac{1}{2}(z + z^{-1})$ and $y \mapsto \frac{1}{2i}(z - z^{-1})$. The inverse map sends z to (the coset of) $x + iy$. If we transfer the C/R-involution $*$ from $C[z, z^{-1}]$ to $C[x, y]/\langle x^2 + y^2 - 1 \rangle$ using (2.11), the elements x and y are fixed under $*$. Therefore, the fixring of $*$ in $C[z, z^{-1}]$ is isomorphic to the R-algebra $R[X] = R[x, y]/\langle x^2 + y^2 - 1 \rangle$. If $p \in R[X]$ is non-negative on the unit circle $X(R) = S$, the Laurent polynomial $f = \varphi(p)$ satisfies $f|_S \geq 0$ and $f = f^*$. Given any Laurent polynomial g with $f = gg^*$, we have $f = g_0^2 + g_1^2$ where $g_0 = \frac{1}{2}(g + g^*)$ and $g_1 = \frac{1}{2i}(g - g^*)$ are fixed under $*$. Transforming back to $R[X]$ via (2.11), we get a decomposition of p as a sum of two squares in $R[X]$.

Exercises

2.3.1 Let $f \in \mathbb{R}[x,y]$ be a polynomial with $f(\cos t, \sin t) \ge 0$ for every $t \in \mathbb{R}$, and let $d = \deg(f)$. Prove that there exist polynomials $f_1, f_2 \in \mathbb{R}[x,y]$ with

$$f(\cos t, \sin t) = f_1(\cos t, \sin t)^2 + f_2(\cos t, \sin t)^2$$

for every $t \in \mathbb{R}$, and such that $\deg(f_i) \le \lceil \frac{d}{2} \rceil$ $(i = 1, 2)$.

2.3.2 Show that the Laurent polynomial $f = -6z^2 - 5z + 38 - 5z^{-1} - 6z^{-2}$ has strictly positive real values on the complex unit circle $|z| = 1$, and find a Laurent polynomial $g \in \mathbb{C}[z, z^{-1}]$ with $f = gg^*$ and $g(z) \ne 0$ for $z \in \mathbb{C}$, $|z| \le 1$. (*Hint*: $z = 2$ is a root of f.)

2.3.3 Let R be a real closed field, let $C = R(\sqrt{-1})$. On $C[z, z^{-1}]$ consider the C/R-involution $*$ as in 2.3.3. Let $f \in C[z, z^{-1}]$ with $f = f^*$, and let $\alpha \in C$ with $f(\alpha) = 0$ and $|\alpha| = 1$. Prove that the function f changes sign on $\{z \in C \colon |z| = 1\}$ at $z = \alpha$ if, and only if, the multiplicity of α as a root of f is odd. (*Hint*: The argument gets easier for $\alpha = 1$.)

2.4 Hilbert's 1888 Theorems

We now present the results from Hilbert's famous 1888 paper [91]. We give complete proofs for almost all the main statements, the only exception being the quantitative theorem on ternary quartics, of which we only show a weakened version. The full version will be proved later in Chapter 7 (Theorem 7.2.7), as part of a more general theory that is much more recent.

Here and later we often switch freely between homogeneous and non-homogeneous polynomials. We start by observing that this is harmless, as far as the psd or sos property is concerned. In the following let R be a real closed field, and let $x = (x_1, \dots, x_n)$ with $n \ge 1$.

2.4.1 Lemma. *If $f_1, \dots, f_r \in R[x]$ are such that $f = f_1^2 + \cdots + f_r^2$ is homogeneous of degree d, then d is even and each f_i is homogeneous of degree $\frac{d}{2}$.*

Of course the lemma is true for polynomials with coefficients in an arbitrary base field k, as long as k is real. Over a non-real field the lemma becomes false, as the identity $4x = (x+1)^2 - (x-1)^2$ shows.

Proof. By assumption $\mathrm{New}(f) \subseteq \{\alpha \in \mathbb{Z}_+^n \colon |\alpha| = d\}$, and so the lemma follows directly from Corollary 2.2.8. □

2.4.2 Lemma. *Any psd polynomial $f \ne 0$ in $R[x]$ has even total degree* $\deg(f)$.

Proof. Let $\deg(f) = d$, write $f = f_d + g$ with f_d homogeneous of degree d and $\deg(g) < d$. Assuming that d is odd, there exists $\xi \in R^n$ with $f_d(\xi) < 0$. So $f(t\xi)$, as a polynomial in the variable t, has negative leading coefficient $f(\xi)$, which implies $f(t\xi) < 0$ for sufficiently large $t > 0$ in R, contradiction. □

2.4.3 Definition. The *homogenization* of a polynomial $f \neq 0$ in $R[x] = R[x_1, \dots, x_n]$ is $f^h = x_0^{\deg(f)} \cdot f(\frac{x_1}{x_0}, \dots, \frac{x_n}{x_0})$. For $f = 0$ we define $0^h = 0$.

So if $f = \sum_{|\alpha| \le d} c_\alpha x^\alpha$ has total degree d, then $f^h = \sum_{|\alpha| \le d} c_\alpha x_0^{d-|\alpha|} x^\alpha$ is homogeneous of degree d.

2.4.4 Lemma. *Let $f \in R[x_0, x] = R[x_0, \dots, x_n]$ be a homogeneous polynomial and let $\tilde{f} = f(1, x_1, \dots, x_n) \in R[x]$ be its dehomogenization. Then f is psd if and only if $\tilde{f}$ is psd and* $\deg(f)$ *is even. The same is true with psd replaced by sos.*

Proof. The forward implications are clear. Conversely assume that $\tilde{f}$ is psd and $\deg(f) = 2d$ is even. Then $\deg((\tilde{f})^h) = \deg(\tilde{f}) = 2e$ is even by Lemma 2.4.2, and $(\tilde{f})^h$ is psd since $(\tilde{f})^h(t, \xi) = t^{2e} \tilde{f}(\xi/t) \ge 0$ for $t \neq 0$ (and by a limit argument). Moreover, $f = x_0^m \cdot (\tilde{f})^h$ where x_0^m is the maximal x_0-power that divides f. So it follows that $m = 2(d - e)$ is even as well, and hence f is psd. If $\tilde{f}$ is sos then f sos follows by homogenizing an sos representation of $\tilde{f}$. □

Note that strict positivity is usually not preserved by homogenization. That is, if $f \in R[x]$ is strictly positive on R^n, its homogenization $f^h \in R[x_0, x]$ may have non-trivial zeros in R^{n+1}. (Example?)

2.4.5 For discussing whether psd polynomials are sums of squares, we may therefore freely choose between the homogeneous and the inhomogeneous setting. We'll take the homogeneous point of view, as Hilbert already did. So we consider homogeneous polynomials (forms) $f \in R[x]$ in n variables $x = (x_1, \dots, x_n)$. A form $p \in R[x]$ is *positive definite* if $p(\xi) > 0$ holds for every $\xi \in R^n$, $\xi \neq 0$. Recall that $R[x]_d$ is the R-vector space of all forms of degree d, and $\dim R[x]_d = \binom{n+d-1}{n-1}$. The set of all *positive semidefinite (psd) forms* of degree d is denoted

$$P_{n,d} = \{f \in R[x]_d \colon \forall\, \xi \in R^n \ f(\xi) \ge 0\},$$

the set of all *sum of squares (sos) forms* of degree is

$$\Sigma_{n,d} = \Big\{ f \in R[x]_d \colon \exists\, r \ge 0,\ \exists\, f_1, \dots, f_r \in R[x] \text{ with } f = \sum_{i=1}^{r} f_i^2 \Big\}.$$

Obviously $\Sigma_{n,d} \subseteq P_{n,d}$ holds. We only consider the case where d is even, since otherwise both sets are reduced to $\{0\}$.

2.4.6 Proposition. *If d is even, both $P_{n,d}$ and $\Sigma_{n,d}$ are closed and R-convex semialgebraic cones with non-empty interior in $R[x]_d$.*

Proof. Both sets $P_{n,d}$ and $\Sigma_{n,d}$ are R-convex cones, since with f_1, f_2 they also contain $f_1 + f_2$ and $a f_1$ for $a \ge 0$. The cone $\Sigma_{n,d}$ has non-empty interior by Exercise 2.4.1. A fortiori, $P_{n,d}$ has non-empty interior.

The set $P_{n,d}$ can be described by a formula in the coefficients of the polynomial, so it is a semialgebraic subset of $R[x]_d$. Clearly $P_{n,d}$ is a closed set since $P_{n,d} = \bigcap_{\xi\in R^n}\{f \in R[x]_d\colon f(\xi) \ge 0\}$ is an intersection of sets that are obviously closed. Let $d = 2e$, and let $N = \dim R[x]_e = \binom{n-1+e}{e}$. According to 2.1.16, $\Sigma_{n,d}$ is the image set of the map

$$\phi\colon (R[x]_e)^N \to R[x]_d, \quad (g_1,\dots,g_N) \mapsto g_1^2 + \cdots + g_N^2.$$

The map ϕ is polynomial in the coefficients of the g_i, so $\Sigma_{n,d}$ is a semialgebraic set. Moreover ϕ is homogeneous of degree two and satisfies $\phi^{-1}(0) = \{0\}$. Therefore $\mathrm{im}(\phi) = \Sigma_{n,d}$ is a closed subset of $R[x]_d$, by the following lemma: □

2.4.7 Lemma. *Let $f\colon R^m \to R^n$ be a polynomial map that is homogeneous of even degree d, i.e. f satisfies $f(a\xi) = a^d f(\xi)$ for $\xi \in R^m$ and $a \in R$. Moreover, assume that $f^{-1}(0) = \{0\}$. If $M \subseteq R^m$ is any closed semialgebraic set such that $R_+\xi \subseteq M$ for every $\xi \in M$, the image set $f(M)$ is closed in R^n.*

Proof. Let $S = S^{n-1}$ be the unit sphere in R^n. The radial projection map

$$p\colon R^n \smallsetminus \{0\} \to S, \quad p(\xi) = \frac{\xi}{|\xi|} \quad (0 \ne \xi \in R^n)$$

is continuous. For any closed subset $B \subseteq S$, the set $c(B) := p^{-1}(B) \cup \{0\}$ is therefore closed in R^n. The semialgebraic set $M \cap S$ is closed and bounded and does not contain 0. Therefore $(p\circ f)(M\cap S)$ is a closed subset of S, by (a mild generalization of) Exercise 1.6.4. On the other hand $(p \circ f)(M \cap S) = f(M) \cap S$, and so $f(M) = c(f(M) \cap S)$ is closed by the first remark. □

The condition that f is polynomial can be weakened to f being a (continuous) semialgebraic map, as introduced in 4.3.1 later. The proof is the same, using 4.5.22 instead of Exercise 1.6.4.

Presumably inspired by Minkowski, Hilbert wondered whether every psd form in $R[x]_d$ might be a sum of squares of forms. In some cases this is true for elementary reasons:

2.4.8 Lemma. *Let $x = (x_1,\dots,x_n)$, and let $f \in P_{n,d} \subseteq R[x]_d$.*

(a) *If $n = 1$ then f is a square.*
(b) *If $n = 2$ then f is a sum of two squares.*
(c) *If $d = 2$ then f is a sum of n squares.*

Proof. (a) is obvious, (b) was proved in Lemma 2.3.2, and (c) is well-known from linear algebra. □

These are the "trivial" cases. For any other pair (n,d) with d even, the answer is not obvious. Hilbert settled them all at once:

2.4.9 Theorem. (Hilbert 1888) *Let $n \ge 1$ and $d \ge 0$ be integers with d even. If $n \le 2$ or $d \le 2$, or if $(n,d) = (3,4)$, then $\Sigma_{n,d} = P_{n,d}$. In all other cases the inclusion $\Sigma_{n,d} \subseteq P_{n,d}$ of cones is strict.*

So, apart from the trivial cases 2.4.8, there is precisely one additional (non-trivial) case where the answer is positive, namely $(n, d) = (3, 4)$, the case of ternary quartics.

The obvious cases $n \leq 2$ or $d \leq 2$ were just discussed. The Motzkin form

$$f = x_1^2 x_2^2 (x_1^2 + x_2^2 - 3x_3^2) + x_3^6 \tag{2.12}$$

lies in $P_{3,6} \setminus \Sigma_{3,6}$, see Exercise 1.5.5 or Example 2.2.10. Hence f lies in $P_{n,6} \setminus \Sigma_{n,6}$ for any $n \geq 3$. Moreover, for any $k \geq 0$, the form $x_1^{2k} f$ is psd and not a sum of squares, by Exercise 1.1.10. Therefore, $x_1^{2k} f$ lies in $P_{n,6+2k} \setminus \Sigma_{n,6+2k}$ for any $k \geq 0$ and $n \geq 3$. The Choi–Lam form (Example 2.2.10.2) lies in $P_{4,4} \setminus \Sigma_{4,4}$, and hence in $P_{n,4} \setminus \Sigma_{n,4}$ for any $n \geq 4$.

It remains to settle the case $(n, d) = (3, 4)$. Hilbert proved an even stronger statement here:

2.4.10 Theorem. (Hilbert 1888) *Any psd ternary form of degree four is a sum of three squares of quadratic forms.*

Hilbert's original proof is not easy to understand. There exist several alternative and more explicit approaches to the three-squares theorem, e.g. [174], [206], [151]. In Chapter 7 (see Theorem 7.2.7 and Remark 7.2.8) we'll reconsider the question from a more general point of view and prove a general theorem that encompasses both Hilbert theorems 2.4.9 and 2.4.10. For now, we are going to use elementary methods to show a statement that is just slightly weaker than Theorem 2.4.10:

2.4.11 Proposition. *Let $f \in R[x, y, z]$ be a psd form of degree four.*

(a) *If f has a non-trivial real zero then f is a sum of three squares.*
(b) *f is always a sum of four squares.*

Proof. The proof is due to Pfister [150]. First note that (b) follows from (a). Indeed, let $a := \min f(S^2)$ be the minimal value taken by f on the unit sphere S^2 (note that a exists, as follows from the case $R = \mathbb{R}$ by an application of Tarski transfer). The quartic form

$$g(x, y, z) := f(x, y, z) - a(x^2 + y^2 + z^2)^2$$

is non-negative on S^2, and hence on all of R^3. Since g has a non-trivial zero by construction, g is a sum of three squares by (a). This implies that f is a sum of four squares. So it remains to prove (a).

2.4.12 Lemma. *Let $q \in R[s, t]$ be a binary form of degree two that is positive definite. Then for any psd form $f \in R[s, t]$ there exist forms g, h in $R[s, t]$ with*

$$f = g^2 + qh^2. \tag{2.13}$$

Proof. We prove the lemma in the dehomogenized setting. So let $q \in R[t]$ satisfy $\deg(q) = 2$ and $q(t) > 0$ for all $t \in R$. Then $q = (at + b)^2 + c^2$ where a, b, $c \in R$ and $ac \neq 0$, and so $q = c^2(u^2 + 1)$ with $u = c^{-1}(at + b)$. We may change variables since $R[u] = R[t]$, and so we can assume $q = t^2 + 1$.

Let $f \in R[t]$ be psd, and assume first that $\deg(f) = 2$, say $f = (t + a)^2 + b^2$ where $a, b \in R$. If $a = 0$ then $f = (b^2 - 1) + q$ (if $b^2 \geq 1$), or $f = (1 - b^2)t^2 + b^2 q$ (if $b^2 \leq 1$) gives an identity as desired. So assume that $a \neq 0$. We want to find $\lambda \in R$, $\lambda \geq 0$ such that

$$f - \lambda q \;=\; f - \lambda(t^2 + 1) \;=\; (1 - \lambda)t^2 + 2at + (a^2 + b^2 - \lambda)$$

is a perfect square. This means we are looking for $0 \leq \lambda < 1$ such that the discriminant

$$\delta(\lambda) \;:=\; 4a^2 - 4(1 - \lambda)(a^2 + b^2 - \lambda) \;=\; -4\big(\lambda^2 - (a^2 + b^2 + 1)\lambda + b^2\big)$$

vanishes. Regarding δ as a polynomial function of λ, it follows from the mean value theorem that such λ exists, since $\delta(0) = -4b^2 \leq 0$ and $\delta(1) = 4a^2 > 0$.

An arbitrary psd polynomial $f \in R[t]$ is a product of quadratic psd polynomials. From the identity

$$(g_1^2 + h_1^2 q)(g_2^2 + h_2^2 q) \;=\; (g_1 g_2 + h_1 h_2 q)^2 + (g_1 h_2 - g_2 h_1)^2 q$$

it therefore follows that f has a representation (2.13) as desired. □

2.4.13 We now prove part (a) of Proposition 2.4.11. Let $f = f(x, y, z)$ be a psd ternary form of degree four that has a non-trivial zero in R^3. After a suitable linear change of coordinates we can assume $f(0, 0, 1) = 0$, and so

$$f \;=\; f_2(x, y) \cdot z^2 + f_3(x, y) \cdot z + f_4(x, y) \tag{2.14}$$

where $f_j = f_j(x, y)$ is a binary form of degree j (or $f_j = 0$), for $j = 2, 3, 4$. (Note that $\deg_z(f)$ cannot be 3.) Since f is a psd ternary form, each of the three binary forms f_2, f_4 and $4f_2 f_4 - f_3^2$ is psd by itself, and hence is a sum of two squares. If $f_2 = 0$, this implies $f_3 = 0$, and then $f = f_4$ is a sum of two squares. If $f_2 = l^2$ is the square of a linear form l, then l divides f_3, say $f_3 = 2lg_2$ for some quadratic binary form g_2. Since $4f_2 f_4 - f_3^2 = 4l^2(f_4 - g_2^2)$ is a sum of two squares, $f_4 - g_2^2$ is a sum of two squares as well. Hence $f = (lz + g_2)^2 + (f_4 - g_2^2)$ is a sum of three squares.

It remains to consider the case where f_2 is positive definite. By Lemma 2.4.12 there exist binary forms $p = p(x, y)$ and $q = q(x, y)$ with $4f_2 f_4 - f_3^2 = q^2 + p^2 f_2$, i.e.

$$q^2 + f_3^2 \;=\; f_2(4f_4 - p^2), \tag{2.15}$$

and here $\deg(p) = 2$ and $\deg(q) = 3$. Since f_2 is psd, there exist linear forms l_1, $l_2 \in R[x, y]$ with $f_2 = l_1^2 + l_2^2 = (l_1 + il_2)(l_1 - il_2)$, where $i = \sqrt{-1}$ as usual. So (2.15) implies that $l_1 + il_2$ divides one of the two forms $q \pm if_3$ (in the "complex" polynomial ring $C[x, y]$ where $C = R(i)$ and $i = \sqrt{-1}$). We may replace l_2 by $-l_2$ if needed, and so we can assume that $l_1 + il_2$ divides $q + if_3$. Multiplying both sides by $l_1 - il_2$, we see that f_2 divides $(q + if_3)(l_1 - il_2) = (ql_1 + f_3 l_2) + i(f_3 l_1 - ql_2)$ (in $C[x, y]$). Hence f_2 divides both the real and the imaginary part of this form (in $R[x, y]$). So the rational functions

$$h_1 := \frac{f_3 l_1 - q l_2}{2f_2}, \qquad h_2 := \frac{q l_1 + f_3 l_2}{2f_2}$$

are in fact binary quadratic forms with R-coefficients. From (2.15) we conclude

$$h_1^2 + h_2^2 = \frac{(q^2 + f_3^2)(l_1^2 + l_2^2)}{4f_2^2} = \frac{q^2 + f_3^2}{4f_2} = f_4 - \frac{1}{4}p^2.$$

Since

$$h_1 l_1 + h_2 l_2 = \frac{f_3(l_1^2 + l_2^2)}{2f_2} = \frac{1}{2} f_3,$$

we finally see that

$$f = \left(\frac{p}{2}\right)^2 + (h_1 + l_1 z)^2 + (h_2 + l_2 z)^2$$

is a sum of three squares of quadratic forms. □

This also completes the proof of Theorem 2.4.9. □

2.4.14 Remarks.

1. Starting from a non-trivial zero of f in R^3, the proof of 2.4.11(a) was entirely explicit and constructive. See Exercise 2.4.2 for a particular example.

2. The minimal cases where the psd cone is strictly larger than the sos cone are $(n, d) = (3, 6)$ and $(4, 4)$. We proved $P \neq \Sigma$ in these cases by presenting explicit forms, for which we could show that they are psd but not sums of squares. It is interesting to note that Hilbert's proof was very different. He did not bother to look for explicit forms in $P \setminus \Sigma$. Instead he used heavier tools from algebraic geometry to give an abstract proof for the existence of such forms. His construction is nicely explained in Reznick [167]. From Hilbert's arguments, it would have easily been possible to extract explicit examples. But it was not before 1967 that the first explicit example of a psd non-sos form was published [141]. This was Motzkin's form (2.12).

3. The Motzkin form has real zeros, and the same is true for many other prominent standard examples of psd, non-sos forms (Choi–Lam, Robinson etc.). On the other hand note that, for any pair (n, d) with $\Sigma_{n,d} \neq P_{n,d}$, there do exist forms in $P_{n,d} \setminus \Sigma_{n,d}$ that are strictly positive definite. Indeed, given $f \in P_{n,d} \setminus \Sigma_{n,d}$, the form $f_t := f + t(x_1^d + \cdots + x_n^d)$ is positive definite for every positive scalar $t > 0$. But for sufficiently small $t > 0$, f_t cannot be a sum of squares since the sos cone is closed (Proposition 2.4.6). It is less straightforward, though, to give explicit examples; see Rudin [174] for one particulary nice construction.

4. For ternary quartics, Hilbert's three-squares theorem 2.4.10 is a significant improvement of the general upper bound from Corollary 2.1.17. In fact, better bounds are known for all pairs $(n, 2d)$. Write

$$p(n, 2d) := \max\{\ell(f) \colon f \in R[x_1, \dots, x_n] \text{ sos}, \ \deg(f) = 2d\}$$

for the maximal sos length of an sos form in n variables of degree $2d$. Apart from the elementary cases $p(2, 2d) = 2$ and $p(n, 2) = n$, the precise value of $p(n, 2d)$ is known in the three cases $p(3, 4) = 3$ (Theorem 2.4.10), $p(3, 6) = 4$ and $p(4, 4) = 5$ [185]. Moreover it is known that $p(3, 2d) \in \{d + 1, d + 2\}$ for all d, and for every $n \geq 4$ there exist upper and lower bounds for $p(n, 2d)$ that are considerably smaller than 2.1.17 (see [185]). From $p(3, 2d) \geq d + 1$ it follows that there exist ternary sos forms of arbitrarily large sos length. Dehomogenizing, we see for every $n \geq 2$ that the polynomial ring $R[x_1, \dots, x_n]$ has infinite Pythagoras number. This fact was already proved in [41].

5. With Theorem 2.4.9, Hilbert determined the complete list of all pairs $(n, 2d)$ for which every psd form of degree $2d$ in n variables is a sum of squares of forms. As indicated before, this is not yet the end of the story. In Chapter 7, we will consider projective $\mathbb{R}$-varieties V on which every psd quadratic form on V is a sum of squares of linear forms. We will arrive at an essentially complete classification of the varieties with this property, and will see in what precise sense Hilbert's results from this section are particular cases of this theorem. The generalization goes even further since it also extends to the quantitative side. In particular, a proof of Hilbert's three-squares theorem 2.4.10 will be given there.

2.4.15 Remark. A natural way of generalizing the questions studied by Hilbert is to consider psd polynomials with a given Newton polytope. More concretely, consider lattice polytopes $P \subseteq \mathbb{R}^n_+$, i.e. polytopes that are the convex hull of finitely many points with non-negative integers as coordinates. Given the preceding examples, it is natural to ask for a characterization of all lattice polytopes P for which every psd polynomial $f \in \mathbb{R}[x_1, \dots, x_n]$ with $\mathrm{New}(f) \subseteq 2P$ is a sum of squares of polynomials $f = \sum_i f_i^2$ (necessarily satisfying $\mathrm{New}(f_i) \subseteq P$ for all i). From the results proved in Chapter 7, a classification of these polytopes is indeed possible. This is remarked upon in 7.2.19.

Exercises

2.4.1 Let $n, d \geq 1$ and $x = (x_1, \dots, x_n)$, and let $\Sigma_{n,2d} \subseteq R[x]_{2d}$ be the sos cone of forms of degree $2d$. Prove that the interior of $\Sigma_{n,2d}$ relative to $R[x]_{2d}$ is non-empty.

2.4.2 Find an explicit representation of the ternary quartic

$$f = x^4 + x^2y^2 + y^4 + x^2z^2 + y^2z^2$$

as a sum of three squares of quadratic forms over $\mathbb{R}$. (*Hint*: f has a real zero.)

2.4.3 Show that a polynomial $f = ax^2y + bxy^2 + cxy + d$ with $a, b, c, d \in \mathbb{R}$ is non-negative on $\mathbb{R}^2_+$ if, and only if, $a, b, d \geq 0$ and $c^3 + 27abd \geq 0$.

2.4.4 Show that the polynomial $1 + x + (y^2 - x^3)^2$ is strictly positive on $\mathbb{R}^2$, but that it is not a sum of squares in $\mathbb{R}[x, y]$.

2.4.5 Let $f(x, y, z) = x^4y^2 + x^2y^4 - 3x^2y^2z^2 + z^6$ be the Motzkin form. Show that $f(x^2, y^2, z^2)$ is a sum of squares of sextic forms.

2.4.6 Let $q \in \mathbb{R}[x_1, x_2, x_3, x_4, x_5]$ be the quadratic form

$$q = (x_1 + \cdots + x_5)^2 - 4(x_1x_2 + x_2x_3 + x_3x_4 + x_4x_5 + x_5x_1)$$

(a) Check the identities $q = (x_1 - x_2 + x_3 + x_4 - x_5)^2 + 4x_2x_4 + 4x_3(x_5 - x_4) = (x_1 - x_2 + x_3 - x_4 + x_5)^2 + 4x_2x_5 + 4x_1(x_4 - x_5)$, and use them to show that $q(\xi) \geq 0$ for every $\xi \in \mathbb{R}^5_+$.
(b) Determine the zero set of q in $\mathbb{R}^5_+$.
(c) Show that the psd quartic form $f = q(x_1^2, x_2^2, x_3^2, x_4^2, x_5^2)$ is not a sum of squares of forms. (*Hint*: Show that no non-zero quadratic form vanishes on the real zeros of f.)

The quartic form f is known as the *Horn form*.

2.4.7 (International Mathematical Olympiad 1971) Let $n \geq 1$ and consider the polynomial

$$f_n = \sum_{i=1}^{n} \prod_{\substack{j=1 \\ j \neq i}}^{n} (x_i - x_j)$$

in $\mathbb{R}[x_1, \ldots, x_n]$. Show that f_n is psd if and only if $n \in \{1, 2, 3, 5\}$. (The argument is elementary but somewhat tricky.) In fact, the form f_5 fails to be sos. Can you show this as well?

2.5 Notes

Hilbert's interest in sum of squares representations of non-negative forms was probably inspired by Minkowski, as he records in his obituary for Minkowski [95]. It was already mentioned that Motzkin [141] contains the first published example of a psd polynomial that is not a sum of squares of polynomials. Soon after, many more constructions of such polynomials were found, by Robinson [170], Choi, Lam and Reznick (e.g. [40], [42], [164], [43]), Lax and Lax [125] and others. We refer to [166] for an excellent and detailed historical account with an extensive bibliography.

Newton polytopes play a central role in algebra and geometry, for example in Gröbner bases, elimination theory, toric geometry or tropical geometry. We refer to the books [72], [49] and [205] for more background.

Corollary 2.2.8 was proved by Reznick [164]. The correspondence between sums of squares representations and Gram matrices is due to Choi, Lam and Reznick [44]. The Fejér–Riesz theorem 2.3.4 was originally proved in [67] and [168]. This theorem plays an important role in Riesz's proof of the spectral theorem for bounded self-adjoint or unitary operators. For an overview with far-reaching generalizations in operator theory, see [57].

Chapter 3
The Real Spectrum

We now generalize the concept of orderings from fields to arbitrary commutative rings. The central object is the real spectrum Sper(A) of a ring A, introduced in the late 1970s by Coste and Roy. As a first application we prove various "stellensätze" (Sections 3.2 and 3.3). In their geometric versions, they assert that positivity of a polynomial on a given closed set can always be certified by an identity from which the positivity is obvious. We then discuss the topology of the real spectrum in some detail (Sections 3.4 to 3.6). Krull valuations play a key role here since they are strongly related to specializations in the real spectrum.

3.1 The Real Spectrum of a Ring

Always let A be a ring in the following. We keep the general convention that all rings are commutative and have a unit, and that every ring homomorphism sends 1 to 1. If $\mathfrak{p}$ is a prime ideal of A, the residue field of $\mathfrak{p}$ is denoted

$$\kappa(\mathfrak{p}) := \mathrm{qf}(A/\mathfrak{p}) = A_{\mathfrak{p}}/\mathfrak{p}A_{\mathfrak{p}},$$

and $\rho_{\mathfrak{p}}\colon A \to \kappa(\mathfrak{p})$ is the natural residue map.

3.1.1 Definition. The *real spectrum* Sper(A) of A is the set of all pairs $\alpha = (\mathfrak{p}, \le)$, where $\mathfrak{p} \in \mathrm{Spec}(A)$ and $\le$ is an ordering of the residue field $\kappa(\mathfrak{p})$. The prime ideal $\mathfrak{p}$ is called the *support* of α, denoted $\mathrm{supp}(\alpha) = \mathfrak{p}$. The elements of Sper($A$) will be called the *orderings* of the ring A.

3.1.2 Notation. Given an ordering $\alpha = (\mathfrak{p}, \le)$ of A, we alternatively refer to the residue field of $\mathfrak{p}$ as the residue field of α and write $\kappa(\alpha) := \kappa(\mathfrak{p})$. The residue map is denoted $\rho_\alpha\colon A \to \kappa(\alpha)$. By $R(\alpha)$ we denote the real closure of the ordered field $(\kappa(\alpha), \le)$, and $r_\alpha\colon A \xrightarrow{\rho_\alpha} \kappa(\alpha) \subseteq R(\alpha)$ is the composite homomorphism. For $f \in A$ we call

$$\mathrm{sign}_\alpha(f) := \mathrm{sign}_{\le} \rho_\alpha(f) = \mathrm{sign}_{R(\alpha)} r_\alpha(f)$$

C. Scheiderer, *A Course in Real Algebraic Geometry*, Graduate Texts in Mathematics 303,
https://doi.org/10.1007/978-3-031-69213-0_3

the *sign* of f at α. For elements $f, g \in A$ we briefly write $f >_\alpha g$ and $f \geq_\alpha g$ instead of $\operatorname{sign}_\alpha(f-g) = 1$ and $\operatorname{sign}_\alpha(f-g) \geq 0$, respectively, and $f <_\alpha g$, $f \leq_\alpha g$ are defined similarly.

We often like to think of the elements $f \in A$ as generalized "functions" on the set (or rather space, see 3.1.6) $\operatorname{Sper}(A)$, in such a way that the "value" that f takes at $\alpha \in \operatorname{Sper}(A)$ is the element $r_\alpha(f)$ in the real closed field $R(\alpha)$. Guided by this point of view, we may alternatively write $f(\alpha) > 0$, $f(\alpha) = 0$ or $f(\alpha) < 0$ instead of $f >_\alpha 0$, $f =_\alpha 0$ or $f <_\alpha 0$, respectively. Notation $f(\alpha) \geq 0$ or $f(\alpha) \leq 0$ is defined accordingly. If Y is any subset of $\operatorname{Sper}(A)$, we will also write $f|_Y > 0$ instead of $f >_\alpha 0$ for every $\alpha \in Y$, etc.

Generalizing the case of fields, there is an equivalent description of orderings by positive cones:

3.1.3 Definition. A *positive cone* of the ring A is a subset $P \subseteq A$ with $P + P \subseteq P$, $PP \subseteq P$ and $P \cup (-P) = A$, such that the ideal $P \cap (-P)$ of A is prime. This prime ideal is called the *support* of P, denoted $\operatorname{supp}(P)$.

3.1.4 Remarks.

1. Any positive cone of A contains the set ΣA^2 of sums of squares in A. When $A = K$ is a field, the notions of orderings or positive cones just introduced agree with the respective notions from Chapter 1 (see 1.1.2 and 1.1.4).

2. In Definition 3.1.3, the condition that $P \cap (-P)$ is a prime ideal can be replaced by the two conditions $-1 \notin P$ and

$$\forall\, a, b \in A \left(a \notin P \wedge b \notin P \Rightarrow -ab \notin P\right),$$

see Exercise 3.1.1.

3.1.5 Proposition. *The map*

$$\alpha = (\mathfrak{p}, \leq) \longmapsto P_\alpha := \{f \in A : f(\alpha) \geq 0\}$$

is a bijection from $\operatorname{Sper}(A)$ *to the set of all positive cones of A. Moreover this map respects supports.*

In view of this proposition, we feel free to confuse orderings of A with positive cones whenever this is convenient, in the same way as we already did for fields. In particular, we'll use notation like $f >_P g$, $f \geq_P g$ etc., whenever $f, g \in A$ and P is a positive cone of A.

Proof. Given $\alpha = (\mathfrak{p}, \leq)$, the set P_α clearly is a positive cone of A, and $\operatorname{supp}(P_\alpha) = P_\alpha \cap (-P_\alpha) = \mathfrak{p}$. Conversely, if P is a positive cone of A and $\mathfrak{p} := \operatorname{supp}(P)$, then P defines an ordering $\leq_P$ of the residue field $\kappa(\mathfrak{p}) = \operatorname{qf}(A/\mathfrak{p})$. It is characterized by $\overline{a}/\overline{b} \geq_P 0 \Leftrightarrow ab \in P$ for $a, b \in A \smallsetminus \mathfrak{p}$ (where $\overline{a} := \rho_\mathfrak{p}(a)$ etc). The map $P \mapsto (\mathfrak{p}, \leq_P)$ is the inverse of $\alpha \mapsto P_\alpha$. □

3.1.6 Definition. Given a subset $M \subseteq A$, we write

$$\begin{aligned} U(M) &= U_A(M) = \{\alpha \in \mathrm{Sper}(A)\colon \forall f \in M\ f(\alpha) > 0\}, \\ X(M) &= X_A(M) = \{\alpha \in \mathrm{Sper}(A)\colon \forall f \in M\ f(\alpha) \geq 0\}, \\ Z(M) &= Z_A(M) = \{\alpha \in \mathrm{Sper}(A)\colon \forall f \in M\ f(\alpha) = 0\}. \end{aligned}$$

When $M = \{f_1, \dots, f_n\}$ is finite we use simplified notation $U(M) = U(f_1, \dots, f_n)$ etc. A subset of Sper(A) of the form $U(M)$ or $X(M)$, where $M \subseteq A$ is a finite set, is called *basic open* or *basic closed*, respectively.

3.1.7 Definition. The *Harrison topology* on Sper(A) is the topology whose open sets are the unions of basic open sets.

In other words, the basic open sets in Sper(A) are a basis of open sets for the Harrison topology on Sper(A). Unless otherwise mentioned, Sper(A) will always be considered with this topology.

3.1.8 Remarks.

1. Rewriting the definition of $U(M)$ and $X(M)$ in terms of positive cones, we see that a positive cone P of A lies in $U(M)$ if and only if $(-M) \cap P = \varnothing$. Similarly, $P \in X(M)$ if and only if $M \subseteq P$.

2. Recall that the Zariski topology on the Zariski spectrum Spec(A) has the sets $D(f) = \{\mathfrak{p} \in \mathrm{Spec}(A)\colon f \notin \mathfrak{p}\}$ ($f \in A$) as a basis of open sets. The support map $\mathrm{Sper}(A) \to \mathrm{Spec}(A)$, $\alpha \mapsto \mathrm{supp}(\alpha)$ is continuous since the preimage of $D(f)$ is the open set $U(f^2)$ in Sper(A).

3. The real spectrum is functorial in a straightforward way: Any ring homomorphism $\varphi\colon A \to B$ induces a (pull-back) map $\varphi^*\colon \mathrm{Sper}(B) \to \mathrm{Sper}(A)$. In terms of positive cones, φ^* is given by $\varphi^*(Q) = \varphi^{-1}(Q)$, for $Q \subseteq B$ a positive cone. Given $f \in A$ and $\beta \in \mathrm{Sper}(B)$ we have

$$\mathrm{sign}_{\varphi^*(\beta)}(f) = \mathrm{sign}_\beta\, \varphi(f)$$

by definition of φ^*. In particular, $(\varphi^*)^{-1}(U_A(f)) = U_B(\varphi(f))$ ($f \in A$), so φ^* is a continuous map. Note also that φ^* commutes with support, i.e. $\varphi^{-1}(\mathrm{supp}(\beta)) = \mathrm{supp}(\varphi^*(\beta))$ for $\beta \in \mathrm{Sper}(B)$. If $\psi\colon B \to C$ is a second ring homomorphism then $(\psi \circ \varphi)^* = \varphi^* \circ \psi^*$.

4. If $\varphi\colon A \to B$ is a ring homomorphism, and if $\beta \in \mathrm{Sper}(B)$, $\alpha := \varphi^*(\beta)$, then φ extends uniquely to an embedding $R(\alpha) \to R(\beta)$ of the real closed fields:

$$\begin{array}{ccc} A & \xrightarrow{\ \varphi\ } & B \\ {\scriptstyle r_\alpha}\downarrow & & \downarrow{\scriptstyle r_\beta} \\ R(\alpha) & \dashrightarrow & R(\beta) \end{array}$$

Indeed, φ induces a map $\kappa(\alpha) \to \kappa(\beta)$ between the residue fields that is order-compatible with respect to α and β. So this map extends uniquely to the real closures (Theorem 1.4.6).

The proof of the next proposition is straightforward (Exercise 3.1.2):

3.1.9 Proposition. *Let A be a ring, let $S \subseteq A$ be a multiplicative subset and $I \subseteq A$ an ideal, and let $\varphi\colon A \to A_S$ and $\pi\colon A \to A/I$ denote the natural homomorphisms.*

(a) *φ^* is a homeomorphism from* $\mathrm{Sper}(A_S)$ *onto the subspace* $U_A(s^2\colon s \in S) = \{\alpha\colon S \cap \mathrm{supp}(\alpha) = \varnothing\}$ *of* $\mathrm{Sper}(A)$.
(b) *π^* is a homeomorphism from* $\mathrm{Sper}(A/I)$ *onto the closed subspace* $Z_A(I) = \{\alpha\colon I \subseteq \mathrm{supp}(\alpha)\}$ *of* $\mathrm{Sper}(A)$.

3.1.10 Corollary. *If $\mathfrak{p}$ is a prime ideal of A the homomorphism $\rho_\mathfrak{p}\colon A \to \kappa(\mathfrak{p})$ induces a homeomorphism from* $\mathrm{Sper}\,\kappa(\mathfrak{p})$ *onto* $\{\alpha \in \mathrm{Sper}(A)\colon\ \mathrm{supp}(\alpha) = \mathfrak{p}\}$.

Proof. Write $\rho_\mathfrak{p}$ as the composition $A \to A_\mathfrak{p} \to A_\mathfrak{p}/\mathfrak{p}A_\mathfrak{p} = \kappa(\mathfrak{p})$ and apply 3.1.9 to both maps. □

We will frequently identify orderings of the residue field $\kappa(\mathfrak{p})$ with points in $\mathrm{Sper}(A)$ that have support $\mathfrak{p}$.

3.1.11 Let X be a topological space. Given $x, y \in X$, we write $x \rightsquigarrow y$ if $y \in \overline{\{x\}}$. In this case one says that y is a *specialization* of x, or that x is a *generalization* of y. The space X has the T_0 property (A.1.2) if and only if the relation $x \rightsquigarrow y$ is antisymmetric, meaning that $x \rightsquigarrow y$ and $y \rightsquigarrow x$ imply $x = y$. Hence $x \rightsquigarrow y$ is a (partial) order relation on X in this case. In general, the T_0 property is much weaker than the Hausdorff (T_2) property. For example, the Zariski topology on $\mathrm{Spec}(A)$ is T_0, but it usually fails to be Hausdorff, like already for $A = \mathbb{Z}$.

The next observation shows that the Harrison topology on the real spectrum is T_0 as well, and that it has an additional property that is quite particular:

3.1.12 Proposition. *Let P, Q, Q' be positive cones of A.*

(a) *$P \rightsquigarrow Q$ if and only if $P \subseteq Q$. In particular,* $\mathrm{Sper}(A)$ *is a T_0-space.*
(b) *If $P \subseteq Q$ then $Q = P + \mathrm{supp}(Q) = P \cup \mathrm{supp}(Q)$.*
(c) *If $P \subseteq Q$ and $P \subseteq Q'$ then one of $Q \subseteq Q'$ or $Q' \subseteq Q$ holds.*

Note that (c) says that any two specializations of a point in $\mathrm{Sper}(A)$ are comparable with respect to $\rightsquigarrow$. In other words, the specializations of α form a chain with respect to $\rightsquigarrow$.

Proof. (a) $P \rightsquigarrow Q$ means that "$f \ge_P 0 \Rightarrow f \ge_Q 0$" holds for every $f \in A$. This simply says $P \subseteq Q$. In (b) the inclusions $P \cup \mathrm{supp}(Q) \subseteq P + \mathrm{supp}(Q) \subseteq Q$ are obvious. Conversely let $f \in Q$. Since $f \in P \cup (-P)$ we may assume $f \in -P$. But then $f \in \mathrm{supp}(Q)$ since $-P \subseteq -Q$. For the proof of (c) assume $Q \not\subseteq Q'$, so there is $f \in Q$ with $f \notin Q'$. To show $Q' \subseteq Q$ let $g \in Q'$ and assume $g \notin Q$. So $g \ge_{Q'} 0$ and $g <_Q 0$. Since also $f \ge_Q 0$ and $f <_{Q'} 0$ it follows that $f - g <_{Q'} 0$ and $g - f <_Q 0$. From $P \rightsquigarrow Q$ and $P \rightsquigarrow Q'$ it follows that $f - g <_P 0$ and $g - f <_P 0$, contradicting $P \cup (-P) = A$. □

3.1.13 Corollary. *If* $\alpha \in \mathrm{Sper}(A)$ *and* $\beta, \gamma \in \overline{\{\alpha\}}$, *then* $\mathrm{supp}(\beta) = \mathrm{supp}(\gamma)$ *implies* $\beta = \gamma$. *In particular, if* $\mathrm{supp}(\alpha)$ *is a maximal ideal of A then* α *is a closed point of* $\mathrm{Sper}(A)$. □

The converse to the last remark is usually false, as already shown by the ring $A = \mathbb{Z}$ of integers and the unique element α in $\mathrm{Sper}(\mathbb{Z})$.

3.1.14 Examples. Here are first examples of real spectra.

1. For every unit $f \in A^*$, the real spectrum $\mathrm{Sper}(A)$ is the disjoint union of $U(f)$ and $U(-f)$. So the sets $U(\pm f)$ are open and closed in $\mathrm{Sper}(A)$. In particular, the real spectrum of a field is a totally disconnected Hausdorff space.

2. A topological space X is said to be Noetherian if every descending sequence $X_1 \supseteq X_2 \supseteq \cdots$ of closed subsets of X becomes stationary, i.e. if there is $N \geq 1$ with $X_n = X_N$ for all $n \geq N$. While the Zariski spectrum of every Noetherian ring is a Noetherian topological space, this usually fails for the real spectrum (Exercise 3.1.3).

3. Let R be a real closed field and let $R[t]$ be the polynomial ring in one variable. The complete list of all positive cones and specializations in $\mathrm{Sper}\, R[t]$ is as follows (Exercise 3.1.4):

(1) For every $c \in R$, a closed point P_c with support $\langle t - c\rangle$, plus two generalizations $P_{c,\pm}$ of P_c with support $\{0\}$;
(2) two closed points $P_{\pm\infty}$ with support $\{0\}$;
(3) for every free Dedekind cut ξ of R, a closed point P_ξ with support $\{0\}$.

Note that points of type (3) exist if and only if $R \neq \mathbb{R}$ (see 1.1.12 and Exercises 1.1.2, 1.1.3).

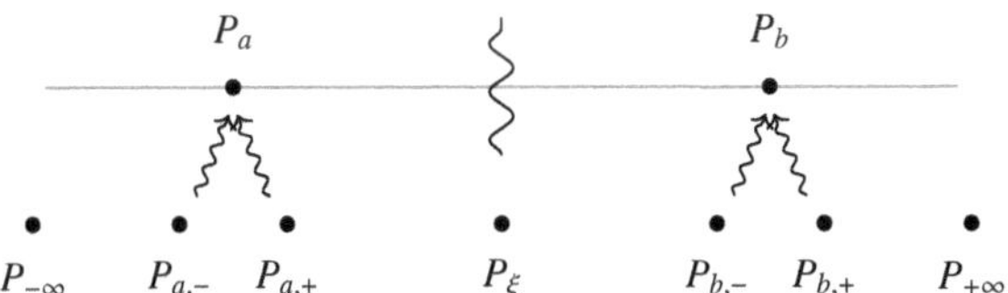

As the schematic picture suggests, there exists a natural total ordering on the set $\mathrm{Sper}\, R[t]$ that extends the ordering of R (Exercises 3.1.5 and 1.1.4). It is easy to see that the topological space $\mathrm{Sper}\, R[t]$ is quasi-compact, and that it contains R with its order topology as a dense topological subspace. We are soon going to prove that these properties hold in much greater generality.

4. Let k be a field and let $k[\![t]\!]$ be the ring of formal power series in the variable t. This is a discrete valuation ring with maximal ideal $\mathfrak{m} = \langle t\rangle$ and quotient field $k(\!(t)\!)$. Every ordering of k extends to $k(\!(t)\!)$ in precisely two ways, which are distinguished by the sign of t. In more detail, if $f = \sum_{\nu\geq 0} a_\nu t^\nu \in k[\![t]\!]$, $f \neq 0$, let $\mathrm{ord}(f) = \min\{\nu\colon a_\nu \neq 0\}$ be the vanishing order and $a_f := a_{\mathrm{ord}(f)}$ the leading coefficient of f.

If P is a positive cone of k, let $\overline{P} = \{\sum_{n\geq 0} a_n x^n \colon a_0 \in P\}$, which is a positive cone of A with $\operatorname{supp}(\overline{P}) = \mathfrak{m}$. Moreover,

$$\begin{aligned} P_+ &= \{0\} \cup \left\{f \in k[\![t]\!] \colon f \neq 0,\ a_f >_P 0\right\}, \\ P_- &= \{0\} \cup \left\{f \in k[\![t]\!] \colon f \neq 0,\ (-1)^{\operatorname{ord}(f)} a_f >_P 0\right\} \end{aligned}$$

are two positive cones of $k[\![t]\!]$ with support $\{0\}$ that specialize to $\overline{P}$. When P ranges over all orderings of k, this gives the complete lists of points in $\operatorname{Sper} k[\![t]\!]$ and specializations between them.

5. Let $(A, \mathfrak{m})$ be a local domain, with residue field $k = A/\mathfrak{m}$ and field of fractions $K = \operatorname{qf}(A)$. Let orderings α of K and β of k be given, and consider α, β as elements of $\operatorname{Sper}(A)$ as in 3.1.10 (with $\operatorname{supp}(\alpha) = \langle 0 \rangle$ and $\operatorname{supp}(\beta) = \mathfrak{m}$). It follows directly from the definitions that the specialization $\alpha \rightsquigarrow \beta$ holds in $\operatorname{Sper}(A)$ if, and only if, $\operatorname{sign}_\alpha(u) = \operatorname{sign}_\beta(\overline{u})$ for every unit $u \in A^*$. Here $\overline{u} \in k^*$ denotes the residue class of u.

3.1.15 We have seen two different ways of representing elements of $\operatorname{Sper}(A)$. Depending on the situation, a third possibility is often convenient. Every ring homomorphism $\varphi \colon A \to R$ into a real closed field R defines a unique point in $\operatorname{Sper}(A)$, namely the image point α of the induced map $\varphi^* \colon \operatorname{Sper}(R) \to \operatorname{Sper}(A)$. We'll write $\alpha = [\varphi]$ and say that φ *represents* α. Here are two characterizations for when two such homomorphisms represent the same point:

3.1.16 Lemma. *Let $\varphi_i \colon A \to R_i$ ($i = 1, 2$) be two ring homomorphisms into real closed fields R_1, R_2. The following conditions are equivalent:*

(i) *$[\varphi_1] = [\varphi_2]$, i.e. φ_1 and φ_2 represent the same point in* $\operatorname{Sper}(A)$*;*
(ii) *there exists a homomorphism $\rho \colon A \to R_0$ into a real closed field R_0 together with homomorphisms $\psi_i \colon R_0 \to R_i$ ($i = 1, 2$), such that $\psi_i \circ \rho = \varphi_i$ for $i = 1, 2$;*
(iii) *there exists a real closed field R together with homomorphisms $\psi_i \colon R_i \to R$ ($i = 1, 2$), such that $\psi_1 \circ \varphi_1 = \psi_2 \circ \varphi_2$.*

Proof. Condition (ii) requires that $\varphi_1(A)$ and $\varphi_2(A)$ are contained in a common real closed subfield of R_1 and R_2, whereas condition (iii) requires that R_1 and R_2 can be amalgamated over A into a real closed field R:

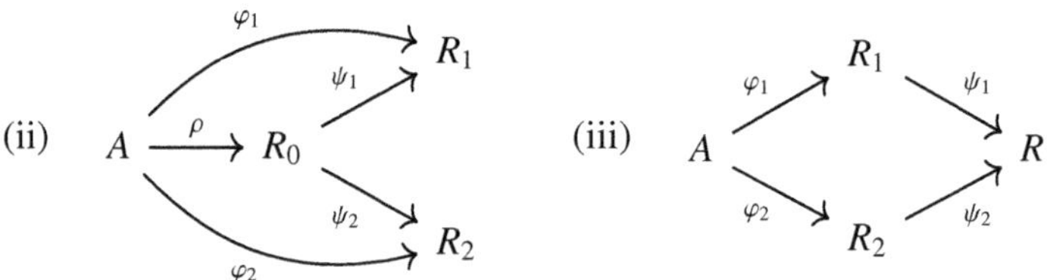

If $[\varphi_1] = [\varphi_2] = \alpha$, it suffices to take $R_0 = R(\alpha)$ and $\rho = r_\alpha$, then ψ_1, ψ_2 will (uniquely) exist by Theorem 1.4.6. This proves (i) $\Rightarrow$ (ii). The implication (ii) $\Rightarrow$ (iii) follows from Exercise 1.4.3, and (iii) $\Rightarrow$ (i) is trivial. □

3.1.17 Remark. We have seen several ways of representing elements of Sper(A). To summarize, an element of Sper(A) may be given in any of the following ways:

1. As a pair $(\mathfrak{p}, \le)$ where $\mathfrak{p}$ is a prime ideal of A and $\le$ is an ordering of the residue field $\kappa(\mathfrak{p}) = \mathrm{qf}(A/\mathfrak{p})$ of $\mathfrak{p}$;
2. as a positive cone P of A (Definition 3.1.3);
3. as the equivalence class $[\varphi]$ of a homomorphism $\varphi\colon A \to R$ into a real closed field R, the equivalence relation being described in Lemma 3.1.16.

From the algebraic geometry perspective, it is natural to ask whether the definition of the real spectrum can be extended from (commutative) rings A to arbitrary schemes X. This is indeed possible in a natural way, by gluing the real spectra of open affine subsets of X along their intersections. We'll outline the construction later in some detail in 4.1.15, in the case of algebraic varieties over fields.

Exercises

3.1.1 Let A be a ring, and let $P \subseteq A$ be a subset with $P + P \subseteq P$, $PP \subseteq P$ and $P \cup (-P) = A$. Show that P is a positive cone of A if, and only if, $-1 \notin P$ and

$$a \notin P \wedge b \notin P \Rightarrow -ab \notin P$$

holds for all $a, b \in A$.

3.1.2 Prove the two assertions in Proposition 3.1.9.

3.1.3 Show that the topological spaces Sper $\mathbb{R}[t]$ and Sper $\mathbb{R}(t)$ fail to be Noetherian (cf. Remark 3.1.14.2).

3.1.4 Let $R[t]$ be the polynomial ring in the variable t over the real closed field R. Verify the assertions made in Example 3.1.14.3, that is, determine all positive cones of Sper $R[t]$, together with all specializations between them.

3.1.5 Let R be a real closed field. Show that the set of positive cones of the univariate polynomial ring $R[t]$ is naturally identified with the set of generalized Dedekind cuts (Exercise 1.1.4) of the totally ordered set $(R, \le)$.

3.1.6 For any ring A, the following are equivalent:

(i) Every $f \in A$ with $f(\alpha) \neq 0$ for all $\alpha \in \mathrm{Sper}\,A$ is a unit in A;
(ii) the residue field of every maximal ideal of A is real;
(iii) $1 + \Sigma A^2 \subseteq A^*$.

If B is an arbitrary ring, the localization $A := B_{1+\Sigma B^2}$ of B in the multiplicative set $1 + \Sigma B^2$ has the above properties.

3.1.7 If K/k is a finite field extension, show that the image set of the restriction map Sper(K) $\to$ Sper(k) is open in Sper(k). (A more general statement will be proved in Exercise 3.4.5.)

3.1.8 Let $\varphi_1 : A \to B_1$, $\varphi_2 : A \to B_2$ be ring homomorphisms, and consider the natural commutative square of ring homomorphisms

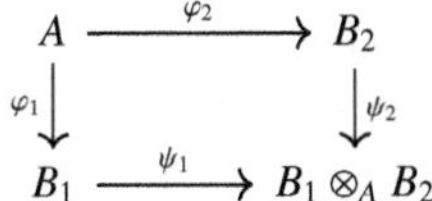

Show that image of the map

$$\mathrm{Sper}(B_1 \otimes_A B_2) \to \mathrm{Sper}(B_1) \times \mathrm{Sper}(B_2), \quad \gamma \mapsto (\psi_1^*(\gamma),\, \psi_2^*(\gamma)) \tag{3.1}$$

is the set of pairs (β_1,β_2) for which $\varphi_1^*(\beta_1) = \varphi_2^*(\beta_2)$. Give an example to show that (3.1) need not be injective.

3.1.9 (For readers who have seen inductive and projective limits) Let $(I, \leq)$ be a directed set and let $\{A_i, \varphi_{ij} \colon A_i \to A_j\ (i, j \in I, i \leq j)\}$ be an inductive system of rings and ring homomorphisms. Show that the natural map

$$\mathrm{Sper}(\varinjlim_{i\in I} A_i) \to \varprojlim_{i\in I} \mathrm{Sper}(A_i)$$

is a homeomorphism.

3.1.10 Let k be a field and let $k(t)$ be the rational function field over k in one variable. Show that the topological space $X := \mathrm{Sper}\, k(t)$ is self-similar: For every integer $m \geq 1$ there exists a homeomorphism between X and the m-fold disjoint topological sum $X \amalg \cdots \amalg X$.

Using this observation it can be shown [76] that if F_1, F_2 are any two real function fields of dimension one over a real closed field R, the topological spaces $\mathrm{Sper}(F_1)$ and $\mathrm{Sper}(F_2)$ are homeomorphic.

3.2 Preorderings and Abstract Stellensätze

Always let A be a ring.

3.2.1 Definition. A *preordering* of A is a subset $T \subseteq A$ that satisfies $T + T \subseteq T$, $TT \subseteq T$ and $a^2 \in T$ for every $a \in A$. The preordering T is *proper* if $-1 \notin T$.

3.2.2 Remarks. This obviously generalizes the definition given for fields in 1.1.23. Here are some immediate remarks.

1. Any positive cone of A is a preordering of A. Conversely, a preordering T is a positive cone if and only if $T \cup (-T) = A$ and $T \cap (-T)$ is a prime ideal.

2. Any intersection of preorderings of A is a preordering of A. For any subset $S \subseteq A$, we may therefore consider

$$PO(S) = PO_A(S) := \bigcap \bigl\{T \colon T \text{ preordering of } A,\ S \subseteq T\bigr\},$$

the preordering of A *generated by* S. Explicitly, $PO(S)$ consists of all finite sums of products $a^2 f_1 \cdots f_r$ with $a \in A$, $r \geq 0$ and $f_1, \ldots, f_r \in S$. In particular,

$$PO(f_1, \ldots, f_r) = \Bigl\{ \sum_{e\in\{0,1\}^r} s_e f_1^{e_1} \cdots f_r^{e_r} \colon s_e \in \Sigma A^2\ (e \in \{0,1\}^r) \Bigr\}.$$

3. Any intersection of positive cones in A is a preordering. If A is a field then, conversely, every proper preordering is an intersection of positive cones (Proposition 1.1.28). But for more general rings this usually fails. For example, the preordering $T = PO(t^3)$ in $\mathbb{R}[t]$ is not an intersection of positive cones (why not?). The question will be discussed more systematically in Section 6.2.

4. Every preordering of A contains ΣA^2, the set of sums of squares in A. Clearly, ΣA^2 is the unique smallest preordering of A.

5. If A contains $\frac{1}{2}$ then every element in A is a difference of two squares. Therefore $T - T = A$ holds for every preordering T in this case, and $T = A$ is the unique improper preordering of A.

6. If T is a preordering of A then $\mathrm{supp}(T) := T \cap (-T)$ is called the *support* of T. This is an additive subgroup of A, and is an ideal of A if $T - T = A$ (for example if $\frac{1}{2} \in A$). Indeed, this follows from $T \cdot \mathrm{supp}(T) \subseteq \mathrm{supp}(T)$.

We are going to state several versions of an abstract "stellensatz" for the real spectrum. For more geometric versions in the polynomial setting, see the next section. We start with the following "ur-stellensatz":

3.2.3 Theorem. *Let A be a ring. Then every proper preordering of A is contained in a positive cone of A.*

When A is a field this was proved in Proposition 1.1.26. Here are two equivalent statements:

3.2.4 Corollary. *Let A be any ring.*

(a) *Every maximal proper preordering of A is a positive cone of A.*
(b) *If T is a preordering of A with $X(T) = \varnothing$, then $-1 \in T$.*

Note that 3.2.4(a) was proved for fields in Proposition 1.1.27. However, the proof given there does not carry over to general rings (Exercise 3.2.3). The statement of 3.2.4(a) implies (and hence is equivalent to) Theorem 3.2.3, by Zorn's lemma, since an ascending union of a family of proper preorderings is again a proper preordering. Statement 3.2.4(b) is directly equivalent to 3.2.3. Altogether it suffices to prove 3.2.4(a).

3.2.5 Lemma. *Let T be a proper preordering of A and let $a \in A$. Then at least one of the two preorderings $T + aT$ and $T - aT$ is again proper.*

Proof. Assume $-1 \in (T + aT) \cap (T - aT)$. Then there exist $s_1, s_2, t_1, t_2 \in T$ with $-1 = s_1 + as_2$ and $-1 = t_1 - at_2$, which means $-as_2 = 1 + s_1$ and $at_2 = 1 + t_1$. Multiplying both identities gives $-a^2 s_2 t_2 = 1 + s_1 + t_1 + s_1 t_1$ and therefore $-1 = s_1 + t_1 + s_1 t_1 + a^2 s_2 t_2 \in T$. This is a contradiction since the right-hand side lies in T. □

3.2.6 We now give the proof of 3.2.4(a), and hence of Theorem 3.2.3. Let T be a maximal proper preordering of A. Then $A = T \cup (-T)$ by Lemma 3.2.5, and so $\mathrm{supp}(T)$ is an ideal of A. It remains to show that this is a prime ideal. Assume to the contrary that there exist $a, b \in A$ with $ab \in \mathrm{supp}(T)$ and $a, b \notin \mathrm{supp}(T)$. We may assume $a, b \in T$, and so $-a, -b \notin T$ and $-ab \in T$. From maximality of T it follows that $-1 \in (T - aT) \cap (T - bT)$. Hence there are identities $as_2 = 1 + s_1$ and $bt_2 = 1 + t_1$ with $s_1, s_2, t_1, t_2 \in T$. Multiplying both identities gives $-1 = s_1 + t_1 + s_1 t_1 - abs_2 t_2$, and hence $-1 \in T$, a contradiction. □

From Theorem 3.2.3 we deduce a number of important consequences.

3.2.7 Corollary. (Positivstellensatz, Krivine–Stengle) *Let T be a preordering of A. For $f \in A$ the following properties are equivalent:*

(i) $f > 0$ *on* $X(T)$*;*
(ii) *there exist* $s, t \in T$ *with* $sf = 1 + t$*;*
(iii) *there exist* $s, t \in T$ *with* $(1 + s)f = 1 + t$.

Proof. (i) ⇒ (ii): Let $f > 0$ on $X(T)$, then $X(T - fT) = X(T) \cap X(-f) = \varnothing$. It follows from Theorem 3.2.3 that $-1 \in T - fT$, which is (ii). To prove (ii) ⇒ (iii), multiply $1+t = sf$ by f to get $(1+t)f = sf^2$. Then addition of both equalities gives $(1 + s + t)f = 1 + (t + sf^2)$. The implication (iii) ⇒ (i) is obvious: Evaluating both sides of identity (iii) at any given point $\alpha \in X(T)$ immediately gives $f(\alpha) > 0$, since both $1 + s$ and $1 + t$ are strictly positive in α. □

Statement (iii) of Corollary 3.2.7 is just a variant of (ii), and it is obvious that both (ii) and (iii) imply (i). The essential part of 3.2.7 is therefore the implication (i) ⇒ (ii). We may informally rephrase it by saying that, whenever a strict inequality $f > 0$ holds on the closed set $X(T)$, there is a *certificate* for this inequality, in the form of an identity that makes this strict positivity obvious. A similar remark applies to each of the next two results:

3.2.8 Corollary. (Nichtnegativstellensatz) *Let T be a preordering of A and let $f \in A$. The following are equivalent:*

(i) $f \ge 0$ *on* $X(T)$*;*
(ii) *there exist* $m \ge 0$ *and* $s, t \in T$ *with* $sf = f^{2m} + t$.

Proof. To show (i) ⇒ (ii), assume $f \ge 0$ on $X(T)$ and let $T_f = \{\frac{t}{f^{2m}} : m \ge 0, t \in T\}$, the preordering generated by T in the ring of fractions A_f. The element f is a unit in A_f, and it is strictly positive on $X_{A_f}(T_f)$ by hypothesis (i) (see 3.1.9). Therefore Corollary 3.2.7, applied to A_f and T_f, implies an identity $s_1 f = 1 + s_2$ in A_f with $s_1, s_2 \in T_f$. Multiplication by a suitable even power f^{2m} gives an identity (ii). The implication (ii) ⇒ (i) is again obvious. □

3.2.9 Remark. In the situation of Theorem 3.2.8, one could think of other identities that imply $f \ge 0$ on $X(T)$, like $(1 + s)f = t$ with $s, t \in T$. However it is *not true* in general that $f \ge 0$ on $X(T)$ implies the existence of such an identity. Examples showing this will later be constructed in Exercise 6.1.6, but see also Remark 6.5.29.1.

3.2.10 Corollary. (Abstract real nullstellensatz, first version) *Let T be a preordering of A and let $f \in A$. The following are equivalent:*

(i) $f \equiv 0$ *on* $X(T)$*;*
(ii) *there is* $m \ge 0$ *with* $f^m \in \operatorname{supp}(T)$*;*
(iii) *there is* $m \ge 0$ *with* $-f^{2m} \in T$.

When supp(T) is an ideal of A (which is the case if $\frac{1}{2} \in A$), condition (ii) says that $f \in \sqrt{\text{supp}(T)}$.

Proof. (i) implies $-f^2 \geq 0$ on $X(T)$, so 3.2.8 implies that $-sf^2 = f^{4m} + t$ for suitable $m \geq 0$ and $s, t \in T$. This is an identity of type (iii). The implications (iii) ⇒ (ii) ⇒ (i) are trivial. □

All three stellensätze 3.2.7, 3.2.8 and 3.2.10 were deduced from Theorem 3.2.3. Conversely, each of the three contains 3.2.3 as a particular case, namely for $f = -1$. This means that all four are essentially just different incarnations of the same result. It is also possible to combine the three stellensätze into a single one:

3.2.11 Theorem. (General real stellensatz) *Let F, G, H be subsets of A. The subset*

$$U(F) \cap X(G) \cap Z(H) \;=\; \bigcap_{f \in F} \{f > 0\} \cap \bigcap_{g \in G} \{g \geq 0\} \cap \bigcap_{h \in H} \{h = 0\}$$

of Sper(A) *is empty if, and only if, there is an identity $s + t + a = 0$ in A with $s \in S$, $t \in T$ and $a \in I$. Here S denotes the multiplicative semigroup generated by $F \cup \{1\}$ in A, while $T = PO(F \cup G)$ and $I = \langle H \rangle = \sum_{h \in H} Ah$.*

Proof. Put $W = U(F) \cap X(G) \cap Z(H)$. An identity $s + t + a = 0$ as above implies $W = \varnothing$ since $s(\alpha) > 0$, $t(\alpha) \geq 0$ and $a(\alpha) = 0$ hold for every $\alpha \in W$. To prove the converse let $B = A_S / IA_S$ and let

$$T' \;:=\; PO_B(T) \;=\; \left\{ \frac{t}{s} + IA_S : t \in T,\ s \in S \right\},$$

the preordering generated by T in B. If we identify Sper(B) with a subset of Sper(A) as in 3.1.9, then $X_B(T') = W$. Therefore $W = \varnothing$ implies that $-1 \in T'$ (Corollary 3.2.4(b)), say $-1 = \frac{t}{s} + \frac{a}{s'}$ in A_S with $s, s' \in S$, $t \in T$ and $a \in I$. This means $s_0(ss' + s't + sa) = 0$ in A with $s_0 \in S$. Now the three summands lie in S, T and I, respectively. □

3.2.12 Remarks.

1. Usage of the German terms "positivstellensatz" or "nichtnegativstellensatz" (literally meaning "theorem of the positivity locus" or "of the nonnegativity locus", respectively) has become customary in English and French texts as well. This may be a continued, and extended, reference to Hilbert, whose famous nullstellensatz (A.3.1) is fundamental in modern algebraic geometry. Since there is a small zoo of such theorems in real algebraic geometry, the general term *stellensatz* (plural *stellensätze*) is used to refer to any of them, or to the whole group.

2. In remarkable contrast to their importance, the proof for the various stellensätze (essentially not more than 3.2.5 and 3.2.6 above) did not use anything beyond the definition of orderings and preorderings, combined with Zorn's lemma.

3.2.13 Definition. A prime ideal $\mathfrak{p}$ of A is said to be *real* if its residue field $\kappa(\mathfrak{p}) = \text{qf}(A/\mathfrak{p})$ is a real field, i.e. can be ordered. Equivalently, $\mathfrak{p}$ is real if and only if $\sum_i a_i^2 \in \mathfrak{p}$ (with $a_i \in A$) implies $a_i \in \mathfrak{p}$ for all i. For any ideal I of A, the ideal

$$\sqrt[\mathrm{re}]{I} := \bigcap \{\mathfrak{p} \in \operatorname{Spec}(A) \colon \mathfrak{p} \text{ real}, I \subseteq \mathfrak{p}\}$$

of A is called the *real radical* of I.

3.2.14 Remarks.

1. Taking $I = \{0\}$ we get the *real nilradical* $N = \sqrt[\mathrm{re}]{\langle 0\rangle}$ of A. The natural map $\operatorname{Sper}(A/N) \to \operatorname{Sper}(A)$ is a homeomorphism, and N is the largest ideal of A with this property (immediate from Proposition 3.1.9).

2. Recall that the radical $\sqrt{I}$ of an ideal I is the intersection of all prime ideals of A that contain I. So the inclusion $\sqrt{I} \subseteq \sqrt[\mathrm{re}]{I}$ always holds, and in general it is strict. The alternative description $\sqrt{I} = \{a \in A \colon \exists n \geq 1\ a^n \in I\}$ of the usual radical has the following analogue for the real radical:

3.2.15 Corollary. (Abstract real nullstellensatz, second version) *For any ideal I of A,*

$$\sqrt[\mathrm{re}]{I} = \left\{f \in A \colon \exists m \geq 0\ \exists s \in \Sigma A^2\ f^{2m} + s \in I\right\}.$$

Moreover, if A contains $\frac{1}{2}$ then $\sqrt[\mathrm{re}]{I} = \sqrt{\operatorname{supp}(I + \Sigma A^2)}$.

Proof. $T = I + \Sigma A^2$ is a preordering of A and satisfies $X(T) = Z(I)$ (the zero set of I in $\operatorname{Sper}(A)$). Given $f \in A$, the condition $f \in \sqrt[\mathrm{re}]{I}$ means that f is contained in every real prime ideal $\mathfrak{p} \supseteq I$, i.e. that $f \equiv 0$ on $Z(I) = X(T)$. By 3.2.10 it is equivalent that $f^{2m} \in -T$ for some $m \geq 0$, which proves the first claim. If $\frac{1}{2} \in A$ then the support of any preordering is an ideal. Since $-f^{2m} \in I + \Sigma A^2$ means $f^{2m} \in \operatorname{supp}(I + \Sigma A^2)$, the existence of some $m \geq 1$ with this property means that $f \in \sqrt{\operatorname{supp}(I + \Sigma A^2)}$. □

Recall that a field K is real, i.e., can be ordered, if and only if it satisfies the equivalent conditions

(1) $-1 \notin \Sigma K^2$,
(2) $a_1, \dots, a_n \in K$ and $a_1^2 + \cdots + a_n^2 = 0$ implies $a_1 = \cdots = a_n = 0$.

For rings more general than fields, the implication (1) ⇒ (2) usually fails. This is why, for rings, we find two different natural concepts of being real:

3.2.16 Lemma and Definition. *A ring A is said to be* real *if it satisfies the following equivalent conditions:*

(i) $\operatorname{Sper}(A) \neq \emptyset$;
(ii) *A has a real prime ideal;*
(iii) $-1 \notin \Sigma A^2$.

Otherwise the ring A is non-real.

Proof. It is obvious that (i), (ii) are equivalent and both imply (iii). Conversely, if $-1 \notin \Sigma A^2$ then apply Theorem 3.2.3 for $T = \Sigma A^2$ to get (i). □

Note the following remarkable consequence of the previous lemma: If -1 is a sum of squares in every residue field of a ring A, then -1 is a sum of squares in A itself.

The next lemma defines a stronger property:

3.2.17 Lemma and Definition. *A ring A is said to be* real reduced *if the following equivalent conditions hold:*

(i) *If* $a_1, \ldots, a_n \in A$ *and* $a_1^2 + \cdots + a_n^2 = 0$ *then* $a_1 = \cdots = a_n = 0$;
(ii) *A is reduced, and every minimal prime ideal of A is real;*
(iii) $\sqrt[\mathrm{re}]{\langle 0\rangle} = \{0\}$, *i.e. the real nilradical of A is trivial.*

Proof. (i) $\Rightarrow$ (ii): It is obvious that A is reduced. If $\mathfrak{p}$ is a minimal prime ideal of A then $A_\mathfrak{p}$ is a field. Assuming that the prime ideal $\mathfrak{p}$ is not real would mean that -1 is a sum of squares in $A_\mathfrak{p}$. Clearing denominators would give an identity $s^2 + a_1^2 + \cdots + a_r^2 = 0$ in A where $s, a_1, \ldots, a_r \in A$ and $s \notin \mathfrak{p}$, contradicting (i).

(ii) $\Rightarrow$ (iii): Since A is reduced, the intersection of all minimal prime ideals of A is $\{0\}$. Since all these prime ideals are real it follows that $\sqrt[\mathrm{re}]{\langle 0\rangle} = \{0\}$.

(iii) $\Rightarrow$ (i): If $a_1, \ldots, a_n \in A$ satisfy $a_1^2 + \cdots + a_n^2 = 0$ then each a_i lies in every real prime ideal of A, and therefore in $\sqrt[\mathrm{re}]{\langle 0\rangle} = \{0\}$. □

Note that a ring A is real reduced if and only if A is isomorphic to a subring of a direct product of real fields. Clearly, if A is real reduced and $A \neq \{0\}$, then A is real. The converse isn't true in general. For example, the rings $\mathbb{R}[x]/\langle x^2\rangle$ or $\mathbb{R}[x, y]/\langle x^2 + y^2\rangle$ are real but not real reduced. The nullring $\{0\}$ is real reduced but not real.

Using the real spectrum, it is clear which elements of a ring A should be considered to be non-negative everywhere.

3.2.18 Definition. If A is a ring we write

$$A_+ := \{f \in A\colon f \geq 0 \text{ on } \mathrm{Sper}(A)\}.$$

This is the preordering of *positive semidefinite elements* (or *psd elements*, for short) of A.

3.2.19 Remarks.

1. The nichtnegativstellensatz 3.2.8 allows us to characterize the psd elements without mentioning the real spectrum. Indeed, an element $f \in A$ lies in A_+ if and only if f satisfies an identity $sf = f^{2m} + t$ with $m \geq 0$ and sums of squares $s, t \in \Sigma A^2$.

2. It is obvious that $\Sigma A^2 \subseteq A_+$ holds for every ring: *Every sum of squares is psd.* Equality holds when A is non-real and $\frac{1}{2} \in A$, by the identity $x = (\frac{x+1}{2})^2 - (\frac{x-1}{2})^2$. Less trivial examples where equality holds are fields of characteristic $\neq 2$ (Theorem 1.1.30), or the rings $R[x]$ or $R[x, y]/\langle 1 - x^2 - y^2\rangle$ (Fejér–Riesz) over a real closed field R. In Chapter 6 we will see families of rings with $A_+ = \Sigma A^2$ that are substantially more interesting. However, for most real rings A, the inclusion $\Sigma A^2 \subsetneq A_+$ is strict. For example, this is so for polynomial rings $A = R[x_1, \ldots, x_n]$ in $n \geq 2$ variables, by Hilbert's results (2.4.9) and since A_+ consists of the polynomials that are non-negative on R^n (Proposition 3.3.2 below).

Exercises

3.2.1 Let T be a preordering of a ring A and let $f \in A$. Show that '$f \geq 0$ on $X(T)$' is also equivalent to each of the following two conditions (cf. Corollary 3.2.8):

(i) There exist $m \geq 0$ and $s, t \in T$ with $f(f^{2m} + s) = t$;
(ii) there exist $m \geq 0$ and $s, t \in T$ with $f(f^{2m} + s) = f^{2m} + t$.

3.2.2 Let R be a real closed field and let $p \in R[x] = R[x_1, \dots, x_n]$ be an irreducible polynomial. Show that the following properties are equivalent:

(i) The (principal) prime ideal $\langle p \rangle$ in $R[x]$ is real;
(ii) the set of R-points on the hypersurface $\mathcal{V}(p) \subseteq \mathbb{A}^n$ is Zariski dense in $\mathcal{V}(p)$;
(iii) p is indefinite, i.e. there exist $u, v \in R^n$ with $p(u) < 0 < p(v)$.

(*Hint*: Section 1.7)

3.2.3 The statement of Lemma 3.2.5 is slightly weaker than its cousin 1.1.25, that we proved for fields. Show that Lemma 1.1.25 does not extend to arbitrary rings. That is, construct a ring A, a proper preordering T of A and an element $f \in A$ such that $f \notin T$ and $T - fT = A$.

3.2.4 Let $\varphi\colon A \to B$ be a ring homomorphism, let P be a positive cone of A. Using Theorem 3.2.11, show that there exists a positive cone Q of B with $\varphi^{-1}(Q) = P$ if, and only if,

$$\varphi(f) + \sum_{i=1}^{r} \varphi(a_i) b_i^2 \neq 0$$

holds in B whenever $f \in P \setminus (-P)$, $r \in \mathbb{N}$ and $a_i \in P$, $b_i \in B$ $(i = 1, \dots, r)$.

3.2.5 Let $(A, \mathfrak{m})$ be a Henselian local ring (see A.4.8 for the definition).

(a) Assuming $\frac{1}{2} \in A$, show that $\operatorname{supp}(\alpha) = \mathfrak{m}$ for every closed point α of $\operatorname{Sper}(A)$.
(b) Show that (a) remains true if 2 is not a unit in A, i.e. show that $\operatorname{Sper}(A)$ is empty then.

Hints: To prove (a), show $P + \mathfrak{m} \neq A$ for every positive cone P of A. For (b), show that $2 \in \mathfrak{m}$ implies that -7 is a square in A.

3.2.6 Let A be a ring that contains a real closed field R, and let $U \subseteq A$ be an R-linear subspace of finite dimension. If A is real reduced, show that ΣU^2 is a closed semialgebraic subset of UU, where UU is the R-linear span of all products $u_1 u_2$ $(u_1, u_2 \in U)$ in A. (*Hint*: Lemma 2.4.7)

3.3 Geometric Stellensätze

We now transfer the "abstract" stellensätze (Section 3.2) from the real spectrum setting to a "geometric" and more concrete situation, namely polynomials over $\mathbb{R}$ (or over a real closed field). These results will allow us to derive a series of non-obvious strengthenings and generalizations of (the solution of) Hilbert's 17th problem. The results of this section may be seen as a first evidence for the usefulness of the real spectrum.

As usual, R denotes a real closed field.

3.3.1 Given an affine R-variety V, the set $V(R)$ of R-points can be naturally identified with a subset of the real spectrum of $R[V]$. Indeed, an R-point $\xi \in V(R)$ gives the element $\alpha_\xi = [\varphi_\xi]$ in $\operatorname{Sper} R[V]$ (see 3.1.15), where $\varphi_\xi \colon R[V] \to R$, $f \mapsto f(\xi)$ denotes evaluation at ξ. The positive cone of $R[V]$ associated with α_ξ is $P_\xi := \{f \in R[V] \colon f(\xi) \ge 0\}$, the support is $\operatorname{supp}(\alpha_\xi) = \ker(\varphi_\xi) = \mathfrak{m}_\xi$, the maximal ideal of $R[V]$ at ξ. Note that, by definition, $\operatorname{sign} f(\xi) = \operatorname{sign}_{\alpha_\xi}(f)$ holds for every $f \in R[V]$. We'll denote the map $\xi \mapsto \alpha_\xi$ by $\iota \colon V(R) \to \operatorname{Sper} R[V]$.

3.3.2 Proposition. *Let $x = (x_1, \dots, x_n)$, and let finitely many polynomials f_i ($i = 1, \dots, r$), g_j ($j = 1, \dots, s$) and h_k ($k = 1, \dots, t$) in $R[x]$ be given. The following are equivalent:*

(i) *There is $\alpha \in \operatorname{Sper} R[x]$ with $f_i(\alpha) > 0$, $g_j(\alpha) \ge 0$ and $h_k(\alpha) = 0$ for all i, j, k;*
(ii) *there is $\xi \in R^n$ with $f_i(\xi) > 0$, $g_j(\xi) \ge 0$ and $h_k(\xi) = 0$ for all i, j, k.*

Proof. (ii) $\Rightarrow$ (i) is obvious by 3.3.1. The forward implication (i) $\Rightarrow$ (ii) is a consequence of Tarski's transfer principle: Given α as in (i), let $r_\alpha \colon R[x] \to R(\alpha)$ be the natural homomorphism, see 3.1.1. Then $r_\alpha(f_i) > 0$, $r_\alpha(g_j) \ge 0$ and $r_\alpha(h_k) = 0$ hold in $R(\alpha)$ for all i, j, k, by (i). So the R-sentence

$$\exists\, y = (y_1, \dots, y_n)\ \Big(\bigwedge_i f_i(y) > 0 \ \wedge\ \bigwedge_j g_j(y) \ge 0 \ \wedge\ \bigwedge_k h_k(y) = 0\Big)$$

holds in the real closed field $R(\alpha)$, since $y = r_\alpha(x) \in R(\alpha)^n$ is such a tuple. By Tarski 1.6.17, the sentence is true in R itself, which means (ii). □

We extend the notion of semialgebraic sets from R^n (1.5.2) to arbitrary affine R-varieties:

3.3.3 Definition. Let V be an affine R-variety with coordinate ring $R[V]$.

(a) If $f_1, \dots, f_r \in R[V]$, write

$$\mathcal{U}_V(f_1, \dots, f_r) := \big\{\xi \in V(R) \colon f_i(\xi) > 0\ (i = 1, \dots, r)\big\},$$

$$\mathcal{S}_V(f_1, \dots, f_r) := \big\{\xi \in V(R) \colon f_i(\xi) \ge 0\ (i = 1, \dots, r)\big\},$$

$$\mathcal{Z}_V(f_1, \dots, f_r) := \big\{\xi \in V(R) \colon f_i(\xi) = 0\ (i = 1, \dots, r)\big\}.$$

The subsets $\mathcal{U}_V(f_1, \dots, f_r)$ and $\mathcal{S}_V(f_1, \dots, f_r)$ of $V(R)$ are called *basic open* and *basic closed*, respectively.

(b) A subset M of $V(R)$ is called *semialgebraic* (with respect to V, but see below) if M is a finite Boolean combination of sets $\mathcal{U}_V(f)$ with $f \in R[V]$.

When V is a closed subvariety of another affine R-variety W, a set $M \subseteq V(R)$ is semialgebraic with respect to V, if and only if it is semialgebraic with respect to W (check this!). Hence there is no need to mention a reference variety. Since this remark applies in particular when $W = \mathbb{A}^n$, the projection theorem 1.5.9 implies:

3.3.4 Proposition. *Images and preimages of semialgebraic sets under morphisms of affine R-varieties are again semialgebraic.* □

The identification of $V(R)$ with a subset of $\operatorname{Sper} R[V]$ is compatible with the respective topologies:

3.3.5 Proposition. *Let V be an affine R-variety. The map $\iota\colon V(R) \to \operatorname{Sper} R[V]$, $\xi \mapsto \alpha_\xi$ is a dense topological embedding.*

In more detail, the proposition says that ι is a homeomorphism from $V(R)$, equipped with the order topology, onto its image set, equipped with the topology induced from $\operatorname{Sper} R[V]$, and that this image set is dense in $\operatorname{Sper} R[V]$.

Proof. It is obvious that ι is injective. For $f_1, \dots, f_r \in R[V]$ consider the basic open set $U := U_{R[V]}(f_1, \dots, f_r)$ in $\operatorname{Sper} R[V]$ (3.1.6). Then $\iota^{-1}(U) = \mathcal{U}_V(f_1, \dots, f_r)$, a basic open semialgebraic set in $V(R)$. Since basic open sets form a basis for the topology of either $\operatorname{Sper} R[V]$ or $V(R)$, it follows that ι is a homeomorphism onto its image. Moreover $\iota(V(R))$ is dense in $\operatorname{Sper} R[V]$ since $U \neq \emptyset$ implies $\iota^{-1}(U) \neq \emptyset$, according to Proposition 3.3.2. □

3.3.6 Remark. With Proposition 3.3.5 in mind, we may think of the real spectrum $\operatorname{Sper} R[V]$ as arising from the topological space $V(R)$ by adding certain "ideal points". Usually, we will identify $V(R)$ with its ι-image in $\operatorname{Sper} R[V]$. In doing so, no topological information on $V(R)$ is lost. In fact, there is something to be gained, since the real spectrum has notable advantages compared to $V(R)$. For $R \neq \mathbb{R}$, the space $V(R)$ has poor topological properties, it is totally disconnected and fails to be locally compact (as long as $|V(R)| = \infty$). Whereas $\operatorname{Sper} R[V]$ is a quasi-compact space that has only finitely many connected components, as will be shown in Section 4.4. Therefore $\operatorname{Sper} R[V]$ is an object of a much more geometric nature.

The connection between $V(R)$ and $\operatorname{Sper} R[V]$ will be made even stronger in Section 4.1 (the tilda operator).

3.3.7 Corollary. *Let V be an affine R-variety. Then:*

(a) *The ring $R[V]$ is real $\Leftrightarrow$ $V(R) \neq \emptyset$;*
(b) *the ring $R[V]$ is real reduced $\Leftrightarrow$ $V(R)$ is Zariski dense in V.*

Proof. (a) By definition, the ring $R[V]$ is real if and only if $\operatorname{Sper} R[V] \neq \emptyset$. This is equivalent to $V(R) \neq \emptyset$ by Proposition 3.3.5.

(b) Since the ring $R[V]$ is always reduced (by our conventions on varieties, Appendix A.6), it is real reduced if and only if the function field of every irreducible component V_i of V is real (Proposition 3.2.17). By Corollary 1.7.9, it is equivalent that $V_i(R)$ is Zariski dense in V_i for every i, and hence that $V(R)$ is Zariski dense in V. □

Combining the "abstract" stellensätze with the density properties of Proposition 3.3.2, we obtain "concrete" geometric versions as follows:

3.3.8 Theorem. (Geometric real stellensätze) *Let $f_1, \dots, f_r \in R[V]$ where V is an affine R-variety, and put $K = \mathcal{S}_V(f_1, \dots, f_r) \subseteq V(R)$ and $T = PO_{R[V]}(f_1, \dots, f_r)$. For $f \in R[V]$, the following equivalences hold:*

(a) $f > 0$ *on* $K \Leftrightarrow \exists\, s, t \in T$ *with* $sf = 1 + t$;
(b) $f \geq 0$ *on* $K \Leftrightarrow \exists\, m \geq 0, \exists\, s, t \in T$ *with* $sf = f^{2m} + t$;
(c) $f \equiv 0$ *on* $K \Leftrightarrow \exists\, m \geq 0, \exists\, t \in T$ *with* $f^{2m} + t = 0$.

Proof. If an inequality $f > 0$ (or ≥ 0, or $= 0$) holds on K, the same inequality will hold on the subset $X(T)$ of $\operatorname{Sper} R[V]$, by Proposition 3.3.2. Therefore, the implications "$\Rightarrow$" are immediate consequences of the abstract stellensätze 3.2.7, 3.2.8 and 3.2.10. The reverse implications "$\Leftarrow$" are trivial thanks to the embedding ι (3.3.1). □

3.3.9 Corollary. (Real nullstellensatz) *Let V be an affine R-variety, let $I \subseteq R[V]$ be an ideal and let $W \subseteq V$ be its vanishing subvariety. A polynomial $f \in R[V]$ vanishes identically on $W(R)$ if, and only if, $f \in \sqrt[\mathrm{re}]{I}$.*

Compare this to the usual Hilbert nullstellensatz (A.6.3) from algebraic geometry, which corresponds to replacing R by an algebraically closed field k and the real radical of an ideal by the usual radical.

Proof. Assume that $f \equiv 0$ on $W(R)$. Let $\mathfrak{p}$ be a real prime ideal of $R[V]$ with $I \subseteq \mathfrak{p}$, we have to show $f \in \mathfrak{p}$. If $Z \subseteq W$ is the irreducible subvariety belonging to $\mathfrak{p}$, then $Z(R)$ is Zariski dense in Z by Corollary 1.7.9. Since $f \equiv 0$ on $Z(R)$, this implies $f \in \mathfrak{p}$ as desired. The reverse implication is obvious since $\sqrt[\mathrm{re}]{I} \subseteq \mathfrak{m}_\xi$ for every $\xi \in W(R)$. □

3.3.10 Remark. The role of the geometric stellensätze is that they characterize the (strict or non-strict) positivity of a polynomial f on a basic closed set by the existence of an identity of suitable shape. In each case, the sign behavior in question is trivial from the identity, whereas the converse is not obvious at all. A similar (but simpler) situation is well known from Hilbert's nullstellensatz in algebraic geometry.

We record a few consequences. The following is at the same time a generalization and a strengthening of Artin's solution to Hilbert 17:

3.3.11 Theorem. *Let V be an affine R-variety and let $K = \mathcal{S}_V(f_1, \dots, f_r)$ be a basic closed semialgebraic set in $V(R)$, where $f_1, \dots, f_r \in R[V]$. If a polynomial $f \in R[V]$ is non-negative on K, there exists $h \in R[V]$ with $fh^2 \in PO_{R[V]}(f_1, \dots, f_r)$, and such that $K \cap \mathcal{Z}_V(h) \subseteq \mathcal{Z}_V(f)$.*

Proof. Write $T := PO_{R[V]}(f_1, \dots, f_r)$. According to Theorem 3.3.8(b), there exist $s, t \in T$ and $m \geq 0$ such that $sf = f^{2m} + t$. The claim follows (with $h = s$) by multiplying the identity by s. Indeed, if $\xi \in K$ satisfies $s(\xi) = 0$, the first identity implies $f(\xi) = 0$ since $t(\xi) \geq 0$. □

3.3.12 Remarks.

1. In the situation of Hilbert's 17th problem ($V = \mathbb{A}^n$, $r = 1$ and $f_1 = 1$), Theorem 3.3.11 asserts in particular that every psd polynomial $f \in R[x_1, \dots, x_n]$ has a sum of squares representation $f = \sum_i g_i^2$, in which the g_i are rational functions whose denominators vanish only in zeros of f (in R^n).

2. The geometric stellensätze 3.3.8 are purely existential statements. Their proofs made essential use of Zorn's lemma, and so these proofs are completely non-constructive. For example, if $f_1, \dots, f_r \in R[x_1, \dots, x_n] = R[x]$ satisfy $\mathcal{S}(f_1, \dots, f_r) = \emptyset$, the positivstellensatz asserts the existence of an identity

$$-1 = \sum_{e \in \{0,1\}^r} s_e \cdot f_1^{e_1} \cdots f_r^{e_r}$$

with sums of squares $s_e \in R[x]$. One would like to know an upper bound for the degrees of the s_e, depending only on n, r and the $\deg(f_i)$. While the existence of such a bound follows rather easily from general principles (Exercise 3.3.5), it is a very difficult problem to prove concrete bounds, or even just to estimate their magnitude. The question is directly related to the problem of bounding degrees in Hilbert 17, see Remark 1.5.25.

3. As we have seen, the geometric stellensätze 3.3.8 are consequences of their "abstract" counterparts (3.2.7 etc), the proofs of which were essentially immediate. On the other hand, Theorem 3.3.8 implies Hilbert 17, and even far-reaching generalizations thereof. This approach did not use the statement of Hilbert 17, so have we found an easier approach to this theorem?

The answer is no: To deduce the geometric stellensätze from their abstract versions, we had to use Proposition 3.3.2, which in turn depends on the Tarski transfer principle in an essential way.

We conclude with another geometric application of the stellensätze. First here is an abstract version that holds in the real spectrum of an arbitrary ring:

3.3.13 Proposition. *Let T be a preordering in the ring A, and let elements f, $g \in A$ be given such that $Z(f) \cap X(T) \subseteq Z(g)$ holds (in* Sper(A)*). Then there exist $h \in A$ and $m \geq 0$ such that $|g|^m \leq |fh|$ holds on $X(T)$.*

Conversely, if an inequality $|g|^m \leq |fh|$ holds on $X(T)$, it is obvious that $Z(f) \cap X \subseteq Z(g)$.

Proof. By assumption, g vanishes identically on $X(T) \cap Z(f) = X(T - f^2T)$. So Corollary 3.2.10 implies that $-g^{2m} \in T - f^2T$ for some $m \geq 0$. In particular, there are s, $t \in T$ with $g^{2m} = sf^2 - t$. On $X(T)$ this implies the inequality $g^{2m} \leq sf^2 \leq (1+s)^2 f^2$, and so the inequality $|g|^m \leq |(1+s)f|$ is satisfied on $X(T)$. □

For polynomials we state the following more precise version. Let R be a real closed field as before, and let $x = (x_1, \dots, x_n)$.

3.3.14 Proposition. (Łojasiewicz inequality, first version) *Let $M \subseteq R^n$ be a basic closed semialgebraic set and let $f, g \in R[x]$ be polynomials such that g vanishes on $\mathcal{Z}(f) \cap M$. Then there are a positive constant $c > 0$ in R and integers $m, p \geq 1$, such that*

$$|g(\xi)|^m \leq c \cdot |f(\xi)| \cdot (1 + |\xi|)^p$$

holds for all $\xi \in M$.

Proof. Transferring the proof of Proposition 3.3.13 to the geometric setting (and invoking Theorem 3.3.8(c) instead of Corollary 3.2.10), we get $m \geq 0$ and a polynomial $h \in R[x]$ such that $|g|^m \leq |fh|$ holds on M. It only remains to show that there exist $c > 0$ in R and an integer $p \geq 1$ with $|h(\xi)| \leq c(1 + |\xi|)^p$ for all $\xi \in R^n$. It suffices to treat the case where $h = x^\alpha$ is a monomial (with $\alpha \in \mathbb{Z}_+^n$). In this case, the inequality $|\xi^\alpha| = \prod_i |\xi_i|^{\alpha_i} \leq (1 + |\xi|)^{|\alpha|}$ holds on R^n. □

The qualitative essence of Proposition 3.3.14 is that if a polynomial g vanishes on the (real) zero set of another polynomial f, then the growth of $|g|$ in a neighborhood of the zeros of f is polynomially bounded in terms of $|f|$. Later we'll prove a much stronger version of this result (Section 4.5).

Exercises

Let R be a real closed field.

3.3.1 Give examples of semialgebraic sets in R^2 that are (a) open but not basic open, (b) closed but not basic closed.

3.3.2 Given $n \geq 1$, prove that the sets of positive semidefinite and of positive definite symmetric $n \times n$ matrices are basic closed and basic open in $\mathrm{Sym}_n(R)$, respectively.

3.3.3 Let $f \in R[x] = R[x_1, \dots, x_n]$ be a homogeneous polynomial with $f \geq 0$ on R^n. Prove the following homogeneous version of Corollary 3.3.11: There exists a non-zero form $h \in R[x]$ such that fh^2 is a sum of squares of forms in $R[x]$, and such that f vanishes in every real zero of h.

3.3.4 Let V be an affine R-variety and W a closed subvariety of V, and let $I \subseteq R[V]$ be an ideal with real zero set $W(R)$. Show that a polynomial $f \in R[V]$ has no zero in $W(R)$ if, and only if, f divides some element of the form $1 + g + h$ with $g \in I$ and $h \in \Sigma R[V]^2$.

3.3.5 Let non-negative integers n, r, d be fixed. Show that there exists a non-negative integer $N = N(n, r, d)$ such that, for every real closed field R and arbitrary polynomials $f_1, \dots, f_r \in R[x] = R[x_1, \dots, x_n]$ with $\deg(f_i) \leq d$ $(i = 1, \dots, r)$, the following is true:

Whenever $\mathcal{S}(f_1, \dots, f_r) = \varnothing$, there exist sums of squares $s_e \in R[x]$ (for $e \in \{0, 1\}^r$) with

$$-1 = \sum_{e \in \{0,1\}^r} s_e \cdot f_1^{e_1} \cdots f_r^{e_r} \tag{3.2}$$

and with $\deg(s_e) \leq N$ for every multi-index e.

Hint: For $N \geq 1$ consider the set X_N of all tuples $(f_1, \dots, f_r)$ in $R[x]^r$ with $\deg(f_i) \leq d$, for which an identity (3.2) exists with $\deg(s_e) \leq N$ for all e. First show that the sets X_N and $\bigcup_{N \geq 1} X_N$ are semialgebraic, then use Exercise 4.1.6 (equivalence (i) ⇔ (ii)) from Chapter 4.

3.4 The Constructible Topology

Always let A be a ring. We now introduce a secondary topology on the real spectrum of A. Although the Harrison topology remains the primary one, the constructible topology is a highly useful auxiliary tool.

3.4.1 Definition.

(a) A subset of $\mathrm{Sper}(A)$ is said to be *constructible* if it is a Boolean combination (unions, intersections, complements) of finitely many sets of the form $U_A(f) = \{\alpha \in \mathrm{Sper}(A)\colon f(\alpha) > 0\}$ (with $f \in A$).

(b) The *constructible topology* on $\mathrm{Sper}(A)$ is the topology that has the constructible sets as a basis of open sets. The topological space defined in this way is denoted $\mathrm{Sper}(A)_{\mathrm{con}}$.

3.4.2 Remarks.

1. The constructible subsets of $\mathrm{Sper}(A)$ are precisely the sets that can be described by sign conditions on finitely many elements of A. Also, they are precisely the finite unions of sets of the form $U_A(f_1, \dots, f_r) \cap Z_A(g)$ with $f_1, \dots, f_r, g \in A$.

2. The constructible topology is finer than the Harrison topology. If $A = K$ is a field then both topologies coincide, since every constructible set is Harrison-open in this case.

3. If $\varphi\colon A \to B$ is a ring homomorphism, preimages of constructible subsets under the induced map $\varphi^*\colon \mathrm{Sper}(B) \to \mathrm{Sper}(A)$ are constructible. Therefore, φ^* is continuous also with respect to the constructible topologies.

Let R be a real closed field, let V be an affine R-variety. We have seen that the topological space $V(R)$ is naturally identified, via the map $\iota\colon V(R) \to \mathrm{Sper}\, R[V]$, with a dense subspace of $\mathrm{Sper}\, R[V]$, equipped with the Harrison topology (Proposition 3.3.5). In fact, we already proved a stronger statement:

3.4.3 Proposition. *$\iota(V(R))$ is dense in $\mathrm{Sper}\, R[V]$ with respect to the constructible topology.*

Proof. Immediate from Proposition 3.3.2 and Remark 3.4.2.1. □

From Section 1.6 recall the concepts of A-formulas and A-sentences. In particular, an A-sentence is an A-formula with no free variables. Given a ring homomorphism $\varphi\colon A \to B$ and an A-formula ϕ, the B-formula ϕ^φ arises from ϕ by applying φ to all constants appearing in ϕ. Recall also for $\alpha \in \mathrm{Sper}(A)$ that $r_\alpha\colon A \to R(\alpha)$ denotes the natural homomorphism into the real closed field $R(\alpha)$ associated with α (3.1.2).

3.4.4 Definition. If ϕ is an A-sentence, the subset

$$K_A(\phi) := \{\alpha \in \mathrm{Sper}(A)\colon R(\alpha) \models \phi^{r_\alpha}\}$$

of $\mathrm{Sper}(A)$ is called the *solution set* of ϕ in the real spectrum of A.

Thus, $K_A(\phi)$ is the set of all $\alpha \in \mathrm{Sper}(A)$ for which the A-sentence ϕ, read via r_α in the real closed field $R(\alpha)$, is true. By definition, the constructible sets in Sper(A) are precisely the solution sets of quantifier-free A-sentences. Quantifier elimination tells us that the word "quantifier-free" may be dropped here:

3.4.5 Proposition.

(a) *A subset* $K \subseteq \mathrm{Sper}(A)$ *is constructible if and only if there is an* A*-sentence* ϕ *with* $K = K_A(\phi)$.
(b) *Two* A*-sentences* ϕ_1, ϕ_2 *are* A*-equivalent (see 1.6.13) if and only if* $K_A(\phi_1) = K_A(\phi_2)$.

Proof. (b) Every homomorphism $\varphi\colon A \to R$ into a real closed field factors as $\varphi = \psi \circ r_\alpha$ for unique $\alpha \in \mathrm{Sper}(A)$ and a unique homomorphism $\psi\colon R(\alpha) \to R$. Therefore, if $K_A(\phi_1) = K_A(\phi_2)$, the transfer principle 1.6.17 implies the equivalence $(R \models \phi_1^\varphi)$ $\Leftrightarrow (R \models \phi_2^\varphi)$ for every such homomorphism. This implies (b), and (a) follows from quantifier elimination (Theorem 1.6.15). □

Therefore, the constructible subsets of Sper(A) are in natural bijection with the A-equivalence classes of A-sentences. The obvious rules

$$\begin{aligned} K_A(\phi_1 \wedge \phi_2) &= K_A(\phi_1) \cap K_A(\phi_2), \\ K_A(\phi_1 \vee \phi_2) &= K_A(\phi_1) \cup K_A(\phi_2), \\ K_A(\neg\, \phi) &= \mathrm{Sper}(A) \setminus K_A(\phi) \end{aligned}$$

hold for any A-sentences ϕ, ϕ_1, ϕ_2.

3.4.6 Remark. As a consequence of quantifier elimination (Proposition 3.4.5(a)), one shows easily (Exercise 3.4.4): If A is a Noetherian ring and $\varphi\colon A \to B$ is a finitely generated A-algebra, the induced map $\varphi^*\colon \mathrm{Sper}(B) \to \mathrm{Sper}(A)$ sends constructible sets to constructible sets. The analogous statement is true for the Zariski spectrum as well, and is known as Chevalley's theorem (cf. Remark 1.5.7). This is not an accident: Chevalley's theorem can be proved in formally exactly the same way as we proved the result for the real spectrum, the only difference being that the Tarski principle (that is used here) has to be replaced by the *Lefschetz principle*. Loosely speaking, the Lefschetz principle states that quantifier elimination holds for algebraically closed fields of fixed characteristic.

3.4.7 Theorem. *For every ring* A*, the topological space* $\mathrm{Sper}(A)_{\mathrm{con}}$ *is compact (Hausdorff) and totally disconnected. The constructible subsets of* Sper(A) *are precisely the sets that are open and closed in the constructible topology.*

Proof. Let $X := \mathrm{Sper}(A)_{\mathrm{con}}$. Given $\alpha \neq \beta$ in X, there is $f \in A$ with $\operatorname{sign} f(\alpha) \neq \operatorname{sign} f(\beta)$. In particular, there is a constructible set $U \subseteq \mathrm{Sper}(A)$ with $\alpha \in U$ and $\beta \notin U$. So U and $X \setminus U$ are complementary open subsets of X that separate α and β. Hence X is a totally disconnected Hausdorff space.

The last assertion follows easily once it is known that X is compact. Indeed, if $U \subseteq X$ is open and closed, U is a union of constructible subsets since U is open. Since U is closed in X it is compact, and so it is a finite such union. Hence U is constructible. It therefore only remains to show that X is compact.

Let $2^A = \prod_{a \in A}\{0, 1\}$, equipped with the product topology. By Tikhonov's theorem, this is a compact topological space. We identify 2^A with the set of subsets S of A, identifying S with the indicator (characteristic) function of S. By regarding points in Sper(A) as positive cones of A, we consider Sper(A) as a subset of 2^A. The topology induced on Sper(A) in this way is the constructible topology, so it remains to show that the complement of Sper(A) is open in 2^A. For this let S be a subset of A that is not a positive cone of A. Then one of the following properties *fails* for S (Exercise 3.1.1):

(1) $S + S \subseteq S$,
(2) $SS \subseteq S$,
(3) $S \cup (-S) = A$,
(4) $a \notin S \wedge b \notin S \Rightarrow -ab \notin S \quad (a, b \in A)$,
(5) $-1 \notin S$.

In each case, this immediately implies that the same property fails for all subsets $S' \subseteq A$ in some neighborhood of S. For example, suppose that (4) fails for S, which means there are $a, b \in A \setminus S$ with $-ab \in S$. Then the set U of subsets $S' \subseteq A$ with $a, b \notin S'$ and $-ab \in S'$ is an open neighborhood of S in 2^A, and $U \cap \text{Sper}(A) = \emptyset$. For the other four properties the argumentation is similar, and so the theorem is proved. □

A topological space that is compact (Hausdorff) and totally disconnected is called a *Boolean space* (see A.1.3). A number of equivalent characterizations of such spaces can be found in [53], for example. By Theorem 3.4.7, $\text{Sper}(A)_{\text{con}}$ is always a Boolean space.

Compactness of the constructible topology has important consequences for the Harrison topology:

3.4.8 Corollary. *Every constructible subset of* Sper(A) *is quasi-compact in the Harrison topology. In particular,* Sper(A) *is quasi-compact.*

Proof. The Harrison topology is coarser than the constructible topology. Every constructible set is closed in $\text{Sper}(A)_{\text{con}}$, therefore it is compact in $\text{Sper}(A)_{\text{con}}$ by Theorem 3.4.7. A fortiori, it is quasi-compact in the Harrison topology. □

3.4.9 Corollary. *An open subset of* Sper(A) *is quasi-compact (in the Harrison topology) if and only if it is constructible.*

Proof. If U is open and quasi-compact in the Harrison topology, then U is a union of finitely many basic open sets. Therefore U is constructible. The converse is contained in 3.4.8. □

3.4.10 Proposition. *Equip* Sper(A) *with the Harrison topology. For any closed irreducible subset X of* Sper(A), *there exists $\alpha \in X$ with $X = \overline{\{\alpha\}}$.*

Note that the point α is uniquely determined by X, since Sper(A) is a T_0-space (3.1.12(a)). One says that α is the *generic point* of X.

Proof. Let

$$Z := \bigcap \big\{U \cap X\colon\ U \subseteq \mathrm{Sper}(A) \text{ open constructible},\ U \cap X \neq \varnothing\big\}.$$

Every intersection of finitely many of these sets $U \cap X$ is non-empty since X is irreducible. Moreover, X and all sets $U \cap X$ are closed in $\mathrm{Sper}(A)_{\mathrm{con}}$, and are therefore compact. It follows that $Z \neq \varnothing$. Let $\alpha \in Z$, then $X = \overline{\{\alpha\}}$. Indeed, if there existed $\beta \in X \setminus \overline{\{\alpha\}}$ then β would have an open constructible neighborhood U in Sper(A) with $\alpha \notin U$, contradicting $\alpha \in Z$. □

3.4.11 Remark. Let A be any ring and $X = \mathrm{Sper}(A)$, equipped with the Harrison topology. As we have seen, the topological space X has the following properties:

(1) X is a quasi-compact T_0-space,
(2) the topology of X has a basis of open quasi-compact sets that is stable under finite intersections,
(3) every closed irreducible subset of X has a generic point.

(For (2), one may take all basic open sets.) A topological space X with properties (1)–(3) is called a *spectral space*. If X is any spectral space, a subset of X is called *constructible* if it is a finite Boolean combination of open quasi-compact subsets. The constructible topology on X is defined to be the topology that has all constructible sets as a basis of open sets. Note that this generalizes the case of real spectra, by Corollary 3.4.9. The analogue of Theorem 3.4.7 holds for all spectral spaces: Every spectral space is a Boolean space when equipped with its constructible topology. Examples of spectral spaces, other than real spectra, are Zariski spectra of arbitrary (commutative) rings. A remarkable theorem of Hochster [96] states that, conversely, every spectral space is homeomorphic to the Zariski spectrum of some ring.

A map $f\colon X \to Y$ between spectral spaces X and Y is said to be a *spectral map* if f is continuous with respect to both the original (spectral) topologies and the constructible topologies. In other words, f is a spectral map if, and only if, the preimage of every open quasi-compact subset of Y is open in X and quasi-compact. The category of spectral spaces and spectral maps is dual (anti-equivalent) to the category of distributive lattices: This is the famous Stone duality. We will discuss a particular instance of this duality in Section 4.1, when the real spectrum of semialgebraic sets is considered.

All this and much more can be found in the book [53] by Dickmann, Schwartz and Tressl.

We continue with some consequences of Theorem 3.4.7.

3.4.12 Definition. A subset $X \subseteq \mathrm{Sper}(A)$ is *pro-constructible* if it is an intersection of constructible subsets of Sper(A).

3.4.13 Proposition.

(a) *A subset* $X \subseteq \mathrm{Sper}(A)$ *is pro-constructible if and only if* X *is closed (equivalently, compact) in the constructible topology of* $\mathrm{Sper}(A)$.
(b) *Every covering of a pro-constructible set by constructible sets has a finite subcovering.*
(c) *Every family* $(X_i)_{i \in I}$ *of pro-constructible sets has the finite intersection property: If* $\bigcap_{j \in J} X_j \neq \emptyset$ *for every finite subset* $J \subseteq I$*, then* $\bigcap_{i \in I} X_i \neq \emptyset$.

Proof. (a) Pro-constructible sets are clearly closed in $\mathrm{Sper}(A)_{\mathrm{con}}$. Conversely, if X is closed then the complement of X is open. Therefore it is a union of constructible sets, which means that X is pro-constructible. The other statements follow from (a) and from compactness of $\mathrm{Sper}(A)_{\mathrm{con}}$. □

3.4.14 Proposition. *Let* X *be a pro-constructible subset of* $\mathrm{Sper}(A)$.

(a) $\overline{X} = \bigcup_{\alpha \in X} \overline{\{\alpha\}}$*: The (Harrison) closure of* X *consists of all specializations of points of* X.
(b) *If* $Y \subseteq X$ *is an irreducible subset that is (Harrison) closed relative to* X*, then there is* $\alpha \in Y$ *with* $Y = X \cap \overline{\{\alpha\}}$.

Proof. (a) Given $\beta \in \overline{X}$, every finite sub-intersection of

$$\bigcap \{X \cap U : U \subseteq \mathrm{Sper}(A) \text{ open constructible}, \ \beta \in U\}$$

is non-empty. So the total intersection is non-empty by 3.4.13(c), which means that there exists $\alpha \in X$ with $\alpha \rightsquigarrow \beta$.

(b) The relative closure $\overline{Y}$ of Y in X is irreducible, so it has a generic point $\alpha \in \overline{Y}$ by 3.4.10. Since Y is pro-constructible, there exists $\beta \in Y$ with $\beta \rightsquigarrow \alpha$, by (a). But also $\alpha \rightsquigarrow \beta$, and so $\alpha = \beta \in Y$. □

3.4.15 Remarks.

1. Proposition 3.4.14(a) is very useful. It implies that a pro-constructible set in $\mathrm{Sper}(A)$ is (Harrison) closed as soon as it is stable under specialization. Passing to the complement, a constructible set is (Harrison) open as soon as it is stable under generalization.

2. Every finite subset of $\mathrm{Sper}(A)$ is pro-constructible (by 3.4.13(a)).

3. Every pro-constructible subset X of $\mathrm{Sper}(A)$ is itself a spectral space (in the Harrison topology). Indeed, properties (1) and (3) from Remark 3.4.11 hold by 3.4.13(a) and 3.4.14(b), respectively, and (2) is a direct consequence of (1).

4. If X is a quasi-compact subset of $\mathrm{Sper}(A)$, the set

$$\mathrm{Gen}(X) := \{\alpha \in \mathrm{Sper}(A) : \overline{\{\alpha\}} \cap X \neq \emptyset\}$$

of all generalizations of elements of X is pro-constructible (Exercise 3.4.10).

5. Let $\varphi: A \to B$ be a ring homomorphism, let $f = \varphi^*: \mathrm{Sper}(B) \to \mathrm{Sper}(A)$ be the induced map. Both preimages *and images* of pro-constructible sets under f are again pro-constructible. (For images this follows from 3.4.13(a) by the compactness theorem 3.4.7.) In particular, the fibre $f^{-1}(\alpha)$ of any $\alpha \in \mathrm{Sper}(A)$ is a pro-constructible set in $\mathrm{Sper}(B)$. Images of constructible sets are not in general constructible, but they are if A is Noetherian and B is finitely generated as an A-algebra (Exercise 3.4.4).

6. If A is a general ring, the closure of a constructible set in $\mathrm{Sper}(A)$ need not be constructible. An example where this fails can be found in [3], p. 199. Under quite general assumptions on A, though, it is true that closures of constructible sets are again constructible. For example, this is so when A is an excellent ring ([3] Proposition VII.6.1). In 4.2.5 we'll see this property in the case where A is a finitely generated algebra over a real closed field.

7. The definition of pro-constructible sets, together with Propositions 3.4.13, 3.4.14 and Remarks 1–5 above, generalizes to arbitrary spectral spaces.

Recall the general definition of Krull dimension for topological spaces (A.1.4). For real spectra, and more generally for all spectral spaces, we have:

3.4.16 Proposition. *If X is a spectral space, the Krull dimension* $\dim(X)$ *is the supremum of all lengths d of finite specialization chains $\alpha_0 \rightsquigarrow \cdots \rightsquigarrow \alpha_d$ in X (with $\alpha_{i-1} \neq \alpha_i$ for $i = 1, \dots, d$).*

Proof. The irreducible closed subsets of X are precisely the closures $\overline{\{\alpha\}}$ of singletons, and $\overline{\{\alpha\}}$ is a proper subset of $\overline{\{\beta\}}$ if and only if $\beta \rightsquigarrow \alpha$ and $\beta \neq \alpha$. This implies the proposition immediately. □

From Corollary 3.1.13 we see:

3.4.17 Corollary. $\dim(\mathrm{Sper}\, A) \le \dim(A)$ *for every ring A.* □

So far, everything that we discussed for real spectra is in fact true for arbitrary spectral spaces (3.4.11). Now we'll see properties that are particular for real spectra.

3.4.18 Definition. Let X be a topological space. By $X_{\min}$ (resp. $X^{\max}$) we denote the set of all points $x \in X$ that are minimal (resp. maximal) in X with respect to specialization. In other words, $X^{\max}$ consists of the closed points of X, while $X_{\min}$ consists of those points of X that are not contained in the closure of any other point.

3.4.19 Proposition. *Let $X \subseteq$* $\mathrm{Sper}(A)$ *be a pro-constructible set.*

(a) *Every point $\alpha \in X$ has a unique specialization in $X^{\max}$.*
(b) *The subset $X^{\max}$ of X is compact (and in particular, Hausdorff) in the relative Harrison topology.*

Proof. Let $\alpha \in X$. The intersection

$$Z := \bigcap_{\beta \in X \cap \overline{\{\alpha\}}} (X \cap \overline{\{\beta\}})$$

is non-empty. Indeed, the sets $X \cap \overline{\{\beta\}}$ are pro-constructible, and the intersection of any finite subsystem is non-empty since $\overline{\{\alpha\}}$ is a chain with respect to specialization (Proposition 3.1.12(c)). If $\beta \in Z$ then $\beta \in X^{\max} \cap \overline{\{\alpha\}}$. Uniqueness of β follows again from 3.1.12(c).

This proves (a). Note that (a) implies that X is the only neighborhood of $X^{\max}$ in X. Indeed, there cannot be a closed subset $Y \neq \emptyset$ of X with $Y \cap X^{\max} = \emptyset$ since such Y would be pro-constructible, and so $Y^{\max}$ (which is a subset of $X^{\max}$) would be non-empty by (a), a contradiction. Therefore, $X^{\max}$ is quasi-compact since X is. It remains to show that $X^{\max}$ is Hausdorff, so let $\alpha \neq \beta$ in $X^{\max}$. Neither of them specializes to the other, so there are $f, g \in A$ with $f(\alpha) > 0$, $f(\beta) \leq 0$ and $g(\beta) > 0$, $g(\alpha) \leq 0$. Therefore $U_A(f - g)$ and $U_A(g - f)$ are disjoint open neighborhoods of α and β, respectively. □

3.4.20 Remarks.

1. In every spectral space X, it is true that $X^{\max} \cap \overline{\{x\}} \neq \emptyset$ for every $x \in X$ (one has to use an argument that is different from the one in the above proof). But the other statements in 3.4.19 (uniqueness in (a), the Hausdorff property in (b)) usually fail completely for Zariski spectra, even of decent rings like $\mathbb{C}[x]$.

2. For a spectral space X, property (a) of Proposition 3.4.19 is equivalent to X being a normal topological space (Exercise 3.4.12). When X is a pro-constructible set in Sper(A), we deduced this property from the fact that $\overline{\{\alpha\}}$ is a chain under specialization, for $\alpha \in X$. A spectral space with this latter property has been called a *spectral root system* in [53]. This property is much stronger than being normal. It turns out that real spectra of rings are still not characterized by this property, i.e. there exist spectral root systems that are not real spectra (Delzell–Madden [52]).

3. The set $\mathrm{Sper}(A)^{\max}$ of closed points usually fails to be pro-constructible (Exercise 3.4.11).

4. Every compact topological space is homeomorphic to the maximal real spectrum $\mathrm{Sper}(A)^{\max}$ of some commutative ring A, as we will see in Exercise 3.6.16.

Exercises

3.4.1 Let A be a ring and let M be a symmetric matrix with coefficients in A. For $\alpha \in \mathrm{Sper}(A)$ let $\mathrm{sign}_\alpha(M)$ denote the signature of the matrix $r_\alpha(M)$ over $R(\alpha)$. Show for $k \in \mathbb{Z}$ that the set $\{\alpha \in \mathrm{Sper}(A)\colon\ \mathrm{sign}_\alpha(M) = k\}$ is constructible in Sper(A).

3.4.2 For any ring A, show that the support map $\mathrm{Sper}(A) \to \mathrm{Spec}(A)$ is spectral (Remark 3.4.11). If A is Noetherian and X is any pro-constructible subset of Sper(A), conclude that there exists a finite subset X_0 of X with the following property: For every $\alpha \in X$ there is $\beta \in X_0$ with $\mathrm{supp}(\beta) \subseteq \mathrm{supp}(\alpha)$.

3.4.3 Let R be a real closed field, let $x = (x_1, \ldots, x_n)$ with $n \geq 1$, and let $\iota\colon R^n \to \mathrm{Sper}\, R[x]$ be the natural embedding (3.3.1). A finite subset of $\mathrm{Sper}\, R[x]$ is constructible in $\mathrm{Sper}\, R[x]$ if and only if it is contained in $\iota(R^n)$.

3.4.4 Let A be a Noetherian ring and let $\varphi\colon A \to B$ be a finitely generated A-algebra. Then the induced map $\varphi^*\colon \operatorname{Sper}(B) \to \operatorname{Sper}(A)$ sends constructible sets to constructible sets. Give an example of a ring homomorphism φ for which φ^* does not have this property. (*Hint* on the first part: If $Y \subseteq \operatorname{Sper}(B)$ is constructible, write $\varphi^*(Y)$ as the solution set of a suitable A-sentence.)

3.4.5 If K/k is a finitely generated field extension, show that the restriction map $\operatorname{Sper}(K) \to \operatorname{Sper}(k)$ is an open map. This generalizes Exercise 3.1.7.

3.4.6 Let R be a real closed field, let K/R be a finitely generated field extension of transcendence degree $n \geq 1$, and assume that the field K is real. For any real closed field extension S/R with $\operatorname{trdeg}_R(S) \geq n$, prove that there exists an R-embedding $K \to S$. *Hint*: Use a transcendence basis for K/R and observe Exercise 3.4.5.

3.4.7 Show that the conclusion of Proposition 3.3.13 remains true when the set $X(T)$ is replaced by an arbitrary closed subset X of $\operatorname{Sper}(A)$.

3.4.8 Let A be a ring, let P be a positive cone of A, and let Q be the specialization of P in $\operatorname{Sper}(A)^{\max}$ (Proposition 3.4.19). Show that an element $a \in A$ lies in $\operatorname{supp}(Q)$ if and only a does not divide any element of $1 + P$.

3.4.9 For $\alpha, \beta \in \operatorname{Sper}(A)$, show that the following are equivalent:

(i) α and β are incomparable with respect to specialization, i.e. $\alpha \notin \overline{\{\beta\}}$ and $\beta \notin \overline{\{\alpha\}}$;
(ii) there is $f \in A$ with $f(\alpha) > 0$ and $f(\beta) < 0$;
(iii) there exist open neighborhoods U of α and V of β with $U \cap V = \varnothing$.

3.4.10 Let A be a ring, let $X \subseteq \operatorname{Sper}(A)$ be a pro-constructible set, and let

$$\operatorname{Gen}(X) = \left\{\alpha \in \operatorname{Sper}(A)\colon \overline{\{\alpha\}} \cap X \neq \varnothing\right\},$$

the set of all generalizations of elements in X.

(a) Every neighborhood of X in $\operatorname{Sper}(A)$ contains a constructible neighborhood of X.
(b) The set $\operatorname{Gen}(X)$ is pro-constructible in $\operatorname{Sper}(A)$, and is the intersection of all open neighborhoods of X in $\operatorname{Sper}(A)$.
(c) If a constructible subset K of $\operatorname{Sper}(A)$ contains $\operatorname{Gen}(X)$, then K contains a neighborhood of X in $\operatorname{Sper}(A)$.

3.4.11 Let R be a real closed field and let V be an affine R-variety. Show that the set $(\operatorname{Sper} R[V])^{\max}$ of closed points of $\operatorname{Sper} R[V]$ is pro-constructible in $\operatorname{Sper} R[V]$, (if and) only if $V(R)$ is a finite set. (An easier way of reasoning will be available after Section 4.6, see Exercise 4.6.4 for a more general formulation.)

3.4.12 For every spectral topological space X, the following are equivalent:

(i) $\left|\overline{\{x\}} \cap X^{\max}\right| = 1$ for every $x \in X$;
(ii) X is a *normal* topological space, i.e., any two disjoint closed subsets of X have disjoint open neighborhoods in X.

Hint: Use Exercise 3.4.10.

3.4.13 Let X be a normal spectral space (Exercise 3.4.12). The map $\rho\colon X \to X^{\max}$ defined by $\rho(x) \in \overline{\{x\}} \cap X^{\max}$ $(x \in X)$ is called the *canonical retraction* of X to $X^{\max}$.

(a) The map ρ is continuous and closed.
(b) $Y \mapsto \rho(Y)$ defines a bijection between the connected components Y of X and the connected components of $X^{\max}$.

3.5 Convex Subrings of Ordered Fields and Valuations

Convex subrings of ordered rings or fields are strongly related to valuations and valuation rings. In this section we'll be working at the level of ordered fields, in the next section we consider general ordered rings. The central result here is the Baer–Krull theorem 3.5.11. We refer to the appendix (Section A.5) for terminology and basic facts on (Krull) valuations and valuation rings. We isolate the following basic fact for its importance:

3.5.1 Proposition. *If v is a valuation of a field K, and if the residue field of v is real, then $v(\sum_i a_i^2) = 2 \min_i v(a_i)$ for arbitrary $a_i \in K$.*

See Exercise 3.5.2 for the proof of a more general version.

3.5.2 Definition. Let (K, P) be an ordered field. A subset $M \subseteq K$ is said to be *P-convex* if $a, b \in M$, $c \in K$ and $a <_P c <_P b$ imply $c \in M$. The smallest P-convex set that contains a given set $M \subseteq K$ is called the *P-convex hull* of M in K.

In Section 3.6 we'll consider notions of convexity that are more general.

Valuation rings play a key role in many areas of real algebra and geometry. One main reason for their importance lies in the following fact. Taken by itself it is almost trivial:

3.5.3 Proposition. *If (K, P) is an ordered field, every P-convex subring of K is a valuation ring of K.*

Proof. Let $B \subseteq K$ be a P-convex subring. Then $[-1, 1]_P \subseteq B$. So if $a \in K^*$ satisfies $|a| \le_P 1$ then $a \in B$. If $|a| >_P 1$ then $|a^{-1}| <_P 1$, and hence $a^{-1} \in B$. □

3.5.4 Proposition. *Let (K, P) be an ordered field and let A be a subring of K.*

(a) *The P-convex hull of A in K is a subring (and hence a valuation ring) of K.*
(b) *A is P-convex in K if and only if $[0, 1]_P \subseteq A$.*
(c) *If A is P-convex in K then so is every A-submodule of K (in particular, every ideal of A and every overring of A in K).*

Proof. (a) Let $A \subseteq K$ be an additive subgroup, let B be its P-convex hull. Then $B = \bigcup_{0 \le a \in A} [-a, a]_P$, which is again an additive subgroup of K. If A is a subring of K then so is B. To prove (b) and (c) assume $[0, 1]_P \subseteq A$, and let M be an A-submodule of K. If $x \in M$ and $x \ge_P 0$, then $[-x, x]_P = \{ax\colon a \in K, a \in [-1, 1]_P\}$ is contained in $Ax \subseteq M$, which shows that M is P-convex. □

3.5.5 Proposition. *Let (K, P) be an ordered field and let B be a valuation ring of K. The following are equivalent:*

(i) *B is P-convex in K;*
(ii) *the maximal ideal $\mathfrak{m} = \mathfrak{m}_B$ is P-convex in B;*
(iii) *$-1 <_P a <_P 1$ for every $a \in \mathfrak{m}$.*

If these conditions hold, B and P are said to be compatible *with each other.*

Proof. (i) ⇒ (ii) holds by 3.5.4(c), and (ii) ⇒ (iii) is trivial. Assuming (iii), it suffices to show $[0,1]_P \subseteq B$ to get (i) (3.5.4(b)). So let $x \in K$ with $0 <_P x <_P 1$. If $x \notin B$ we would get $\frac{1}{x} \in \mathfrak{m}$, contradicting (iii) since $\frac{1}{x} >_P 1$. □

3.5.6 Construction. Let (K,P) be an ordered field. Let B be a P-convex subring of K, let $\mathfrak{m} = \mathfrak{m}_B$ and $k = B/\mathfrak{m}$, and write $\overline{a} = a + \mathfrak{m}$ for $a \in B$. The subset

$$\overline{P} := \{\overline{a}\colon\ a \in B,\ a \geq_P 0\}$$

of k is a positive cone of k. Indeed, $\overline{P} + \overline{P} \subseteq \overline{P}$, $\overline{P} \cdot \overline{P} \subseteq \overline{P}$ and $\overline{P} \cup (-\overline{P}) = k$ are immediate. Assuming $-1 \in \overline{P}$, there would be an element $a \in B$ with $a \geq_P 0$ and $1 + a \in \mathfrak{m}$, contradicting 3.5.5. The positive cone $\overline{P}$, or the corresponding ordering of k, is called the *residue ordering* of k induced by P. Note that, by construction,

$$\operatorname{sign}_P(u) = \operatorname{sign}_{\overline{P}}(\overline{u})$$

holds for every unit u of B. We see in particular:

3.5.7 Corollary. *Every convex subring of an ordered field K is a valuation ring of K, and its residue field is real.* □

3.5.8 Example. For any ordered field (K,P), the P convex hull $O(P) = \{a \in K\colon \exists n \in \mathbb{N}\ -n \leq_P a \leq_P n\}$ of $\mathbb{Z}$ in K is a valuation ring of K, with maximal ideal $I(P) = \{a \in K\colon \forall n \in \mathbb{N}\ -1 \leq_P na \leq_P 1\}$. The ordering $\overline{P}$ induced by P on the residue field k of $O(P)$ is an Archimedean ordering of k. In particular, there exists a natural field embedding $k \to \mathbb{R}$. The elements of $I(P)$ are called the *infinitesimal elements* of K with respect to P.

3.5.9 Example. Let R be a real closed field and let $S = R(\!(t^{1/\infty})\!)$, the field of formal Puiseux series over R (see 1.4.10). The field S has a natural Krull valuation v with value group $\mathbb{Q}$, given by

$$v\left(\sum_{k=m}^{\infty} a_k t^{k/d}\right) = \frac{m}{d}$$

if $d \geq 1$, $a_k \in R$ and $a_m \neq 0$. The valuation ring $B = O_v$ of v is the convex hull of R in S.

The following result on convex subrings of real closed fields will be useful in the next chapter.

3.5.10 Proposition. *Let R be a real closed field. For every convex subring B of R, the following are true:*

(a) *The residue field* $k = B/\mathfrak{m}_B$ *of B is real closed.*
(b) *The natural homomorphism* $\pi\colon B \to k$ *has a section, i.e. there is a homomorphism* $s\colon k \to B$ *such that* $\pi \circ s = \mathrm{id}_k$.

Proof. (a) Let $f \in B[t]$ be a monic polynomial of odd degree. Then f has a root α in R, and $\alpha \in B$ since B is integrally closed. Therefore k has no proper field extensions of odd degree. Moreover, the set of squares in k is a positive cone of k, corresponding to the residue ordering induced by the ordering of R (3.5.6). So k is real closed by Proposition 1.2.7.

(b) By Zorn's lemma there exists a subfield F of R that is contained in B and that is maximal with respect to this property. Since B is integrally closed, F is relatively algebraically closed in R, and so F is real closed itself. To prove (b) it suffices to show that $\pi(F) = k$. Assume to the contrary that there is $b \in B$ with $\pi(b) \notin \pi(F)$. Since $\pi(F)$, being real closed, is relatively algebraically closed in k, it follows that $\pi(b)$ is transcendental over $\pi(F)$. This means $F[b] \cap \mathfrak{m}_B = \{0\}$. But then the subfield $F(b)$ of R is contained in B, contradicting the maximal choice of F. □

The next theorem provides a strong converse to Corollary 3.5.7:

3.5.11 Theorem. (Baer–Krull) *Let B be a valuation ring of a field K and let Q be a positive cone of the residue field $k = B/\mathfrak{m}$. Then there is a natural bijection between the following two sets:*

(1) *The set of all positive cones P of K that are compatible with B and satisfy $\overline{P} = Q$;*
(2) *the set of all group homomorphisms $\chi\colon K^* \to \{\pm 1\}$ with $\chi(u) = \mathrm{sign}_Q(\overline{u})$ for all $u \in B^*$.*

If $\Gamma = K^*/B^*$ denotes the value group of B, both sets (1), (2) are in non-canonical bijection with the group $\mathrm{Hom}(\Gamma, \{\pm 1\})$. *In particular, they are non-empty.*

Proof. If P is a positive cone as in (1) then $\chi_P(a) := \mathrm{sign}_P(a)$ ($a \in K^*$) is a character as in (2) (see 3.5.6), and it is obvious that P is determined by χ_P. Conversely let χ be as in (2), we show that $P := \{0\} \cup \ker(\chi)$ is a positive cone of K. Clearly $PP \subseteq P$ and $P \cup (-P) = K$ hold, and also $-1 \notin P$ since $\chi(-1) = -1$. Let $a, b \in K^*$ with $\chi(a) = \chi(b) = 1$. After switching a and b if necessary we may assume $\frac{b}{a} \in B$. Note that $\chi(\frac{b}{a}) = 1$. Therefore, if $\frac{b}{a} = u$ is a unit of B then $\overline{u} >_Q 0$ in k, hence also $1 + \overline{u} >_Q 0$. This implies $\chi(1 + u) = 1$, and so $\chi(a + b) = \chi(a(1 + u)) = 1$. On the other hand, if $\frac{b}{a} \in \mathfrak{m}$ then $v = 1 + \frac{b}{a}$ is a unit of B with $\overline{v} >_Q 0$, and so again $\chi(v) = 1$ and $\chi(a + b) = \chi(av) = 1$. So we have proved that P is a positive cone of K. In fact, P is compatible with B. Indeed, for $a \in \mathfrak{m}$ we have $\chi(1 \pm a) = 1$, hence $1 \pm a >_P 0$, which shows $\mathfrak{m} \subseteq]-1, 1[_P$, and an application of Proposition 3.5.5 shows that P and B are compatible. By construction we have $\overline{P} = Q$.

To prove the last assertion, consider the exact sequence of abelian groups $1 \to B^* \xrightarrow{i} K^* \xrightarrow{v} \Gamma \to 0$, together with the homomorphism $B^* \xrightarrow{\pi} k^*$. Tensorizing the sequence with $\mathbb{Z}/2\mathbb{Z}$ leaves it exact since the value group Γ has no torsion. So we get the exact sequence

$$1 \to B^*/B^{*2} \to K^*/K^{*2} \to \Gamma/2\Gamma \to 0. \tag{3.3}$$

Dualizing both sequence (3.3) and the homomorphism π, we get the exact sequence

$$1 \to \mathrm{Hom}(\Gamma, \{\pm 1\}) \to \mathrm{Hom}(K^*, \{\pm 1\}) \xrightarrow{i^*} \mathrm{Hom}(B^*, \{\pm 1\}) \to 1 \qquad (3.4)$$

plus the inclusion $\pi^*\colon \mathrm{Hom}(k^*, \{\pm 1\}) \subseteq \mathrm{Hom}(B^*, \{\pm 1\})$. The characters χ in (2) are precisely the preimages of $\pi^*(\mathrm{sign}_Q)$ under i^*. Since i^* is surjective, the set (2) is non-empty and the last assertion follows. □

3.5.12 Remark. By fixing a basis of the $\mathbb{Z}/2\mathbb{Z}$-vector space $\Gamma/2\Gamma = K^*/B^*K^{*2}$, we find a family x_i $(i \in I)$ of elements of K^* with the property that every $x \in K^*$ has a representation $x = uy^2 \prod_{j \in J} x_j$ with $u \in B^*$ and $y \in K^*$, where $J \subseteq I$ is a finite set that is uniquely determined by x. In these terms we may state the Baer–Krull theorem as follows: For any given family of signs $\varepsilon_i \in \{\pm 1\}$ $(i \in I)$ there exists a unique ordering P of K that is compatible with B and satisfies $\overline{P} = Q$ and $\mathrm{sign}_P(x_i) = \varepsilon_i$ for every $i \in I$. In particular, $\mathrm{sign}_P(x) = \mathrm{sign}_Q(\overline{u}) \prod_{j \in J} \varepsilon_j$ for x as above.

3.5.13 Example. Let B be a discrete valuation ring with prime element $t \in \mathfrak{m}$. The value group $\Gamma = K^*/B^*$ is infinite cyclic, generated by $v(t) = tB^*$. Let Q be an ordering of $k = B/\mathfrak{m}$. According to Theorem 3.5.11, there are precisely two orderings P_1, P_2 of K that are compatible with B and induce Q on k. Note that P_1, P_2 are distinguished by the sign of t. In the case of a formal power series ring $B = k[\![x]\!]$, we have already seen this in 3.1.14.4 by a direct argument.

3.5.14 Example. Let $n \geq 1$ and $x = (x_1, \ldots, x_n)$, let $\Gamma = \mathbb{Z}^n_{\mathrm{lex}}$ denote the additive group $\mathbb{Z}^n$, ordered lexicographically. Let k be a field, let $v\colon k(x)^* \to \Gamma$ be the valuation defined by $v(x^\alpha) = \alpha$ for $\alpha \in \Gamma$ and $v(c) = 0$ for $c \in k^*$. If $f = \sum_\alpha c_\alpha x^\alpha$ is a non-zero polynomial with support $\mathrm{supp}(f) = \{\alpha\colon c_\alpha \neq 0\}$, then $v(f) = \min_{\mathrm{lex}} \mathrm{supp}(f)$, the *lex*-smallest multi-index α with $c_\alpha \neq 0$. Given an ordering Q of k, there are exactly 2^n orderings P_ε of $k(x)$, parametrized by $\varepsilon = (\varepsilon_1, \ldots, \varepsilon_n) \in \{\pm 1\}^n$, that are compatible with the valuation v and induce the residue ordering Q on k. The ordering P_ε is characterized by $\mathrm{sign}_{P_\varepsilon}(x_i) = \varepsilon_i$ $(i = 1, \ldots, n)$ and by $c|x^\alpha| <_{P_\varepsilon} |x^\beta|$ for every $c \in k$ whenever the first non-zero entry of $\alpha - \beta$ is positive. For $0 \neq f \in k[x]$ as above, the sign of f with respect to P_ε is therefore

$$\mathrm{sign}_{P_\varepsilon}(f) = \varepsilon^{v(f)} \,\mathrm{sign}_Q(c_{v(f)}) \qquad (3.5)$$

using multinomial notation $\varepsilon^\alpha = \prod_i \varepsilon_i^{\alpha_i}$. Note that v extends to a valuation of $k(\!(x)\!)$, the quotient field of the ring of formal power series $k[\![x]\!]$, with the same value group Γ. Similarly, the orderings P_ε extend to orderings of $k(\!(x)\!)$, with the same characterization (3.5) for non-zero power series $f = \sum_\alpha c_\alpha x^\alpha \in k[\![x]\!]$.

Exercises

3.5.1 Let B be a convex subring of a real closed field. Show that the local ring B is Henselian.

3.5.2 Let K be a field, let v be a valuation of K and let $f_1, \dots, f_r \in K$. Assume that there exists an ordering $\leq$ of K that is compatible with v and such that $f_i \geq 0$ for $i = 1, \dots, r$. Show that

$$v(f_1 + \cdots + f_r) = \min\{v(f_1), \dots, v(f_r)\}.$$

This generalizes Exercise 1.1.10. (A "geometric" version of this fact will be stated in Exercise 4.6.3.)

3.5.3 Let K be a field. Show that K has a proper valuation ring $B \neq K$ with real residue field if, and only if, K admits a non-Archimedean ordering.

3.5.4 Let B be a valuation ring of a field K, with value group Γ_B, and let C be a subring of K that contains B.

(a) Show that $\mathfrak{m}_C \subseteq \mathfrak{m}_B$ and that $\overline{B} := B/\mathfrak{m}_C$ is a valuation ring of $C/\mathfrak{m}_C = k_C$.
(b) There is a natural exact sequence $0 \to \Gamma_{\overline{B}} \xrightarrow{i} \Gamma_B \to \Gamma_C \to 0$ of the value groups. Moreover $i(\Gamma_{\overline{B}})$ is a convex subgroup of Γ_B, and the maps are order-preserving in the sense that they send non-negative elements to non-negative elements.
(c) Conclude that the overrings of B in K are in natural bijective correspondence with the convex subgroups of Γ_B, and also with the prime ideals of B. Make these correspondences explicit.

3.5.5 Let B be a convex subring of a real closed field. Show that the support map $\mathrm{Sper}(B) \to \mathrm{Spec}(B)$ is a homeomorphism.

3.6 Specialization in the Real Spectrum

We now relate specializations in the real spectrum of a ring A to convex subrings of ordered residue fields of A, and also to prime ideals of A that are convex in a suitable sense. We start by setting up a general framework of convexity in rings or abelian groups. With an eye on later applications (Chapter 5), this will be done in greater generality than needed at this moment.

3.6.1 Let G be an abelian group, written additively, and let $M \subseteq G$ be a subsemigroup (always containing 0). For $a, b \in G$ write

$$a \leq_M b \;:\Leftrightarrow\; b - a \in M.$$

The relation $\leq_M$ on G is transitive and reflexive, and is compatible with addition, i.e. $a \leq_M b$ and $c \in G$ imply $a+c \leq_M b+c$. The subgroup $\mathrm{supp}(M) := M \cap (-M)$ of G is called the *support* of M. Note that M induces a partial ordering (again denoted $\leq_M$) on the quotient group $\overline{G} := G/\,\mathrm{supp}(M)$, again compatible with addition. The latter is a total ordering, i.e. makes $\overline{G}$ an ordered abelian group, if and only if $M \cup (-M) = G$.

3.6.2 Lemma. *Let $M \subseteq G$ be a semigroup. For any subgroup $H \subseteq G$, the following are equivalent:*

(i) $\forall\, a, b \in H\ \forall\, c \in G\ \ (a \le_M c \le_M b \Rightarrow c \in H)$;
(ii) $\forall\, a, b \in M\ \ (a + b \in H \Rightarrow a, b \in H)$;
(iii) $\operatorname{supp}(M + H) = H$.

*If these conditions hold, the subgroup H of G is said to be M-*convex.

Proof. The proofs are straightforward: Suppose that (i) holds, and let $a, b \in M$ with $a + b \in H$. Since $0 \le_M a, b \le_M a + b$, (i) implies $a, b \in H$, proving (ii). Assuming (ii) let $a \in \operatorname{supp}(M + H)$, which means $a = x_1 + h_1 = -(x_2 + h_2)$ where $x_i \in M$ and $h_i \in H$ $(i = 1, 2)$. Since $x_1 + x_2 \in H$, (ii) implies that $x_1, x_2 \in H$, and hence $a \in H$. This proves $\operatorname{supp}(M + H) \subseteq H$, and the opposite inclusion is trivial. Now assume (iii), let $a, b \in H$ and $c \in G$ such that $a \le_M c \le_M b$, which means $c - a \in M$ and $b - c \in M$. Then $c = (c - a) + a \in M + H$ and $-c = (b - c) - b \in M + H$, and so $c \in \operatorname{supp}(M + H) = H$ by (iii). □

3.6.3 Example. Let $(G, \le)$ be a totally ordered abelian group, and let $M = \{a \in G : a \ge 0\}$, the semigroup of non-negative elements in G. Then the M-convex subgroups of G are just the convex subgroups of M in the usual sense.

3.6.4 Lemma. *Let $M \subseteq G$ be a semigroup.*

(a) *Every M-convex subgroup of G contains $S = \operatorname{supp}(M) = M \cap (-M)$. The map $H \mapsto H/S$ defines a bijective correspondence between the M-convex subgroups H of G and the M/S-convex subgroups of G/S.*
(b) *Assume that $M \cup (-M) = G$. Then the M-convex subgroups H of G form a chain under inclusion (i.e., any two of them are comparable with respect to inclusion), and $M + H = M \cup H$ holds for each of them.*

Proof. (a) Let $H \subseteq G$ be a M-convex subgroup. Then $H = \operatorname{supp}(M + H)$ by 3.6.2, from which $\operatorname{supp}(M) \subseteq H$ is clear. If H is any subgroup of G with $S \subseteq H$, then condition 3.6.2(ii) holds for G, H and M if and only if it holds for G/S, H/S and M/S, respectively.

(b) Assume that $M \cup (-M) = G$. The first assertion follows from the general fact that, in any (totally) ordered abelian group, the convex subgroups form a chain under inclusion. In more detail, let $H_1, H_2 \subseteq G$ be M-convex subgroups and assume $H_1 \not\subseteq H_2$. So there is $h_1 \in H_1 \setminus H_2$, and we may assume $h_1 \in M$. For every $h_2 \in H_2$ it follows that $-h_1 \le_M h_2 \le_M h_1$ (otherwise we get a contradiction to H_2 being M-convex), and therefore $h_2 \in H_1$ since H_1 is M-convex, showing that $H_2 \subseteq H_1$. To prove the last assertion, let $x = p + h$ with $p \in M$, $h \in H$ and assume $x \notin M$. Then $-x \in M$, and so $-h = (-x) + p \in H$ implies $-x \in H$ by 3.6.2(ii)). Hence also $x \in H$, which shows $M + H = M \cup H$. □

Now we return to real spectra of rings. If A is a ring and $\alpha \in \operatorname{Sper}(A)$ has associated positive cone $P_\alpha \subseteq A$, the notion of P_α-convex ideals I of A is defined by 3.6.2. Such an ideal I will simply be called *α-convex*, so an ideal $I \subseteq A$ is α-convex iff $a, b \ge_\alpha 0$ and $a + b \in I$ implies $a, b \in I$. By Lemma 3.6.4, the α-convex ideals I of A form a chain under inclusion, and $\operatorname{supp}(\alpha) \subseteq I$ holds for each of them.

3.6.5 Corollary. *Let* $\alpha \in \mathrm{Sper}(A)$. *The support map*

$$\mathrm{supp}\colon \overline{\{\alpha\}} \to \mathrm{Spec}(A), \quad \beta \mapsto \mathrm{supp}(\beta)$$

defines a bijective correspondence between the specializations β *of* α *in* $\mathrm{Sper}(A)$ *and the* α*-convex prime ideals* $\mathfrak{q}$ *of* A. *The inverse map is given by*

$$\mathfrak{q} \mapsto P_\alpha + \mathfrak{q} = P_\alpha \cup \mathfrak{q}.$$

Proof. Let $\beta \in \overline{\{\alpha\}}$ and $\mathfrak{q} = \mathrm{supp}(\beta)$. Then $P_\beta = P_\alpha + \mathfrak{q}$ by 3.1.12(b), and so $\mathfrak{q} = \mathrm{supp}(P_\alpha + \mathfrak{q})$, which means that $\mathfrak{q}$ is α-convex (3.6.2). Conversely, if $\mathfrak{q}$ is an α-convex prime ideal of A, then $\mathrm{supp}(P_\alpha + \mathfrak{q}) = \mathfrak{q}$, again by 3.6.2. Therefore the preordering $P_\alpha + \mathfrak{q}$ is a positive cone, see 3.2.2.1. □

Note that, in view of Corollary 3.6.5, the key properties 3.1.12(b) and (c) of the real spectrum can be seen as particular cases of Lemma 3.6.4(b).

3.6.6 Remark. Let $(B, \mathfrak{m}, k)$ be a local domain with quotient field $K = \mathrm{qf}(B)$. If $\alpha, \beta \in \mathrm{Sper}(B)$ satisfy $\mathrm{supp}(\alpha) = \{0\}$ and $\mathrm{supp}(\beta) = \mathfrak{m}$, the following are equivalent:

(i) $\alpha \rightsquigarrow \beta$ in $\mathrm{Sper}(B)$;
(ii) $\mathrm{sign}_\alpha(u) = \mathrm{sign}_\beta(u)$ holds for every unit $u \in B^*$.

This is nothing but the definition of specialization. Now let B be a valuation ring. Observing that α (resp. β) is naturally identified with an ordering of K (resp. k), another equivalent condition is:

(iii) α is compatible (cf. 3.5.5) with the valuation ring B, and $\beta = \overline{\alpha}$, the residue ordering induced by α.

Indeed, (ii) implies $1 + b >_\alpha 0$ for all $b \in \mathfrak{m}$, which means that B is α-convex in K (Proposition 3.5.5). Conversely, if α and B are compatible then (ii) holds for $\beta = \overline{\alpha}$, see 3.5.6.

3.6.7 Remark. In particular, the valuation ring B of K is compatible with the ordering α of K if, and only if, α has a specialization β in $\mathrm{Sper}(B)$ with $\mathrm{supp}(\beta) = \mathfrak{m}$. If so, then $\beta = \overline{\alpha}$ is the residue ordering of α. Hence we may rephrase the statement of the Baer–Krull theorem as follows: Let B be a valuation ring with value group Γ, and let $\beta \in \mathrm{Sper}(B)$ with $\mathrm{supp}(\beta) = \mathfrak{m}_B$. Then the set of generalizations α of β in $\mathrm{Sper}(B)$ with $\mathrm{supp}(\alpha) = \{0\}$ is in bijection with $\mathrm{Hom}(\Gamma, \{\pm 1\})$.

By inductively applying the Baer–Krull theorem for discrete valuation rings, we deduce an important consequence:

3.6.8 Theorem. *Let* $(A, \mathfrak{m})$ *be a regular local ring of dimension* d. *Then for any* $\beta \in \mathrm{Sper}(A)$ *with* $\mathrm{supp}(\beta) = \mathfrak{m}$, *there exists a sequence* $\alpha_0 \rightsquigarrow \alpha_1 \rightsquigarrow \cdots \rightsquigarrow \alpha_d = \beta$ *of length* d *of proper specializations in* $\mathrm{Sper}(A)$, *that ends with* β.

Proof. If $d = 0$ there is nothing to be shown. If $d = 1$ then A is a discrete valuation ring, and the assertion is a consequence of the Baer–Krull theorem, cf. Remark 3.6.7. For $d > 1$ we proceed by induction on d. Let $a \in \mathfrak{m} \setminus \mathfrak{m}^2$, then the principal ideal $\mathfrak{p} = Aa$ of A is prime, and the quotient ring $A/\mathfrak{p}$ is regular of dimension $d - 1$ (see A.4.5). The localized ring $A_\mathfrak{p}$ is a local domain, not a field, and its maximal ideal is generated by a. Therefore $A_\mathfrak{p}$ is a discrete valuation ring, with residue field $A_\mathfrak{p}/\mathfrak{p}A_\mathfrak{p} = \mathrm{qf}(A/\mathfrak{p})$.

By the inductive hypothesis, applied to $A/\mathfrak{p}$, there exists a chain $\alpha_1 \rightsquigarrow \cdots \rightsquigarrow \alpha_d = \beta$ in $\mathrm{Sper}(A)$ of length $d - 1$ such that $\mathrm{supp}(\alpha_1) = \mathfrak{p}$. In particular, α_1 is an ordering of the residue field of $A_\mathfrak{p}$. Therefore, by the first step (case of discrete valuation rings), there exists $\alpha_0 \in \mathrm{Sper}(A)$ with $\mathrm{supp}(\alpha_0) = \{0\}$ and with $\alpha_0 \rightsquigarrow \alpha_1$. This completes the proof. □

The localization of a regular local ring at an arbitrary prime ideal is again regular, see A.4.5. So we conclude:

3.6.9 Corollary. *If A is a regular local ring with quotient field K, any point in* $\mathrm{Sper}(A)$ *has a generalization that lies in* $\mathrm{Sper}(K)$*, i.e. has support* $\{0\}$. □

3.6.10 Remark. When we proved the Artin–Lang theorem (Theorem 1.7.8), we used an *ad hoc* argument to settle the implication (iii) ⇒ (ii) there. Using Theorem 3.6.8, this step becomes immediate. Indeed, in the situation at hand, X is an irreducible variety over a real closed field R, with a non-singular R-point $\xi \in X(R)$. The local ring $\mathcal{O}_{X,\xi} = \mathcal{O}_\xi$ of X at ξ is a regular local domain of dimension $\dim(X)$, with residue field R. The ordering of the residue field, considered as an element of $\mathrm{Sper}(\mathcal{O}_\xi)$, has a generalization α in $\mathrm{Sper}(\mathcal{O}_\xi)$ with support $\{0\}$, as a consequence of Theorem 3.6.8. This means that α is an ordering of $\mathrm{qf}(\mathcal{O}_\xi) = R(X)$ that satisfies $\mathrm{sign}_\alpha(f) = \mathrm{sign}\, f(\xi)$ for every $f \in \mathcal{O}_\xi$ with $f(\xi) \neq 0$. The existence of an ordering α with this property is exactly what had to be proved.

3.6.11 Example. Let R be a real closed field, let $n \geq 1$ and $x = (x_1, \dots, x_n)$. For every point $\xi \in R^n$ there is a specialization chain $\alpha_n \rightsquigarrow \alpha_{n-1} \rightsquigarrow \cdots \rightsquigarrow \alpha_1 \rightsquigarrow \alpha_0 = \xi$ in $\mathrm{Sper}\, R[x]$ of length n, that ends in ξ. This follows from Theorem 3.6.8 since the local ring $\mathcal{O}_\xi = R[x]_{\mathfrak{m}_\xi}$ at ξ is regular of dimension n. It is easy to construct such chains explicitly. After a coordinate translation assume that ξ is the origin in R^n. Then take any of the orderings P_ε of $R(x)$ that were discussed in 3.5.14, and consider it as a positive cone of $R[x]$ with support $\{0\}$. If we do this for $\varepsilon = (1, \dots, 1)$, we get the positive cone

$$P = \{0\} \cup \Big\{ f = \sum_\alpha c_\alpha x^\alpha : f \neq 0,\ c_{\nu(f)} > 0 \Big\}$$

of $R[x]$ (with $\nu(f) = \min_{\mathrm{lex}}(\mathrm{supp}(f))$ as in 3.5.14). The closure of the singleton $\{P\}$ in $\mathrm{Sper}\, R[x]$ is a chain of length n as above, namely

$$P \rightsquigarrow P + \langle x_1 \rangle \rightsquigarrow P + \langle x_1, x_2 \rangle \rightsquigarrow \cdots \rightsquigarrow P + \langle x_1, \dots, x_n \rangle,$$

cf. 3.1.12(b). The support of the i-th ordering $P_i = P + \langle x_1, \ldots, x_i \rangle$ is $\langle x_1, \ldots, x_i \rangle$, and P_n is identified with the point ξ under the embedding ι (3.3.1).

3.6.12 Remark. Conversely, if we start from an arbitrary ordering $\alpha \in \operatorname{Sper} R[x]$ with $\operatorname{supp}(\alpha) = \{0\}$, then α has *at most* n proper specializations. In fact there may be less, for two different reasons. First, the closed specialization $\overline{\alpha}$ of α need not be an R-rational point. For example, $\overline{\alpha}$ may lie "at infinity", meaning that $|x_i| >_\alpha c$ for some index i and every $c \in R$. If $R \neq \mathbb{R}$, it may also happen that $|x_i| <_\alpha c$ for some $c \in R$ and all i, but still the prime ideal $\operatorname{supp}(\overline{\alpha})$ of $R[x]$ is not maximal. Both cases occur for $n = 1$ already (Example 3.1.14.3). If $\overline{\alpha} \notin R^n$ then α will have less than n proper specializations since the prime ideal $\operatorname{supp}(\overline{\alpha})$ has height $< n$.

Second, gaps may occur in the specialization chain, or rather, in the chain of their supports: If $\alpha = \alpha_0 \rightsquigarrow \alpha_1 \rightsquigarrow \cdots \rightsquigarrow \alpha_m = \overline{\alpha}$ is the full chain of specializations of α, and if $\mathfrak{p}_i = \operatorname{supp}(\alpha_i)$, it may happen that $\dim R[x]/\mathfrak{p}_{i-1} > 1 + \dim R[x]/\mathfrak{p}_i$ for one or several indices i. Examples are given in Exercise 3.6.15.

3.6.13 Example. The explicit construction of specialization chains of maximal length in 3.6.11 extends from polynomial rings over R to polynomial rings over any field k, after fixing an ordering of k. It also extends to the rings $k[\![x]\!]$ of formal power series, essentially without change (cf. 3.5.14). In fact, we may even go one step further and extend the construction to any regular local ring with a real residue field. This is carried out in Exercise 3.6.8.

Back to specializations in general real spectra. For their study it is convenient to represent orderings of rings by homomorphisms into real closed fields, as we did in 3.1.15. First consider a particular situation:

3.6.14 Lemma. *Let A be a subring of a real closed field R, let $\alpha \in \operatorname{Sper}(A)$ be the point with positive cone $P_\alpha = A \cap R_+$.*

(a) *If B is a convex subring of R with $A \subseteq B$, then $\mathfrak{m}_B \cap A$ is an α-convex prime ideal of A.*
(b) *Conversely, for every α-convex prime ideal $\mathfrak{q}$ of A, there exists a convex subring B of R that contains A such that $\mathfrak{q} = \mathfrak{m}_B \cap A$.*

Proof. (a) is clear since $\mathfrak{m}_B$ is convex in R (3.5.5). Conversely let $\mathfrak{q}$ be an α-convex prime ideal of A, let B be the convex hull of $A_\mathfrak{q}$ in R, a (convex) subring of R (3.5.4(a)). We show $\mathfrak{m}_B \cap A = \mathfrak{q}$, with "$\subseteq$" being clear. Let $a \in \mathfrak{q}$, and assume $a \notin \mathfrak{m}_B$, which means $\frac{1}{a} \in B$. By definition of B there exists $\frac{b}{s} \in A_\mathfrak{q}$ with $\left|\frac{1}{a}\right| < \left|\frac{b}{s}\right|$, hence $0 < s^2 < (ab)^2$. Since $b \in \mathfrak{q}$ and $s \notin \mathfrak{q}$, this contradicts the assumption that $\mathfrak{q}$ is α-convex in A. □

In (b) of the previous lemma, there will in general exist several convex overrings B that dominate $A_\mathfrak{q}$. The ring constructed in the proof above is the smallest among them, see Exercise 3.6.9 for more details.

Now consider the general case. If $\alpha \in \operatorname{Sper}(A)$, recall that $r_\alpha \colon A \to R(\alpha)$ denotes the canonical homomorphism into the real closed field associated with α (3.1.2).

3.6.15 Proposition. *Let A be a ring, and let $\alpha \in \mathrm{Sper}(A)$. If B is a convex subring of $R(\alpha)$ with $r_\alpha(A) \subseteq B$, the composite homomorphism*

$$A \xrightarrow{r_\alpha} B \to B/\mathfrak{m}_B = k_B \tag{3.6}$$

represents a specialization β of α in $\mathrm{Sper}(A)$. Conversely, for every $\beta \in \overline{\{\alpha\}}$ there exists a convex subring $B \subseteq R(\alpha)$ with $r_\alpha(A) \subseteq B$ such that β is represented by (3.6).

Proof. Recall that the residue field of B is real closed (Proposition 3.5.10). If $f \in A$ satisfies $r_\alpha(f) \geq 0$ in $R(\alpha)$, the residue class of $r_\alpha(f)$ in k_B is ≥ 0, so the first statement is clear. For the converse let $\mathfrak{p} = \mathrm{supp}(\alpha)$. Replacing A by $A/\mathfrak{p}$ we may assume $\mathfrak{p} = \{0\}$ and identify A with the subring $r_\alpha(A)$ of $R(\alpha)$. Then we are in the situation of Lemma 3.6.14. Let $\beta \in \overline{\{\alpha\}}$, then $\mathfrak{q} := \mathrm{supp}(\beta)$ is an α-convex prime ideal of A (Corollary 3.6.5). Therefore Lemma 3.6.14(b) gives a convex overring B of A in $R(\alpha)$ with $\mathfrak{q} = \mathfrak{m}_B \cap A$, and so the composite homomorphism $A \subseteq B \to B/\mathfrak{m}_B = k_B$ represents a specialization β' of α with $\mathrm{supp}(\beta') = \mathfrak{q} = \mathrm{supp}(\beta)$. This implies $\beta' = \beta$ by 3.6.5. □

Altogether we have natural maps

$$\left\{\begin{matrix}\text{convex overrings} \\ \text{of } r_\alpha(A) \text{ in } R(\alpha)\end{matrix}\right\} \twoheadrightarrow \left\{\begin{matrix}\alpha\text{-convex} \\ \text{prime ideals of } A\end{matrix}\right\} \xrightarrow{\sim} \left\{\begin{matrix}\text{specializations} \\ \text{of } \alpha \text{ in } \mathrm{Sper}(A)\end{matrix}\right\}$$

The left-hand map $B \mapsto r_\alpha^{-1}(\mathfrak{m}_B)$ is surjective by Lemma 3.6.14, the right-hand map is bijective by Corollary 3.6.5.

3.6.16 Corollary. *Let $\varphi \colon A \to R$ be a homomorphism into a real closed field R, let B be the convex hull of $\varphi(A)$ in R and let $\varphi_0 \colon A \to B$ be the induced homomorphism. Then the image of the map $\varphi_0^* \colon \mathrm{Sper}(B) \to \mathrm{Sper}(A)$ consists precisely of all specializations of $\alpha := [\varphi]$ in $\mathrm{Sper}(A)$.*

Proof. This follows directly from Lemma 3.6.14. □

Finally we give a direct characterization of the closed points in the real spectrum. We use the following terminology: If (K, P) is an ordered field and A is a subring of K, we say that K is *relatively Archimedean* over A with respect to P if, for every $b \in K$, there exists $a \in A$ with $b \leq_P a$. It is equivalent that K is the P-convex hull of A in K.

3.6.17 Proposition. *Let A be a ring, let $\alpha \in \mathrm{Sper}(A)$ and $\mathfrak{p} = \mathrm{supp}(\alpha)$. The following are equivalent:*

(i) *α is a closed point of $\mathrm{Sper}(A)$;*
(ii) *the residue field $\kappa(\alpha) = \mathrm{qf}(A/\mathfrak{p})$ is relatively Archimedean over its subring $A/\mathfrak{p}$ with respect to α;*
(iii) *the real closed field $R(\alpha)$ is relatively Archimedean over its subring $r_\alpha(A)$.*

Proof. (i) $\Rightarrow$ (ii): Let α be a closed point. By Proposition 3.6.5, $\{0\}$ is the only α-convex prime ideal of $A/\mathfrak{p}$. Let B denote the α-convex hull of $A/\mathfrak{p}$ in its quotient field $\kappa = \kappa(\mathfrak{p})$. The maximal ideal $\mathfrak{m}_B$ is α-convex in B (Proposition 3.5.5), and so $\mathfrak{m}_B \cap (A/\mathfrak{p})$ is an α-convex prime ideal of $A/\mathfrak{p}$. It follows that $\mathfrak{m}_B \cap (A/\mathfrak{p}) = \{0\}$, which means that every non-zero element of $A/\mathfrak{p}$ is a unit in B. Therefore $B = \kappa$, which is condition (ii).

The implication (ii) $\Rightarrow$ (iii) follows from Exercise 1.2.4, and (iii) $\Rightarrow$ (ii) is trivial. Suppose that (ii) holds, and assume that α admits a proper specialization β in $\mathrm{Sper}(A)$. Then $\mathfrak{q} := \mathrm{supp}(\beta)$ is an α-convex prime ideal of A that strictly contains $\mathfrak{p}$ (3.6.5), so there exists $a \in \mathfrak{q} \setminus \mathfrak{p}$. By (ii) there is $b \in A$ with $\frac{1}{a^2} <_\alpha b$ in κ, hence $a^2 b >_\alpha 1$. But $a^2 b \in \mathfrak{q}$, which contradicts the fact that $\mathfrak{q}$ is an α-convex prime ideal of $A/\mathfrak{p}$. □

Exercises

3.6.1 Let G be an abelian group and H a subgroup of G, and let M be a semigroup in G.

(a) Show that $\mathrm{supp}(M + H) = \{x \in G \colon \exists h, h' \in H \text{ with } h \le_M x \le_M h'\}$, and that $\mathrm{supp}(M + H)$ is the smallest M-convex subgroup of G that contains H. We write $O_M(H) := \mathrm{supp}(M + H)$ and call $O_M(H)$ the *M-convex hull* of H.

(b) The set $O'_M(H) = \{x \in G \colon \exists h \in H \cap M \text{ with } h \pm x \in M\}$ is an M-convex subgroup of G, and is contained in $O_M(H)$.

(c) Show that $O'_M(H) = O_M(H)$ if and only if the group H is generated by $M \cap H$.

3.6.2 Let A be a ring and let $\alpha \in \mathrm{Sper}(A)$ with associated positive cone P_α.

(a) If I is an ideal of A, show that the P_α-convex hull of I (Exercise 3.6.1) is $\{f \in A \colon \exists g \in I$ with $|f(\alpha)| \le g(\alpha)\}$, and that this is an ideal of A.

(b) Arbitrary sums and intersections of α-convex ideals of A are again α-convex.

(c) Let $I \ne \langle 1 \rangle$ be an α-convex ideal of A. Show that $\sqrt{I}$ is a prime ideal of A.

(d) Let I, J be α-convex ideals of A. Is the product ideal IJ again α-convex? Same question for the radical $\sqrt{I}$ of I.

3.6.3 Let R be a real closed field and let V be an affine R-variety in which $V(R)$ is Zariski dense. Show that the topological spaces $\mathrm{Sper}\, R[V]$ and $\mathrm{Spec}\, R[V]$ have the same Krull dimension. (In Section 4.6, a much more general result will be proved.)

3.6.4 Prove the real analogue of going-up: If $\varphi \colon A \to B$ is an integral ring homomorphism, the induced map $\varphi^* \colon \mathrm{Sper}\, B \to \mathrm{Sper}\, A$ sends closed sets to closed sets. (*Hint*: Use Proposition 3.6.15.)

3.6.5 The going-down theorem does not hold in real algebra: Find a finite flat homomorphism $\varphi \colon A \to B$ of integrally closed domains such that there are $\beta \in \mathrm{Sper}(B)$ and $\alpha \in \mathrm{Sper}(A)$ with $\alpha \rightsquigarrow \varphi^*(\beta)$, but $\alpha \ne \varphi^*(\beta')$ for every generalization β' of β in $\mathrm{Sper}(B)$. (There exist very easy examples.)

3.6.6 Let $\varphi \colon A \to B$ be a ring homomorphism and let $\alpha \in \mathrm{Sper}(A)$. Show that the preimage of α under $\varphi^* \colon \mathrm{Sper}(B) \to \mathrm{Sper}(A)$ is naturally homeomorphic to the real spectrum of $B \otimes_A R(\alpha)$. (*Hint*: Start by reducing to the case where A is a field.)

3.6.7 Let R_0 be the field of real algebraic numbers. As an application of Exercise 3.6.6, show that the real spectra of $\mathbb{Z}[x_1, \ldots, x_n]$ and of $R_0[x_1, \ldots, x_n]$ are homeomorphic.

3.6.8 Prove the following refinements of Theorem 3.6.8 and Proposition A.5.5. Let $(A, \mathfrak{m})$ be a regular local ring with field of fractions K, let $a_1, \dots, a_n$ be a regular system of parameters in A, and let $\mathfrak{p}_i = \langle a_1, \dots, a_i \rangle$ for $i = 0, \dots, n$, a prime ideal in A. By following the proof of A.5.5 and using Exercise 3.5.4, show that there exists a valuation ring B of K with the following properties:

(a) B dominates A, and the map $k_A \to k_B$ of the residue fields is an isomorphism;
(b) the value group Γ_B is isomorphic to $\mathbb{Z}^n_{\text{lex}}$ (as an ordered abelian group);
(c) if $\{0\} = \mathfrak{q}_0 \subsetneq \mathfrak{q}_1 \subsetneq \cdots \subsetneq \mathfrak{q}_n = \mathfrak{m}_B$ are the prime ideals of B, then $\mathfrak{q}_i \cap A = \mathfrak{p}_i$ for $i = 0, \dots, n$;
(d) if $P \subseteq \text{Sper}(A)$ is a positive cone with $\text{supp}(P) = \{0\}$, and if P (considered as an ordering of K) is compatible with B, show that P has exactly $n + 1$ different specializations in $\text{Sper}(A)$, namely the positive cones $P + \mathfrak{p}_i$ for $i = 0, \dots, n$.

3.6.9 In the situation of Lemma 3.6.14, this exercise provides more detailed information. Let A be a subring of a real closed field R, let $\alpha \in \text{Sper}(A)$ be the point with support $\{0\}$ and with positive cone $P_\alpha = A \cap R_+$, and let $\mathfrak{p}$ be an α-convex prime ideal of A. Prove:

(a) There exist convex overrings $B \subseteq C$ of A in R such that, for every convex overring D of A in R, one has $\mathfrak{m}_D \cap A = \mathfrak{p}$ if and only if $B \subseteq D \subseteq C$.
(b) B is the convex hull of $A_\mathfrak{p}$ in R, and the field extension $\kappa(\mathfrak{p}) \subseteq k_B$ is relatively Archimedean.
(c) $C = B_\mathfrak{q}$ where $\mathfrak{q} = \sqrt{I}$ and I is the convex hull of $\mathfrak{p}$ in B.

Hint: Note that the residue field k_B of B is real closed, and that every radical ideal $\neq \langle 1 \rangle$ in a valuation ring is prime.

3.6.10 Let A be a ring and let $\alpha, \beta \in \text{Sper}(A)$. Show that α and β have no common specialization in $\text{Sper}(A)$ if, and only if, there exists $f \in A$ with $f(\alpha) > 1$ and $f(\beta) < -1$.

3.6.11 Let A be a semilocal ring (that is, A has only finitely many maximal ideals). Show that the (compact) topological space $\text{Sper}(A)^{\max}$ is totally disconnected.

3.6.12 Let A be a semilocal ring, and let $\alpha, \beta \in \text{Sper}(A)$ be such that $\text{sign}_\alpha(u) = \text{sign}_\beta(u)$ holds for every unit u of A. Then α and β have a common specialization in $\text{Sper}(A)$. (*Hint*: Give an indirect proof using Exercise 3.6.10)

3.6.13 Let $\varphi\colon A \to B$ be a ring homomorphism, let $X \subseteq \text{Sper}(A)$ be the image set of the induced map $\varphi^*\colon \text{Sper}(B) \to \text{Sper}(A)$.

(a) X is a pro-constructible subset of $\text{Sper}(A)$.
(b) X is specialization-convex in $\text{Sper}(A)$, i.e. given $\beta_0 \rightsquigarrow \beta_1$ in $\text{Sper}(B)$ and $\alpha \in \text{Sper}(A)$ with $\varphi^*(\beta_0) \rightsquigarrow \alpha \rightsquigarrow \varphi^*(\beta_1)$, there exists $\beta \in \text{Sper}(B)$ with $\beta_0 \rightsquigarrow \beta \rightsquigarrow \beta_1$ and $\varphi^*(\beta) = \alpha$.
(c) Does the analogue of assertion (b) hold for Zariski spectra as well?

3.6.14 Let A be a ring. For every subset X of $\text{Sper}(A)$, let the *specialization-convex hull* $c(X)$ of X be defined by

$$c(X) := \left\{\alpha \in \text{Sper}(A)\colon \exists \beta, \gamma \in X \text{ with } \beta \rightsquigarrow \alpha \rightsquigarrow \gamma\right\}.$$

If X is pro-constructible, show that $c(X) = \overline{X} \cap \text{Gen}(X)$, and that $c(X)$ is again pro-constructible.

3.6.15 Let $A = \mathbb{R}[x, y]$. In both (a) and (b) below, an ordering $\alpha \in \text{Sper}(A)$ with $\text{supp}(\alpha) = \{0\}$ is given. Show in either case that the only specialization of α in $\text{Sper}(A)$ is the origin:

(a) Let $\beta \in \text{Sper}(A)$ be defined by $\text{supp}(\beta) = \{0\}$ and $0 < x \ll y \ll 1$ (see Example 3.5.14 for the $\ll$ notation), let $\varphi\colon A \to A$ be the homomorphism with $\varphi(x) = x$, $\varphi(y) = xy$, and let $\alpha = \varphi^*(\beta)$.
(b) Let α correspond to the positive cone P of $R(x, y)$ from Exercise 1.2.3.

3.6.16 Every compact topological space X is homeomorphic to the maximal real spectrum of some ring. In fact, let $A = \mathcal{C}(X, \mathbb{R})$ be the ring of continuous $\mathbb{R}$-valued functions on X. Prove that the natural map $X \to \mathrm{Sper}(A)^{\max}$ is a homeomorphism.

3.7 Notes

The real spectrum of a ring was constructed and investigated by Coste and Roy, beginning around 1979 [47], [48], [46]. Originally its introduction was motivated by topos-theoretic considerations, but soon the usefulness for real algebraic geometry became apparent. Before, preorderings of general rings, and the set of maximal (proper) preorderings, were introduced as early as 1964 by Krivine [113]. Note that maximal preorderings in A are just the closed points in the real spectrum of A, by Corollary 3.2.4. Krivine deduced the (abstract) positivstellensatz 3.2.7 and the real nullstellensatz 3.2.10 in much the same way as we did, and also a version of the geometric real nullstellensatz, using Tarski's principle. However, Krivine's paper went largely unnoticed at the time, and anyway, there didn't yet exist anything like a community in real algebraic geometry. The (geometric) real nullstellensatz was reproved by Dubois and Risler [59], [169] in 1969–1970, the geometric positivstellensatz was rediscovered by Stengle [202] in 1974. Neither of them was aware of Krivine's earlier work.

Spectral spaces were introduced in the 1930s by M. H. Stone. They appear naturally in many fields of mathematics, like lattice theory, algebraic and real algebraic geometry, valuation theory, logic and others. The term *spectral space* was coined by Hochster [96], who proved that spectral spaces are precisely the topological spaces homeomorphic to the Zariski spectrum of a commutative ring. The monograph [53] by Dickmann, Schwartz and Tressl offers a comprehensive treatment of all aspects of spectral spaces.

The Baer–Krull theorem goes back to work of Baer [9] and Krull [114] from 1927 and 1932, respectively. The relation between specializations in the real spectrum on the one hand, and convex subrings of ordered residue fields on the other, is in [15].

Chapter 4
Semialgebraic Geometry

This chapter takes a closer look at semialgebraic sets. After introducing the basic correspondence between semialgebraic sets and constructible subsets of the real spectrum, we use the real spectrum to prove the finiteness theorem (Section 4.2). Next we discuss cylindrical algebraic decomposition (CAD), which is a key technique for the study of the geometry of semialgebraic sets. CAD is also important for computational questions, since it can be obtained effectively. We only prove a basic version, but much more refined approaches are possible. In the remaining chapter, semialgebraic notions of dimension and connected components are introduced, and semialgebraic paths are discussed as another useful technique. Throughout we try to emphasize the usefulness of the real spectrum for semialgebraic geometry.

In the entire chapter, R denotes a real closed field.

4.1 Semialgebraic Sets and the Real Spectrum

Let V be an affine R-variety. We start by showing that the semialgebraic subsets of $V(R)$ are in natural bijective correspondence with the constructible sets in the real spectrum of $R[V]$, the affine coordinate ring of V. This correspondence, called the tilda operator, is the key to making real spectrum techniques available for semialgebraic geometry. We'll see many instances in the sequel where this principle is at work.

From an algebraic geometry perspective, the restriction to affine varieties is artificial. In fact, the entire setup (semialgebraic sets, real spectrum and tilda operator) can be generalized to arbitrary R-varieties. In 4.1.15–4.1.16 below we'll sketch how to achieve this, using the language of schemes (at a very modest level). The reader who is not familiar with this background may safely skip these paragraphs without any loss for the sequel. Starting with 4.1.18, we explain an alternative approach to the real spectrum that works for all varieties (affine or not), and that doesn't need schemes. It is a particular instance of Stone duality between distributive lattices and spectral spaces.

C. Scheiderer, *A Course in Real Algebraic Geometry*, Graduate Texts in Mathematics 303,
https://doi.org/10.1007/978-3-031-69213-0_4

As explained in Appendix A.6, the reader may assume that all varieties considered are quasi-projective.

4.1.1 Let V be an affine R-variety. Recall (3.3.3) that a set $M \subseteq V(R)$ is semialgebraic if M is a finite Boolean combination of principal open sets $\mathcal{U}_V(f) = \{\xi \in V(R)\colon f(\xi) > 0\}$ with $f \in R[V]$. The system $\mathfrak{S}(V)$ of all semialgebraic sets in $V(R)$ is closed under finite unions and intersections, and under taking complements. In other words, it is a Boolean lattice of subsets of $V(R)$.

For systematic reasons we introduce the alternative notation $V_r := \operatorname{Sper} R[V]$, and call V_r the real spectrum of V. Recall (3.3.5) that $\iota\colon V(R) \to V_r$ denotes the natural inclusion map, and that the image of ι is dense in V_r with respect to the constructible topology on V_r (Proposition 3.4.3). Let $\mathcal{K}(V_r)$ denote the system of all constructible subsets of V_r. Again, $\mathcal{K}(V_r)$ is closed under finite unions and intersections and under taking complements.

4.1.2 Proposition. *For every affine R-variety V, the assignment $K \mapsto \iota^{-1}(K)$ defines a bijective mapping $\mathcal{K}(V_r) \to \mathfrak{S}(V)$.*

Proof. For $K \in \mathcal{K}(V_r)$ it is clear that $\iota^{-1}(K)$ is a semialgebraic set in $V(R)$. Conversely, every semialgebraic set in $V(R)$ has the form $\iota^{-1}(K)$ for some $K \in \mathcal{K}(V_r)$. Let $K_1,\ K_2 \subseteq V_r$ be constructible with $\iota^{-1}(K_1) = \iota^{-1}(K_2)$. Then the symmetric difference $K := (K_1 \cup K_2) \smallsetminus (K_1 \cap K_2)$ in V_r is constructible and satisfies $\iota^{-1}(K) = \varnothing$. Since the image of ι is constructibly dense in V_r, this implies $K = \varnothing$, hence $K_1 = K_2$. □

4.1.3 Definition. Given an affine R-variety V and a semialgebraic set M in $V(R)$, we let $\widetilde{M}$ denote the unique constructible subset of V_r with $M = \iota^{-1}(\widetilde{M})$. The topological space $\widetilde{M}$, equipped with the (relative) Harrison topology, will be called the *real spectrum* of M.

4.1.4 Corollary. *The tilda operator $M \mapsto \widetilde{M}$ is a bijective map $\mathfrak{S}(V) \to \mathcal{K}(V_r)$ that commutes with finite unions and intersections and with taking complements. For $M \in \mathfrak{S}(V)$, the set $\widetilde{M}$ is the closure of $\iota(M)$ in V_r with respect to the constructible topology.*

Proof. It only remains to prove the last assertion. Let $M \subseteq V(R)$ be a semialgebraic set and let M' be the closure of $\iota(M)$ in V_r with respect to the constructible topology. Then $M' \subseteq \widetilde{M}$ since $\iota(M) \subseteq \widetilde{M}$ and $\widetilde{M}$ is constructible. For the reverse inclusion we need to show, for any $\alpha \in \widetilde{M}$ and any $K \in \mathcal{K}(V_r)$ with $\alpha \in K$, that $\iota^{-1}(\widetilde{M} \cap K) \neq \varnothing$. But this is clear since $\widetilde{M} \cap K$ is constructible and non-empty, and since the image of ι is constructibly dense in V_r. □

For ease of notation, we will often identify M with a subset (or rather, topological subspace) of $\widetilde{M}$, via the map ι.

4.1.5 Remarks.

1. If V is affine and $M \in \mathfrak{S}(V)$ is described by finitely many polynomial sign conditions, say $M = \bigcup_{i=1}^m (\mathcal{Z}_V(f_i) \cap \mathcal{S}_V(g_{i1}, \dots, g_{ir_i}))$ with $f_i,\ g_{ij} \in R[V]$, then $\widetilde{M}$ is described by the same sign conditions in V_r, namely $\widetilde{M} = \bigcup_{i=1}^m (Z_{R[V]}(f_i) \cap X_{R[V]}(g_{i1}, \dots, g_{ir_i}))$. This is true independently of the chosen description of M.

2. Let $x = (x_1, \dots, x_n)$. With every R-formula $\phi = \phi(x)$ we associated the relation $\mathcal{S}_R(\phi) \subseteq R^n$, consisting of all points $\xi \in R^n$ for which $\phi(\xi)$ is true (1.6.9). Assume that ϕ contains no quantifiers, or more generally, that every occurrence of $x_1, \dots, x_n$ in ϕ is free. Then we may regard ϕ as an $R[x]$-sentence (no free variables). Associated with ϕ we therefore have the constructible set $K_{R[x]}(\phi)$ in $\operatorname{Sper} R[x]$, see 3.4.4. Both sets are related by

$$\widetilde{\mathcal{S}_R(\phi)} = K_{R[x]}(\phi).$$

Indeed, $K_{R[x]}(\phi)$ is a constructible set in $\operatorname{Sper} R[x]$ that intersects R^n precisely in the semialgebraic set $\mathcal{S}_R(\phi)$.

4.1.6 Remark. Let V be an affine R-variety, let $M \subseteq V(R)$ be a semialgebraic set. If $\widetilde{M}$ is (Harrison) open in $V_r = \operatorname{Sper} R[V]$, then $M = \iota^{-1}(\widetilde{M})$ is open in $V(R)$ since ι is continuous. In fact $\widetilde{M}$, being quasi-compact, is then a *finite* union of basic open sets in V_r. So it follows that M is a *finite* union of basic open semialgebraic sets in $V(R)$.

Conversely let M be open in $V(R)$. Does it follow that $\widetilde{M}$ is open in $\operatorname{Sper} R[V]$? Equivalently, is M a union of *finitely* many basic open semialgebraic sets? The answer is yes, but this is a non-trivial theorem. We'll prove it in the next section.

We need to discuss another operation on semialgebraic sets. Let $R \subseteq R'$ be an extension of real closed fields, and let V be an affine R-variety. Extending the base field from R to R' gives the affine R'-variety $V_{R'}$, with affine coordinate ring $R[V] \otimes_R R'$. On $V_{R'}$, we have the notion of (R'-) semialgebraic subsets of $V_{R'}(R') = V(R')$. These sets form the Boolean lattice $\mathfrak{S}(V_{R'})$.

4.1.7 Definition. Let V be an affine R-variety, let $M \subseteq V(R)$ be a semialgebraic set in $V(R)$. We define $M_{R'}$, the *base field extension* of M from R to R', to be the subset of $V(R')$ that consists of all $\eta \in V(R') = \operatorname{Hom}_R(R[V], R')$ for which the point $[\eta] \in \operatorname{Sper} R[V]$ represented by $\eta\colon R[V] \to R'$ (see 3.1.15) lies in $\widetilde{M}$.

This definition, albeit of quite abstract nature, has the advantage of being free of choices, so there is no question of well-definedness. The following formulation is equivalent and is much more intuitive:

4.1.8 Proposition. *Let $M \subseteq R^n$ be a semialgebraic set and let $\phi(x) = \phi(x_1, \dots, x_n)$ be an R-formula with $M = \mathcal{S}_R(\phi)$. Then $M_{R'}$ coincides with $\mathcal{S}_{R'}(\phi)$, the relation defined over R' by the same formula. In particular, $M_{R'}$ is an R'-semialgebraic subset of $(R')^n$.*

Proof. Let $\psi(x)$ be a quantifier-free R-formula that is R-equivalent to ϕ (such ψ exists by Theorem 1.6.15). Then ϕ and ψ define the same relation, both in R^n and in $(R')^n$. So we may assume that the formula ϕ has no quantifiers. If $\eta \in (R')^n$ then, by definition of the constructible set $K_{R[x]}(\phi)$, the equivalences $\eta \in \mathcal{S}_{R'}(\phi) \Leftrightarrow (R' \models \phi^\eta) \Leftrightarrow [\eta] \in K_{R[x]}(\phi)$ holds. Since $K_{R[x]}(\phi) = \widetilde{M}$ by Remark 4.1.5.2, if follows that $\eta \in \mathcal{S}_{R'}(\phi)$ if and only if $\eta \in M_{R'}$. □

4.1.9 Remarks.

1. Still more concretely, if V is an affine R-variety and $M \subseteq V(R)$ is described by a fixed finite system of polynomial sign conditions on V, the base field extension $M_{R'}$ is the subset of $V(R')$ that is described by exactly the same sign conditions.

2. For any affine R-variety V, the operation $M \mapsto M_{R'}$ commutes with finite Boolean set operations. It also commutes with taking closures. Indeed, it suffices to prove this for a semialgebraic set M in R^n. If $\varphi(x) = \varphi(x_1, \ldots, x_n)$ is an R-formula (quantifier-free, if we want) that describes M, the closure $\overline{M}$ of M in R^n is defined by the R-formula

$$\psi(x)\colon\ \forall\, \varepsilon > 0\ \exists\, y = (y_1, \ldots, y_n)\ \big(|y - x|^2 < \varepsilon\ \wedge\ \varphi(y)\big)$$

By 4.1.8, the base field extension of this set is $(\overline{M})_{R'} = \mathcal{S}_{R'}(\psi)$, the relation defined by ψ in $(R')^n$. But this set is the closure of $\mathcal{S}_{R'}(\varphi) = M_{R'}$ in $(R')^n$, showing that $(\overline{M})_{R'} = \overline{M_{R'}}$. In a completely similar way one sees that base field extension commutes with taking interior, boundary or (R-) convex hull (cf. Corollary 1.6.18).

The following remark relates the operators tilda and base field extension. In spite of its "abstractness", it is very useful:

4.1.10 Proposition. *Let $M \subseteq R^n$ be a semialgebraic set, let $x = (x_1, \ldots, x_n)$. For $\alpha \in \widetilde{R^n} = \operatorname{Sper} R[x]$ let $r_\alpha\colon R[x] \to R(\alpha)$ be the ring homomorphism associated with α (see 3.1.2). Then the equivalence*

$$\alpha \in \widetilde{M} \quad\Leftrightarrow\quad r_\alpha(x) \;=\; (r_\alpha(x_1), \ldots, r_\alpha(x_n)) \;\in\; M_{R(\alpha)}$$

holds.

Proof. Indeed, under the natural identification $R(\alpha)^n = \operatorname{Hom}_R(R[x], R(\alpha))$, the point $r_\alpha(x) \in R(\alpha)^n$ corresponds to the homomorphism $r_\alpha\colon R[x] \to R(\alpha)$. Since the latter represents the point α in $\operatorname{Sper} R[x]$, the claim follows directly from our definition of $M_{R(\alpha)}$ (Definition 4.1.7). □

4.1.11 Remark. Let $M \subseteq R^n$ be a semialgebraic set and let $\alpha \in \widetilde{M}$. We may think of the point

$$\xi_\alpha \;:=\; (r_\alpha(x_1), \ldots, r_\alpha(x_n)) \;\in\; M_{R(\alpha)}$$

as the "canonical $R(\alpha)$-rational point" of M that is associated with α. Tautologically we have $r_\alpha(f) = f(\xi_\alpha)$ for every polynomial $f \in R[x]$.

We now extend the notion of semialgebraic sets to R-varieties that are not necessarily affine. (See Section A.6 for conventions on algebraic varieties over a field.) The following observation is obvious:

4.1.12 Lemma. *Let V be an affine R-variety and let $V = \bigcup_{i=1}^r U_i$ be a covering of V by Zariski open affine sets U_i. A subset M of $V(R)$ is semialgebraic (with respect to V) if, and only if, $M \cap U_i(R)$ is semialgebraic (with respect to U_i) for $i = 1, \ldots, r$.*

□

4.1.13 Proposition. *Let V be an arbitrary R-variety. For any subset M of $V(R)$, the following two conditions are equivalent:*

(i) *There exists a covering $V = \bigcup_{i=1}^r U_i$ of V by open affine subsets U_i such that $M \cap U_i(R)$ is semialgebraic (with respect to U_i) for $i = 1, \dots, r$;*
(ii) *for every open affine subset U of V, the intersection $M \cap U(R)$ is semialgebraic (with respect to U).*

M is said to be a semialgebraic subset of $V(R)$ if either condition holds.

As for affine varieties, we write $\mathfrak{S}(V)$ to denote the set of all semialgebraic subsets of $V(R)$. Of course, in the case of affine varieties, the definition just given agrees with the one given before (see 3.3.3), by Lemma 4.1.12.

Proof. The implication (ii) $\Rightarrow$ (i) is obvious. To prove the converse, recall that the intersection of any two open affine subsets of V is again affine (A.6.12). If M satisfies (i), and if $U \subseteq V$ is an open affine subset, it follows that $(U \cap U_i)_{1 \le i \le r}$ is an open affine covering of U. Since $(M \cap U(R)) \cap U_i(R) = (M \cap U_i(R)) \cap U(R)$ is semialgebraic with respect to U_i for each i, $M \cap U(R)$ is semialgebraic with respect to U by Lemma 4.1.12. □

The system of all semialgebraic sets has the expected properties:

4.1.14 Proposition. *Let V be an R-variety.*

(a) *$\mathfrak{S}(V)$ is closed under the finite Boolean set operations ($\cap$, $\cup$, complement in $V(R)$).*
(b) *If $M \in \mathfrak{S}(V)$ then closure, interior and boundary of M (relative to $V(R)$) are again in $\mathfrak{S}(V)$.*
(c) *If $f\colon V \to W$ is a morphism of R-varieties and $M \in \mathfrak{S}(V)$, $N \in \mathfrak{S}(W)$, then $f(M) \in \mathfrak{S}(W)$ and $f^{-1}(N) \in \mathfrak{S}(V)$.*

Proof. All statements reduce to the case of affine varieties, in which they are clear (cf. Corollary 1.6.18 for (b) and Proposition 3.3.4 for (c)). □

To extend the tilda correspondence $M \leftrightarrow \widetilde{M}$ beyond the case of affine R-varieties, there exist several equivalent ways. We will first achieve this by constructing the real spectrum for an arbitrary R-variety, assuming familiarity with not more than the definition of a scheme. Thereafter, an alternative and equivalent construction will be presented that works for arbitrary R-varieties and does not rely on the notion of schemes. If the reader feels uneasy with schemes, he or she may directly pass to 4.1.17 below.

4.1.15 Recall the notation $V_r = \operatorname{Sper} R[V]$ if V is an affine R-variety. For an arbitrary R-variety V, we define the real spectrum V_r of V by glueing the real spectra of an open affine covering of V. In more detail, let $(V_i)_{i \in I}$ be a family of (Zariski) open affine subsets of V with $V = \bigcup_{i \in I} V_i$. For each pair i, j of indices, the inclusion $V_i \cap V_j \subseteq V_i$ induces a natural open topological embedding $\varphi_{ij}\colon (V_i \cap V_j)_r \to (V_i)_r$ whose image is constructible in $(V_i)_r$. (Note that $V_i \cap V_j$ is again affine, see A.6.12.)

The *real spectrum* V_r of V is defined as the topological space that results from glueing the spaces $(V_i)_r$ along the maps φ_{ij} $(i, j \in I)$. As a set, therefore, V_r consists of equivalence classes of pairs (i, α) with $i \in I$ and $\alpha \in (V_i)_r$, where two pairs (i, α) and (j, β) are considered equivalent iff there exists $\gamma \in (V_i \cap V_j)_r$ with $\varphi_{ij}(\gamma) = \alpha$ and $\varphi_{ji}(\gamma) = \beta$. By definition, a subset $U \subseteq V_r$ is open in V_r if and only $\varphi_i^{-1}(U)$ is open in $(V_i)_r$ for every $i \in I$, where $\varphi_i \colon (V_i)_r \to V_r$ denotes the natural map.

The topological space V_r so defined is again a spectral space (3.4.11) and does not depend on the choice of the covering $(V_i)_{i\in I}$, up to natural homeomorphism. Each of the inclusion maps $\varphi_i \colon (V_i)_r \to V_r$ $(i \in I)$ is a homeomorphism onto its image, and this image is an open constructible subset of V_r. A subset Y of V_r is constructible in V_r if and only if $\varphi_i^{-1}(Y)$ is constructible in $(V_i)_r$ for every $i \in I$. The real spectrum V_r is functorial in the expected sense: Every morphism $f \colon V \to W$ of R-varieties induces a spectral map $f_r \colon V_r \to W_r$. Moreover, there is a natural support map $V_r \to V$ that generalizes the support map in the affine case. (Note that we consider V as a scheme.) For the proofs of these assertions we refer to Exercise 4.1.7; this only needs easy standard arguments.

In fact, most of the above generalizes from R-varieties to arbitrary schemes X: By glueing the real spectra $\operatorname{Sper} \mathcal{O}_X(U_i)$, for (U_i) an open-affine covering of X, one defines the topological space X_r, the real spectrum of the scheme X. This space depends functorially on X and comes with a natural support map $X_r \to X$. In general however, X_r may fail to be quasi-compact (and in particular, to be a spectral space).

4.1.16 Remarks.

1. The natural inclusion map $\iota \colon V(R) \to V_r$, previously discussed for affine V, extends naturally to arbitrary R-varieties. Propositions 3.3.5 and 3.4.3 carry over immediately from the affine case: The map ι is a topological embedding, and the image $\iota(V(R))$ is dense in V_r with respect to the constructible topology.

The tilda correspondence continues to hold, and the proof reduces directly to the affine case. So, for each $M \in \mathfrak{S}(V)$ there exists a unique constructible set $\widetilde{M} \subseteq V_r$ with $\iota^{-1}(\widetilde{M}) = M$, and $\widetilde{M}$ is the closure of $\iota(M)$ in the constructible topology of V_r.

2. The operation of extending the real closed base field (4.1.7) passes naturally to semialgebraic subsets on arbitrary R-varieties. We skip over the details.

3. When V is quasi-projective, there exists an open affine subset V' of V with $V'(R) = V(R)$ (Proposition 1.7.6). Both for semialgebraic sets and real spectrum, as well as for the tilda correspondence and for base field extension of semialgebraic sets, the variety V may be replaced by the affine variety V'.

Finally, we'll sketch an alternative way of constructing real spectrum and tilda correspondence for arbitrary R-varieties. The construction is a particular case of Stone duality, see the remarks in 4.1.20 below. It will come in handy in Section 4.6.

4.1.17 Recall that a (bounded) *lattice* is a triple $(L, \vee, \wedge)$ consisting of a set L and two binary operations $\vee$ and $\wedge$ on L (called join and meet, respectively) that are associative and commutative and satisfy the absorption laws $x \vee (x \wedge y) = x$ and $x \wedge (x \vee y) = x$ for all $x, y \in L$. Moreover it is required that there exist elements

$0,\ 1 \in L$ with $x \vee 0 = x$ and $x \wedge 1 = x$ for all $x \in L$. The lattice L is *distributive* if the distributive laws $x \vee (y \wedge z) = (x \vee y) \wedge (x \vee z)$ and $x \wedge (y \vee z) = (x \wedge y) \vee (x \wedge z)$ hold. A distributive lattice L in which every $x \in L$ has a complement x' (i.e., an element satisfying $x \wedge x' = 0$ and $x \vee x' = 1$) is called a *Boolean lattice*. In a Boolean lattice, the complement x' of x is uniquely determined.

4.1.18 Let V be an R-variety and let M be a semialgebraic subset of $V(R)$. We consider $\mathfrak{S}(M)$, the set of all semialgebraic subsets of M, and also $\mathring{\mathfrak{S}}(M)$, the set of all $U \in \mathfrak{S}(M)$ that are (relatively) open in M. Both $\mathfrak{S}(M)$ and $\mathring{\mathfrak{S}}(M)$ are (bounded) distributive lattices, with meet ($\wedge$) and join ($\vee$) operations corresponding to $\cap$ and $\cup$, respectively. Moreover, both have a smallest ($\varnothing$) and a largest (M) element. The lattice $\mathfrak{S}(M)$ is even Boolean. In the following let L denote either $\mathfrak{S}(M)$ or $\mathring{\mathfrak{S}}(M)$. We write $\cup$ and $\cap$ instead of $\vee$ and $\wedge$, respectively.

A *filter* in L is a non-empty subset F of L that is stable under $\cap$ and is upward closed, i.e. satisfies ($A \in F,\ B \in L,\ A \subseteq B \Rightarrow B \in F$). The filter F is a *prime filter* of L if $\varnothing \notin F$, and if $A,\ B \in L$ and $A \cup B \in F$ imply $A \in F$ or $B \in F$. Let $\mathrm{St}(L)$ denote the set of all prime filters of L. For any $A \in L$ let $\mathfrak{U}_L(A) := \{F \in \mathrm{St}(L)\colon A \in F\}$. We make $\mathrm{St}(L)$ a topological space by declaring the family of all subsets $\mathfrak{U}_L(A)$ (with $A \in L$) to be a basis of open sets for $\mathrm{St}(L)$. The topological space $\mathrm{St}(L)$ is called the *Stone space* of L.

For every $\alpha \in \widetilde{M}$, the set $F_\alpha^L := \{A \in L\colon \alpha \in \widetilde{A}\}$ is a prime filter in L. Conversely let F be a prime filter in L. We'll show that $F = F_\alpha^L$ for a unique $\alpha \in \widetilde{M}$. Given $A,\ B \in L$ with $A \in F$ and $B \notin F$, the set-theoretic difference $A \setminus B$ is non-empty. Therefore, the intersection

$$Z_F := \bigcap_{A \in F} \bigcap_{B \in L \setminus F} \widetilde{A \setminus B} \tag{4.1}$$

in $\widetilde{M}$ is non-empty, by compactness of $(\widetilde{M})_{\mathrm{con}}$ and since every finite partial intersection is non-empty. Let α be an element in Z_F, and let $\beta \in \widetilde{M}$ be any point different from α. Then either $\beta \not\rightsquigarrow \alpha$, in which case there is $A \in \mathring{\mathfrak{S}}(M)$ with $\alpha \in \widetilde{A}$ (hence $A \in F$) and $\beta \notin \widetilde{A}$. Or else $\alpha \not\rightsquigarrow \beta$, and then there is $B \in \mathring{\mathfrak{S}}(M)$ with $\beta \in \widetilde{B}$ and $\alpha \notin \widetilde{B}$ (hence $B \in L \setminus F$). In either case, we see that $\beta \notin Z_F$. Therefore the set $Z_F = \{\alpha\}$ is a singleton, and hence $F = F_\alpha^L$ as asserted. In the following we'll write α_F for α.

4.1.19 Theorem. *Let V be an arbitrary R-variety and let $M \subseteq V(R)$ be a semialgebraic subset.*

(a) *For $L = \mathring{\mathfrak{S}}(M)$, the map $\widetilde{M} \to \mathrm{St}(L)$, $\alpha \mapsto F_\alpha^L$ is a homeomorphism with respect to the Harrison topology on $\widetilde{M}$.*
(b) *For $L = \mathfrak{S}(M)$, the map $\widetilde{M} \to \mathrm{St}(L)$, $\alpha \mapsto F_\alpha^L$ is a homeomorphism with respect to the constructible topology on $\widetilde{M}$.*

In either case, the inverse map is $F \mapsto \alpha_F$ with α_F defined as above.

Proof. In both cases let ϕ and ψ denote the maps $\alpha \mapsto F^L_\alpha$ and $F \mapsto \alpha_F$, respectively. For $\alpha \in V_r$ it is clear that $\alpha \in Z_{F^L_\alpha}$, and therefore $\alpha = \alpha_{F^L_\alpha}$, which means $\psi \circ \phi = \mathrm{id}$. Conversely, let F be a prime filter in L and let $\alpha := \alpha_F$. Then $\alpha \in \widetilde{A}$ for every $A \in F$, which means $F \subseteq F^L_\alpha$. This inclusion is an equality, because otherwise there would exist $B \in F^L_\alpha \smallsetminus F$, meaning that $\alpha \in \widetilde{B}$ but $Z_F \cap \widetilde{B} = \varnothing$, contradicting $\alpha \in Z_F$. Therefore $\phi \circ \psi = \mathrm{id}$, which shows that both maps are bijective and are inverses of each other. Moreover both maps are continuous, since for $A \in L$ the subsets $\widetilde{A}$ of $\widetilde{M}$ and $\mathcal{U}_L(A)$ of $\mathrm{St}(L)$ correspond to each other under ϕ and ψ. □

4.1.20 Remarks.

1. A maximal prime filter F in L is called an *ultrafilter*. By Zorn's lemma, every prime filter is contained in some ultrafilter. In case (b) of Theorem 4.1.19 ($L = \mathfrak{S}(M)$), the lattice $L = \mathfrak{S}(M)$ is Boolean, which means that it contains the complement $A^c = M \smallsetminus A$ of each $A \in L$. Hence, for every prime filter F and every $A \in L$, exactly one of $A \in L$ and $A^c \in L$ holds. So every prime filter is an ultrafilter in this case.

2. If $L = (L, \vee, \wedge)$ is any distributive lattice, the Stone space $\mathrm{St}(L)$ of L is defined in a way completely analogous to 4.1.18 (with $\subseteq$, $\cup$, $\cap$ replaced by $\leq$, $\vee$, $\wedge$). It is a general fact that the topological space $\mathrm{St}(L)$ is always spectral (Remark 3.4.11). In addition, the lattice L can be recovered from $\mathrm{St}(L)$, up to isomorphism, as the lattice of open quasi-compact subsets of $\mathrm{St}(L)$. These facts extend to a full-fledged duality, i.e., an anti-equivalence between the category of distributive lattices (and maps preserving finite infs and sups) and the category of spectral spaces (and spectral maps). This is the famous *Stone duality*. The ultrafilters of L are precisely the specialization-minimal points in $\mathrm{St}(L)$, cf. Exercise 4.1.9.

3. The distributive lattice L is Boolean (i.e., has complements) if and only if the spectral space $\mathrm{St}(L)$ is Hausdorff (i.e., a Boolean space), if and only if every prime filter is maximal (i.e., an ultrafilter). Therefore, Stone duality restricts to a duality between Boolean lattices and Boolean topological spaces. For proofs of the assertions just made, and for full details on Stone duality, we refer to [53], Chap. 3.

4. Theorem 4.1.19 may thus be phrased as follows. For every semialgebraic set M on an R-variety V, the real spectrum $\widetilde{M}$ of M (with its Harrison topology) is the Stone dual of the lattice of all open semialgebraic subsets of M. With its constructible topology, $\widetilde{M}$ is the Stone dual of the Boolean lattice of all semialgebraic subsets of M. In particular, any point in $\widetilde{M}$ may be identified either with a prime filter of open semialgebraic subsets of M, or with an ultrafilter of arbitrary semialgebraic subsets of M. The case $M = V(R)$ recovers the full real spectrum V_r of the variety V.

4.1.21 Example. Let $\alpha_2 \rightsquigarrow \alpha_1 \rightsquigarrow \alpha_0$ be the specialization chain in $\mathrm{Sper}\, R[x, y] = \widetilde{R^2}$ that was described in Example 3.6.11 (with $n = 2$ and $\xi = 0$ there). The (ultra) filters corresponding to the α_i have the following description. Given a semialgebraic set $M \subseteq R^2$, we have

(1) $\alpha_2 \in \widetilde{M} \Leftrightarrow (0,0) \in M$,
(2) $\alpha_1 \in \widetilde{M} \Leftrightarrow \exists 0 < \varepsilon \in R$ such that $(t,0) \in M$ for all $0 < t < \varepsilon$,
(3) $\alpha_0 \in \widetilde{M} \Leftrightarrow \exists 0 < \varepsilon \in R, \exists n \in \mathbb{N}$ with

$$\{(a,b) \in R^2 : 0 < a < \varepsilon,\ 0 < b < a^n\} \subseteq M.$$

For the proof see Exercise 4.1.3. Can you generalize to R^n for $n = 3$? for arbitrary n?

Exercises

Let R always be a real closed field.

4.1.1 Let V be an irreducible R-variety and let $M \subseteq V(R)$ be a semialgebraic set. The following are equivalent:

(i) M is Zariski dense in V;
(ii) M contains a non-empty open subset of $V_{\mathrm{reg}}(R)$.

4.1.2 Let V be an affine R-variety and let M be a semialgebraic set in $V(R)$. Show that M is Zariski dense in V if, and only if, every minimal prime ideal of $R[V]$ is the support of some point in $\widetilde{M}$.

4.1.3 Prove the assertions made in Example 4.1.21.

4.1.4 Let $\alpha \in \operatorname{Sper} \mathbb{R}[x,y]$ be the ordering with positive cone

$$P_\alpha = \left\{ f \in \mathbb{R}[x,y] : \exists \varepsilon \geq 0\ f(t,e^t) > 0 \text{ for } 0 < t < \varepsilon \right\},$$

cf. Exercise 1.2.3. Given a semialgebraic set $M \subseteq \mathbb{R}^2$, show that $\widetilde{M}$ contains α if and only if $(t,e^t) \in M$ for all sufficiently small real numbers $t > 0$.

4.1.5 Generalize the remarks in 4.1.9.2 as follows: Let V be an R-variety, let $N \subseteq M$ be semialgebraic subsets of $V(R)$, and let R' be a real closed field extension of R. Show that N is open (or closed, or dense) in M if, and only if, $M_{R'}$ is open (or closed, or dense) in $N_{R'}$, respectively.

4.1.6 Let V be an R-variety, let $(M_i)_{i \in I}$ be a family of semialgebraic sets in $V(R)$. Prove that the following are equivalent:

(i) There is a finite subset $J \subseteq I$ with $\bigcap_{j \in J} M_j = \varnothing$;
(ii) $\bigcap_{i \in I} (M_i)_{R'} = \varnothing$ for every real closed field extension R' of R;
(iii) $\bigcap_{i \in I} \widetilde{M}_i = \varnothing$.

Here $(M_i)_{R'}$ denotes the base field extension of M_i from R to R'.

4.1.7 Assuming that you are familiar with the notion of schemes, prove the claims made in 4.1.15.

4.1.8 Let A be a ring and let $\mathring{\mathcal{K}}(A)$ denote the distributive lattice of all open quasi-compact subsets of $\operatorname{Sper}(A)$. Prove that the Stone space of $\mathring{\mathcal{K}}(A)$ is naturally homeomorphic to the real spectrum $\operatorname{Sper}(A)$ (cf. Remark 4.1.20.2).

4.1.9 Let M be a semialgebraic set, F, F' be two prime filters in the lattice $\mathring{\mathfrak{S}}(M)$ and let $\alpha = \alpha_F$, $\alpha' = \alpha_{F'}$ be the corresponding points in $\widetilde{M}$ (Theorem 4.1.19). Show that $F' \subseteq F$ if and only if $\alpha \rightsquigarrow \alpha'$. Conclude that the ultrafilters in $\mathring{\mathfrak{S}}(M)$ correspond to the specialization-minimal points in $\widetilde{M}$.

4.2 The Finiteness Theorem

R is always a real closed field. We are going to prove the finiteness theorem, mentioned already in 4.1.6, using the real spectrum. As a consequence we'll see that the tilda operator commutes with taking closures and (relative) interiors.

4.2.1 Theorem. (Finiteness Theorem) *Let V be an affine R-variety. Every open semialgebraic set in $V(R)$ is a union of finitely many basic open semialgebraic sets.*

Taking complements, we get the following equivalent version:

4.2.2 Corollary. *If V is an affine R-variety, every closed semialgebraic set in $V(R)$ is a union of finitely many basic closed semialgebraic sets.*

Proof. Let $M \subseteq V(R)$ be closed and semialgebraic. Assuming 4.2.1 we have $V(R) \setminus M = \bigcup_{i=1}^{m} \bigcap_{j=1}^{s_i} \mathfrak{U}_V(f_{ij})$ for suitable m, $s_i \geq 0$ and $f_{ij} \in R[V]$. Passing to the complement gives $M = \bigcap_{i=1}^{m} \bigcup_{j=1}^{s_i} \mathcal{S}_V(-f_{ij})$. Using the distributive law we can rearrange and get $M = \bigcup_{j_1=1}^{s_1} \cdots \bigcup_{j_m=1}^{s_m} \bigcap_{i=1}^{m} \mathcal{S}_V(-f_{ij_i})$. □

The following statement concerns arbitrary R-varieties V. For affine V, it is equivalent to Theorem 4.2.1, as was noticed in Remark 4.1.6:

4.2.3 Theorem. *Let V be an R-variety and let $M \subseteq V(R)$ be an open semialgebraic set. Then the constructible subset $\widetilde{M}$ of V_r is open as well.*

Proof. We prove Theorem 4.2.3, which will also imply Theorem 4.2.1. Passing to the complement, we show that when $M \in \mathfrak{S}(V)$ is closed, the set $\widetilde{M}$ in V_r is closed as well. The proof reduces immediately to the case where V is affine, and after choosing a closed embedding of V into affine space we may replace V by $\mathbb{A}^n$. So let $M \subseteq R^n$ be a closed semialgebraic set and let $x = (x_1, \ldots, x_n)$. We have to show that the constructible set $\widetilde{M}$ is closed in $\operatorname{Sper} R[x]$. By Proposition 3.4.14(a), it suffices to prove that $\widetilde{M}$ is stable under specialization, see Remark 3.4.15.1.

Let $\alpha \in \widetilde{M}$ and let $\beta \in \operatorname{Sper} R[x]$ be a specialization of α. By Proposition 3.6.15 there exists a convex subring B of the real closed field $R(\alpha)$ with $\operatorname{im}(r_\alpha) \subseteq B$, such that β is represented by the composite homomorphism

$$R[x] \xrightarrow{r_\alpha} B \xrightarrow{\pi} k,$$

with $\pi\colon B \to k := B/\mathfrak{m}_B$ the residue homomorphism. Since the field k is real closed (Proposition 3.5.10), this means (see Lemma 3.1.16) that there is a field embedding $R(\beta) \to k$ such that the diagram (solid arrows)

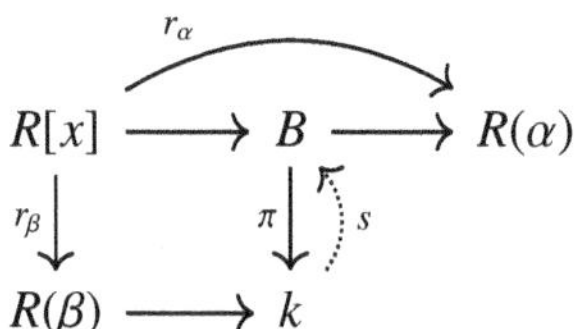

commutes. By Proposition 3.5.10, $\pi\colon B \to k$ has a homomorphic section $s\colon k \to B$.

Let $\xi := (r_\alpha(x_1), \dots, r_\alpha(x_n))$. Then $\xi \in B^n$. Seen as an element of $R(\alpha)^n$, the point ξ lies in $M_{R(\alpha)}$ since $\alpha \in \widetilde{M}$ (Proposition 4.1.10). Assume that $\beta \notin \widetilde{M}$. By the same token, this means that the point $\overline{\xi} := \pi(\xi) \in k^n$ does not lie in M_k. The semialgebraic set M_k is closed in k^n since extension of the base field commutes with taking closures (Remark 4.1.9.2). Therefore there exists $\delta > 0$ in k such that

$$\forall \overline{\eta} \in k^n \ \big(|\overline{\eta} - \overline{\xi}| < \delta \ \Rightarrow \ \overline{\eta} \notin M_k\big).$$

We may formulate this as a k-sentence ϕ that holds in k. By Tarski's transfer principle (Corollary 1.6.17), ϕ remains true if we extend the real closed field via $s\colon k \to R(\alpha)$. Therefore, writing $\varepsilon := s(\delta) \in B \subseteq R(\alpha)$ and $\xi' := s(\overline{\xi}) \in B^n \subseteq R(\alpha)^n$, the following $R(\alpha)$-sentence holds:

$$\forall \eta \in R(\alpha)^n \ \big(|\eta - \xi'| < \varepsilon \ \Rightarrow \ \eta \notin M_{R(\alpha)}\big). \tag{4.2}$$

If we now take $\eta := \xi$ we get a contradiction. Indeed, on one hand we have $\xi \in M_{R(\alpha)}$. On the other, $\xi - \xi' \in \mathfrak{m}_B^n$ since $\pi(\xi') = \pi(\xi)$. Hence $|\xi - \xi'| < \varepsilon$ since $\varepsilon \notin \mathfrak{m}_B$, and therefore $\xi \notin M_{R(\alpha)}$ according to (4.2). This contradiction completes the proof of the finiteness theorem. □

We record a few direct consequences. Let V be an R-variety.

4.2.4 Corollary. *Let $N \subseteq M$ be semialgebraic subsets of $V(R)$. Then N is relatively open (or relatively closed) in M if, and only if, $\widetilde{N}$ is relatively open (or relatively closed) in $\widetilde{M}$, respectively.*

Proof. Assume that N is relatively closed in M. There exists a closed semialgebraic set A in $V(R)$ with $N = A \cap M$, for example the closure of N. By Theorem 4.2.3, $\widetilde{A}$ is closed in V_r, and so $\widetilde{N} = \widetilde{A} \cap \widetilde{M}$ is relatively closed in $\widetilde{M}$. The converse is clear anyway, using ι. □

4.2.5 Corollary. *The tilda operator commutes with taking closure, interior and boundary.*

Proof. Let M be a semialgebraic subset of $V(R)$, let $\overline{M}$ be the closure of M in $V(R)$ and $\overline{\widetilde{M}}$ the closure of $\widetilde{M}$ in V_r. From $\iota(M) \subseteq \widetilde{M}$ it follows that $\iota(\overline{M}) \subseteq \overline{\widetilde{M}}$. On the other hand, the constructible set $\widetilde{\overline{M}}$ is the closure of $\iota(\overline{M})$ in V_r with respect to the constructible topology (Corollary 4.1.4). This implies $\widetilde{\overline{M}} \subseteq \overline{\widetilde{M}}$. Conversely, the set $\widetilde{\overline{M}}$ is closed in V_r by Theorem 4.2.3. Since it contains $\widetilde{M}$, it follows that $\overline{\widetilde{M}} \subseteq \widetilde{\overline{M}}$, and altogether that $\widetilde{\overline{M}} = \overline{\widetilde{M}}$. Since interior or boundary can be expressed in terms of closures, this implies all statements of the corollary. □

4.2.6 Remark. We digress to mention a few important results on the complexity of semialgebraic sets, that otherwise are outside the scope of this book. By the finiteness theorem, every open semialgebraic set U in R^n can be written as a finite union

$$U = \bigcup_{i=1}^{t} \mathcal{U}(f_{i1}, \ldots, f_{is_i})$$

of basic open semialgebraic sets, with polynomials $f_{ij} \in R[x]$ and suitably chosen numbers s_i, $t \geq 1$. Surprisingly, the numbers s_i and t can be bounded uniformly, only in terms of n. In fact, the following more precise and more general results hold:

4.2.7 Theorem. (Bröcker) *For every integer $n \geq 1$ there exist integers s_n, $t_n \geq 1$ such that the following are true, for every n-dimensional affine R-variety V:*

(a) *Every basic open semialgebraic set in $V(R)$ has the form $\mathcal{U}_V(f_1, \ldots, f_r)$ with $r \leq s_n$ and $f_1, \ldots, f_r \in R[V]$;*
(b) *every open semialgebraic set in $V(R)$ is a union of at most t_n many basic open semialgebraic sets.*

4.2.8 Theorem. *Let $s(n)$ and $t(n)$ denote the smallest integers s_n and t_n, respectively, for which the previous theorem holds.*

(a) (Bröcker–Scheiderer) $s(n) = n$.
(b) $t(1) = 1,\ t(2) = 2,\ t(3) \leq 1719,\ t(4) < 1.51 \cdot 10^{16}, \ldots$

There exists an explicit recursive upper bound for $t(n)$ and all n, but very likely this bound is way too large for $n \geq 3$.

The same questions may be asked for closed instead of open semialgebraic sets. Every closed semialgebraic set can be expressed as a finite union of basic closed sets (Corollary 4.2.2). And again, there exist universal upper bounds for the number of polynomials necessary for such a representation:

4.2.9 Theorem. *For every $n \geq 1$ there exist integers $\bar{s}_n$, $\bar{t}_n \geq 1$ such that the following hold for every n-dimensional affine R-variety V:*

(a) *Every basic closed semialgebraic set in $V(R)$ has the form $\mathcal{S}_V(f_1, \ldots, f_r)$ with $r \leq \bar{s}_n$ and $f_1, \ldots, f_r \in R[V]$;*
(b) *every closed semialgebraic set in $V(R)$ is a union of at most $\bar{t}_n$ many basic closed semialgebraic sets;*
(c) *the smallest number $\bar{s}(n)$ satisfying (a) is $\bar{s}(n) = \frac{n}{2}(n+1)$.*

There also exists an explicit recursive upper bound for $\bar{t}(n)$, which is even larger than the above mentioned bound for $t(n)$.

4.2.10 Remarks.

1. A typical basic open set in R^n that requires $s(n) = n$ strict inequalities is the open orthant $\mathcal{U}(x_1, \ldots, x_n)$, or an open hypercube, see Exercise 4.2.4. A typical basic closed set that requires at least $\bar{s}(n) = \frac{n}{2}(n+1)$ non-strict inequalities is

$$S = \bigcup_{i=1}^{n} \big\{\xi \in R^n \colon \xi_1 \geq i-1,\ \xi_2 \geq i-2,\ \ldots,\ \xi_i \geq 0,\ \xi_{i+1} = \cdots = \xi_n = 0\big\}.$$

The reader may check that S is indeed basic closed and can be described by $\frac{n}{2}(n+1)$ simultaneous non-strict inequalities.

Similar examples exist in $V(R)$, for every affine R-variety V of dimension n in which $V(R)$ is Zariski dense.

2. The theoretical background for these results is the reduced theory of quadratic forms over fields. Unfortunately we don't have the space here to go into details. This theory was given a very powerful extension by Marshall, in his theory of spaces of orderings. Theorems 4.2.8 and 4.2.9 were later generalized from affine R-algebras to arbitrary excellent rings and constructible sets in their real spectrum. In particular, this allows for theorems similar to the above ones, in the context of global semianalytic sets. For fully detailed accounts we refer to the books [135] by Marshall and [3] by Andradas, Bröcker and Ruiz.

3. It seems that basic closed sets of dimension n that need the full number $\frac{n}{2}(n+1)$ of non-strict inequalities, tend to be somewhat pathological, as in the previous example. In contrast, Averkov and Bröcker [8] proved that n non-strict inequalities suffice for every polyhedron in n-space.

Exercises

Let R always be a real closed field.

4.2.1 Let N, $N' \subseteq M$ be semialgebraic sets in R^n.

(a) Show that N is dense in M if and only the set $\widetilde{M}_{\min}$ (of specialization-minimal points of $\widetilde{M}$) is contained in $\widetilde{N}$.
(b) Conclude that if N and N' are dense in M, then so is $N \cap N'$.

4.2.2 Show that Proposition 3.3.14 (the Łojasiewicz inequality) remains true for arbitrary closed semialgebraic sets $M \subseteq R^n$ (not necessarily basic closed).

4.2.3 Let $\phi = \phi(x_1, \dots, x_n)$ be a $\mathbb{Z}$-formula and let R be a real closed field. Assume that $\mathcal{S}_R(\varphi)$, the relation in R^n defined by φ (1.6.9), is open in R^n. For any ring A and any tuple $a \in A^n$, show that the constructible set $K_A(\varphi(a))$ in Sper(A) (3.4.4) is open in Sper(A). (*Hint*: Show that the hypothesis does not depend on the real closed field R. Then use the finiteness theorem.)

4.2.4 This exercise outlines the reasoning for why the positive orthant in R^n cannot be described by less than n polynomial inequalities. If K is a field, we use the shorthand notation $X_K :=$ Sper(K) in this exercise. Identify X_K with the set of all characters $\chi\colon K^* \to \{\pm 1\}$ of the multiplicative group that come from an ordering of K, i.e. for which $\ker(\chi) = \{a \in K^* : \chi(a) = 1\}$ is additively closed.

(a) A non-empty subset $F \subseteq X_K$ is a *fan* if there exists a subgroup H of K^* with $a^2 \in H$ for every $a \in K^*$, and such that

$$F = \big\{\chi \in \mathrm{Hom}(K^*, \{\pm 1\})\colon \chi|_H = 1,\ \chi(-1) = -1\big\}.$$

The cardinality $|F|$ is called the *order* of the fan F. Show that every subset $F \subseteq X_K$ with $|F| = 1$ or $|F| = 2$ is a fan. These are the *trivial fans*.
(b) Show that every fan has a natural structure of affine space over the field with two elements. In particular, the order of every finite fan is a power of 2.

(c) Let $F \subseteq X_K$ be a fan, and let $Y \subseteq X_K$ be a basic closed set that is described by r inequalities, say $Y = X_K(f_1, \dots, f_r)$ with $f_i \in K^*$ (notation as in 3.1.6). Show that $F \cap Y$ is an affine subspace of F of index at most 2^r.

(d) Let B be a valuation ring of K, with residue field k. If G is a fan in X_k, show that the full pull-back

$$F := \{\chi \in X_K : \exists \eta \in G \text{ with } \chi(u) = \eta(\overline{u}) \text{ for all } u \in B^*\}$$

of G is a fan in X_K, of order $|F| = |G| \cdot |\Gamma/2\Gamma|$ where Γ is the value group of B. (Use the Baer–Krull theorem)

(e) Show that the positive open orthant Q in R^n cannot be expressed with less than n simultaneous strict inequalities. Find other examples of basic open sets in R^n with the same property. (*Hint*: Use (d) to construct a fan of size 2^n in $R(x_1, \dots, x_n)$ that meets $\widetilde{Q}$ in a single point. Then apply (c).)

4.3 Cylindrical Algebraic Decomposition

R is always a real closed field, as before. We start by introducing the concepts of definable and semialgebraic maps, before we turn to the question of decomposing semialgebraic sets into elementary pieces.

4.3.1 Definition. Let V, W be R-varieties and let $M \subseteq V(R)$, $N \subseteq W(R)$ be semialgebraic sets. A map $f : M \to N$ is *definable* if $\operatorname{graph}(f) = \{(\xi, f(\xi)) : \xi \in M\}$ is a semialgebraic subset of $V(R) \times W(R) = (V \times W)(R)$. If f is definable and in addition continuous, we say that f is a *semialgebraic map*. Definable or semialgebraic maps $M \to R$ are also called *definable* or *semialgebraic functions* on M, respectively.[1] The set of all semialgebraic functions on M will be denoted by $\mathcal{A}(M)$.

4.3.2 Remarks. Let $M \subseteq V(R)$, $N \subseteq W(R)$ be semialgebraic sets.

1. For every morphism $f : V \to W$ of R-varieties, the induced map $f : V(R) \to W(R)$ is semialgebraic.

2. Given R-varieties W_1, W_2, a map $f = (f_1, f_2) : M \to W_1(R) \times W_2(R)$ is definable (or semialgebraic) if and only if each component $f_i : M \to W_i(R)$ is definable (or semialgebraic, respectively).

3. Given a definable map $f : M \to N$ between semialgebraic sets, and given a real closed overfield $R' \supseteq R$, the base field extension of $\operatorname{graph}(f) \subseteq M \times N$ from R to R' is the graph of a map $M_{R'} \to N_{R'}$. This map is definable (over R') and is called the *base field extension* of f to R', denoted $f_{R'}$. Note that f is continuous if and only if $f_{R'}$ is continuous.

4. Semialgebraic functions $R^n \to R$ need not be piecewise polynomial, as the square root $[0, \infty[\to R$, $x \mapsto \sqrt{x}$ shows. In fact, a semialgebraic function need not be expressible by iterated (higher) roots, not even piecewise. This follows from the well-known fact that the solutions of a general equation of degree ≥ 5 cannot be expressed by iterated roots. A concrete example is given by the equation

[1] What we call definable functions are called semialgebraic functions in [26].

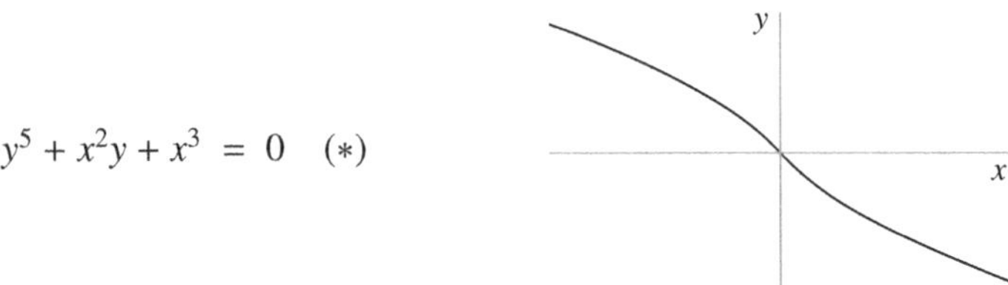

For every $x \in R$, $(*)$ has a unique solution $y = f(x)$ in R. The function $f: R \to R$ defined in this way is semialgebraic, but a small Galois-theoretic argument shows that $f(x)$ is not expressible by iterated roots.

4.3.3 *Convention*: In the sequel, when we say that $M, N \dots$ are semialgebraic sets, it is always understood that $M \subseteq V(R)$, $N \subseteq W(R) \dots$ are semialgebraic subsets of suitable R-varieties $V, W \dots$

4.3.4 Lemma.

(a) *If $f: M \to N$ and $g: N \to L$ are semialgebraic maps (between semialgebraic sets on R-varieties), the composite map $g \circ f: M \to L$ is semialgebraic as well.*
(b) *If $f: M \to K$ and $g: N \to L$ are semialgebraic maps, the map $f \times g: M \times N \to K \times L$, $(\xi, \eta) \mapsto (f(\xi), g(\eta))$ is semialgebraic as well.*
(c) *Images and preimages of semialgebraic sets under semialgebraic maps are again semialgebraic sets.*
(d) *The set $\mathcal{A}(M)$ of all semialgebraic functions on a semialgebraic set M is a commutative ring under pointwise addition and multiplication.*

The same are true with semialgebraic maps replaced by definable maps.

Proof. The proofs are straightforward using Proposition 4.1.14(c). For instance, (a) follows from $\text{graph}(g \circ f) = \text{pr}_{13}\big((\text{graph}(f) \times L) \cap (M \times \text{graph}(g))\big)$, where pr_{13} denotes projection to the first and third component of $M \times N \times L$. The other cases are left to the reader. □

4.3.5 Definition. A map $f: M \to N$ between semialgebraic sets is called a

(a) *semialgebraic (s.a.) homeomorphism* if f is semialgebraic and bijective and the inverse map $f^{-1}: N \to M$ is continuous (hence semialgebraic);
(b) *semialgebraic (s.a.) embedding* if f is a semialgebraic homeomorphism from M onto $f(M)$; if in addition $f(M)$ is (relatively) open or closed in N, then f is an *open* or *closed embedding*, respectively.

Semialgebraic sets M and N are *semialgebraically (s.a.) homeomorphic*, denoted $M \approx N$, if there exists a s.a. homeomorphism $M \to N$.

4.3.6 Examples. Let M be a semialgebraic set.

1. If f, g are semialgebraic functions on M then so are $|f|$, $\max(f, g)$, $\min(f, g)$, and also $1/f$ (if $f \neq 0$ on M) and $\sqrt[n]{f}$ (if n is odd or $f \geq 0$ on M). (If $n \geq 0$ is even and $0 \leq a \in R$, the notation $\sqrt[n]{a}$ always stands for the non-negative n-th root of a in R.)

2. Let $M \subseteq R^n$ be a non-empty semialgebraic set. The infimum

$$d_M(\xi) := \operatorname{dist}(\xi, M) = \inf\{|\eta - \xi| \colon \eta \in M\}$$

exists in R for every $\xi \in R^n$, even though bounded infima won't exist in R in general. Indeed, the set $\{|\eta - \xi| \colon \eta \in M\}$ is semialgebraic in R, hence it is a finite union of intervals and has an infimum in R. It is easy to see that the distance function $d_M \colon R^n \to R$ is semialgebraic, and that $d_M^{-1}(0) = \overline{M}$ (Exercise 4.3.3).

3. For $n \geq 1$, the sets R^n, $]0, \infty[^n$, $]0, 1[^n$ and $B_n := \{x \in R^n \colon |x| < 1\}$ are all s.a. homeomorphic to each other (Exercise 4.3.3). A semialgebraic set that is s.a. homeomorphic to any of the above is called an *open n-cell*. Note in particular that every semialgebraic set in R^n is s.a. homeomorphic to a bounded semialgebraic set.

To state the main result of this section, we introduce the following terminology:

4.3.7 Notation. Let M be a set. For any two maps $f, g \colon M \to R \cup \{\pm\infty\}$, the set

$$\operatorname{band}(f, g) := \left\{(\xi, t) \in M \times R \colon f(\xi) < t < g(\xi)\right\} \subseteq M \times R$$

will be called the *(open) band* between f and g.

4.3.8 Lemma. *Let M be a semialgebraic set.*

(a) *For any semialgebraic function $f \colon M \to R$, the first projection* $\operatorname{pr}_1 \colon \operatorname{graph}(f) \to M$, $(\xi, f(\xi)) \mapsto \xi$ *is a s.a. homeomorphism.*
(b) *If $f, g \colon M \to R$ are semialgebraic maps that satisfy $f < g$ on M, there exists a s.a. homeomorphism* $\phi \colon \operatorname{band}(f, g) \xrightarrow{\approx} M \times]0, 1[$ *over M. The same is true if $f \equiv -\infty$ or $g \equiv +\infty$.*

By saying that ϕ is a homeomorphism *over M* we mean that the triangle

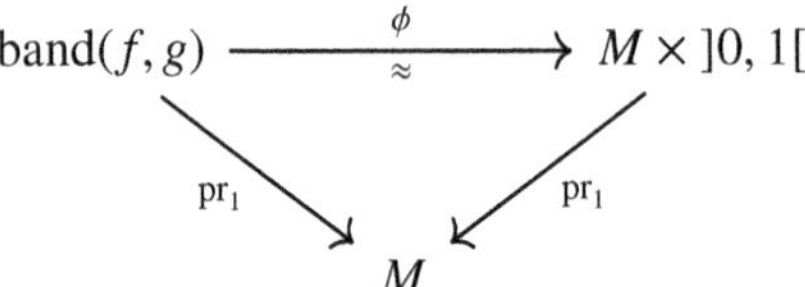

commutes.

Proof. (a) The inverse map is $\xi \mapsto (\xi, f(\xi))$. (b) If $f < g$ are R-valued functions, define ϕ to be the map that is inverse to

$$M \times]0, 1[\xrightarrow{\approx} \operatorname{band}(f, g), \quad (\xi, t) \mapsto (\xi,\ (1 - t)f(\xi) + tg(\xi)).$$

If $f \equiv -\infty$ and $g < +\infty$ then

$$M \times]-\infty, 0[\to \operatorname{band}(-\infty, g), \quad (\xi, t) \mapsto (\xi,\ g(\xi) + t)$$

is a s.a. homeomorphism over M that can be combined with a s.a. homeomorphism $]0,1[\ \stackrel{\approx}{\longrightarrow}\ ]-\infty,0[$ (see 4.3.6.3). Similarly if $f > -\infty$ and $g \equiv +\infty$. If $f \equiv -\infty$ and $g \equiv +\infty$, (b) follows from $R \approx\ ]0,1[$. □

Given any semialgebraic set $M \subseteq R^{n+1}$, the following theorem gives an effective way of decomposing M into a finite disjoint union of open cells (of varying dimensions):

4.3.9 Theorem. (Cylindrical algebraic decomposition, CAD) *Let finitely many polynomials $f_1, \dots, f_r \in R[x_1, \dots, x_n, t]$ be given. There exists a finite decomposition*

$$R^n = K_1 \cup \cdots \cup K_s$$

of R^n into pairwise disjoint semialgebraic sets K_i, together with finitely many semialgebraic functions

$$z_{ij} \colon K_i \to R$$

($j = 1, \dots, m_i$, where $m_i \ge 0$) for every $i = 1, \dots, s$, such that the following hold for every $i = 1, \dots, s$:

(1) $z_{i1} < z_{i2} < \cdots < z_{im_i}$ *(pointwise on K_i);*
(2) *for every $\xi \in K_i$ one has*

$$\bigl\{z_{ij}(\xi) \colon 1 \le j \le m_i\bigr\} = \bigcup_{\substack{\nu=1,\dots,r \\ f_\nu(\xi,t)\not\equiv 0}} \bigl\{t \in R \colon f_\nu(\xi,t) = 0\bigr\};$$

(3) *each polynomial f_ν ($\nu = 1, \dots, r$) has constant sign on each of the sets* graph(z_{ij}) *($1 \le j \le m_i$) and* band$(z_{ij}, z_{i,j+1})$ *($0 \le j \le m_i$). Here we put $z_{i0} := -\infty$ and $z_{i,m_i+1} := +\infty$.*

4.3.10 Remark. Over each piece $K_i \subseteq R^n$, the "cylinder" $K_i \times R$ is decomposed into the graphs $G_{ij} :=$ graph(z_{ij}) ($j = 1, \dots, m_i$) and the bands $B_{ij} :=$ band$(z_{ij}, z_{i,j+1})$ ($j = 0, \dots, m_i$) between them:

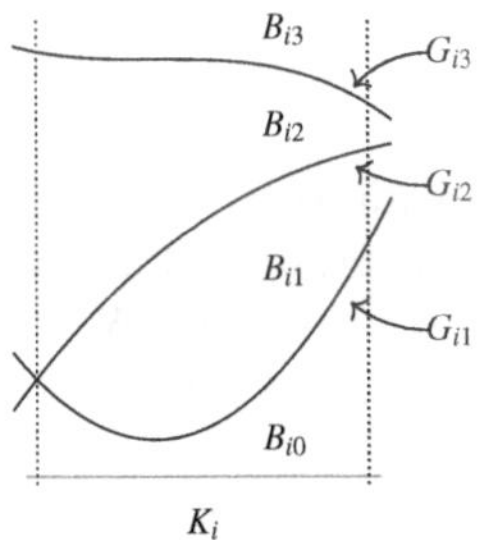

This explains the name CAD. Every semialgebraic set $M \subseteq R^{n+1}$ that can be described by a Boolean combination of sign conditions on the given polynomials $f_1, \dots, f_r$, is the union of some of the graphs G_{ij} and some of the bands B_{ij}, by property (3). In this way, the set M has been decomposed into simpler pieces $G_{ij} \approx K_i$

and $B_{ij} \approx K_i \times]0,1[\approx K_i \times R$ (Lemma 4.3.8). In a next step one can proceed in the same way with the sets $K_i \subseteq R^n$, using a projection $R^n \to R^{n-1}$, and then iterate. In this way we see:

4.3.11 Corollary. *If V is an R-variety, every semialgebraic set $M \subseteq V(R)$ can be expressed as a finite disjoint union $M = M_1 \dot\cup \cdots \dot\cup M_r$ of semialgebraic cells $M_i \approx R^{n_i}$ (with suitable $n_i \ge 0$).*

Proof. Indeed, for V affine this is explained in the previous remark. The general case ensues immediately. □

If $M \subseteq R^n$ is semialgebraic then $n_i \le n$ holds for all i in 4.3.11. A refined analysis of dimensions will be made shortly (Section 4.6).

4.3.12 We now give the proof of Theorem 4.3.9. To the given list $f_1, \dots, f_r$ of polynomials we may add an arbitrary finite number of further polynomials. If the theorem has been proved for the extended list, simply drop all those z_{ij} for which none of $f_1, \dots, f_r$ vanishes on graph(z_{ij}) without vanishing identically on $K_i \times R$. By the intermediate value theorem 1.3.3, property (3) will remain true for $f_1, \dots, f_r$ and the remaining z_{ij}. In this way we can assume that the set $\{f_1, \dots, f_r\}$ is stable under forming the partial derivative $\partial/\partial t$. (The proof will show that it suffices to add $\partial f_1/\partial t, \dots, \partial f_r/\partial t$ to the initial list.)

For $\nu = 1, \dots, r$ and $\xi \in R^n$ put $Z_\nu(\xi) = \{b \in R\colon f_\nu(\xi, b) = 0\}$, and let

$$m(\xi) = \Bigl| \bigcup_{\substack{\nu=1,\dots,r \text{ with} \\ |Z_\nu(\xi)| < \infty}} Z_\nu(\xi) \Bigr|.$$

Then $0 \le m(\xi) \le d := \sum_{\nu=1}^r \deg_t f_\nu(x,t)$ for all ξ. On the other hand, put

$$K(m) = \{\xi \in R^n\colon m(\xi) = m\}$$

for $m = 0, \dots, d$. The sets $K(m)$ are semialgebraic: There exists a formula in the coefficients of a univariate polynomial $p(t)$ of degree $\le d$, that expresses that p has exactly m distinct zeros in R. So $R^n = K(0) \cup \cdots \cup K(d)$ is a (disjoint) decomposition of R^n in semialgebraic sets.

Fix a number $m \in \{0, \dots, d\}$ with $K(m) \ne \emptyset$ and put $K := K(m)$. For $\xi \in K$ let $z_1(\xi) < \cdots < z_m(\xi)$ be the distinct zeros of all those univariate polynomials $f_\nu(\xi, t)$ $(\nu = 1, \dots, r)$ that do not vanish identically. We further write $z_0(\xi) := -\infty$ and $z_{m+1}(\xi) := +\infty$. For $\nu = 1, \dots, r$ and $\xi \in K$ let

$$\varepsilon_j^\nu(\xi) := \operatorname{sign} f_\nu(\xi, b) \ \text{ for } z_j(\xi) < b < z_{j+1}(\xi)$$

$(j = 0, 1, \dots, m)$ and

$$\delta_j^\nu(\xi) := \operatorname{sign} f_\nu(\xi, z_j(\xi))$$

$(j = 1, \dots, m)$. Note that the $\varepsilon_j^\nu(\xi)$ are well-defined by the intermediate value theorem 1.3.3. Let further

$$s^\nu(\xi) := \Big(\varepsilon_0^\nu(\xi),\ \delta_1^\nu(\xi),\ \varepsilon_1^\nu(\xi), \ldots, \delta_m^\nu(\xi),\ \varepsilon_m^\nu(\xi)\Big) \ \in\ \{-1, 0, 1\}^{2m+1}$$

($\nu = 1, \ldots, r$). We call $s^\nu(\xi)$ the *sign pattern* of f_ν at ξ. If sign patterns $s^1, \ldots, s^r \in \{-1, 0, 1\}^{2m+1}$ are given, let

$$K(s^1, \ldots, s^r) := \Big\{\xi \in K\colon s^1(\xi) = s^1,\ \ldots,\ s^r(\xi) = s^r\Big\}.$$

This is the set of points in K at which f_ν has sign pattern s^ν, for $\nu = 1, \ldots, r$. The set $K(s^1, \ldots, s^r)$ is semialgebraic: The property that a polynomial $p(t)$ has prescribed signs both in the m real roots of another polynomial $q(t)$ and in the intervals between them, can be expressed by a formula in the coefficients of p and q. The set K is the disjoint union of all the sets $K(s^1, \ldots, s^r)$. We fix a sign pattern $(s^1, \ldots, s^r)$ and replace K by $K(s^1, \ldots, s^r)$. Then we have:

There exist functions $z_1, \ldots, z_m\colon K \to R$ satisfying $z_1 < \cdots < z_m$ on K, such that the following are true:

(1) *For every $\xi \in K$, the set $\{z_1(\xi), \ldots, z_m(\xi)\}$ consists precisely of all real zeros of those $f_\nu(\xi, t)$ that are not identically zero;*

(2) *each f_ν has constant sign on each of the sets* $\mathrm{graph}(z_j)$ *($j = 1, \ldots, m$) and* $\mathrm{band}(z_j, z_{j+1})$ *($j = 0, \ldots, m$).*

The functions $z_1, \ldots, z_m$ are definable since they can be described by R-formulas. We show that they are continuous, which will complete the proof of Theorem 4.3.9.

For this fix $\xi_0 \in K$ and put $b_j := z_j(\xi_0)$ for $j = 1, \ldots, m$. Let $\delta > 0$ in R with $2\delta < b_{j+1} - b_j$ for $1 \le j < m$, and fix $j \in \{1, \ldots, m\}$. By hypothesis, the family $f_1, \ldots, f_r$ is stable under $\partial/\partial t$. Therefore there exists an index $\nu = \nu(j) \in \{1, \ldots, r\}$ such that $t = b_j$ is a zero of $f_\nu(\xi_0, t)$ of odd multiplicity. For this index ν we have

$$f_\nu(\xi_0,\ b_j - \delta) \cdot f_\nu(\xi_0,\ b_j + \delta)\ <\ 0.$$

Let $U_{\nu j}$ be a neighborhood of ξ_0 in K such that

$$f_\nu(\xi,\ b_j - \delta) \cdot f_\nu(\xi,\ b_j + \delta) < 0$$

for every $\xi \in U_{\nu j}$:

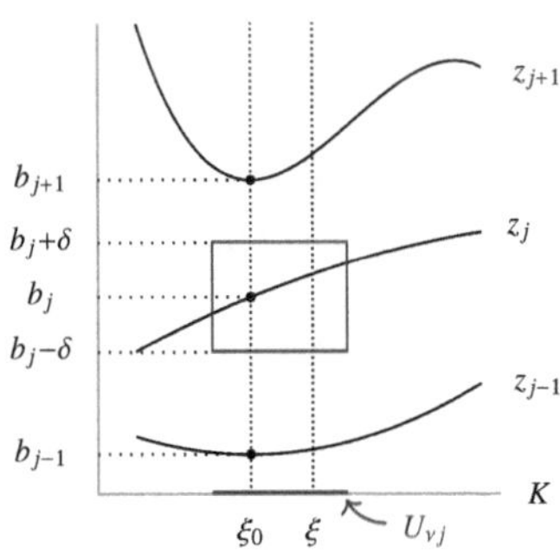

For every $\xi \in U_{\nu j}$, therefore, $f_\nu(\xi, t) \not\equiv 0$, and $f_\nu(\xi, t)$ has a zero between $b_j - \delta$ and $b_j + \delta$ (intermediate value theorem 1.3.3). So there exists an index $k = k_j(\xi)$ with

$$z_k(\xi) \in \;]b_j - \delta, \, b_j + \delta[.$$

Let $U := \bigcap_{j=1}^m U_{\nu(j),j}$, a neighborhood of ξ_0 in K. Since the intervals $]b_j - \delta, \, b_j + \delta[$ $(j = 1, \ldots, m)$ do not overlap, it follows that $k_j(\xi) = j$ for $j = 1, \ldots, m$ and $\xi \in U$. This means $\left|z_j(\xi) - z_j(\xi_0)\right| < \delta$ for all $j = 1, \ldots, m$ and all $\xi \in U$, so the functions z_j are continuous in ξ_0. Since $\xi_0 \in K$ was arbitrary, the z_j are continuous on K. □

4.3.13 Definition. Let $\pi \colon R^{n+1} \to R^n$, $\pi(\xi, t) = \xi$ denote projection to the first n components, and write $x = (x_1, \ldots, x_n)$.

(a) A *cylindrical algebraic decomposition* (with respect to π), or *CAD* for short, is a finite decomposition $R^n = K_1 \cup \cdots \cup K_s$ into pairwise disjoint semialgebraic sets K_i, together with finitely many semialgebraic functions $z_{i1} < \cdots < z_{im_i}$ on K_i for every $i = 1, \ldots, s$ (with $m_i \geq 0$).
(b) Let $f_1, \ldots, f_r \in R[x, t]$. The CAD is said to be *adapted to* $f_1, \ldots, f_r$ if, for every $\nu = 1, \ldots, r$ and $i = 1, \ldots, s$, the polynomial f_ν has constant sign on each of the sets $\operatorname{graph}(z_{ij})$ $(1 \leq j \leq m_i)$ and $\operatorname{band}(z_{ij}, z_{i,j+1})$ $(0 \leq j \leq m_i)$.
(c) Let $M_1, \ldots, M_r \subseteq R^n \times R$ be semialgebraic sets. The CAD is said to be *adapted to* $M_1, \ldots, M_r$ if every M_ν is the union of some of the sets $\operatorname{graph}(z_{ij})$ $(1 \leq j \leq m_i)$ and $\operatorname{band}(z_{ij}, z_{i,j+1})$ $(0 \leq j \leq m_i, \; 1 \leq i \leq s)$.

(Here again we put $z_{i,0} = -\infty$ and $z_{i,m_i+1} = +\infty$ in (b) and (c).)

Theorem 4.3.9 implies:

4.3.14 Corollary. *For any finite number of polynomials in $R[x, t]$, or of semialgebraic sets in R^{n+1}, there exists an adapted CAD.* □

The CAD theorem has the following consequence (see Exercise 4.3.2 for the proof):

4.3.15 Corollary. *Let M be a semialgebraic set on an R-variety V, let $f \colon M \to R^k$ be a definable map. There is a finite covering $M = M_1 \cup \cdots \cup M_s$ of M by semialgebraic sets M_i, such that the restriction $f|_{M_i}$ is continuous for every $i = 1, \ldots, s$.* □

In particular:

4.3.16 Corollary. *Let $I \subseteq R$ be an interval, let $f \colon I \to R^k$ be a definable map. Then f is continuous at all but a finite number of points.* □

4.3.17 Example. We use an explicit example to illustrate the construction in the proof of Theorem 4.3.9. Let

$$f \;=\; f(x, t) \;=\; (t^2 - x)(xt - 1)^2 \;=\; x^2t^4 - 2xt^3 + (1 - x^3)t^2 + 2x^2t - x.$$

Let $f_1 = f$ and put

$$f_2 = \frac{\partial f_1}{\partial t} = 4x^2t^3 - 6xt^2 + 2(1 - x^3)t + 2x^2 = (xt - 1)(4xt^2 - 2t - 2x^2)$$

Fixing $x \in R$, the real zeros of $f_2(x, -)$ are $\frac{1}{x}$ (if $x \neq 0$) and

$$h_1(x) := \frac{1}{4x}(1 - \sqrt{1 + 8x^3}), \quad h_2(x) := \frac{1}{4x}(1 + \sqrt{1 + 8x^3})$$

(if $x \geq -\frac{1}{2}$, $x \neq 0$), together with 0 if $x = 0$. Proceeding as in the proof of Theorem 4.3.9, the x-axis gets subdivided by the points $x = -\frac{1}{2}$, $x = 0$ and $x = 1$. So we get the picture below (with the curves f_1 and f_2 drawn in blue and red, respectively, and the black hyperbola belonging to both):

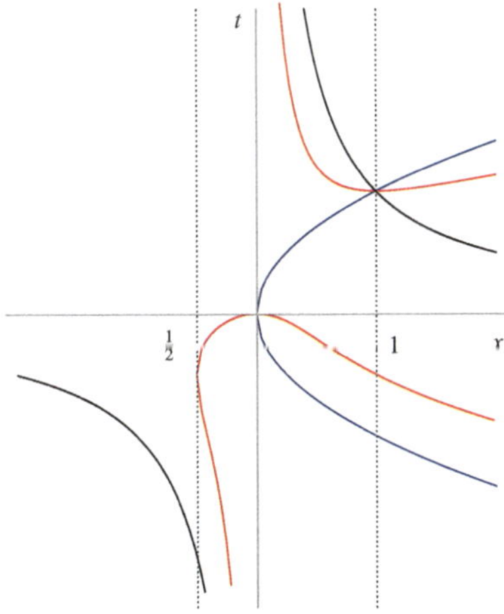

Note that it is necessary to add the partial derivative $f_2 = \partial f / \partial t$, in order to get a CAD that is adapted to f. Otherwise the subdivision constructed in the proof would be too coarse. Note also that some of the functions $z_{ij} \colon K_i \to R$ extend continuously to the closure of K_i, while others do not.

4.3.18 Remarks.

1. Theorem 4.3.9 decomposes the space into open semialgebraic cells, subject to refining a given finite semialgebraic partition. This is still a very coarse first version of CAD, that can be refined in several ways. One problem with this version is that it doesn't give any information on the closures of the cells. This can be fixed by choosing the direction of the projection $R^{n+1} \to R^n$ properly. An important role for such a refinement is played by Thom's lemma below.

2. Implementations of CAD have been written for major software packages, both commercial and free ones.

As a key step for establishing an improved version of CAD, we mention Thom's lemma. Write $\Sigma := \{-1, 0, 1\}$ in the following. For every tuple $e = (e_1, \dots, e_r) \in \Sigma^r$ let

$$\overline{e} = \Big\{\varepsilon \in \Sigma^r : \varepsilon_i \in \{0, e_i\} \text{ for } i = 1, \dots, r\Big\}.$$

Let $f_1, \dots, f_r \in R[t]$ be a finite sequence of univariate polynomials satisfying $f_i' = df_i/dt \in \{f_1, \dots, f_r\} \cup R$ for $i = 1, \dots, r$. For $e \in \Sigma^r$ let

$$S(e) := \Big\{x \in R\colon \operatorname{sign} f_i(x) = e_i \text{ for } i = 1, \dots, r\Big\},$$

and put $S(E) := \bigcup_{e \in E} S(e)$ for every subset $E \subseteq \Sigma^r$. Then the following hold:

4.3.19 Proposition. (Thom's Lemma) *Let $f_1, \dots, f_r \in R[t]$ be as above.*

(a) *If $|S(e)| > 1$ then $S(e)$ is an open interval;*
(b) *if $S(e) \neq \varnothing$ then $\overline{S(e)} = S(\overline{e})$;*
(c) *if $S(e) = \varnothing$ then $|S(\overline{e})| \le 1$.*

If $S(e)$ is non-empty, the remarkable point is that the closure of $S(e)$ is obtained by simply relaxing all defining inequalities. When $S(e)$ is empty this may fail (take $f_1 = t^2$, $f_2 = 2t$ and $e = (-1, 0)$, for example), but according to (c), the relaxed system of inequalities describes at most a singleton set in this case.

4.3.20 Corollary. *Let $f \in R[t]$ be of degree $n \ge 1$. If $\alpha, \beta \in R$ are roots of f such that* $\operatorname{sign} f^{(i)}(\alpha) = \operatorname{sign} f^{(i)}(\beta)$ *for $i = 1, \dots, n-1$, then $\alpha = \beta$.*

Thus a real root α of f may be uniquely encoded by giving the signs of all the iterated derivatives of f at α. The corollary is an immediate consequence of Thom's lemma.

4.3.21 Remark. Thom's lemma in general, and Corollary 4.3.20 in particular, is heavily used in algorithmic real algebraic geometry (see [13] for much more details). For one example, a real algebraic number α can be represented by the string of coefficients of an integer polynomial $f(t)$ with $f(\alpha) = 0$, together with the sequence $\operatorname{sign} f^{(i)}(\alpha)$ $(i = 1, \dots, \deg(f) - 1)$ of signs (this is the so-called *Thom encoding* of α). There exist implementations of an exact arithmetic of real algebraic numbers that are based on this fact.

Proof of Proposition 4.3.19. We'll show parts (a) and (b) by induction on r, and refer to Exercise 4.3.12 for (c).

If $r = 1$ then $\deg(f_1) \le 1$, and the assertion is obvious. Therefore let $r > 1$. We may assume that $\deg(f_r) \ge \deg(f_i)$ for every i, and that $\deg(f_r) \ge 1$. Let $e = (e_1, \dots, e_r) \in \Sigma^r$ be given and put $e' := (e_1, \dots, e_{r-1})$. By the inductive hypothesis applied to $f_1, \dots, f_{r-1}$, the set

$$S(e') := \bigcap_{i=1}^{r-1} \{x \in R\colon \operatorname{sign} f_i(x) = e_i\}$$

is either an open interval or contains at most one point. In addition, the closure of $S(e')$ is $S(\overline{e'})$ if $S(e') \neq \emptyset$.

Case 1: $\left|S(e')\right| \leq 1$. Then clearly $|S(e)| \leq 1$ as well. Therefore, if $S(e)$ is non-empty then $S(e) = S(e')$, and so

$$\overline{S(e)} = S(\overline{e'}) \supseteq S(\overline{e}) \supseteq \overline{S(e)}.$$

Here the first equality holds by the inductive hypothesis, while the last one is trivial. This proves $\overline{S(e)} = S(\overline{e})$.

Case 2: $S(e')$ is a non-empty open-interval. By hypothesis we have $f_r' \in R \cup \{f_1, \ldots, f_{r-1}\}$. Therefore f_r' has constant sign ± 1 on $S(e')$, which means that f_r is strictly monotonic on $S(e')$ (Corollary 1.3.10). Clearly, this implies that

$$S(e) = S(e') \cap \{x\colon\ \text{sign}\, f_r(x) = e_r\}$$

is either an open interval, or $|S(e)| \leq 1$. We may assume that $S(e) \neq \emptyset$. Using the inductive we have to show that

$$\overline{S(e)} = \overline{S(e')} \cap \{x\colon\ \text{sign}\, f_r(x) \in \{0, e_r\}\}.$$

Only the inclusion

$$\overline{S(e)} \supseteq \{x \in \overline{S(e')}\colon\ f_r(x) = 0\}.$$

is not clear a priori. But it follows immediately from the fact that f_r is strictly monotonic on $S(e')$. □

Exercises

Let R always be a real closed field.

4.3.1 Show that each of the following properties of a definable map is preserved under extension of the base field (cf. Remark 4.3.2.3): Being injective, surjective, bijective, continuous, a semialgebraic homeomorphism, an open map, a map that sends closed semialgebraic sets to closed sets.

4.3.2 Give the proof of Corollary 4.3.15.

4.3.3 Prove the claims made in Examples 2 and 3 of 4.3.6.

4.3.4 Let $B = \{\xi \in R^n : |x| \leq 1\}$ be the closed unit ball. Then

$$f\colon B \to \mathbb{P}^n(R), \quad f(\xi) = \Big(1 - \sum_i \xi_i^2 : \xi_1 : \cdots : \xi_n\Big)$$

defines a semialgebraic map. The restriction of f to $\partial B = S^{n-1}$ is a two-to-one covering of the hyperplane $H = \{u\colon u_0 = 0\}$ in $\mathbb{P}^n(R)$, while f restricted to the interior of B is a homeomorphism onto the complement of H in $\mathbb{P}^n(R)$.

4.3.5 Let $S = \{\xi \in R^{n+1} : |\xi| = 1\}$, the n-dimensional sphere over R, and let $\infty = (1, 0, \dots, 0)$ be the north pole.

(a) The stereographic projection $p: S^n \smallsetminus \{\infty\} \to R^n$ is a s.a. homeomorphism.
(b) A subset $M \subseteq R^n$ is unbounded if and only if ∞ lies in the closure of $p^{-1}(M)$.

4.3.6 Let $\pi: V \to \mathbb{A}^n$ be the blowing-up of affine n-space in the origin. Show that $V(R)$ is s.a. homeomorphic to $\mathbb{P}^n(R) \smallsetminus \{(1 : 0 : \cdots : 0)\}$.

4.3.7 Let $f: M \to N$ be a definable map between semialgebraic sets. Show that the points of M where f fails to be continuous form a semialgebraic subset of M.

4.3.8 Let $f: M \to R$ be a definable function where $M \subseteq R^n$ is a semialgebraic set. Show that there are semialgebraic sets $M_1, \dots, M_r$ with $M = \bigcup_{i=1}^r M_i$, together with polynomials $p_1, \dots, p_r$ in $R[x_1, \dots, x_n, t]$ such that, for every index i and every $\xi \in M_i$, the univariate polynomial $p_i(\xi, t)$ is not identically zero and $p_i(\xi, f(\xi)) = 0$.

4.3.9 Construct a CAD for the projection $\pi: R^3 \to R^2$, $\pi(x, y, z) = (x, y)$ that is adapted to $f = x^2 z - y^2$, and use it to find a decomposition of the Whitney umbrella

$$W = \left\{(u, v, w) \in R^3 : v^2 = u^2 w\right\}$$

into finitely many disjoint semialgebraic cells.

4.3.10 Let $f: \,]0,1[\, \to R$ be a definable function. Show that there exists an integer $n \geq 1$ such that $|f(t)| < t^{-n}$ for all sufficiently small $t > 0$. If f doesn't vanish identically on $]0, c[$ for any $0 < c < 1$, there also exists an integer $m \geq 1$ with $|f(t)| > t^m$ for all sufficiently small $t > 0$. (*Hint*: f satisfies an algebraic identity.)

4.3.11 With a refined reasoning, the statement of Exercise 4.3.10 can be made more precise. Let $f: \,]0,1[\, \to R$ be a definable function, and assume that f does not vanish identically on $]0, \delta[$ for any $0 < \delta < 1$. Show that there exist $q \in \mathbb{Q}$ and $0 \neq c \in R$ such that $\lim_{t \to 0} t^{-q} f(t) = c$. Moreover, prove that both q and c are uniquely determined. (*Hint*: Remark 1.4.10)

4.3.12 Prove statement (c) in Thom's lemma 4.3.19.

4.4 Connected Components

R is always a real closed field. Following the convention in 4.3.3, every semialgebraic set is assumed to be a semialgebraic subset of $V(R)$ for some R-variety V.

With its order topology, the field R is totally disconnected unless $R = \mathbb{R}$ (Exercise 1.2.2). Therefore the same is true for every semialgebraic set. Still there exists a reasonable concept of connectedness for semialgebraic sets:

4.4.1 Definition. Let V be an R-variety. A set $M \subseteq V(R)$ is called *semialgebraically (s.a.) connected* if M is a semialgebraic subset of $V(R)$ and the following holds: Whenever M_1, M_2 are two disjoint semialgebraic subsets of M that are (relatively) open in M and satisfy $M = M_1 \cup M_2$, either $M_1 = \varnothing$ or $M_2 = \varnothing$.

This is almost the usual definition of connectedness, except that only decompositions $M = M_1 \uplus M_2$ into *semialgebraic* sets are considered.

4.4.2 Remark. It is immediately clear that s.a. connectedness is invariant under s.a. homeomorphism. It is also true that M is s.a. connected if and only if the topological space $\widetilde{M}$ is connected in the usual sense, Exercise 4.4.2.

Recall (1.6.18) that a subset $K \subseteq R^n$ is R-convex if $u, v \in K$ implies $(1-t)u+tv \in K$ for every $0 \le t \le 1$ in R.

4.4.3 Lemma. *Every semialgebraic and R-convex subset of R^n is s.a. connected. In particular, R^n is s.a. connected.*

Proof. The interval $[0, 1]$ in R is s.a. connected, hence so is every closed line segment $[u, v]$ between two points in R^n. From this, the general case follows immediately. □

4.4.4 Proposition. *Let M, $N \subseteq V(R)$ be s.a. connected sets.*

(a) *The closure $\overline{M}$ is s.a. connected.*
(b) *If $M \cap N \neq \varnothing$ then $M \cup N$ is s.a. connected.*
(c) *If $L \subseteq W(R)$ is s.a. connected, then so is $M \times L \subseteq V(R) \times W(R)$.*
(d) *If $f\colon M \to R^k$ is a semialgebraic map then $f(M)$ is s.a. connected.*

It suffices to remark that the usual proofs carry over to the semialgebraic setting.

4.4.5 Proposition. *Let M be a non-empty semialgebraic set. There exists a decomposition $M = M_1 \uplus \cdots \uplus M_r$ into finitely many semialgebraic sets M_i that are relatively open in M, s.a. connected and pairwise disjoint. The sets $M_1, \ldots, M_r$ are uniquely determined up to permutation, and are called the* semialgebraic connected components *of M.*

Proof. We first prove the existence of such a decomposition. Let $M = N_1 \uplus \cdots \uplus N_s$ be a disjoint decomposition into non-empty and s.a. connected sets N_i. Such a decomposition exists (with the N_i being open semialgebraic cells) by Corollary 4.3.11. By induction on s, we show that such a decomposition can be found for which the N_i are relatively closed in M. The start $s = 1$ is trivial, let $s > 1$. If N_1 is open and closed in M, the inductive hypothesis can be applied to $M \setminus N_1 = N_2 \uplus \cdots \uplus N_s$. Otherwise there is an index $i \in \{2, \ldots, s\}$ with $\overline{N_1} \cap N_i \neq \varnothing$ (if N_1 fails to be relatively closed) or with $N_1 \cap \overline{N_i} \neq \varnothing$ (if N_1 fails to be relatively open). In either case, $N_1 \cup N_i$ is s.a. connected by Proposition 4.4.4(b). Therefore, M is a disjoint union of $s - 1$ s.a. connected sets, and we are done by induction.

To prove uniqueness, let $M = M_1 \cup \cdots \cup M_r = M'_1 \cup \cdots \cup M'_s$ be two decompositions as in the proposition. Each set M_i is contained in M'_j for precisely one index j, and vice versa. Therefore $r = s$ and $M'_i = M_i$ after relabelling. □

4.4.6 Corollary. *Every semialgebraic set M is locally s.a. connected: Every point $\xi \in M$ has arbitrarily small neighborhoods in M that are s.a. connected.*

Proof. Let U be a semialgebraic neighborhood of ξ in M, and let U_0 be the s.a. connected component of U that contains ξ. By Proposition 4.4.5, U_0 is relatively open in U, and so U_0 is a semialgebraic connected neighborhood of ξ in M. □

4.4.7 Remark. The proof of Proposition 4.4.5 shows the following. Let $M = M_1 \cup \cdots \cup M_r$ with s.a. connected sets M_i, not necessarily disjoint. Let $\sim$ be the equivalence relation on the set $\{1, \ldots, r\}$ that is generated by

$$i \sim j \quad \text{if} \quad M_i \cap \overline{M_j} \neq \varnothing.$$

Then the s.a. connected components of M are the unions $M_\alpha = \bigcup_{i \in \alpha} M_i$ where α ranges over the equivalence classes with respect to $\sim$.

4.4.8 Corollary. *($R = \mathbb{R}$) A semialgebraic set M over the real numbers $\mathbb{R}$ is s.a. connected if and only if it is connected in the usual topological sense.*

Proof. Topological connectedness implies s.a. connectedness. Conversely assume that M is s.a. connected. We can write $M = N_1 \cup \cdots \cup N_r$ with semialgebraic sets N_i that are topologically connected (homeomorphic to $\mathbb{R}^{n_i}$). By Remark 4.4.7 we have $i \sim j$ for all $1 \leq i, j \leq r$. Since 4.4.4(b) holds for topological connectedness as well, it follows that M is topologically connected. □

In the next section we will see that every s.a. connected set is semialgebraically path-connected.

Exercises

4.4.1 Let $f\colon M \to N$ be a surjective semialgebraic map between semialgebraic sets. Assume that N is s.a. connected and that $f^{-1}(\eta)$ is s.a. connected for every $\eta \in N$.

(a) If f maps open semialgebraic subsets of M to open subsets of N, then M is s.a. connected. The same is true if one replaces "open" by "closed".
(b) Give an example to show that (a) usually becomes false without further assumptions on f.

4.4.2 A semialgebraic set M is s.a. connected if and only if the topological space $\widetilde{M}$ is connected.

4.5 Semialgebraic Paths

We start by extending the elementary concepts of differentiability to semialgebraic functions over an arbitrary real closed field R.

4.5.1 Let $U \subseteq R^n$ be an open semialgebraic set and let $f = (f_1, \dots, f_p)\colon U \to R^p$ be a definable map. The notion of differentiability carries over from $\mathbb{R}$ to R without change. So f is *differentiable at* $\xi \in U$ if there exists an R-linear map $L\colon R^n \to R^p$ such that

$$\lim_{u \to 0} \frac{1}{|u|} \Big(f(\xi + u) - f(\xi) - L(u) \Big) = 0.$$

If this holds then f is continuous at ξ and the linear map L is uniquely determined. As usual, L is called the *derivative* of f at ξ, denoted $Df(\xi)$. Viewing this linear map as a matrix, we have $Df(\xi) = (\frac{\partial f_i}{\partial x_j}(\xi))$, the $p \times n$ matrix of partial derivatives of the f_i at ξ. As usual, f is called *differentiable* (on U) if f is differentiable at every point $\xi \in U$.

4.5.2 Lemma. *The set* $U' := \{\xi \in U : f \text{ is differentiable in } \xi\}$ *is semialgebraic, and the map* $\xi \mapsto Df(\xi)$, $U' \to \mathrm{Hom}(R^n, R^p) = M_{p\times n}(R)$ *is definable.*

Proof. We may assume $p = 1$. Then U' can be described by the following R-formula $\varphi(x) = \varphi(x_1, \dots, x_n)$:

$$\exists z_1 \cdots \exists z_n \, \forall \varepsilon > 0 \, \exists \delta > 0 \, \forall y_1 \cdots \forall y_n \left(|y| < \delta \to \Big| f(x+y) - f(x) - \sum_i y_i z_i \Big| \le \varepsilon |y| \right).$$

Similarly, graph$(\partial f / \partial x_j)$ is characterized by the following R-formula $\psi(x, t)$:

$$\forall \varepsilon > 0 \, \exists \delta > 0 \, \forall s \left(|s| < \delta \;\to\; \big| f(x + s e_j) - f(x) - st \big| \le \varepsilon |s| \right).$$

(You may have observed that we have slightly cheated here, since in general these are not formulas unless f is a polynomial. Rather, one has to express the relation $f(x) = u$ by using a formula $\omega(x, u)$ that describes the graph of f. Also, one has to intersect the descriptions above with the condition $x \in U$.) □

4.5.3 Remarks.

1. k-fold differentiability is defined inductively for $k \ge 1$, as well as the higher derivatives $D^k f = D(D^{k-1} f)$. If f is k-times continuously differentiable, f is called a *semialgebraic* C^k*-function.*

2. All formal rules for derivatives (linearity, product rule, chain rule) hold in the semialgebraic context as well. *However there is no general semialgebraic analogue of integration*! Already the function $\int_1^x \frac{dt}{t} = \log(x)$ over $R = \mathbb{R}$ fails to be semialgebraic.

3. A semialgebraic map $f : U \to R^p$ is said to be *Nash* (or a *Nash map*, or a *semialgebraic* C^∞*-map*) if f is C^k for every $k \ge 1$. So f is Nash if and only if all iterated partial derivatives exist on U. A Nash map $f : U \to R$ is called a *Nash function* on U.

4.5.4 Proposition. (Inverse functions) *Let* $U \subseteq R^n$ *be an open semialgebraic set and let* $f : U \to R^n$ *be a semialgebraic* C^k*-map (with* $1 \le k \le \infty$*). Let* $\xi \in U$ *be such that* $\det Df(\xi) \ne 0$. *Then there exists an open semialgebraic neighborhood* $V \subseteq U$ *of* ξ *such that* $f(V)$ *is open, the restriction* $f|_V : V \to R^n$ *of* f *is injective and the inverse map* $f(V) \to V$ *of* f *is again a semialgebraic* C^k*-map.*

Proof. For $R = \mathbb{R}$ this is known from first year calculus. In the semialgebraic setting, the result can be proved over arbitrary R in the same way as over $\mathbb{R}$, see for example [26], Section 2.9. To save time and space we use Tarski instead, to derive the result over R from the case $R = \mathbb{R}$. Fixing f, the graph of f has a semialgebraic description by a certain number of polynomial inequalities of certain degrees. Replace each coefficient of each of these polynomials by a separate variable. Then there exists a formula that expresses that the set so described is the graph of a C^k-map f (for fixed $k < \infty$), that $\det Df(\xi) \ne 0$, and that the assertion of the proposition holds. The $\mathbb{Z}$-sentence constructed in this way holds over $\mathbb{R}$, and therefore it holds over every real closed field R. The case $k = \infty$ follows from the cases $k < \infty$. □

4.5.5 Proposition. (Implicit functions) *Let $W \subseteq R^n \times R^p$ be an open semialgebraic set and let $f\colon W \to R^p$ be a semialgebraic C^k-map ($1 \le k \le \infty$). Let $(\xi_0, \eta_0) \in W$ with $f(\xi_0, \eta_0) = 0$, and assume that the matrix*

$$(D_y f)(\xi_0, \eta_0) = \left(\frac{\partial f_i}{\partial y_j}(\xi_0, \eta_0)\right)_{1 \le i, j \le p}$$

is invertible. Then there are open semialgebraic neighborhoods $U \subseteq R^n$ of ξ_0 and $V \subseteq R^p$ of η_0 satisfying $U \times V \subseteq W$, together with a semialgebraic C^k-map $g\colon U \to V$, such that for all $(\xi, \eta) \in U \times V$ one has

$$\eta = g(\xi) \quad \Leftrightarrow \quad f(\xi, \eta) = 0.$$

Moreover, g is uniquely determined by this condition, locally around ξ_0.

Proof. Same as in calculus: Apply Proposition 4.5.4 to the semialgebraic C^k-map $F\colon W \to R^n \times R^p$, $F(\xi, \eta) = (\xi, f(\xi, \eta))$. Since $F(\xi_0, \eta_0) = (\xi_0, 0)$ and

$$DF = \begin{pmatrix} I & 0 \\ D_x f & D_y f \end{pmatrix}$$

is invertible in (ξ_0, η_0), there exist neighborhoods U of ξ_0 and V of η_0 such that $F|_{U \times V}$ is a semialgebraic open embedding with C^k-inverse map $G\colon F(U \times V) \to U \times V$. Define $g\colon U \to V$ by $(\xi, g(\xi)) = G(\xi, 0)$ ($\xi \in U$). □

4.5.6 Example. Let $p \in R[x_1, \dots, x_n, t]$, let $\xi_0 \in R^n$, and let $a \in R$ be a simple root of the polynomial $p(\xi_0, t) \in R[t]$. Then there are an open semialgebraic neighborhood $U \subseteq R^n$ of ξ_0 and a Nash function $f\colon U \to R$, unique locally around ξ_0, such that $f(\xi_0) = a$ and $p(\xi, f(\xi)) = 0$ for all $\xi \in U$:

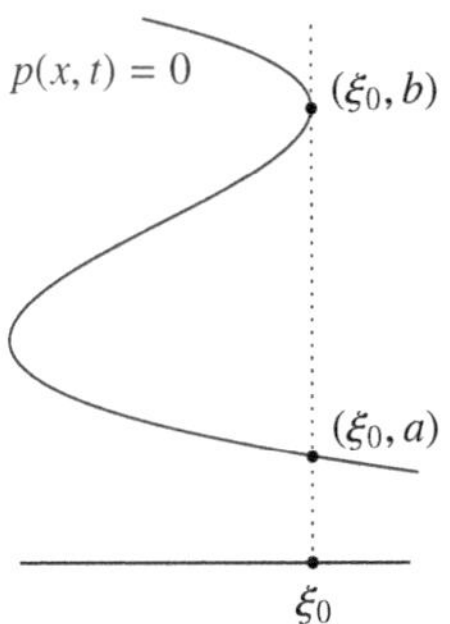

A simple root a and a non-simple root b of $p(\xi_0, t)$

Indeed, $\frac{\partial p}{\partial t}(\xi_0, a) \ne 0$ since a is a simple root, so the assertion follows from Proposition 4.5.5.

4.5.7 Theorem. *Let $U \subseteq R^n$ be an open semialgebraic set and let $f: U \to R^p$ be a definable map. Then there is an open dense semialgebraic subset V of U such that $f|_V$ is Nash.*

Proof. Let $x = (x_1, \dots, x_n)$. We may assume $p = 1$. The graph of f can be written as a union $\operatorname{graph}(f) = G_1 \cup \cdots \cup G_r$, where $G_i = \mathcal{Z}(p_i) \cap \mathcal{U}(q_{i1}, \dots, q_{is_i})$ with polynomials p_i, $q_{ij} \in R[x, t]$. Let $\pi: R^n \times R \to R^n$ be projection to the first n components. Since U is the union of the semialgebraic sets $\operatorname{pr}_1(G_i)$, the union of their interiors is open and dense in U (Exercise 4.5.3). So we may assume $r = 1$, which means

$$\operatorname{graph}(f) = \{(\xi, t): p(\xi, t) = 0, \ q_1(\xi, t) > 0, \dots, q_s(\xi, t) > 0\}$$

with p, $q_1, \dots, q_s \in R[x, t]$. We can also assume that $p(x, t)$ does not have multiple factors. Therefore the discriminant $\Delta(x) \in R[x]$ of $p(x, t)$ with respect to t is a non-zero polynomial, and the set $V := \{\xi \in U: \Delta(\xi) \neq 0\}$ is open and dense in U. Let $d = \deg_t(p)$ and put

$$V_r := \{\xi \in V: p(\xi, t) \text{ has exactly } r \text{ zeros in } R\}$$

for $1 \leq r \leq d$. Then $V = \bigcup_{r=1}^d V_r$. Again $\bigcup_{r=1}^d \operatorname{int}(V_r)$ is open dense in V, hence in U, and we may replace U by $\operatorname{int}(V_r)$ for some fixed $r \geq 1$. This means:

For every $\xi \in U$, the polynomial $p(\xi, t)$ has exactly r real zeros, and they are all simple roots.

Let $z_1(\xi) < \cdots < z_r(\xi)$ be the real zeros of $p(\xi, t)$, for $\xi \in U$. Then $z_1, \dots, z_r$ are Nash functions on U, by the implicit function theorem (Example 4.5.6). Finally, let $U_i := \{\xi \in U: f(\xi) = z_i(\xi)\}$ for $i = 1, \dots, r$. Then $U = \bigcup_{i=1} U_i$, the set $U' := \bigcup_{i=1}^r \operatorname{int}(U_i)$ is open dense in U, and $f|_{U'}$ is Nash. □

4.5.8 Remarks.

1. Let $F \in R[x_1, \dots, x_n, t]$. If $U \subseteq R^n$ is an open semialgebraic set and $f: U \to R$ is a continuous map such that $f(\xi)$ is a simple root of $F(\xi, t)$ for every $\xi \in U$, then f is a Nash function (Example 4.5.6).

2. The converse is usually false: Given a Nash function $f: U \to R$, there need not exist a polynomial $p(x, t)$ with $p(\xi, f(\xi)) = 0$ for all $\xi \in U$, and such that $t = f(\xi)$ is a simple root of $p(\xi, t)$ for every $\xi \in U$. An example is given by the Nash function $f:]-1, 1[\to R$, $f(x) = x\sqrt{1 + x}$ (non-negative square root):

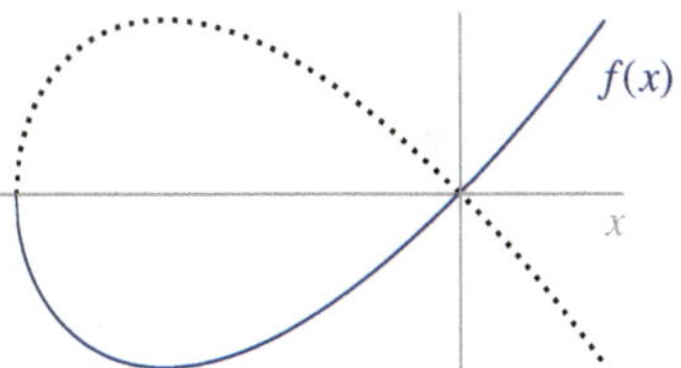

Graph of the Nash function $f(x)$
and its Zariski closure

The minimal polynomial that vanishes on graph(f) is $q(x,t) = t^2 - x^2 - x^3$, and it has a singularity at the origin. Accordingly, $t = 0$ is a two-fold root (at least) of $p(0,t)$, for any polynomial $p(x,t)$ that vanishes on the graph of f.

3. In the exercises it will be shown that every Nash function is locally determined by its Taylor series (see Exercise 4.5.14). When $R = \mathbb{R}$, every Nash function is in fact real analytic, not just C^∞. More precisely, a function $f\colon U \to \mathbb{R}$ defined on an open semialgebraic set $U \subseteq \mathbb{R}^n$ is a semialgebraic C^∞-function if, and only if, f is definable and analytic. The proof can be found in [26] Proposition 8.1.8.

4.5.9 Proposition. (Existence of definable sections) *Let $f\colon M \to N$ be a definable and surjective map between semialgebraic sets. Then f has a definable section: There exists a definable map $s\colon N \to M$ such that $f(s(\eta)) = \eta$ for every $\eta \in N$.*

Proof. We may assume that $M \subseteq R^m$ and $N \subseteq R^n$ are semialgebraic sets. The map f is the composition $M \xrightarrow{\gamma} \text{graph}(f) \xrightarrow{\pi} N$ of definable maps, where $\gamma(\xi) = (\xi, f(\xi))$ and $\pi(\xi,\eta) = \eta$. Since γ is bijective, it suffices to find a definable section for π. So we may assume that $M \subseteq R^m \times R^n$ is a semialgebraic set, and f is the second projection $f(\xi,\eta) = \eta$. By an induction argument one reduces to the case $m = 1$, and this case is settled in Exercise 4.5.1. □

Proposition 4.5.9 is remarkable in that it has no counterpart in classical topology or analysis. The proof itself is very easy.

To prepare for the discussion of semialgebraic paths, we start with a few simple lemmas:

4.5.10 Lemma. *Let $f\colon I \to R$ be a semialgebraic (continuous) function defined on an interval $I \subseteq R$. If f is injective then f is strictly monotonic.*

The proof is obvious using Proposition 4.4.4(d).

4.5.11 Lemma. *Let $f\colon\ ]0,1[\ \to R$ be a definable function. For some $a \in R$ with $0 < a \leq 1$, the restriction of f to $]0,a[$ is continuous, and either strictly monotonic or constant.*

Proof. There is $0 < b \leq 1$ such that f is C^1 on $]0,b[$ (Theorem 4.5.7). If $f' \equiv 0$ on $]0,a[$ for some $0 < a \leq b$ then f is constant on $]0,a[$. Otherwise there is $0 < a \leq b$ such that either $f' > 0$ or $f' < 0$ on $]0,a[$. Accordingly, f is strictly increasing or strictly decreasing on $]0,a[$, respectively. □

4.5.12 Corollary. *Let $f\colon\ ]0,1[\ \to R^n$ be a bounded definable function. Then the limit $\lim_{t\to 0} f(t)$ exists in R^n.*

Proof. We may assume $n = 1$. Then the assertion is immediate from Lemma 4.5.11. For example, if f is strictly increasing on $]0,a[$ (where $0 < a \leq 1$), then $\lim_{t\to 0} f(t) = \inf f(]0,a[)$. The infimum exists in R since $f(]0,a[)$ is a union of finitely many bounded intervals in R. □

4.5.13 Definition. Let M be a semialgebraic set. A *semialgebraic path* in M is a (continuous) semialgebraic map $\alpha\colon I \to M$ where I is a non-degenerate interval in R. Frequently we'll write α_t, rather than $\alpha(t)$, for $t \in I$.

4.5.14 Lemma. *Let $f\colon M \to N$ be a definable map between semialgebraic sets that is surjective, and let $\beta\colon\ ]0,1[\ \to N$ be a semialgebraic path in N. Then, for some $0 < c \leq 1$, there exists a semialgebraic path $\alpha\colon\ ]0,c[\ \to M$ such that $\beta_t = f(\alpha_t)$ for $0 < t < c$.*

Speaking informally, the lemma says that "open" semialgebraic path germs can be lifted under definable maps.

Proof. There exists a definable map $\gamma\colon\ ]0,1[\ \to M$ with $f \circ \gamma = \beta$ (Proposition 4.5.9). Then there is $0 < c \leq 1$ such that γ is continuous on $]0,c[$ (Corollary 4.3.16). □

4.5.15 Proposition. (Curve selection lemma) *Let $M \subseteq R^n$ be a semialgebraic set and let $\xi \in \overline{M}$. Then there is a semialgebraic path $\alpha\colon\ [0,1[\ \to \overline{M}$ such that $\alpha_t \in M$ for $0 < t < 1$ and $\alpha_0 = \xi$.*

Proof. We may assume $\xi \notin M$, otherwise we can take the constant path ξ. For $\eta \in R^n$ let $d_\xi(\eta) := |\eta - \xi|$. The set $d_\xi(M) \subseteq\ ]0,\infty[$ is semialgebraic and contains $]0,a[$ for some $a > 0$, since $\xi \in \overline{M}$. By Lemma 4.5.14, applied to $d_\xi\colon M \to R_+$, there are $0 < c \leq a$ and a semialgebraic path $\beta\colon\ ]0,c[\ \to M$ such that $d_\xi(\beta_t) = t$ for all $0 < t < c$. Let $\alpha\colon\ [0,1[\ \to \overline{M}$ be defined by $\alpha_t := \beta_{tc}$ for $0 < t < 1$ and by $\alpha_0 := \xi$. Then α is continuous and has the desired properties. □

4.5.16 Corollary. *If $M \subseteq R^n$ is an unbounded semialgebraic set, there exists a semialgebraic path $\alpha\colon\ ]0,1[\ \to M$ satisfying* $\lim_{t\to 0} |\alpha_t| = \infty$.

Proof. See Exercise 4.5.4. □

The curve selection lemma is a very useful tool. In the context of semialgebraic maps over arbitrary real closed fields, it can replace the traditional use of convergent sequences in analysis or topology. In the following we present a selection of various applications, starting with a path criterion for continuous maps:

4.5.17 Proposition. *A definable map $f\colon M \to N$ between semialgebraic sets is continuous if, and only if, $f(\alpha_0) = \lim_{t\to 0,\, t>0} f(\alpha_t)$ holds for every semialgebraic path $\alpha\colon\ [0,1[\ \to M$.*

Proof. We only need to prove the "if" part. Let $K \subseteq N$ be a semialgebraic subset that is (relatively) closed in N, and let $\xi \in \overline{f^{-1}(K)}$, we need to show $\xi \in f^{-1}(K)$. By curve selection 4.5.15 there is a semialgebraic path $\alpha\colon\ [0,1[\ \to M$ that satisfies $\alpha_0 = \xi$ and $\alpha_t \in f^{-1}(K)$ for $0 < t \leq 1$. By hypothesis, $f(\xi)$ is the limit of $f(\alpha_t)$ for $t \to 0$, $t > 0$, and so $f(\xi) \in K$ since K is closed in N. □

4.5.18 In general topology, the notion of path-connected spaces is stronger than that of connected spaces. For semialgebraic sets we show that both agree. A semialgebraic set M is said to be *semialgebraically (s.a.) path connected* if, for any two points $\xi, \eta \in M$, there exists a semialgebraic path $\alpha\colon [0,1] \to M$ with $\alpha_0 = \xi$ and $\alpha_1 = \eta$. Such α will be called a *semialgebraic path from ξ to η*.

4.5.19 Corollary. *A semialgebraic set is s.a. connected if and only if it is s.a. path connected.*

Proof. Let M be s.a. connected, write M as a union $M = M_1 \cup \cdots \cup M_r$ of semialgebraic subsets that are s.a. path connected (for example open semialgebraic cells, 4.3.11). Curve selection 4.5.15 implies that the relative closure $\overline{M_i}$ of M_i in M is s.a. path connected for each i. So we may assume that the M_i are relatively closed in M. If $\sim$ denotes the equivalence relation on $\{1, \dots, r\}$ generated by $i \sim j$ if $M_i \cap M_j \neq \varnothing$, we have $i \sim j$ for all i, j since M is s.a. connected (Remark 4.4.7). Being connectible by a s.a. path in M is an equivalence relation on M. Therefore M is s.a. path connected. □

If the real closed field R is different from $\mathbb{R}$, a semialgebraic set will never be locally compact, except when it is finite (Exercise 1.2.2). Still there is a reasonable semialgebraic analogue of the notion of compactness:

4.5.20 Definition. A semialgebraic set M is *semialgebraically (s.a.) compact* if, for any semialgebraic path $\alpha\colon\ ]0,1[\ \to M$, the limit $\lim_{t\to 0} \alpha_t$ exists in M.

The definition can be regarded as analogous to the notion of sequential compactness from general topology. The next few results show that it is the correct generalization of usual compactness to the semialgebraic context:

4.5.21 Proposition. *A semialgebraic subset M of R^n is s.a. compact if, and only if, M is closed and bounded in R^n.*

Proof. If M is s.a. compact then M is closed in R^n by curve selection 4.5.15, and M is bounded by Corollary 4.5.16. Conversely, if M is closed and bounded then M is s.a. compact by Corollary 4.5.12. □

In particular, over $R = \mathbb{R}$, a semialgebraic set is s.a. compact if and only if it is compact.

4.5.22 Proposition. *Let $f\colon M \to N$ be a semialgebraic map between semialgebraic sets, and assume that M is s.a. compact. Then $f(M)$ is s.a. compact as well. In particular, every semialgebraic function defined on M takes its minimum and its maximum on M.*

Proof. Let $\beta\colon\ ]0,1[\ \to f(M)$ be a semialgebraic path. By Lemma 4.5.14 there is $0 < a \le 1$ together with a semialgebraic path $\alpha\colon\ ]0,a[\ \to M$ such that $\beta_t = f(\alpha_t)$ for $0 < t \le a$. The limit $\xi := \lim_{t\to 0} \alpha_t$ exists in M since M is s.a. compact, and so $\lim_{t\to 0} \beta_t = f(\xi)$ exists in $f(M)$. □

For a characterization of semialgebraic compactness via the real spectrum, see Exercise 4.5.12. Further applications of the curve selection lemma are contained in some of the other exercises.

We end this section with an improved version of the Łojasiewicz inequality. It refines and generalizes Theorem 3.3.14 at the same time.

4.5.23 Proposition. *Let $M \subseteq R^n$ be a semialgebraic set that is locally closed in R^n, and let $g\colon M \to R$ be a semialgebraic function. If $f\colon M \setminus \mathcal{Z}(g) \to R$ is another semialgebraic function, there exists an integer $m \geq 1$ such that the function $h_m\colon M \to R$,*

$$h_m(\xi) = \begin{cases} g(\xi)^m f(\xi) & \text{if } g(\xi) \neq 0, \\ 0 & \text{if } g(\xi) = 0, \end{cases}$$

is continuous (hence semialgebraic).

Proof. Write $U = M \setminus \mathcal{Z}(g)$. Note that each function h_m is definable. We first show that, for any given $\xi \in \mathcal{Z}(g)$, there exists $m \geq 1$ such that h_m is continuous in ξ. For this we may assume $\xi \in \overline{U}$. It is easy to see that ξ has a neighborhood K in M that is s.a. compact (Exercise 4.5.10). For any $t > 0$ in R, the subset $K_t = \{\eta \in K\colon |g(\eta)| = t\}$ of $K \cap U$ is semialgebraic and s.a. compact, and $K_t \neq \varnothing$ for small t, say for $0 < t < c$. Let the function $\varphi\colon\ ]0, c[\ \to R$ be defined by

$$\varphi(t) := \sup\{|f(\eta)|\colon \eta \in K_t\},$$

and note that the supremum is finite by Corollary 4.5.22. Since the function φ is definable, there exists an integer $m \geq 1$ such that $\varphi(t) \leq t^{-m}$ for $0 < t < \varepsilon$ and some $\varepsilon \in R$, $0 < \varepsilon \leq c$ (Exercise 4.3.10). It follows that the function h_{m+1} is continuous in ξ: If $\eta \in K$ is such that $t := |g(\eta)|$ satisfies $0 < t < \varepsilon$, we have $|h_{m+1}(\eta)| = t^{m+1} \cdot |f(\eta)| \leq t^{m+1} \cdot \varphi(t) \leq t = |g(\eta)|$.

Given any integer $m \geq 1$, write $M_m = \{\xi \in M\colon h_m \text{ is continuous in } \xi\}$. We have $U \subseteq M_1 \subseteq M_2 \subseteq \cdots$, and $M = \bigcup_{m\geq 0} M_m$ holds by the first part of the proof. In fact, the same is true after arbitrary extension of the real closed base field, i.e., $M_S = \bigcup_{m\geq 0}(M_m)_S$ for every real closed field $S \supseteq R$. Therefore it follows from Exercise 4.1.6 that there exists $m \geq 1$ with $M_m = M$. This proves the proposition. □

We get the following consequence. Together with the next corollary, it is a stronger version of Proposition 3.3.14:

4.5.24 Theorem. (Łojasiewicz inequality, second version) *Let $M \subseteq R^n$ be a semialgebraic set that is locally closed. If f, g are semialgebraic functions defined on M that satisfy $\mathcal{Z}(f) \subseteq \mathcal{Z}(g)$, there exists $m \geq 1$ together with a semialgebraic function $h\colon M \to R$ such that $g^m = fh$.*

Proof. By Proposition 4.5.23, applied to $g\colon M \to R$ and $1/f\colon M \setminus \mathcal{Z}(g) \to R$, there is $m \geq 1$ such that $g^m/f\colon M \setminus \mathcal{Z}(g) \to R$ can be extended to a semialgebraic function h on all of M. □

4.5.25 Corollary. *Let $M \subseteq R^n$ be a closed semialgebraic set. If $f, g\colon M \to R$ are semialgebraic functions with $\mathcal{Z}(f) \subseteq \mathcal{Z}(g)$, there are $c > 0$ in R and integers $m, p \geq 1$ such that*

$$|g(\xi)|^m \leq c \cdot |f(\xi)| \cdot (1 + |\xi|)^p$$

holds for all $\xi \in M$. If in addition M is bounded, one can get $p = 0$.

Proof. Follows immediately from Theorem 4.5.24 combined with Exercise 4.5.13. □

Exercises

Let R always be a real closed field.

4.5.1 Let M be a semialgebraic subset of $R^{n+1} = R^n \times R$, let $\pi\colon R^{n+1} \to R^n$, $\pi(x,t) = x$ (for $x \in R^n$, $t \in R$) be the projection. Complete the proof of Proposition 4.5.9 by showing: There exists a definable map $s\colon \pi(M) \to R$ such that $(\eta, s(\eta)) \in M$ for every $\eta \in \pi(M)$. (*Hint*: This can be proved using the CAD theorem 4.3.9. For an even easier proof define, for every semialgebraic set $J \neq \varnothing$ in R, a "canonical point" $\theta_J \in J$ in a suitable way.)

4.5.2 Let M be a semialgebraic set and let $U \subseteq M$ be a semialgebraic subset. State and prove a path criterion for U to be relatively open in M.

4.5.3 Let V be a semialgebraic set.

(a) If $\alpha \in \widetilde{M}_{\min}$, and if $N \subseteq M$ is a semialgebraic subset with $\alpha \in \widetilde{N}$, show that there is a semialgebraic subset U of N, relatively open in M, with $\alpha \in \widetilde{U}$.
(b) If $M_1, \dots, M_r$ are semialgebraic subsets of M whose union is dense in M, conclude that $\mathrm{int}(M_1) \cup \cdots \cup \mathrm{int}(M_r)$ is dense in M as well. (Here $\mathrm{int}(M_i)$ denotes the relative interior of M_i with respect to M.)
(c) Show that, for any definable map $f\colon M \to N$ into a semialgebraic set N, there exists a dense and relatively open semialgebraic set $M' \subseteq M$ for which $f|_{M'}$ is continuous.

4.5.4 Prove Corollary 4.5.16.

4.5.5 "The complex roots of a non-zero univariate polynomial f depend continuously on the coefficients of f". Use the curve selection lemma to prove the following precise version of this statement. Let $C = R(\sqrt{-1})$, the algebraic closure of R. Fix $d \geq 1$ and let $[0,1] \to C[x]_{\leq d}$, $t \mapsto f_t$ be a (continuous) semialgebraic path in the R-vector space $C[x]_{\leq d}$, satisfying $\deg(f_t) = d$ for $0 < t \leq 1$ and $f_0 \neq 0$. If $m = \deg(f_0)$, show that the complex roots of f_t can be labelled in such a way that the following holds, for some $0 < c \leq 1$ and all $0 < t < c$:

(1) The j-th root of f_t depends continuously and semialgebraically on t ($j = 1, \dots, d$);
(2) for $t \to 0$, the first m roots of f_t converge against the roots of f_0, whereas the last $d - m$ roots diverge to ∞.

4.5.6 As an application of Exercise 4.5.5, do the following: If $1 \leq m \leq d$ and $U \subseteq C$ is an open subset, show that the set of all polynomials $f \neq 0$ in $C[x]_{\leq d}$ that have at least m roots in U (counting with multiplicities) is open in $C[x]_{\leq d}$.

4.5.7 Another application of Exercise 4.5.5: Let $d \geq 1$, let W_d be the set of all univariate polynomials $f \in R[x]_{\leq d}$ all of whose real roots are simple.

(a) Show that W_d is a semialgebraic subset of $R[x]_{\leq d}$, and decide whether W_d is open or closed in $R[x]_{\leq d}$.
(b) For $m \geq 0$ let $W_{d,m} = \{f \in W_d : f \text{ has precisely } m \text{ real roots}\}$. Show that $W_{d,m}$ is a semialgebraic set, and decide whether $W_{d,m}$ is open and/or closed in W_d. Same question for $W_{d,\geq m} := \bigcup_{k \geq m} W_{d,m}$.

4.5.8 Let V be a projective R-variety. Show that the semialgebraic set $V(R)$ is s.a. compact.

4.5.9 For $m \geq 0$ consider the rational function $f_m(x,y) = \frac{x^m}{x^2+y^2}$ defined on $R^2 \setminus \{(0,0)\}$. For which values of m does f_m extend to a continuous (semialgebraic) function $g_m : R^2 \to R$? For which values of k is g_m a semialgebraic C^k-function?

4.5.10 Let $M \subseteq R^n$ be a semialgebraic set. Show that the following conditions are equivalent:

(i) M is locally closed in R^n (i.e. M is the intersection of an open and a closed semialgebraic set in R^n);
(ii) every $\xi \in M$ has a semialgebraic neighborhood in M that is s.a. compact (and hence ξ has arbitrarily small such neighborhoods);
(iii) M is s.a. homeomorphic to a closed semialgebraic set in some R^m;
(iv) M is s.a. homeomorphic to $K \setminus \{\xi\}$ where K is a s.a. compact set and $\xi \in K$ is a point;
(v) $\widetilde{M}$ is specialization-convex in $\widetilde{R^n}$, i.e., $\alpha \rightsquigarrow \gamma \rightsquigarrow \beta$ in $\widetilde{R^n}$ and $\alpha, \beta \in \widetilde{M}$ imply $\gamma \in \widetilde{M}$.

Hint: (v) implies that $\widetilde{M}$ is stable under generalization inside its closure (why?). This can be used to prove (v) $\Rightarrow$ (i).

4.5.11 Let M be a subset of R^n that is s.a. compact, and let $f: M \to M$ be a definable map that satisfies $|f(\xi) - f(\eta)| < |\xi - \eta|$ for all $\xi \neq \eta$ in M. Show that f has a fixed point.

4.5.12 Let M be a semialgebraic set. Prove the following characterizations of semialgebraic compactness via the real spectrum:

(a) M is s.a. compact if and only if, for every closed point α of $\widetilde{M}$, the extension $R \subseteq R(\alpha)$ of real closed fields is relatively Archimedean.
(b) If $R = \mathbb{R}$, it is also equivalent that every element of $\widetilde{M}$ specializes to a point in $\iota(M)$.

4.5.13 Let $M \subseteq R^n$ be a closed semialgebraic set, let $f: M \to R^m$ be a semialgebraic map. Generalizing the statement in the proof of Proposition 3.3.14, show that there exist $c > 0$ in R and an integer $N \geq 1$ satisfying

$$|f(x)| \leq c \cdot (1 + |x|)^N$$

for all $x \in M$.

4.5.14 Let $U \subseteq R^n$ be a semialgebraic open neighborhood of the origin and let $f: U \to R$ be a Nash function. Assume that the formal Taylor series of f at the origin vanishes identically, i.e. that $\frac{\partial^\alpha f}{\partial x^\alpha}(0) = 0$ for all multi-indices $\alpha \in \mathbb{Z}_+^n$. Prove that $f \equiv 0$ in a neighborhood of 0. Informally speaking, this says that there does not exist any non-zero semialgebraic bump function. (*Hint*: Łojasiewicz inequality)

4.6 Dimension of Semialgebraic Sets

We start by showing that every definable map between semialgebraic sets induces a map between the real spectra in a natural way. Given a semialgebraic set M, recall that we may identify the topological space M with the subset $\iota(M)$ of $\widetilde{M}$.

4.6.1 Proposition. *Let $f: M \to N$ be a definable map between semialgebraic sets. There exists a unique map $\widetilde{f}: \widetilde{M} \to \widetilde{N}$ that extends f and that is continuous in the constructible topologies. For every semialgebraic subset T of N one has $\widetilde{f}^{-1}(\widetilde{T}) = \widetilde{f^{-1}(T)}$.*

Proof. Since M is dense in $\widetilde{M}$ and $\widetilde{N}$ is Hausdorff, both with respect to the constructible topologies, there can be at most one such map $\widetilde{f}$. To construct it we identify $\widetilde{M}$ (with its constructible topology) with St $\mathfrak{S}(M)$, the Stone space of the distributive lattice $\mathfrak{S}(M)$ (Theorem 4.1.19(b)), and similarly for $\widetilde{N}$. Consider the map $\widetilde{f}$: St $\mathfrak{S}(M) \to$ St $\mathfrak{S}(N)$ defined by

$$F \mapsto \widetilde{f}(F) := \{T \in \mathfrak{S}(N) : f^{-1}(T) \in F\}$$

and note that the right-hand set $\widetilde{f}(F)$ is indeed an ultrafilter in $\mathfrak{S}(N)$. For $T \in \mathfrak{S}(N)$ we have

$$\widetilde{f}^{-1}(\widetilde{T}) = \{F \in \mathrm{St}\,\mathfrak{S}(M) : f^{-1}(T) \in F\} = \widetilde{f^{-1}(T)}$$

by construction, which implies that $\widetilde{f}$ is continuous. If $\xi \in M$, the map $\widetilde{f}$ sends the principal ultrafilter $F_\xi \in$ St $\mathfrak{S}(M)$ to $F_{f(\xi)} \in$ St $\mathfrak{S}(N)$, and so $\widetilde{f}$ extends f. □

4.6.2 Example. Let $f: V \to W$ be a morphism of affine R-varieties, and let f also denote the induced map $f: V(R) \to W(R)$. Then $\widetilde{f}: \widetilde{V(R)} \to \widetilde{W(R)}$ is the map φ^*: Sper $R[V] \to$ Sper $R[W]$ induced by the ring homomorphism $\varphi = f^*: R[W] \to R[V]$ associated with f. More generally, if f is a morphism of arbitrary R-varieties, then $\widetilde{f} = f_r: V_r \to W_r$, the map between the real spectra induced by f (see 4.1.15).

4.6.3 Proposition. *Let $f: M \to N$ be a definable map between semialgebraic sets, let $\widetilde{f}: \widetilde{M} \to \widetilde{N}$ be the map between the real spectra constructed in 4.6.1.*

(a) *For every semialgebraic set $S \subseteq M$ we have $\widetilde{f(S)} = \widetilde{f}(\widetilde{S})$.*
(b) *If $g: N \to N'$ is another definable map then $\widetilde{g \circ f} = \widetilde{g} \circ \widetilde{f}$.*
(c) *f is continuous if and only if $\widetilde{f}$ is continuous (with respect to the Harrison topologies).*
(d) *f is injective (or surjective, or bijective) if and only if $\widetilde{f}$ is injective (or surjective, or bijective, respectively).*

Proof. We use the notation introduced in the proof of Proposition 4.6.1. For (a) let $\alpha \in \widetilde{S}$. Then $S \in U_\alpha$ and hence $f^{-1}(f(S)) \in U_\alpha$, which implies $Y_\alpha = \{\widetilde{f}(\alpha)\} \subseteq \widetilde{f(S)}$ by step (1) of the previous proof. For the converse note that $f(S) \subseteq \widetilde{f}(\widetilde{S})$ and that $\widetilde{f}(\widetilde{S})$ is constructibly closed in $\widetilde{N}$. Since $\widetilde{f(S)}$ is the constructible closure of $f(S)$ in $\widetilde{N}$, this implies $\widetilde{f(S)} \subseteq \widetilde{f}(\widetilde{S})$. (b) is clear from the uniqueness part of 4.6.1, since $\widetilde{g} \circ \widetilde{f}$ is continuous in the constructible topologies and extends $g \circ f$. In (c) it is clear that $\widetilde{f}$ continuous implies f continuous. Conversely, if f is continuous and $W \subseteq N$ is an open semialgebraic set, then $\widetilde{f}^{-1}(\widetilde{W}) = \widetilde{f^{-1}(W)}$ is open in $\widetilde{M}$ by the finiteness theorem 4.2.3. This implies that $\widetilde{f}$ is continuous. For the proof of (d) let f be injective and $M \neq \varnothing$. Then there exists a definable map $g: N \to M$ with $g \circ f = \mathrm{id}_M$, and so (b) implies that $\widetilde{f}$ is injective. It is trivial that $\widetilde{f}$ injective implies f injective. The equivalence of f or $\widetilde{f}$ being surjective follows from (a). □

4.6.4 Definition. Let V be an R-variety and let $M \subseteq V(R)$ be a semialgebraic set. We define the *(semialgebraic) dimension* of M to be the (Krull) dimension of the Zariski closure $cl_{zar}(M)$ of M in V:

$$\dim(M) \;:=\; \dim cl_{zar}(M).$$

The *local (semialgebraic) dimension* $\dim_\xi(M)$ of M at a point $\xi \in V(R)$ is defined to be the minimal value of $\dim(M \cap U)$, where U ranges over the open semialgebraic neighborhoods of ξ in $V(R)$.

Since semialgebraic dimension applies to semialgebraic sets, while Krull dimension applies to algebraic varieties, there should be no danger of confusion.

4.6.5 Remarks. Here is a number of immediate observations.

1. Let $M \subseteq V(R)$ be a semialgebraic set and let $\overline{M}$ be its closure in $V(R)$. Then $\dim(\overline{M}) = \dim(M)$.

2. If $M' \subseteq M$ is a semialgebraic subset then $\dim(M') \leq \dim(M)$.

3. If $M_1,\ M_2 \subseteq V(R)$ are semialgebraic sets then

$$\dim(M_1 \cup M_2) \;=\; \max\{\dim(M_1),\ \dim(M_2)\}.$$

4. If $M_i \subseteq V_i(R)$ $(i = 1, 2)$ are non-empty semialgebraic sets then

$$\dim(M_1 \times M_2) \;=\; \dim(M_1) + \dim(M_2).$$

Indeed, the Zariski closure of $M_1 \times M_2$ in $V_1 \times V_2$ satisfies $cl_{zar}(M_1 \times M_2) = cl_{zar}(M_1) \times cl_{zar}(M_2)$ (prove this).

5. $\dim(R^n) \;=\; n$. More generally, if $M \subseteq R^n$ is any semialgebraic set, then $\dim(M) = n$ if and only if the interior of M is non-empty. This follows from Exercise 4.1.1.

6. The function field of an irreducible R-variety V can be ordered if, and only if, $\dim V(R) = \dim(V)$. Indeed, by the definition of $\dim V(R)$, this is an immediate consequence of the Artin–Lang theorem 1.7.9.

From Definition 4.6.4 it is not directly clear that semialgebraic dimension is invariant under semialgebraic homeomorphisms. We use the real spectrum to prove that this is true.

4.6.6 Proposition. $\dim(M) = \dim(\widetilde{M})$ *holds for every semialgebraic set* M.

Here, of course, $\dim(M)$ is the (semialgebraic) dimension of M as defined in 4.6.4, while $\dim(\widetilde{M})$ is the Krull dimension of the spectral space $\widetilde{M}$ (see 3.4.16).

Proof. We can assume that M is a Zariski dense semialgebraic set in the R-variety V. Let $n \;=\; \dim(M) \;=\; \dim(V)$. Every specialization chain in $\widetilde{M}$ has length at most $\dim(\widetilde{M}) \leq \dim(V_r) \leq \dim(V)$ (3.4.17), which shows $\dim(M) \geq \dim(\widetilde{M})$. On the other hand we may write M as a union of finitely many open semialgebraic cells

M_i (Corollary 4.3.11), i.e. semialgebraic sets $M_i \approx R^{n_i}$. In particular, M contains a semialgebraic subset M' for which there is a s.a. homeomorphism $f: R^n \to M'$. Since $\widetilde{f}: \widetilde{R^n} \to \widetilde{M'}$ is a homeomorphism (Proposition 4.6.3), the space $\widetilde{M'}$ contains a specialization chain of length n (Example 3.6.11). Therefore $\dim(\widetilde{M}) \geq n$. □

As a direct consequence of Proposition 4.6.6, we record:

4.6.7 Corollary. *If two semialgebraic sets are s.a. homeomorphic, they have the same semialgebraic dimension.*

Proof. If $f: M \to N$ is a s.a. homeomorphism then $\widetilde{f}: \widetilde{M} \to \widetilde{N}$ is a homeomorphism, by 4.6.3. So the corollary follows from Proposition 4.6.6. □

Surprisingly, semialgebraic dimension is invariant even under definable bijective maps that need not be continuous:

4.6.8 Proposition. *Let $f: M \to N$ be a definable map between semialgebraic sets.*

(a) *If f is injective then* $\dim(M) \leq \dim(N)$.
(b) *If f is surjective then* $\dim(M) \geq \dim(N)$.
(c) *If f is bijective then* $\dim(M) = \dim(N)$.

Proof. (a) Choose a finite covering $M = M_1 \cup \cdots \cup M_r$ by semialgebraic subsets such that $f|_{M_i}$ is continuous for $i = 1, \dots, r$ (Corollary 4.3.15). Choose an index i with $\dim(M_i) = \dim(M) =: d$ (Remark 4.6.5.3). By 4.6.6 there is a specialization chain of length d in $\widetilde{M}_i$. Since $\widetilde{f}|_{\widetilde{M}_i}$ is continuous and injective (using Proposition 4.6.3), there is a specialization chain of length d in $\widetilde{f}(\widetilde{M}_i) \subseteq \widetilde{N}$. Therefore $\dim(N) = \dim(\widetilde{N}) \geq d$.

(b) Proposition 4.5.9 gives a definable section $s: N \to M$ of f, and so (b) follows from (a) applied to s. □

In particular, there exists nothing like a definable space-filling (Peano) curve. We conclude by showing that the boundary of any open semialgebraic set in R^n has local dimension $n - 1$ everywhere.

4.6.9 Theorem. *Let U be an open subset of R^n that is s.a. connected, and let $S \subseteq U$ be a semialgebraic subset with* $\dim(S) \leq n-2$. *Then $U \setminus S$ is again s.a. connected.*

An equivalent formulation is:

4.6.10 Corollary. *Let $U \subseteq R^n$ be an open set that is s.a. connected. If V_1, V_2 are non-empty open semialgebraic subsets of U with $V_1 \cap V_2 = \emptyset$, then* $\dim(U \setminus (V_1 \cup V_2)) \geq n - 1$.

4.6.11 We first prove equivalence between the theorem and the corollary. In the situation of 4.6.10 let $S := U \setminus (V_1 \cup V_2)$. Since $U \setminus S = V_1 \cup V_2$ is s.a. disconnected, Theorem 4.6.9 implies $\dim(S) \geq n - 1$. Conversely let $S \subseteq U$ with $\dim(S) \leq n - 2$. The relative closure $S' := \overline{S} \cap U$ of S has $\dim(S') = \dim(S)$ (Remark 4.6.5.1). Moreover $U \setminus S'$ is dense in U, and hence in $U \setminus S$, since otherwise S would contain a non-empty open ball. It therefore suffices to show that $U \setminus S'$ is s.a. connected.

Assume this is false, then there exist open semialgebraic subsets V_1, $V_2 \neq \emptyset$ of U with $V_1 \cap V_2 = \emptyset$ and $V_1 \cup V_2 = U \setminus S'$. Assuming Corollary 4.6.10 it follows that $\dim(S') \geq n-1$, a contradiction.

Proof of Corollary 4.6.10. We may assume $U \subseteq \overline{V_1} \cup \overline{V_2}$, since otherwise the dimension in question is n. Then $U \cap \overline{V_1} \cap \overline{V_2} \neq \emptyset$ since U is s.a. connected. Fix a point $\xi \in U \cap \overline{V_1} \cap \overline{V_2}$. There is an open ball $B \subseteq U$ with $\xi \in B$, and we may replace U by B and V_i by $B \cap V_i$. The set $A = U \setminus (V_1 \cup V_2)$ is relatively closed in U. Choose points $\eta_i \in V_i$ $(i = 1, 2)$, let $H_1 \subseteq R^n$ be a hyperplane with $\eta_1 \in H_1$ and $\eta_2 \notin H_1$, and let H_2 be the hyperplane through η_2 that is parallel to H_1. Finally let $\pi \colon R^n \setminus H_2 \to H_1$ be the linear projection with centre η_2:

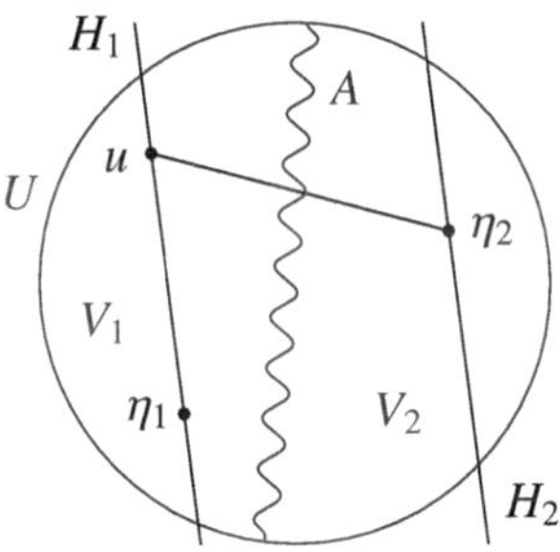

For every point $u \in H_1 \cap V_1$, the line segment $[u, \eta_2]$ is s.a. connected, and so it contains a point in A. Therefore $H_1 \cap V_1 \subseteq \pi(A)$. Since $H_1 \cap V_1$ is a non-empty open subset of H_1 we have $\dim(H_1 \cap V_1) = n-1$. It follows that $\dim(A) \geq \dim \pi(A) \geq \dim(H_1 \cap V_1) = n-1$. □

4.6.12 Corollary. *Let $U \subseteq R^n$ be an open semialgebraic set. Then the boundary $\partial U = \overline{U} \setminus U$ of U has local dimension $n-1$ at each of its points.*

Proof. For $\xi \in \partial U$, the assertion says $\dim(W \cap \partial U) = n-1$ for every sufficiently small neighborhood W of ξ in R^n. The inequality $\leq$ is clear, cf. Remark 4.6.5.5. For the converse we may assume that W is connected (e.g., a ball). Then $W \cap \partial U = W \setminus (W_1 \cup W_2)$ with $W_1 = W \cap U$ and $W_2 = W \cap (R^n \setminus \overline{U})$. Since W_1, W_2 are open, disjoint and non-empty, Corollary 4.6.10 implies $\dim(W \cap \partial U) \geq n-1$. □

4.6.13 Corollary. *Every R-convex semialgebraic set $K \subseteq R^n$ with non-empty interior has purely $(n-1)$-dimensional boundary.*

Proof. This follows from 4.6.12 applied to $U = \operatorname{int}(K)$, since $\overline{K}$ is the closure of U. (Over $R = \mathbb{R}$, this is an elementary fact about convex sets, a proof of which is given in 8.1.4 and Exercise 8.1.3. It extends to R-convex semialgebraic sets over any real closed field R, for example by Tarski's principle.) □

4.6.14 Remark. Let $K \subseteq R^n$, $K \neq R^n$, be a non-empty semialgebraic set which is the closure of its interior. For example, K may be a proper R-convex semialgebraic subset of R^n with non-empty interior. The *algebraic boundary* $\partial_a K := cl_{zar}(\partial K)$ of

K is defined to be the Zariski closure of the ordinary boundary $\partial K \subseteq R^n$. Combined with Exercise 4.6.6, Corollary 4.6.13 shows that $\partial_a K$ is a hypersurface, i.e. it is the zero set of some non-constant polynomial $g \in R[x]$. We may require that g has no multiple factors, then g is unique up to a scalar factor. Every irreducible component of the hypersurface $\partial_a K = \mathcal{V}(g)$ has Zariski dense R-points. Finding this hypersurface explicitly, or just determining its degree, is often an interesting and challenging question. Note also that K is the closure of a union of connected components of $\{\xi \in R^n : g(\xi) \neq 0\}$.

Exercises

Let R always be a real closed field.

4.6.1 Let $N \subseteq M$ be non-empty semialgebraic sets with $\dim(N) = \dim(M)$. Show that the relative interior of N in M is non-empty.

4.6.2 Let $M \neq \varnothing$ be a semialgebraic set and let N be a dense semialgebraic subset of M. Prove that $\dim(M \setminus N) < \dim(M)$. (Use the real spectrum.)

4.6.3 With the notion of semialgebraic dimension available, prove the following geometric version of Exercise 3.5.2: Let polynomials $p, f_1, \dots, f_r \in R[x] = R[x_1, \dots, x_n]$ be given such that p is irreducible and p^m divides $\sum_{i=1}^r f_i$. Assume that there exists an open semialgebraic set $U \subseteq R^n$ with $f_i|_U > 0$ for every i such that $\dim(\overline{U} \cap \mathcal{Z}(p)) = n - 1$. Then p^m divides each f_i. (*Hint*: Show that the field $K = R(x)$ has an ordering P that satisfies the condition in Exercise 3.5.2.)

4.6.4 Let M be a semialgebraic set, let $\widetilde{M}$ be the real spectrum of M. Show that the set $(\widetilde{M})^{\max}$ of closed points of $\widetilde{M}$ is pro-constructible in $\widetilde{M}$ (if and) only if M is a finite set.

4.6.5 Let $f \in R[x] = R[x_1, \dots, x_n]$ be an irreducible polynomial that is indefinite, i.e. takes both positive and negative values on R^n. Prove that the hypersurface $f = 0$ has a real function field (cf. Theorem 1.7.14), as follows:

(a) Reduce to showing that $\mathcal{Z}(f) = \{\xi \in R^n : f(\xi) = 0\}$ has (semialgebraic) dimension $n-1$;
(b) show $\dim \mathcal{Z}(f) = n - 1$ by a topological argument.

4.6.6 Let $M \subseteq R^n$ be a semialgebraic set, let $\xi \in M$, and assume that $\dim_\xi(M) = d$ for every $\xi \in M$. Prove that every irreducible component of the Zariski closure $cl_{zar}(M)$ (of M in $\mathbb{A}^n$) has dimension d.

4.6.7 Let $M \subseteq R^n$ be a semialgebraic set. For any integer $d \geq 0$ let $M_{(d)} := \{\xi \in R^n : \dim_\xi(M) = d\}$ and $M_{(\geq d)} := \bigcup_{e \geq d} M_{(e)}$.

(a) The sets $M_{(\geq d)}$ are closed and semialgebraic in R^n.
(b) The set $M_{(d)}$ is purely d-dimensional, i.e. $\dim_\xi M_{(d)} = d$ holds for every $\xi \in M_{(d)}$.

4.6.8 Let V be an irreducible R-variety of dimension n. For $\xi \in V(R)$, show that the following are equivalent:

(i) ξ lies in the closure of $V_{\mathrm{reg}}(R)$;
(ii) the local dimension of $V(R)$ at ξ is n.

ξ is called a *central point* of V if it satisfies conditions (i) and (ii).

4.6.9 Let $M \subseteq R^n$ be a semialgebraic set, let $\xi \in R^n$ and $d \geq 0$. Show that $\dim_\xi(M) \geq d$ if, and only if, there is a specialization chain $\alpha_d \rightsquigarrow \cdots \rightsquigarrow \alpha_0 = \iota(\xi)$ of length d in $\widetilde{R^n}$ (with proper specializations) for which $\alpha_0, \dots, \alpha_{d-1} \in \widetilde{M}$.

4.6.10 (Marker–Steinhorn Theorem) Let R be a real closed field extension of $\mathbb{R}$ and let $M \subseteq R^n$ be an R-semialgebraic set. Then $M \cap \mathbb{R}^n$ is an $\mathbb{R}$-semialgebraic set. (*Hint*: It suffices to show this for $M = \mathcal{U}(f)$ with $f \in R[x]$. Use the canonical valuation ring of R over $\mathbb{R}$ and argue by induction on n.)

4.6.11 Let V be an $\mathbb{R}$-vector space of finite dimension and let $f\colon V \to R$ be an $\mathbb{R}$-linear map into a real closed overfield R of $\mathbb{R}$. Use the Marker–Steinhorn theorem (Exercise 4.6.10) to show that the subset $H = \{u \in V\colon f(u) \geq 0\}$ of V is semialgebraic. Find examples where H fails to be closed in $\mathbb{R}^n$. Show also that the set H need not be semialgebraic when $\mathbb{R}$ is replaced by a different real closed field.

4.6.12 Let $x = (x_1, \dots, x_n)$, and write $\mathbb{R}[x]_{\leq d} := \{f \in \mathbb{R}[x]\colon \deg(f) \leq d\}$ for $d \geq 0$. Let P be a positive cone of the ring $\mathbb{R}[x]$. Prove for every $d \geq 0$ that $P \cap \mathbb{R}[x]_{\leq d}$ is a semialgebraic subset of $\mathbb{R}[x]_{\leq d}$. Show that this becomes false in general when $\mathbb{R}$ is replaced by a different real closed field. (*Hint*: Exercise 4.6.11)

4.6.13 Let $M \subseteq R^n$ be a semialgebraic set with associated constructible set $\widetilde{M} \subseteq \operatorname{Sper} R[x_1, \dots, x_n]$, and let $\mathcal{A}(M)$ denote the ring of (continuous) semialgebraic functions $M \to R$. If M is locally closed in R^n, show that the real spectrum of $\mathcal{A}(M)$ is naturally homeomorphic to $\widetilde{M}$, as follows:

(a) Every closed constructible subset of $\widetilde{M}$ has the form $\widetilde{\mathcal{Z}(f)}$ for some $f \in \mathcal{A}(M)$.
(b) Prove a similar statement for the closed constructible subsets of $\operatorname{Sper} \mathcal{A}(M)$.
(c) If $f, g \in \mathcal{A}(M)$ satisfy $\mathcal{Z}(f) \subseteq \mathcal{Z}(g)$, and if M is locally closed, show that $Z_{\mathcal{A}(M)}(f) \subseteq Z_{\mathcal{A}(M)}(g)$ (as subsets of $\operatorname{Sper} \mathcal{A}(M)$).
(d) When M is locally closed, conclude that $\operatorname{Sper} \mathcal{A}(M)$ is homeomorphic to $\widetilde{M}$. More precisely, show that the map $\pi_M\colon \operatorname{Sper} \mathcal{A}(M) \to \widetilde{R^n}$ induced by the ring homomorphism $R[x_1, \dots, x_n] \to \mathcal{A}(M)$ induces a homeomorphism $\operatorname{Sper} \mathcal{A}(M) \to \widetilde{M}$.

Hint: Use Theorems 4.1.19 and 4.5.24.

4.7 Notes

A profound systematic study of the geometry of semialgebraic and semianalytic sets was initiated by Łojasiewicz around 1964. The tilda operator was introduced in [46]. Its interpretation in terms of ultrafilters of semialgebraic sets is due to Bröcker [33]. The finiteness theorem figures as "unproved proposition" in [35]. It was given several different proofs later, first in [27]. Theorems 4.2.7 and 4.2.9 (parts (a) and (b)) are due to Bröcker [34]. The precise values of $s(n)$ and of $\bar{s}(n)$ were found by Scheiderer [177]. Easier proofs or more general versions were later given by Mahé [131] and Marshall [134]. Cylindrical algebraic decomposition of semialgebraic sets is originally due to Collins (1975). Collins' algorithm is of fundamental importance in computational real algebraic geometry, and also for applications like robot motion planning, see [13]. Note that our text contains only a very basic version.

A considerable part of the results in this chapter extends to the much more general setting of o-minimal sets (see [56]).

Chapter 5
The Archimedean Property

Let $K \subseteq \mathbb{R}^n$ be a basic closed set, described by finitely many polynomial inequalities $g_1 \geq 0, \ldots, g_r \geq 0$. We consider polynomials $f = f(x)$ that are strictly positive on K. The general Krivine–Stengle positivstellensatz 3.2.7 provides certificates for this positivity in the form of identities involving f. These identities amount to representing f as a weighted sum of squares of *rational* functions, with products of the g_i as weights. Certainly, it would be even more desirable to have *denominator-free* such certificates, representing f as a weighted sum of squares of *polynomials*, rather than of rational functions.

The central notion in this chapter is the Archimedean property, the main result is the Archimedean positivstellensatz 5.3.3. It implies the existence of denominator-free certificates for strictly positive f under very general and weak assumptions. This result has many prominent applications, and some of the most important ones are discussed in Sections 5.4 and 5.5. Generally we are working in a framework of "abstract" rings (rather than rings of polynomials), both for systematic reasons and with the goal of being more flexible in applications. In the last two sections, an elegant alternative approach to the main results is sketched that uses classical tools from locally convex vector spaces. These sections are optional and are not needed for the rest of the book.

5.1 Semirings and Modules

Always let A be a ring, commutative and unital as usual. *We always assume that A contains $\frac{1}{2}$.* For many of the general results this restriction could be removed, at the cost of more technical statements and proofs. The most important applications are to $\mathbb{R}$-algebras anyway.

Given subsets X, Y of A, we write $X + Y = \{x + y \colon x \in X, y \in Y\}$ and $XY = \{xy \colon x \in X, y \in Y\}$.

C. Scheiderer, *A Course in Real Algebraic Geometry*, Graduate Texts in Mathematics 303,
https://doi.org/10.1007/978-3-031-69213-0_5

5.1.1 Definition. Let A be a ring.

(a) A *semiring*[1] in A is a subset $S \subseteq A$ that contains 0, 1 and is closed under addition and multiplication, i.e. satisfies $S + S \subseteq S$ and $SS \subseteq S$.
(b) Let S be a semiring in A. An S*-module* is a subset M of A for which $1 \in M$, $M + M \subseteq M$ and $SM \subseteq M$.
(c) A *quadratic module* of A is a subset $M \subseteq A$ such that $1 \in M$, $M + M \subseteq M$ and $a^2 M \subseteq M$ for every $a \in A$.

Note that every semiring in A is a module over itself.

5.1.2 Definition. Let M be a module over the semiring $S \subseteq A$.

(a) M is said to be *proper* if $-1 \notin M$, and *improper* otherwise.
(b) M is called *generating* if $M - M = A$.
(c) The *support* of M is $\mathrm{supp}(M) := M \cap (-M)$. This is an additive subgroup of A.

5.1.3 Lemma. *Let A be a ring and let $S \subseteq A$ be a semiring that is generating.*

(a) *The only improper S-module is $M = A$.*
(b) *The support* $\mathrm{supp}(M)$ *of any S-module M is an ideal of A.*

Proof. Let M be an S-module, then clearly $S \cdot \mathrm{supp}(M) \subseteq \mathrm{supp}(M)$. Therefore $S - S = A$ implies that $\mathrm{supp}(M)$ is an ideal of A. If $-1 \in M$ then $1 \in \mathrm{supp}(M)$, and so $\mathrm{supp}(M) = A$, which means $M = A$. □

5.1.4 Remarks.

1. A preordering of A (Definition 3.2.1) is the same as a semiring in A that contains all squares. Every preordering is generating, by our general assumption $\frac{1}{2} \in A$ and since $x = (\frac{x+1}{2})^2 - (\frac{x-1}{2})^2$. Quadratic modules of A are the same as modules over $S = \Sigma A^2$, the preordering of sums of squares. Every quadratic module is generating, and so the support of every quadratic module is an ideal.

2. Every preordering is a quadratic module, but not vice versa. For an example let $A = \mathbb{R}[x, y]$. The quadratic module $M := \Sigma A^2 + x\Sigma A^2 + y\Sigma A^2$ of A fails to be a preordering since $xy \notin M$, which means that there is no identity $xy = s_0 + s_1 x + s_2 y$ in A with $s_i \in \Sigma A^2$. See Exercise 5.1.3 for this and other (non-) examples.

3. The smallest semiring in A is $\mathbb{Z}_+ = \{0, 1, 2, \dots\}$. Note that $\mathbb{Z}_+$-modules in A are just additive subsemigroups M of A that contain 1. This already indicates that arbitrary semirings (or their modules) are far too general to be of interest. Usually only semirings will play a role that are generating.

4. Let R be a real closed field an A an R-algebra. Given any family $(p_i)_{i \in I}$ of elements of A, we may consider the semiring $S \subseteq A$ that is generated by R_+ and by the p_i $(i \in I)$. So S is the R-convex cone in A that is generated by all products of the form $p_{i_1}^{e_1} \cdots p_{i_r}^{e_r}$ with $i_1, \dots, i_r \in I$, where $r \geq 0$ and $e_1, \dots, e_r$ are positive integers. Clearly, S is generating as a semiring of A if, and only if, the R-algebra A is generated by the p_i $(i \in I)$.

[1] also called *preprime* in some of the older literature

The next two examples of semirings are more interesting.

5.1.5 Example. Let A be a ring. For any integer $n \geq 1$ let ΣA^n denote the semiring of all finite sums of n-th powers in A. If $2, 3, \dots, n$ are invertible in A then this semiring is generating, by the identity

$$n!\,x = \sum_{j=0}^{n-1} (-1)^{n-1-j} \binom{n-1}{j} \cdot \big((x+j)^n - j^n\big) \tag{5.1}$$

(see Exercise 5.1.5). If moreover n is odd, we see that every element of A is a sum of n-th powers in A. Therefore only the case where $n = 2m$ is even is of interest. A semiring S in A with $\Sigma A^{2m} \subseteq S$ is called a *preordering of level* $2m$, a module over ΣA^{2m} is called a *module of level* $2m$. See [159] ch. 7 for much more information and background.

5.1.6 Example. Let $n \geq 1$ and let $z = (z_1, \dots, z_n)$ be a tuple of variables. An expression of the form

$$f = \sum_{j=1}^{r} |p_j(z)|^2 = \sum_{j=1}^{r} p_j(z) \cdot \overline{p_j(z)} \in \mathbb{C}[z, \overline{z}]$$

with complex polynomials $p_1, \dots, p_r \in \mathbb{C}[z]$ is called a *sum of Hermitian squares*. Note that f is a polynomial in the $2n$ independent variables $z = (z_1, \dots, z_n)$ and $\overline{z} = (\overline{z}_1, \dots, \overline{z}_n)$. Writing $z_j = x_j + iy_j$ and $\overline{z}_j = x_j - iy_j$ for $j = 1, \dots, n$, every sum of Hermitian squares is contained in the polynomial subring

$$A := \mathbb{R}[x, y] = \mathbb{R}[x_1, y_1, \dots, x_n, y_n]$$

of $\mathbb{C}[z, \overline{z}]$. The set Σ_h of all sums of Hermitian squares is a generating semiring in A. It is strictly contained in the semiring ΣA^2 of all sums of squares, see Exercise 5.1.6.

5.1.7 Remarks. Let A be a ring.

1. If A contains a proper quadratic module then the ring A is real (see 3.2.16), i.e. A has an ordering. Indeed, A nonreal implies $-1 \in \Sigma A^2$, and so -1 is contained in every quadratic module in this case.

2. If $S \subseteq A$ is a semiring, every proper S-module is contained in a maximal proper S-module. This is a consequence of Zorn's lemma, since an upward directed union of proper S-modules is again a proper S-module.

3. Any intersection of semirings in A is again a semiring. If S is a given semiring and $(M_i)_{i\in I}$ is a family of S-modules, both the intersection $\bigcap_{i\in I} M_i$ and the sum

$$\sum_{i\in I} M_i = \Big\{ \sum_{i\in I} x_i \colon\ x_i \in M_i\ (i \in I),\ x_i = 0 \text{ for almost all } i \in I \Big\}$$

are again S-modules. The S-module generated by a family $(p_i)_{i\in I}$ in A is $M = S + \sum_{i\in I} S p_i$. The case when $S = \Sigma A^2$ is the semiring of sums of squares will be used frequently in the sequel: The quadratic module generated by $f_1, \ldots, f_r \in A$ is denoted

$$QM(f_1, \ldots, f_r) := QM_A(f_1, \ldots, f_r) := \Sigma A^2 + f_1 \Sigma A^2 + \cdots + f_r \Sigma A^2.$$

Note that $QM(f)$ is a preordering for every $f \in A$.

Next we discuss a few generalities on quadratic modules.

5.1.8 Remark. Let $\varphi\colon A \to B$ be a ring homomorphism. For any quadratic module N of B, the preimage $\varphi^{-1}(N)$ is a quadratic module of A. If N is a preordering then so is $\varphi^{-1}(N)$. Conversely, if M is a quadratic module of A, we may consider the quadratic module M^B of B generated by $\varphi(M)$, called the *extension* of M to B. It consists of all finite sums $\sum_i b_i^2 \varphi(x_i)$ with $x_i \in M$ and $b_i \in B$. If M is a preordering in A then M^B is a preordering in B. Even if M is proper, M^B need not be. When φ is surjective, we can be more precise:

5.1.9 Lemma. *Let I be an ideal of A and let $\pi\colon A \to A/I$ be the canonical map. There is a natural bijection between quadratic modules M of A that satisfy $I \subseteq \operatorname{supp}(M)$, and quadratic modules N of A/I, given by $M \mapsto \pi(M)$ and $N \mapsto \pi^{-1}(N)$, respectively. Moreover $\operatorname{supp}(\pi(M)) = \pi(\operatorname{supp}(M))$ and $\operatorname{supp}(\pi^{-1}(N)) = \pi^{-1}(\operatorname{supp}(N))$.*

5.1.10 We skip the proof, which is straightforward. Let $M \subseteq A$ be a quadratic module and S a multiplicative subset of A. The extension of M to the ring of fractions A_S (5.1.8) will be denoted M_S (rather than M^{A_S}). This is the quadratic module

$$M_S = \left\{ \frac{x}{s^2} : x \in M,\ s \in S \right\}$$

of A_S. Again, there is a natural bijection between quadratic modules of A_S and certain quadratic modules of A (Exercise 5.1.4). We just record:

5.1.11 Lemma. *The quadratic module M_S is proper if and only if $S \cap \operatorname{supp}(M) = \emptyset$.*

Proof. If $-1 \in M_S$, say $\frac{x}{s^2} = -1$ with $x \in M$ and $s \in S$, then there is $t \in S$ with $t^2(x+s^2) = 0$ in A, which implies $s^2t^2 \in S \cap \operatorname{supp}(M)$. If conversely $s \in S \cap \operatorname{supp}(M)$, then $-s^2 \in \operatorname{supp}(M)$ since $\operatorname{supp}(M)$ is an ideal (5.1.3), and so $-1 \in M_S$. □

5.1.12 Remark. Let A be a ring, let $S \subseteq A$ be a generating semiring in A and M an S-module. For $f, g \in A$ we write $f \le_M g$ if $g - f \in M$, as in 3.6.1. This defines a partial order relation $\le_M$ on A modulo the ideal $\operatorname{supp}(M)$, that satisfies the following properties:

(1) $a \le_M b$ and $a' \le_M b' \Rightarrow a + a' \le_M b + b'$;
(2) $a \le_M b$ and $s \in S \Rightarrow as \le_M bs$.

An ideal I of A is *M-convex* if the following equivalent conditions hold (see 3.6.2):

(i) $a \leq_M c \leq_M b$ and $a, b \in I$, $c \in A$ implies $c \in I$;
(ii) $a, b \in M$ and $a + b \in I$ implies $a, b \in I$;
(iii) $\mathrm{supp}(M + I) = I$.

Essentially, these concepts will play a role only when $M \cup (-M) = A$. Note that $\leq_M$ is a total ordering on the abelian group $A/\mathrm{supp}(M)$ in this case. Let us issue a warning here: Even though $A/\mathrm{supp}(M)$ is a ring, *the total ordering $\leq_M$ need not be compatible with products*. See Exercise 5.1.9 for an example.

5.1.13 Remark. Semirings, or modules over semirings, generalize preorderings. We are going to explore to what extent one can expect analogues of the geometric stellensätze from Section 3.3 in this greater generality. Usually this won't work, not even for generating semirings or modules. The reason is that, contrary to the case of preorderings, a proper module M over a semiring need not be contained in any positive cone (Exercises 5.1.2, 5.1.9). In this case even the "ur-stellensatz" 3.2.3 fails for S. The key concept that will allow us to avoid such phenomena will be the Archimedean property, to be introduced in the next section.

Every proper S-module is contained in a maximal proper S-module, see Remark 5.1.7.2. Moreover we have:

5.1.14 Proposition. *If $S \subseteq A$ is a generating semiring and M is a maximal proper S-module, then $M \cup (-M) = A$. Hence, in this case, $\leq_M$ induces a total ordering on the abelian group $A/\mathrm{supp}(M)$.*

Proof. Assume that there exists $a \in A$ with $\pm a \notin M$. By maximality of M we have $-1 \in (M + Sa) \cap (M - Sa)$. Hence there are $x, y \in M$ and $s, t \in S$ with $-1 = x + sa$ and $-1 = y - ta$. Multiplying the identities by t resp. s and adding them, we get $-s = t + sy + tx$, which shows $-s \in M$ and therefore $s \in \mathrm{supp}(M)$. Now $\mathrm{supp}(M)$ is an ideal of A since S is generating (Lemma 5.1.3). Therefore $-1 = x + sa$ implies $-1 \in M$, contradiction. □

Conversely, the condition $M \cup (-M) = A$ does not imply that M is maximal (think of positive cones in A). But it implies that the M-convex subgroups of A form a chain (Lemma 3.6.4). In the context of modules over a semiring, this gives:

5.1.15 Proposition. *Let M be a module over a generating semiring S, and assume that $M \cup (-M) = A$. Then there is an inclusion-preserving bijective correspondence between*

(1) *the S-modules N in A with $M \subseteq N$, and*
(2) *the M-convex ideals I of A,*

given by $N \mapsto \mathrm{supp}(N)$ and $I \mapsto M + I = M \cup I$. Each of the sets (1) and (2) forms a chain with respect to inclusion.

Note that the improper S-module $N = A$ corresponds to the unit ideal $I = A$ under this correspondence.

Proof. If I is an M-convex ideal, the set $M + I$ is an S-module and satisfies $\operatorname{supp}(M + I) = I$. Therefore $M + I = M \cup I$ by Lemma 3.6.4(b). Conversely, let N be an S-module with $M \subseteq N$. Then $I := \operatorname{supp}(N)$ is an N-convex (and hence M-convex) ideal of A. Moreover, if $a \in N$ and $a \notin M$, then $a \in -M$ since $M \cup (-M) = A$, and therefore $a \in -N$, which means $a \in \operatorname{supp}(N) = I$. This shows $N \subseteq M \cup I$, and therefore $N = M \cup I = M + I$. Both sets (1) and (2) are totally ordered by inclusion, see Lemma 3.6.4. □

In the situation of Proposition 5.1.15 there exists a largest proper ideal of A that is M-convex. Indeed, any two M-convex ideals are comparable with respect to inclusion, and so the union of all proper M-convex ideals is the largest such ideal. This ideal can be described as follows:

5.1.16 Proposition. *Let S be a generating semiring in A and let M be a proper S-module for which $M \cup (-M) = A$. The largest proper M-convex ideal of A is*

$$I = \{a \in A \colon \forall\, b \in A\ ab \le_M 1\} = \{a \in A \colon \forall\, s \in S\ \ -1 \le_M as \le_M 1\}.$$

The largest proper S-module that contains M is $N = M + I$, and $\operatorname{supp}(N) = I$.

Proof. The second assertion follows from the first by Proposition 5.1.15. We first note for $a, b \in A$ that $a \le_M b$ implies $\frac{a}{2} \le_M \frac{b}{2}$. To see this it suffices to show that $x \in M$ implies $\frac{x}{2} \in M$. Since $M \cup (-M) = A$, we may assume $\frac{x}{2} \in -M$, from which we get $x \in \operatorname{supp}(M)$. Therefore $\frac{x}{2} \in \operatorname{supp}(M)$ since $\operatorname{supp}(M)$ is an ideal.

Let $a \in A$. Since the semiring S is generating, $-1 \le_M as \le_M 1$ for all $s \in S$ implies the same statement for all $s \in A$, using the previous remark. So the two sets in the statement of 5.1.16 are equal. To check that I is an ideal of A, one has to show that $a_1, a_2 \in I$ implies $a_1 + a_2 \in I$. This is true since, for $b \in A$, the inequalities $2a_1 b \le_M 1$ and $2a_2 b \le_M 1$ imply $2(a_1 + a_2)b \le_M 2$, from which we get $(a_1 + a_2)b \le_M 1$ by the remark at the beginning. The ideal I is proper since $-1 \notin M$.

Every proper M-convex ideal J of A is contained in I. Indeed, given $a \in J$ and $b \in A$, then $ab \ge_M 1 \ge_M 0$ would imply $1 \in J$, since $ab \in J$ and J is M-convex. On the other hand, I itself is M-convex. Indeed, given $f, g \in M$ with $f + g \in I$, we have to show $f, g \in I$. For any $s \in S$ we see from $0 \le_M s(f + g) = sf + sg \le_M 1$ and $sf, sg \ge_M 0$ that $0 \le_M sf, sg \le_M 1$, which shows $f, g \in I$. Altogether we have proved that I is the largest proper M-convex ideal in A. □

In the following, it will be important to know whether the ideal I described in Proposition 5.1.16 is prime. We are first going to show that this is true for quadratic modules (Corollary 5.1.18). In the next section, such a statement will be proved in general under Archimedean hypotheses (Theorem 5.2.11).

5.1.17 Proposition. *Let M be a (proper) quadratic module in A, and let $\mathfrak{p} \subseteq A$ be a prime ideal that is minimal with respect to* $\operatorname{supp}(M) \subseteq \mathfrak{p}$. *Then $\mathfrak{p}$ is M-convex. In particular, $\mathfrak{p}$ is real and* $\operatorname{supp}(M + \mathfrak{p}) = \mathfrak{p}$, *so $M + \mathfrak{p}$ is again a proper quadratic module.*

Proof. It suffices to show that $\mathfrak{p}$ is M-convex, since this means $\mathrm{supp}(M + \mathfrak{p}) = \mathfrak{p}$ (see 5.1.12), and since every M-convex prime ideal is clearly real ($\sum_i a_i^2 \in \mathfrak{p}$ implies $a_i^2 \in \mathfrak{p}$ for every i). Write $I = \mathrm{supp}(M)$. This is an ideal of A since the semiring $S = \Sigma A^2$ is generating (5.1.3). Let $f, g \in M$ with $f + g \in \mathfrak{p}$, we have to show $f, g \in \mathfrak{p}$. The ring $A_\mathfrak{p}/IA_\mathfrak{p}$ has only one prime ideal, which is the ideal generated by $\mathfrak{p}$. Therefore $f + g$ is nilpotent in this ring. In particular, there exist $n \geq 1$ and $u \in A$ with $u \notin \mathfrak{p}$ such that $u(f + g)^n \in I$. Expanding the power, we see in particular that

$$u^2 \sum_{i=0}^{n} \binom{n}{i} f^i g^{n-i} \in I.$$

Clearly we may assume that n is odd. Then every single summand lies in M since for each index i, one of i or $n - i$ will be even. Since the sum lies in I, the same is true for each summand. In particular, $u^2 f^n \in I \subseteq \mathfrak{p}$, and so $f \in \mathfrak{p}$. □

We record a few consequences of 5.1.17:

5.1.18 Corollary. *Let $M \subseteq A$ be a proper quadratic module with $M \cup (-M) = A$. Then the largest proper M-convex ideal I (5.1.16) is prime.*

Proof. We have $I = \mathrm{supp}(N)$ for some quadratic module $N \supseteq M$. Let $\mathfrak{p}$ be a minimal prime ideal containing I. Then $\mathfrak{p}$ is N-convex by Proposition 5.1.17, hence also M-convex, and so $\mathfrak{p} = I$ by maximality of I. □

Directly from Proposition 5.1.17 we get:

5.1.19 Corollary. *Every quadratic module M satisfies* $\sqrt[\mathrm{re}]{\mathrm{supp}(M)} = \sqrt{\mathrm{supp}(M)}$. □

5.1.20 Definition. A *semiordering* of A is a quadratic module M of A for which $M \cup (-M) = A$ and $\mathrm{supp}(M)$ is a prime ideal of A.

5.1.21 Corollary. *Let A be any ring. Every maximal proper quadratic module of A is a semiordering of A.*

Proof. Let M be a proper quadratic module of A that is maximal. Then $M \cup (-M) = A$ by Proposition 5.1.14. The support $\mathrm{supp}(M)$ is the only M-convex ideal in A (Proposition 5.1.15), so it is a prime ideal by Corollary 5.1.18. □

5.1.22 Remarks.

1. Every positive cone of A is a semiordering of A. But in many real rings there exist semiorderings that are not positive cones, which means they are not closed under products. See Exercise 5.1.9 for an example, and see Section 5.5 for more results on semiorderings.

2. Real fields in which every semiordering is an ordering have been studied in quadratic form theory, under the name *SAP-fields* (see [118]). There exists a long list of equivalent characterizations of these fields. Every number field is an SAP-field, and so is every one-dimensional function field over a real closed base field R. In the case $R = \mathbb{R}$, we will see this in Section 5.5. Every quadratic module in an SAP-field is a preordering, by Exercise 5.1.8.

Exercises

5.1.1 Let A be a ring and let $I \subseteq A$ be the ideal generated by elements $f_1, \dots, f_r \in A$. If S is a generating semiring in A, show that the S-module generated by $\pm f_1, \dots, \pm f_r$ is $M = S + I$.

5.1.2 Find a proper generating semiring S in $\mathbb{R}[x, y]$ that is not contained in any positive cone of $\mathbb{R}[x, y]$. (*Hint*: Examples where S contains all squares are somewhat tricky to find, see for instance Exercise 5.1.9. For easy examples, you may take a semiring that is generated by suitable linear polynomials.)

5.1.3 Let $A = \mathbb{R}[x]$ for (a), (b) and $A = \mathbb{R}[x, y]$ for (c). Decide for each of the following quadratic modules whether M is a preordering in A:

(a) $M = QM(x, 1 - x)$,
(b) $M = QM(x + 1,\ x^2 - x)$,
(c) $M = QM(x, y)$.

5.1.4 Let $I \subseteq A$ be an ideal and $S \subseteq A$ a multiplicative subset, let $\varphi\colon A \to A_S$, $\varphi(a) = \frac{a}{1}$ be the canonical map.

(a) Prove Lemma 5.1.9.
(b) If $M \subseteq A$ is a quadratic module, let $M_S = \{\frac{x}{s^2} : x \in M,\ s \in S\}$. Show that $\mathrm{supp}(M_S) = \mathrm{supp}(M)_S$, the ideal generated in A_S by $\mathrm{supp}(M)$.
(c) The quadratic module M of A is called S-saturated if $x \in A$, $t \in S$ and $t^2 x \in M$ imply $x \in M$. Show that $M \mapsto M_S$ and $N \mapsto \varphi^{-1}(N)$ define a bijective correspondence between the S-saturated quadratic modules M of A and all quadratic modules N of A_S.
(d) The bijection (c) restricts to a bijective correspondence between S-saturated semiorderings of A and all semiorderings of A_S (see Definition 5.1.20).

5.1.5 For each natural number n, prove the identity (of bivariate polynomials over $\mathbb{Z}$)

$$n!\, xy^{n-1} = \sum_{j=0}^{n-1} (-1)^{n-1-j} \binom{n-1}{j} \Big((x + jy)^n - j^n y^n\Big).$$

Hint: Use the difference operator $\Delta f(x) = f(x + 1) - f(x)$ and show the identity

$$\Delta^e f(x) = \sum_{j=0}^{e} (-1)^{e-j} \binom{e}{j} f(x + j)$$

for $e \geq 0$. Then show, for f a monic polynomial of degree n, that $\Delta^e f(x)$ has degree $n - e$ and leading coefficient $e!\binom{n}{e}$ for $0 \leq e \leq n$.

5.1.6 Let $z = (z_1, \dots, z_n)$ and $w = (w_1, \dots, w_n)$ be tuples of variables, and consider the involutive ring automorphism $f \mapsto f^*$ of the polynomial ring $\mathbb{C}[z, w]$ defined by $z_j^* = w_j$ $(j = 1, \dots, n)$ and $c^* = \overline{c}$ for $c \in \mathbb{C}$ (complex conjugate), see Example 5.1.6.

(a) The fixring of $*$ is $A := \mathbb{R}[x, y]$ where $x = (x_1, \dots, x_n)$, $y = (y_1, \dots, y_n)$ and $x_j = \frac{1}{2}(z_j + w_j)$, $y_j = \frac{1}{2i}(z_j - w_j)$ $(j = 1, \dots, n)$.
(b) Given $f \in \mathbb{C}[z, w]$, let $|f|^2 := ff^*$. The set $\{\sum_{j=1}^r |f_j(z, w)|^2 : r \in \mathbb{N},\ f_j \in \mathbb{C}[z, w]$ for $j = 1, \dots, r\}$ is equal to ΣA^2.
(c) Let $\Sigma_h := \{\sum_{j=1}^r |f_j(z)|^2 : r \in \mathbb{N},\ f_j \in \mathbb{C}[z]\ (j = 1, \dots, r)\}$. Prove that Σ_h is a generating semiring in A.
(d) Show that $\Sigma_h \neq \Sigma A^2$. (*Hint*: Consider polynomials of degree 2.)

5.1.7 With notation as in Exercise 5.1.6, let $f \in \mathbb{C}[z, w]$ with $f = f^*$. Using multinomial notation, write

$$f = \sum_{|\alpha|, |\beta| \le d} c_{\alpha\beta} z^\alpha w^\beta$$

with complex coefficients $c_{\alpha\beta} = \overline{c}_{\beta\alpha}$, where $d = \deg_z(f) = \deg_{\overline{z}}(f)$. Show that $f \in \Sigma_h$ if and only if the Hermitian matrix $H(f) = (c_{\alpha\beta})_{|\alpha|, |\beta| \le d}$ is positive semidefinite.

5.1.8 If K is a field, prove that every quadratic module in K is an intersection of semiorderings in K. Show that this usually fails in rings that are more general than fields, e.g. in the polynomial ring $\mathbb{R}[t]$.

5.1.9 Let $A = R[x, y]$ where R is a real closed field. Given a polynomial $0 \ne f \in A$, write $f = \sum_{i \ge 0} f_i(x) y^i$ where $f_i \in R[x]$ are univariate polynomials. Let $n = \deg_y(f)$ and $m = \deg_x(f_n)$, let $a \ne 0$ be the coefficient of $x^m y^n$ in f and write $c(f) := (-1)^{mn} a$. Let $M \subseteq A$ be the set of all polynomials $f \ne 0$ for which $c(f) > 0$, together with $f = 0$.

(a) M is a semiordering of $R[x, y]$, and $\mathrm{supp}(M) = \{0\}$.
(b) M is not contained in any positive cone of $R[x, y]$.
(c) Let $S = A \smallsetminus \{0\}$. Show that the localization M_S (see Lemma 5.1.11) is a semiordering in the field of fractions $A_S = R(x, y)$ of A that is not an ordering.

5.2 Archimedean Modules

The following definition is of central importance for all that follows. Recall that every subsemigroup of an abelian group is required to contain the neutral element.

5.2.1 Definition. Let A be a ring. An additive subsemigroup M of A is *Archimedean* if $1 \in M$, and if for every $f \in A$ there exists an integer n such that $n - f \in M$.

5.2.2 Remarks.

1. For a semigroup $M \subseteq A$ with $1 \in M$, the Archimedean property requires that for every $f \in A$ there exists an integer $n \ge 1$ with $f \le_M n$ (notation as in Remark 5.1.12). This is the exact analogue of the classical Axiom of Archimedes, which explains the terminology. Equivalently, M is Archimedean if and only if $1 \in M$ and $M + \mathbb{Z} = A$. Clearly, if M is Archimedean then M is generating, i.e. $M - M = A$.

2. Every semigroup in A that contains an Archimedean semigroup is itself Archimedean.

3. A generalization of Archimedean semigroups that doesn't require $1 \in M$ are semigroups with an order unit (Definition 5.6.1). They will be considered in Sections 5.6 and 5.7.

We discuss a few criteria for semirings or modules to be Archimedean.

5.2.3 Definition. If $M \subseteq A$ is an additive semigroup with $1 \in M$, write

$$O(M) = O_A(M) = \{f \in A \colon \exists n \in \mathbb{N} \;\; n \pm f \in M\}.$$

5.2.4 Remark. Note that $O_A(M) = \{f \in A : \exists n \in \mathbb{N}\ -n \leq_M f \leq_M n\}$, with $\leq_M$ defined as in Remark 5.1.12. Therefore the elements of $O_A(M)$ may be considered as the *M-bounded elements* of A. Following the terminology of Exercise 3.6.1, $O_A(M)$ is the M-convex hull of $\mathbb{Z}$ in A and should be denoted $O_M(\mathbb{Z})$. However the notation $O(M) = O_A(M)$ will prove more convenient, which is why we deviate from the general notation in Exercise 3.6.1. We will not use the (easy) results from this exercise.

5.2.5 Lemma. *If $M \subseteq A$ is a semigroup with $1 \in M$, then $O(M) = \operatorname{supp}(M + \mathbb{Z})$. In particular, $O(M)$ is an additive subgroup of A, and M is Archimedean if and only if $O(M) = A$.*

Proof. Since $1 \in M$ we have $O(M) = \{f \in A : \pm f \in M + \mathbb{Z}\} = \operatorname{supp}(M + \mathbb{Z})$. The other assertions follow immediately. □

5.2.6 Proposition. *If M is a semiring or a quadratic module in A, then $O(M)$ is a subring of A.*

Proof. First let M be a semiring. Let $m, n \in \mathbb{N}$ and $f, g \in A$ such that $m \pm f \in M$ and $n \pm g \in M$. For $\varepsilon = \pm 1$ we have

$$(m + f)(n + \varepsilon g) + n(m - f) + m(n - \varepsilon g) = 3mn + \varepsilon fg.$$

The left-hand expression is obviously in M, and so $fg \in O(M)$.

Now let M be a quadratic module in A. Let $f \in O(M)$, so $n \pm f \in M$ for some $n \in \mathbb{N}$. For any $m \in \mathbb{Z}$ we have

$$(m - 4f)^2(n + f) + (m + 4f)^2(n - f) = 2nm^2 - 16(m - 2n)f^2.$$

The left-hand side is in M. This implies $f^2 \in O(M)$ since we may choose $m > 2n$. From $fg = (\frac{f+g}{2})^2 - (\frac{f-g}{2})^2$ we conclude that $O(M)$ is a subring of A, since $O(M)$ is an additive subgroup and $\frac{1}{2}M \subseteq M$. □

Recall that the ring A contains $\frac{1}{2}$ by assumption. For any quadratic module M in A, the subring $O(M)$ of A contains $\frac{1}{2}$ as well, since $1 \pm \frac{1}{2} \in \Sigma A^2 \subseteq M$.

5.2.7 Proposition. *Let A be an $\mathbb{R}$-algebra, finitely generated by $x_1, \dots, x_n$.*

(a) *Let $S \subseteq A$ be a semiring that contains $\mathbb{R}_+$. If there is a real number c with $c \pm x_i \in S$ $(i = 1, \dots, n)$, then S is Archimedean.*
(b) *Let $M \subseteq A$ be a quadratic module. If there is $c \in \mathbb{R}$ with $c - (x_1^2 + \cdots + x_n^2) \in M$, then M is Archimedean.*

Clearly the converses hold as well.

Proof. (a) The set $O(S)$ is an $\mathbb{R}$-subalgebra of A, by 5.2.6. Condition (a) implies $x_1, \dots, x_n \in O(S)$, hence $O(S) = A$.

(b) Again $O(M)$ is an $\mathbb{R}$-subalgebra of A. Let $c \in \mathbb{R}$ be as in (b), then also $c - x_i^2 \in M$ for $i = 1, \dots, n$, and so $(c - x_i^2) + (x_i \pm 1)^2 = (c + 1) \pm 2x_i \in M$. This shows $x_1, \dots, x_n \in O(M)$, and we conclude $O(M) = A$ as before. □

For later use we show that the subring $O(M)$ is integrally closed in A whenever M is a quadratic module. First a simple lemma:

5.2.8 Lemma. *Let A be a ring, let $f(t) \in A[t]$ be a monic polynomial of even degree. Then there exists $c \in A$ such that $f(t) + c$ is a sum of squares in $A[t]$.*

Proof. Let $\deg(f) = 2d$, the proof is by induction on d. If $2d = 2$ then

$$t^2 + at + b = \Big(t + \frac{a}{2}\Big)^2 + \Big(b - \frac{a^2}{4}\Big),$$

and we may take $c = \frac{a^2}{4} - b$, for example. Let now $f = t^{2d} + at^{2d-1} + bt^{2d-2} + \cdots$ where $2d \geq 4$, and assume that the lemma has been proved for all smaller degrees. Put $\alpha := \frac{a}{2}$, and let $\beta \in A$ be such that $\alpha^2 + 2\beta = b - 1$. Then

$$f(t) = (t^d + \alpha t^{d-1} + \beta t^{d-2})^2 + g(t)$$

where $g(t)$ is a monic polynomial of degree $2d-2$. By the inductive hypothesis there is $c \in A$ with $g(t) + c$ sos in $A[t]$. Hence $f(t) + c$ is sos in $A[t]$ as well. □

The proof has shown that if $\deg(f) = 2d$, the constant $c \in A$ can be chosen such that $f + c = g_d^2 + \cdots + g_1^2$, with g_i monic of degree i for $i = 1, \ldots, d$.

5.2.9 Proposition. *Let M be any quadratic module in A. Then the subring $O(M)$ of A is integrally closed in A.*

Proof. Let $a \in A$ be integral over $O(M)$, say $f(a) = 0$ with some monic polynomial $f(t)$ in $O(M)[t]$. We may assume that $\deg(f) = n \geq 2$ is even. By Lemma 5.2.8, applied to the ring $O(M)$, and since $\frac{1}{2} \in O(M)$, there are $u, v \in O(M)$ and sums of squares g, h in $O(M)[t]$ such that $f(t) + t = u + g(t)$ and $f(t) - t = v + h(t)$ hold (note that $f(t) \pm t$ are monic of degree n). By the definition of $O(M)$, there exists an integer $N \geq 1$ with $N+u, N+v \in M$. Substitution $t = a$ gives $N+a = N+f(a)+a = (N+u) + g(a) \in M$ and $N - a = N + f(a) - a = (N + v) + h(a) \in M$. Therefore $a \in O(M)$. □

5.2.10 Corollary. *Let $A \to B$ be an integral ring homomorphism and let M be an Archimedean quadratic module in A. Then the quadratic module M^B generated by M in B is again Archimedean.*

Proof. We have $O_A(M) = A$ since M is Archimedean. The subring $O_B(M^B)$ of B contains the image of $O_A(M) = A$ in B, and it is integrally closed in B by 5.2.9. Hence $O_B(M^B) = B$. □

The next theorem will play a key role in the proof of the Archimedean positivstellensatz (Section 5.3):

5.2.11 Theorem. *Let M be a proper module over a generating semiring S in A. If M is Archimedean then M is contained in a positive cone of A.*

Proof. M is contained in a maximal proper S-module M_1 (see 5.1.7.2), and M_1 is again Archimedean. Replacing M by M_1 we may assume that M itself is maximal as a proper S-module, and have to show that M is a positive cone. By Proposition 5.1.14 we have $M \cup (-M) = A$. Writing

$$I := \{a \in A\colon \forall\, b \in A\ ab \le_M 1\} = \{a \in A\colon \forall\, s \in S\ \ -1 \le_M sa \le_M 1\} \tag{5.2}$$

we claim that $I = \operatorname{supp}(M)$. Indeed, I is the largest proper M-convex ideal of A, and so $M + I$ is a proper S-module itself, both by Proposition 5.1.16). From maximality of M it follows that $I \subseteq M$, which means $I = \operatorname{supp}(M)$. The proof of the theorem is now given in three steps:

(1) *If $a, b \in A$ satisfy $a \ge_M 1$ and $b \ge_M 1$, there is an integer $n \ge 1$ with $nab \ge_M 1$.*

Let $s, t \in S$ with $a = s - t$. Since M is Archimedean, there is an integer $n \ge 1$ with $t \le_M n - 1$, and then there is an integer $N \ge 1$ with $-N \le_M nb \le_M N$. In particular, there exists a largest positive integer m satisfying $m \le_M nb$. Since $\le_M$ is a total ordering modulo I, and since $m+1 \le_M nb$ does not hold, we have $nb \le_M m+1$. Moreover $b \ge_M 1$ implies $n \le_M nb$, and so $m \ge n$. Altogether we get

$$nab = nsb - ntb \ge_M ms - (m+1)t = ma - t \ge_M m - t \ge_M 1.$$

Here the last inequality holds since $m - t \ge_M n - t \ge_M 1$.

(2) *M is a preordering in A.*

It suffices to show that $a, b \in M$ implies $ab \in M$. Indeed, since $M \cup (-M) = A$, this will imply that M contains all squares. So let $a, b \in M$. We can assume $a, b \notin \operatorname{supp}(M)$ since $\operatorname{supp}(M)$ is an ideal. By the above description (5.2) of $\operatorname{supp}(M)$, there exist $s, t \in S$ such that $sa \ge_M 1$ and $tb \ge_M 1$. By (1) there is an integer $n \ge 1$ with $nstab \ge_M 1$. Assuming $ab \notin M$ we would have $ab \le_M 0$, hence also $nstab \le_M 0$, a contradiction. Therefore $ab \in M$.

(3) *M is a positive cone of A.*

It remains to show that the ideal $\operatorname{supp}(M)$ is prime, so let $a, b \in A$ satisfy $ab \in \operatorname{supp}(M)$. Assuming that neither a nor b lies in $\operatorname{supp}(M)$, there exist $u, v \in A$ with $au \ge_M 1$ and $bv \ge_M 1$, again by (5.2). By (1) we find $n \ge 1$ with $nabuv \ge_M 1$. On the other hand $nabuv$ lies in $\operatorname{supp}(M)$, which implies $nabuv \le_M 0$, a contradiction. Alternatively we could first use Proposition 5.1.15 to see that M is maximal as a proper quadratic module, and then apply Corollary 5.1.21 to conclude that $\operatorname{supp}(M)$ is a prime ideal. □

5.2.12 Corollary. *Let $S \subseteq A$ be a generating semiring. If M is a maximal proper S-module, and if M is Archimedean, then M is a positive cone of A. In particular, the support of M is* $\operatorname{supp}(M) = \{a \in A\colon 1 + aA \subseteq M\}$*, and this is a prime ideal of A.*

Proof. See the beginning of the last proof for the description of $\operatorname{supp}(M)$. □

In Exercise 5.1.9 we saw an example of a proper semiordering M in $\mathbb{R}[x,y]$ that is not contained in any positive cone of $\mathbb{R}[x,y]$. According to Theorem 5.2.11, such M cannot be Archimedean. The example also shows that the Archimedean hypothesis in Theorem 5.2.11 is essential.

5.2.13 Corollary. *Let M be a proper module over a generating semiring S in A. If M is Archimedean, there exists a ring homomorphism $\varphi\colon A \to \mathbb{R}$ with $\varphi(f) \ge 0$ for every $f \in M$.*

Proof. Let P be a maximal proper S-module with $M \subseteq P$. Then P is a positive cone in A (Corollary 5.2.12), and is maximal as such. In other words, P is a closed point of Sper(A). Write $\mathfrak{p} = \operatorname{supp}(P)$. By Proposition 3.6.17, the ring extension $A/\mathfrak{p} \subseteq \kappa(\mathfrak{p}) = \operatorname{qf}(A/\mathfrak{p})$ is relatively Archimedean with respect to the ordering $\le_P$. For every $b \in \kappa(\mathfrak{p})$, this means that there is $a \in A$ with $b \le_P a$ in $\kappa(\mathfrak{p})$. On the other hand, there is an integer $N \ge 1$ with $a \le_P N$, since P is Archimedean. Together this shows that the ordered field $(\kappa(\mathfrak{p}), \le_P)$ is Archimedean. By Hölder's theorem 1.1.18 there is an order-preserving field embedding $(\kappa(\mathfrak{p}), \le_P) \to \mathbb{R}$. Combine it with the residue map $A \to \kappa(\mathfrak{p})$ to get φ as desired. □

5.2.14 Definition. Given a ring A and a subset $M \subseteq A$, let

$$X_M := \{\varphi \in \operatorname{Hom}(A,\mathbb{R})\colon \varphi(M) \subseteq \mathbb{R}_+\},$$

the set of all ring homomorphisms $A \to \mathbb{R}$ that take non-negative values on M. We equip X_M with the relative topology that is induced from the inclusion

$$X_M \subseteq \operatorname{Hom}(A,\mathbb{R}) \subseteq \mathbb{R}^A = \prod_{f\in A} \mathbb{R}, \quad \varphi \mapsto (\varphi(f))_{f\in A}$$

and from the product topology on $\mathbb{R}^A$. If $\varphi \in X_M$, let $\alpha_\varphi := [\varphi]$ denote the point in Sper(A) that is represented by φ (3.1.15).

5.2.15 Remarks.

1. The only ring homomorphism $\mathbb{R} \to \mathbb{R}$ is the identity (1.1.21). If A is an $\mathbb{R}$-algebra, this means that $\operatorname{Hom}(A,\mathbb{R}) = \operatorname{Hom}_{\mathbb{R}}(A,\mathbb{R})$. In particular, when $A = \mathbb{R}[V]$ is the coordinate ring of an affine $\mathbb{R}$-variety V, then $\operatorname{Hom}(A,\mathbb{R}) = V(\mathbb{R})$, the set of $\mathbb{R}$-points of V. This identification is correct even as topological spaces, with $V(\mathbb{R})$ being given the order topology (1.7.5). Therefore

$$X_M = \mathcal{S}_V(M) = \{\xi \in V(\mathbb{R})\colon \forall\, f \in M\ f(\xi) \ge 0\}$$

in this case, as topological spaces. In concrete applications, M will usually be a finitely generated quadratic module in $\mathbb{R}[V]$, or just a finite subset of $\mathbb{R}[V]$. Then X_M is a basic closed semialgebraic set in $V(\mathbb{R})$. If A is an "abstract" ring, one should think of the topological spaces $\operatorname{Hom}(A,\mathbb{R})$ and X_M as "abstract" analogues of $V(\mathbb{R})$ and $\mathcal{S}_V(M)$, respectively.

2. Observe the notational difference between the set $X_M \subseteq \operatorname{Hom}(A, \mathbb{R})$ just defined, and the closed subset $X(M)$ of $\operatorname{Sper}(A)$. The relation between both sets is exhibited in 5.2.22 below. Corollary 5.2.13 above states that X_M is non-empty, whenever M is a proper Archimedean module over a generating semiring.

5.2.16 Lemma. *Let $M \subseteq A$ be a subset. Then X_M is a closed subset of $\mathbb{R}^A$, and is compact if M is an Archimedean semigroup in A.*

Proof. An argument similar to the proof of 3.4.7 (compactness of the constructible topology) shows that $\operatorname{Hom}(A, \mathbb{R})$ is a closed subset of $\mathbb{R}^A$. On the other hand, it is clear that X_M is closed in $\operatorname{Hom}(A, \mathbb{R})$. Let M be an Archimedean semigroup. Then for every $f \in A$ there exists an integer $N_f \geq 1$ with $N_f \pm f \in M$. Therefore $\varphi(f) \in [-N_f, N_f]$ for every $\varphi \in X_M$, from which we see that X_M is contained in $\prod_{f \in A} [-N_f, N_f]$. This is a compact topological space by Tikhonov's theorem. □

In the remainder of this section, we are going to relate the topological spaces X_M to the real spectrum. For this we will back up a little and work under assumptions that are more general than the Archimedean property. Being Archimedean is a property of M that is strictly stronger than compactness of X_M. In fact, there exist (finitely generated) quadratic modules M in $\mathbb{R}[x_1, \ldots, x_n]$ that fail to be Archimedean, but for which X_M is compact. We'll see examples a little later, in Section 5.5.

5.2.17 Definition. A subset Y of $\operatorname{Sper}(A)$ is *absolutely bounded* if for every $f \in A$ there exists an integer $N \geq 1$ such that $N \pm f > 0$ on Y, i.e. with $Y \subseteq U(N+f, N-f)$.

5.2.18 Remarks.

1. If R is a real closed field, recall (Exercise 1.6.4) that a subset of R^n is called (semialgebraically) bounded if it is contained in $[-c, c]^n$ for some $c \in R$. Beware not to confuse this notion with the property defined in 5.2.17: If $R = \mathbb{R}$ is the field of real numbers and $S \subseteq \mathbb{R}^n$ is a semialgebraic set, then the constructible set $\widetilde{S} \subseteq \operatorname{Sper} \mathbb{R}[x]$ is absolutely bounded if, and only if, S is semialgebraically bounded. But when the real closed field R is non-Archimedean, no subset of $\operatorname{Sper} R[x]$ whatsoever will be absolutely bounded, except for the empty set. As the terminology indicates, absolute boundedness is an *absolute* concept, meaning boundedness over $\mathbb{Z}$, and not a relative one.

2. For every Archimedean semigroup M in a ring A, the set $X(M) \subseteq \operatorname{Sper}(A)$ is absolutely bounded. Conversely, a set $Y \subseteq \operatorname{Sper}(A)$ is absolutely bounded if and only if the preordering $\{f \in A\colon f \geq 0 \text{ on } Y\}$ is Archimedean. While absolute boundedness of the set $X(M) \subseteq \operatorname{Sper}(A)$ is a "geometric" condition on M, the Archimedean property of M is a stronger condition of "arithmetic" nature.

5.2.19 Proposition. *Let $Y \subseteq \operatorname{Sper}(A)$ be a closed set. The set*

$$Y^{\text{arch}} := \{\alpha \in Y\colon R(\alpha) \text{ is Archimedean}\}$$

of Archimedean orderings in Y is always contained in $Y^{\max}$. If Y is absolutely bounded then equality $Y^{\text{arch}} = Y^{\max}$ holds. In particular, the topological space Y^{arch} is compact in this case.

Proof. Recall that $Y^{\max}$ denotes the set of closed points of Y. This is a compact topological space, see Proposition 3.4.19. The inclusion $Y^{\mathrm{arch}} \subseteq Y^{\max}$ holds in general. Indeed, if $\alpha \in Y^{\mathrm{arch}}$ and $\mathfrak{p} = \mathrm{supp}(\alpha)$, the real closed field $R(\alpha)$ is Archimedean. So it will *a fortiori* be relatively Archimedean over its subring $A/\mathfrak{p}$, which implies that α is a closed point (Proposition 3.6.17). Now assume that Y is absolutely bounded, let $\alpha \in Y^{\max}$ and put $\mathfrak{p} = \mathrm{supp}(\alpha)$. We claim that the real closed field $R(\alpha)$ is Archimedean, which will prove $\alpha \in Y^{\mathrm{arch}}$. Since α is a closed point, the ring extension $A/\mathfrak{p} \subseteq R(\alpha)$ is relatively Archimedean (3.6.17). For $b \in R(\alpha)$, this means that there is $a \in A$ with $|b| < a$ in $R(\alpha)$. On the other hand, Y is absolutely bounded, so there exists an integer N with $a(\alpha) < N$. Together this shows that the real closed field $R(\alpha)$ is Archimedean. Compactness of $Y^{\max}$ was observed in 3.4.19(b). □

5.2.20 Let $Y \subseteq \mathrm{Sper}(A)$ be a closed set. For every $\alpha \in Y^{\mathrm{arch}}$ there is a unique field embedding $R(\alpha) \to \mathbb{R}$ (Hölder's theorem 1.1.18). Hence there exists a unique ring homomorphism $\varphi_\alpha \colon A \to \mathbb{R}$ that represents the point α (in the sense of 3.1.15). This implies that every element $f \in A$ defines a real-valued function $\hat{f}$ on Y^{arch}, namely

$$\hat{f} \colon Y^{\mathrm{arch}} \to \mathbb{R}, \quad \hat{f}(\alpha) = \varphi_\alpha(f) \quad (\alpha \in Y^{\mathrm{arch}}).$$

Clearly $\widehat{f+g} = \hat{f} + \hat{g}$ and $\widehat{fg} = \hat{f}\,\hat{g}$ hold for all $f, g \in A$, and $\hat{1} = 1$. Moreover, the functions $\hat{f}$ are continuous. Indeed, if a, b, n are integers with $n \geq 1$ and $a < b$, the preimage of the open interval $]\frac{a}{n}, \frac{b}{n}[\subseteq \mathbb{R}$ under $\hat{f}$ is $Y^{\mathrm{arch}} \cap U(nf-a, b-nf)$, which is an open subset of Y^{arch}. Since every open set in $\mathbb{R}$ is a union of such intervals, continuity of $\hat{f}$ follows.

If X is any topological space, let $\mathcal{C}(X, \mathbb{R})$ denote the ring of continuous $\mathbb{R}$-valued functions on X. For any closed set $Y \subseteq \mathrm{Sper}(A)$, the previous discussion shows that $f \mapsto \hat{f}$ defines a ring homomorphism $\Phi \colon A \to \mathcal{C}(Y^{\mathrm{arch}}, \mathbb{R})$. When Y is absolutely bounded we have $Y^{\mathrm{arch}} = Y^{\max}$, and so we get:

5.2.21 Corollary. *For every closed and absolutely bounded set $Y \subseteq \mathrm{Sper}(A)$, the map $\Phi \colon A \to \mathcal{C}(Y^{\max}, \mathbb{R})$, $\Phi(f) = \hat{f}$ is a homomorphism from A into the ring of continuous real-valued functions on the compact space $Y^{\max}$.* □

We now identify the sets X_M with suitable subsets of the real spectrum:

5.2.22 Proposition. *Let $M \subseteq A$ be a subset. The map $\psi \colon X_M \to \mathrm{Sper}(A)$, $\varphi \mapsto \alpha_\varphi = [\varphi]$ is a homeomorphism from X_M onto $X(M)^{\mathrm{arch}}$.*

Proof. Recall (5.2.14) that $[\varphi]$ denotes the point in $\mathrm{Sper}(A)$ that is represented by φ. For $\varphi \in X_M$, it is obvious that α_φ lies in $X(M)^{\mathrm{arch}}$. The map ψ is bijective onto $X(M)^{\mathrm{arch}}$, and the inverse map sends a point $\beta \in X(M)^{\mathrm{arch}}$ to the homomorphism $\varphi_\beta \in X_M$ (5.2.20). Moreover ψ is continuous: Given $f \in A$ we have $\psi^{-1}(U(f)) = \{\varphi \colon \varphi(f) > 0\}$, which is an open subset of X_M since $\hat{f}$ is continuous. To show that ψ is a topological embedding, it suffices to show for any $f \in A$ and any *rational* numbers $\varepsilon, t \in \mathbb{Q}$, that the set

$$\{\alpha_\varphi \colon \varphi \in X_M,\ |\varphi(f) - t| < \varepsilon\}$$

is relatively open in $X(M)^{\text{arch}}$. Indeed, this suffices since the finite intersections of sets $\{\beta \in X_M : |\beta(f) - t| < \varepsilon\}$ (with $f \in A$ and $\varepsilon, t \in \mathbb{Q}$) form a basis for the topology of X_M. Now

$$\left|\varphi(f) - \frac{m}{n}\right| < \frac{1}{n} \quad \Leftrightarrow \quad \left|\varphi(nf - m)\right| < 1 \quad \Leftrightarrow \quad \varphi(1 \pm (nf - m)) > 0.$$

So it is enough to show for every $f \in A$ that the set $\{\alpha_\varphi : \varphi \in X_M, \varphi(1 \pm f) > 0\}$ is relatively open in $X(M)^{\text{arch}}$. This set is $X(M)^{\text{arch}} \cap U(1+f,\ 1-f)$, and the proposition is proved. □

Combining the homeomorphism 5.2.22 with 5.2.19, we see in particular:

5.2.23 Corollary. *If $M \subseteq A$ is an Archimedean semigroup, the topological spaces X_M and $X(M)^{\max}$ are canonically homeomorphic.* □

5.2.24 Remarks. Let $M \subseteq A$ be a semigroup.

1. Assume that M is Archimedean. An element $f \in A$ is strictly positive on $X(M)$ if, and only if, f is strictly positive on X_M. Indeed, for the backward ($\Leftarrow$) direction it suffices to note that every point in $X(M)$ specializes to a point in $X(M)^{\max} = X_M$. For non-strict positivity we have $f \geq 0$ on $X(M) \Rightarrow f \geq 0$ on X_M, but not conversely in general.

2. For general rings, the topological space $\operatorname{Hom}(A, \mathbb{R})$ or its subspace X_M may have little or no significance. For example, $\operatorname{Hom}(A, \mathbb{R})$ is empty whenever A contains a non-Archimedean real closed field. In this generality, the real spectrum is the correct object to work with. In the important case of $\mathbb{R}$-algebras however, one may replace the real spectrum by the "less abstract" space $X_M \approx X(M)^{\text{arch}}$ for some purposes.

5.2.25 Remark. Corollary 5.2.21 has an important consequence that we want to record. Before doing so, recall the Stone–Weierstrass approximation theorem (see [172], for example): Let X be a compact topological space and let $A \subseteq \mathcal{C}(X, \mathbb{R})$ be a ring of real-valued continuous functions on X (containing $\frac{1}{2}$). Assume that A separates points, i.e. that for any pair $x \neq y$ of points in X there exists $f \in A$ with $f(x) \neq f(y)$. Then A is dense in $\mathcal{C}(X, \mathbb{R})$ with respect to the norm of uniform convergence: For every continuous map $g : X \to \mathbb{R}$ and every real number $\varepsilon > 0$ there exists $f \in A$ with $|f - g| < \varepsilon$, uniformly on X.

Usually this theorem is stated for subrings $A \subseteq \mathcal{C}(X, \mathbb{R})$ that contain $\mathbb{R}$. It holds in the generality above since $\frac{1}{2} \in A$ implies that the closure of A contains all of $\mathbb{R}$.

5.2.26 Proposition. *Let $Y \subseteq \operatorname{Sper}(A)$ be a closed set that is absolutely bounded. Then the image of the ring homomorphism*

$$\Phi : A \to \mathcal{C}(Y^{\max}, \mathbb{R}), \quad f \mapsto \hat{f}$$

(see 5.2.20) is a dense subring of $\mathcal{C}(Y^{\max}, \mathbb{R})$.

Proof. The topological space $Y^{\max}$ is compact (5.2.19), and the ring $\Phi(A)$ separates points: Given $\alpha \neq \beta$ in $Y^{\max}$, there exists $f \in A$ with $\hat{f}(\alpha) \neq \hat{f}(\beta)$. In fact, there even exists $f \in A$ with $\hat{f}(\alpha) > 1$ and $\hat{f}(\beta) < -1$ (Exercise 3.6.10). □

We will come back to Proposition 5.2.26 in Theorem 5.3.9, and then again in Section 6.2.

Exercises

5.2.1 Let A be a ring that is generated by two subrings A_1, A_2. For $i = 1, 2$ let M_i be an Archimedean semigroup in A_i. Prove that the set of all finite sums $\sum_j b_j c_j$ with $b_j \in M_1$, $c_j \in M_2$ is an Archimedean semigroup in A.

5.2.2 Give an example of a ring A for which the subset $\mathrm{Sper}(A)^{\mathrm{arch}}$ is not closed.

5.2.3 Let K be a field and let $M \subseteq K$ be a proper quadratic module. Show that

$$O_K(M) = \left\{\frac{n}{1+x} - \frac{n}{1+y} \colon x, y \in M,\ n \in \mathbb{N}\right\}.$$

Hint: To prove that $f \in O_K(M)$ lies in the right-hand set, choose $n \in \mathbb{N}$ with $u := n^2 - f^2 \in M$, and note that $f = 2n^2 f(f^2 + u + n^2)^{-1}$. Now try to rewrite the right-hand expression in the desired way.

5.2.4 Let $A = \mathbb{R}[x, y]$ with $x = (x_1, \dots, x_n)$, $y = (y_1, \dots, y_n)$. Using the notation from Exercise 5.1.6, let $\Sigma_h \subseteq A$ be the semiring of sums of Hermitian squares in $\mathbb{C}[z, \overline{z}]$, and let M be a Σ_h-module in A. Prove that M is Archimedean if (and only if) $c - (|z_1|^2 + \cdots + |z_n|^2) \in M$ for some real number $c > 0$. (*Hint*: Show that $\{p \in \mathbb{C}[z] \colon -|p(z)|^2 \in \mathbb{R} + M\}$ is a subring of $\mathbb{C}[z]$.)

5.2.5 Let $S \subseteq \mathbb{R}[t]$ be the semiring generated by $1 - t^2$ and all fourth powers in $\mathbb{R}[t]$.

(a) Show that every element in S has even degree, and conclude that S is not Archimedean.
(b) When fourth powers are replaced by sixth powers, it is known [19] that the semiring becomes Archimedean. In particular, there exists an identity

$$a - t = \sum_{j=0}^{5} (1 - t^2)^j p_j(t) \tag{5.3}$$

where $a > 1$ is a real number and $p_0, \dots, p_5$ are sums of sixth powers of polynomials. Conversely, such an identity implies that the semiring is Archimedean (why?). Can you find an identity (5.3)? (The author does not know an answer.)

5.2.6 Let M be a semiordering of the field K, and write $a \leq_M b \Leftrightarrow b - a \in M$ for $a, b \in K$. Let $a, b \in K$ in (b) and (c), and prove:

(a) The restriction of $\leq_M$ to $\mathbb{Q} \subseteq K$ is the usual ordering of $\mathbb{Q}$;
(b) $0 <_M a <_M b$ implies $0 <_M \frac{1}{b} <_M \frac{1}{a}$;
(c) if $0 <_M a <_M b$, and if $a \in \Sigma K^2$ or $b \in \Sigma K^2$, then $0 <_M a^2 <_M b^2$.

Hint on (b): $\frac{b}{a(b-a)} = \frac{1}{a} + \frac{1}{b-a}$.

5.2.7 Show that every Archimedean semiordering M of a field K is (the positive cone of) an ordering of K. (*Hint*: Use Exercise 5.2.6, and start by showing that $\mathbb{Q}$ is dense in K with respect to the order topology of M.)

5.2.8 (Compare Example 3.5.8) Let M be a semiordering of the field K, and let $O(M) = O_K(M)$ as in Definition 5.2.3.

(a) $O(M)$ is a valuation ring of K, and is M-convex;
(b) the maximal ideal $\mathfrak{m}$ of $O(M)$ is M-convex;
(c) the set $\overline{M} := \{\overline{a}\colon a \in M\}$ is the positive cone of an Archimedean ordering of the residue field $k = O(M)/\mathfrak{m}$.

5.3 The Archimedean Positivstellensatz

The following theorem is one of the central results in this entire course. Although the statement itself is of an abstract nature, it has far-reaching consequences that are very concrete and explicit, as we will see. Recall the definition of X_M from 5.2.14.

5.3.1 Theorem. (Archimedean positivstellensatz) *Let $M \subseteq A$ be an Archimedean quadratic module, or a module over an Archimedean semiring S in A. Given $f \in A$, the following are equivalent:*

(i) *$f > 0$ on X_M;*
(ii) *there exists an integer $n \geq 1$ with $nf \in 1 + M$.*

Proof. When M is an Archimedean quadratic module write $S = \Sigma A^2$. So in any case, S is a generating semiring and M is an S-module. The implication (ii) $\Rightarrow$ (i) is clear. Indeed, from $nf \in 1 + M$ we get $nf \geq 1$ on X_M, and hence $f \geq \frac{1}{n} > 0$ on X_M. The essential part of the theorem is therefore the implication (i) $\Rightarrow$ (ii). Assume $f > 0$ on X_M and write $M' = M - Sf$. Then M' is an Archimedean S-module with $X_{M'} = \varnothing$. Since S is generating we have $-1 \in M'$ by Corollary 5.2.13, so there exists $s \in S$ with $sf - 1 \in M$. Since M is Archimedean, there are integers k, $q \geq 1$ with $2k - 1 - s^2 f \in M$ and $f + q \in M$. By enlarging k we can ensure that also $(k - s)^2 \in S$. Indeed, if S is Archimedean we can even get $k - s \in S$. Otherwise $S = \Sigma A^2$, and $(k - s)^2 \in S$ is automatic. In any case we conclude

$$2k - s = (2k - 1 - s^2 f) + s(sf - 1) + 1 \in M.$$

Writing

$$Q := \{(m, n)\colon m, n \in \mathbb{Z},\ m \geq 1,\ mf + n \in M\}$$

we have $(1, q) \in Q$. Let $(m, n) \in Q$ be any pair with $n \geq 0$. Then also

$$\begin{aligned} &k^2 mf + (k^2 n - m) \\ &\quad = (k - s)^2(mf + n) + 2km(sf - 1) + ns(2k - s) + m(2k - 1 - s^2 f) \end{aligned}$$

lies in M, since $mf + n \in M$ and $(k - s)^2 \in S$. So the implication

$$(m, n) \in Q \text{ and } n \geq 0 \quad \Rightarrow \quad (k^2 m,\ k^2 n - m) \in Q \tag{5.4}$$

holds in general. Repeating the argument and starting with $(1, q) \in Q$, we successively find that $(1, q)$, $(k^2, k^2q - 1)$, $(k^4, k^4q - 2k^2)$, ... lie all in Q, until the second component becomes negative for the first time. More precisely, we find that

$$k^{2j}(k^2, k^2q - j - 1) \in Q \quad \text{for } j = 0, 1, \dots, k^2q.$$

The last possible value j gives $k^{2j}(k^2, -1) \in Q$. This means $k^{2j}(k^2f - 1) \in M$, and therefore $k^{2j+2}f \in 1 + M$. Theorem 5.3.1 has been proved. □

5.3.2 Remarks.

1. Inspecting the proof, we see that the number n in (ii) can be chosen to be a power of 2. Indeed, the choice of k in the beginning may be increased arbitrarily, and in particular, k may be chosen to be a 2-power.

2. The inductive step (5.4) in the previous proof becomes more transparent if we assume $\mathbb{Q} \subseteq A$ and $\mathbb{Q}_+ \subseteq S$. Then $(m, n) \in Q$ is equivalent to $f + \frac{n}{m} \in M$, and (5.4) says that $0 \le q \in \mathbb{Q}$ and $f + q \in M$ imply $f + q - \frac{1}{k^2} \in M$. After finitely many steps, therefore, we have found a rational number $q < 0$ with $f + q \in M$.

We stress the fact that Theorem 5.3.1 is essentially a *denominator-free positivstellensatz*. This becomes obvious from the following reformulation:

5.3.3 Corollary. *Let M be an Archimedean quadratic module, or a module over an Archimedean semiring S with $\frac{1}{2} \in S$. Then M contains every $f \in A$ with $f > 0$ on X_M.*

Proof. In Theorem 5.3.1 there is an integer $r \ge 0$ with $2^r f \in 1 + M$ (use Remark 5.3.2.1). Since we assumed $\frac{1}{2} \in S$, we can divide by 2^r and get $f \in 2^{-r} + M \subseteq M$. □

5.3.4 Remarks.

1. Note how Corollary 5.3.3 gives a much stronger conclusion than the Krivine–Stengle positivstellensatz 3.2.7. Of course, there is a price to pay: 5.3.3 needs the Archimedean condition, whereas 3.2.7 is true in complete generality. On the other hand, 3.2.7 holds for preorderings only, whereas the Archimedean positivstellensatz applies to situations of much more general type. This includes very relevant explicit cases, as we will see in the next sections.

2. Most important for applications will be the following situation. Let V be an affine $\mathbb{R}$-variety and let M be either an Archimedean quadratic module in $\mathbb{R}[V]$, or a module over an Archimedean semiring S in $\mathbb{R}[V]$. Then M contains every $f \in \mathbb{R}[V]$ with $f > 0$ on the subset X_M of $V(\mathbb{R})$. Prominent examples of this sort will be discussed in the next two sections.

3. The proof of Theorem 5.3.1 becomes much simpler if $M = S$ is assumed to be an Archimedean *preordering*. Indeed, the hypothesis $f > 0$ on X_S, together with S Archimedean, implies that $f > 0$ on $X(S)$ (Remark 5.2.24.1). If S is a preordering, one finds $s \in S$ with $sf \in 1 + S$ (first step in the proof of 5.3.1) just from the general Krivine–Stengle positivstellensatz 3.2.7. In particular, there is no need to invoke Theorem 5.2.11 or Corollary 5.2.13, and the proof can directly enter into the inductive argument.

4. Note that the case where M is an Archimedean module over a generating semiring is *not covered* by Theorem 5.3.1. Leaving the quadratic modules case aside, what the theorem requires is that the semiring itself is Archimedean. In the proof there is only one single step where the Archimedean property for S is needed, namely to guarantee for $s \in S$ the existence of an integer $k \geq 1$ with $(k - s)^2 \in S$. In fact, there do exist examples of Archimedean modules over generating semirings for which Theorem 5.3.1 fails, see Exercises 5.3.2 and 5.3.3.

The conclusion of the Archimedean positivstellensatz can also be stated as a nichtnegativstellensatz:

5.3.5 Corollary. (Archimedean nichtnegativstellensatz) *Let M be an Archimedean quadratic module, or a module over an Archimedean semiring $S \subseteq A$ with $\frac{1}{2} \in S$. For any $f \in A$, the following are equivalent:*

(i) $f \geq 0$ *on* X_M,
(ii) $\forall n \in \mathbb{N}\ \ 1 + nf \in M$.

Proof. If $f \geq 0$ on X_M then $1 + nf > 0$ on X_M for every $n \geq 1$, and therefore $1 + nf \in M$ by 5.3.3. Conversely, if $f(\alpha) < 0$ for some $\alpha \in X_M$ then $f(\alpha) < -\frac{1}{n}$ for some integer $n \geq 1$. Therefore $(1 + nf)(\alpha) < 0$, which implies $1 + nf \notin M$. □

5.3.6 Remark. Conversely, Corollary 5.3.5 implies the positivstellensatz 5.3.3: If $f > 0$ on X_M, there exists $m \geq 0$ with $f \geq \frac{1}{2^m}$ on X_M, by compactness of X_M. Applying 5.3.5 to $g := f - 2^{-m}$ with $n = 2^m$ gives $1 + 2^m g = 2^m f \in M$, hence $f \in M$.

5.3.7 Corollary. *A quadratic module $M \subseteq A$ is Archimedean if, and only if, X_M is compact and M contains every $f \in A$ with $f > 0$ on X_M.*

Proof. M Archimedean implies that X_M is compact (5.2.23) and that M contains all elements that are strictly positive on X_M (5.3.3). The converse is true for any semigroup M, not just for quadratic modules: If X_M is compact and $f \in A$, there exists $n \geq 1$ with $n \pm f > 0$ on X_M. So if M contains all elements strictly positive on X_M, it follows that $n \pm f \in M$. Therefore M is Archimedean. □

5.3.8 Remark. The historical genesis of the Archimedean positivstellensatz has many ramifications. An ur-version was proved by Stone [204] in 1940. Generalizations were later found by Kadison [104] in 1951 and Dubois [58] in 1967. Independently, Krivine [113] in 1964 stated essentially the version discussed here, in the case of preorderings. Apparently, neither was Krivine aware of the work of Stone, Kadison or Dubois, nor did Krivine's result get much attention. A purely algebraic proof (for modules over Archimedean semirings) was given by Becker–Schwartz [16] in 1983. There, as well as in other places in the literature, the result goes under the name "Kadison–Dubois theorem". The extension to Archimedean quadratic modules was found by Jacobi [100] in 1999. He also proved the result for Archimedean modules of higher level, a case that is not included in our discussion. A recent approach by Schmüdgen–Schötz [192] derives the semiring case directly from the case of quadratic modules (for $\mathbb{R}$-algebras). We refer to Section 5.6 in [159] for a detailed account of the history of the Archimedean positivstellensatz. In Section 5.7 we will present a very different alternative approach to these results.

In the literature, the Archimedean positivstellensatz is often referred to as the *representation theorem* (for example [104], [16], [159] or [138]). Let us briefly indicate how this can be justified. Let X be a topological space and let $\mathcal{C}(X,\mathbb{R})$ be the ring of continuous $\mathbb{R}$-valued functions on X, as before. If B is a subring of $\mathcal{C}(X,\mathbb{R})$, let $B_+ := \{f \in B\colon \forall\, x \in X\ f(x) \ge 0\}$, which is a semiring in B. The starting point for the representation theorem is the question how to characterize, in purely algebraic terms, when a pair (A,S) of a commutative ring A and a semiring $S \subseteq A$ is isomorphic to a pair (B,B_+), with X a compact topological space and $B \subseteq \mathcal{C}(X,\mathbb{R})$ a subring. This question can now be answered:

5.3.9 Corollary. (Representation Theorem) *Let A be a ring and $S \subseteq A$ a semiring with $\frac{1}{2} \in S$. There exists a compact topological space X and a subring B of $\mathcal{C}(X,\mathbb{R})$ with $(A,S) \cong (B,B_+)$ if, and only if, the following hold:*

(1) *S is Archimedean;*
(2) *$S \cap (-S) = \{0\}$;*
(3) *if $f \in A$ and $1 + nf \in S$ for all $n \in \mathbb{N}$, then $f \in S$.*

If (1)–(3) hold, one can moreover arrange in addition that B is a dense subring of $\mathcal{C}(X,\mathbb{R})$.

Here "dense subring" refers to the norm $\|g\| = \max\{|g(x)|\colon x \in X\}$ of uniform convergence. The condition $(A,S) \cong (B,B_+)$ means that there exists a ring isomorphism $A \to B$ that maps S onto B_+.

Proof. Properties (1)–(3) are obvious if X is a compact space and $(A,S) = (B,B_+)$, with $B \subseteq \mathcal{C}(X,\mathbb{R})$ a subring. Conversely assume that (A,S) satisfies (1)–(3). We take $X := X_S$, which is a compact topological space since S is Archimedean (5.2.23). The ring homomorphism

$$\Phi\colon\ A \to \mathcal{C}(X_S,\mathbb{R}),\quad f \mapsto \hat{f}$$

(see 5.2.20) satisfies $\Phi^{-1}(\mathcal{C}_+(X_S,\mathbb{R})) = S$, by the Archimedean nichtnegativstellensatz 5.3.5 and by (3). So (2) implies that Φ is injective. By Corollary 5.2.26, $\Phi(A)$ is a dense subring of $\mathcal{C}(X_S,\mathbb{R})$. □

Exercises

5.3.1 Let the set $S \subseteq \mathbb{R}[x,y]$ consist of all finite sums of polynomials of the form

$$\big(p_1(x)^2 + p_2(x)^2(1-x^2)\big)\big(q_1(y)^2 + q_2(y)^2(1-y^2)\big)$$

with $p_i(x) \in \mathbb{R}[x]$ and $q_i(y) \in \mathbb{R}[y]$. Show that S is an Archimedean semiring in $\mathbb{R}[x,y]$ whose associated basic closed set is $X_S = [-1,1] \times [-1,1] \subseteq \mathbb{R}^2$. Deduce that $c + (x-y)^2 \in S$ for every real number $c > 0$, and decide whether this remains true for $c = 0$. (A more general statement will be proved in Exercise 5.5.2.)

5.3.2 Let $\mathbb{R}[t]$ be the polynomial ring in one variable, let $S \subseteq \mathbb{R}[t]$ be the semiring generated by $\mathbb{R}_+$ and t, and let $M = S + S(1-t)$.

(a) S is a generating semiring of $\mathbb{R}[t]$, and M is an Archimedean S-module.
(b) If $0 < c < 1$ and $f = c + (1-t^2)^2$ then $f > 0$ on X_M, but $f \notin M$.

Hence the Archimedean positivstellensatz is usually false for Archimedean modules over generating semirings.

5.3.3 For another example in a similar vein, let Σ_h be the set of sums of Hermitian squares in $\mathbb{C}[z, \overline{z}]$ (one complex variable z), a generating semiring in the polynomial ring $A = \mathbb{R}[x, y]$ in two variables (Example 5.1.6). The Σ_h-module $M = \Sigma_h + \Sigma_h(1 - |z|^2)$ in A is Archimedean according to Exercise 5.2.4. When $0 < c < 1$ is a real number, show that the polynomial $f = c + (1 - |z|^2)^2$ is strictly positive on X_M, but $f \notin M$.

5.3.4 In the notation of Exercise 5.1.6, let $z = (z_1, \dots, z_n)$ and $w = (w_1, \dots, w_n)$, let $f(z, w) \in \mathbb{C}[z, w]$ be such that $f = f^*$ and $f(u, \overline{u}) > 0$ for every $u \in \mathbb{C}^n$ with $|u| = 1$. Prove Quillen's theorem [162]: There exist finitely many polynomials $p_1, \dots, p_r \in \mathbb{C}[z]$ such that

$$f(u, \overline{u}) = \sum_{j=1}^{r} |p_j(u)|^2$$

holds for every $u \in \mathbb{C}^n$ with $|u| = 1$. (*Hint*: Use Exercise 5.2.4 for the module $M = \Sigma_h + I$, where I is the principal ideal generated by $1 - \sum_{j=1}^{n} |z_j|^2$.)

5.4 First Applications: Theorems of Pólya and Handelman

With the Archimedean positivstellensatz at our disposal, we start the row of applications, and we begin with semirings. The following theorem is a classical result by Pólya. Using Theorem 5.3.1 it admits an easy proof:

5.4.1 Theorem. (Pólya) *Let $f \in \mathbb{R}[x_1, \dots, x_n]$ be a homogeneous polynomial. The following are equivalent:*

(i) *f is strictly positive on $C := \{\xi \in \mathbb{R}^n : \xi_1 \geq 0, \dots, \xi_n \geq 0\} \setminus \{(0, \dots, 0)\}$;*
(ii) *there is an integer $N \geq 1$ such that all coefficients of the form $(x_1 + \cdots + x_n)^N \cdot f$ are strictly positive.*

Proof. Let $h = x_1 + \cdots + x_n$. Condition (ii) implies that $h^N f$ is strictly positive on C, and hence that $f > 0$ on C since $h > 0$ on C. So (ii) $\Rightarrow$ (i) is clear. We prove (i) $\Rightarrow$ (ii) by applying the Archimedean positivstellensatz to a suitable semiring. Let V be the complement of the hyperplane $h = 0$ in projective space $\mathbb{P}^{n-1}$. Then V is an affine $\mathbb{R}$-variety, isomorphic to $\mathbb{A}^{n-1}$, and has coordinate ring $\mathbb{R}[V] = \mathbb{R}[\frac{x_1}{h}, \dots, \frac{x_n}{h}]$ (see A.6.10). Let $S \subseteq \mathbb{R}[V]$ be the semiring generated by $\mathbb{R}_+$ and $\frac{x_1}{h}, \dots, \frac{x_n}{h}$. From $\frac{x_1}{h} + \cdots + \frac{x_n}{h} = 1$ we see that $1 - \frac{x_i}{h} \in S$ for $i = 1, \dots, n$. So S is Archimedean by Proposition 5.2.7(a). The set $X_S \subseteq V(\mathbb{R})$ consists of the homogeneous tuples $[\xi] \in V(\mathbb{R})$ with $\frac{\xi_j}{\xi_1 + \cdots + \xi_n} \geq 0$ for $j = 1, \dots, n$, and so

$$X_S = \{(\xi_1 : \cdots : \xi_n) \in \mathbb{P}^{n-1}(\mathbb{R}) : \xi_1 \geq 0, \dots, \xi_n \geq 0\}.$$

Let $d = \deg(f)$. Then $\frac{f}{h^d} \in \mathbb{R}[V]$, and $\frac{f}{h^d} > 0$ on X_S holds by (i). So the positivstellensatz 5.3.3 implies $\frac{f}{h^d} \in S$. This means that there is an identity

$$\frac{f}{h^d} = \sum_{e \in \mathbb{Z}_+^n} c_e \cdot \frac{x_1^{e_1} \cdots x_n^{e_n}}{h^{e_1+\cdots+e_n}}$$

in $\mathbb{R}(x)$ with real numbers $c_e \geq 0$, almost all of them zero. Multiplying by a high power of h we see that, for large N, the coefficients of $h^N f$ are non-negative.

We show that all coefficients of $h^N f$ (in degree $d + N$) can even be made strictly positive, by further increasing N. Since $f > 0$ on the standard $(n-1)$-simplex

$$\Delta = \Big\{\xi \in \mathbb{R}^n \colon \xi_1 \geq 0, \ldots, \xi_n \geq 0, \sum_{i=1}^n \xi_i = 1\Big\},$$

and since Δ is compact, there exists $\varepsilon > 0$ with $f > \varepsilon > 0$ on Δ. Hence the form $f_1 := f - \varepsilon h^d$ is strictly positive on Δ, and therefore on C. By the first part of the proof there exists $N \geq 0$ such that $h^N f_1$ has non-negative coefficients. Therefore $h^N f = h^N f_1 + \varepsilon h^{N+d}$ has all coefficients strictly positive. □

5.4.2 Remarks.

1. The essential step in the previous proof was to show that the first condition in 5.4.1 implies

(ii') there is $N \in \mathbb{N}$ such that $(x_1 + \cdots + x_n)^N f$ has non-negative coefficients,

which is a slightly weaker form of (ii). Often the conclusion (i) $\Rightarrow$ (ii') alone is referred to as Pólya's theorem.

2. If we allow f in Theorem 5.4.1 to be inhomogeneous, the conclusion of this theorem becomes false in general. For similar reasons, it fails if f is allowed to have a zero in C. See Exercises 5.4.1 and 5.4.2.

3. Pólya's theorem can be stated over any real closed field R. However it becomes false as soon as R is non-Archimedean. A counterexample is the quadratic form $f = (x + y)^2 + c(x - y)^2$, if $c \in R$ is larger than any integer. This follows from Exercise 5.4.3.

4. It is natural to ask for a quantitative version of Pólya's theorem, i.e. for a bound on the exponent N in terms of the given form f. Writing f with normalized coefficients

$$f = \sum_{|\alpha|=d} \frac{d!}{\alpha_1! \cdots \alpha_n!} c_\alpha x_1^{\alpha_1} \cdots x_n^{\alpha_n}$$

(where $c_\alpha \in \mathbb{R}$), let $c = \max_\alpha |c_\alpha|$ and $\lambda := \min f(\Delta) > 0$, where $\Delta \subseteq \mathbb{R}^n$ is the standard $(n-1)$-simplex. Then if

$$N > \frac{d}{2}(d-1)\frac{c}{\lambda} - d,$$

the form $(x_1 + \cdots + x_n)^N f$ has positive coefficients (Powers-Reznick [156]). Remarkably, this bound doesn't depend on n. For $d = 2$ it is even best possible, see Exercise 5.4.3.

We now discuss another application of Theorem 5.3.1 to semirings.

5.4.3 Theorem. (Handelman) *Let $f_1, \dots, f_r \in \mathbb{R}[x] = \mathbb{R}[x_1, \dots, x_n]$ be linear polynomials such that the polyhedron*

$$K = \mathcal{S}(f_1, \dots, f_r) = \{\xi \in \mathbb{R}^n \colon f_1(\xi) \geq 0, \dots, f_r(\xi) \geq 0\}$$

is compact (i.e., a polytope) and non-empty. Then every polynomial $f \in \mathbb{R}[x]$ with $f|_K > 0$ has a representation

$$f = \sum_{\alpha \in \mathbb{Z}_+^r} c_\alpha f_1^{\alpha_1} \cdots f_r^{\alpha_r} \tag{5.5}$$

with real numbers $c_\alpha \geq 0$ (almost all of them zero).

Our proof via the Archimedean positivstellensatz needs a classical result from polyhedral geometry. A (self-contained) proof will be given later (Corollary 8.1.24), and for the moment we just quote it:

5.4.4 Proposition. *Let $f, f_1, \dots, f_r \in \mathbb{R}[x] = \mathbb{R}[x_1, \dots, x_n]$ be linear polynomials such that the polyhedron $K = \{\xi \in \mathbb{R}^n \colon f_1(\xi) \geq 0, \dots, f_r(\xi) \geq 0\}$ is non-empty. If $f \geq 0$ on K, there exist $a_0, \dots, a_r \geq 0$ in $\mathbb{R}$ with $f = a_0 + a_1 f_1 + \cdots + a_r f_r$.*

Obviously, the converse is true as well. We remark that Theorem 5.4.4 holds over any real closed field, for example by Tarski's principle.

Proof of Theorem 5.4.3. Let $S \subseteq \mathbb{R}[x]$ be the semiring generated by $\mathbb{R}_+$ and $f_1, \dots, f_r$. Clearly $X_S = K$. Since K is compact by hypothesis, there is a positive real number c with $K \subseteq [-c, c]^n$. From Proposition 5.4.4 it follows that $c \pm x_i \in S$ for $i = 1, \dots, n$. Therefore S is Archimedean by 5.2.7(a). Now the claim follows again from Theorem 5.3.3. □

Using the strength of the Archimedean positivstellensatz, it is easy to arrive at versions of Theorem 5.4.3 that are considerably more general, like the following one:

5.4.5 Corollary. *Let $f_1, \dots, f_r \in \mathbb{R}[x]$ be linear polynomials such that the polyhedron $K = \{\xi \in \mathbb{R}^n \colon f_i(\xi) \geq 0 \ (i = 1, \dots, r)\}$ is compact and non-empty. Let again $S \subseteq \mathbb{R}[x]$ be the semiring generated by $\mathbb{R}_+$ and $f_1, \dots, f_r$, and let $g_1, \dots, g_s \in \mathbb{R}[x]$ be arbitrary polynomials. Then every polynomial that is strictly positive on $K' = \{\xi \in K \colon g_j(\xi) \geq 0 \ (j = 1, \dots, s)\}$ lies in $M = S + S g_1 + \cdots + S g_s$.*

Proof. The semiring S is Archimedean as shown in the previous proof, and M is an S-module. Since $X_M = K'$, the assertion follows again from 5.3.3. □

5.4.6 Remarks.

1. The condition $K \neq \emptyset$ cannot be dropped in 5.4.3 (or in 5.4.4), as the example $f_1 = x_1 - 1$, $f_2 = -x_1$, $f = x_2$ in $\mathbb{R}[x_1, x_2]$ shows. Here $K = \mathcal{S}(f_1, f_2)$ is empty, but f is not contained in the semiring generated by f_1, f_2 and $\mathbb{R}_+$.

2. In Handelman's theorem 5.4.3, it is clear that a non-zero polynomial f can never have a representation (5.5) if it vanishes somewhere in the interior of K. Moreover, it is not hard to see that the theorem fails if the field $\mathbb{R}$ is replaced by a non-Archimedean field, see Exercise 5.4.4. On the other hand, Theorem 5.4.3 remains true if $\mathbb{R}$ is replaced by an Archimedean real closed field.

3. Linearity of the polynomials f_i in Theorem 5.4.3 is essential. For an illustration, let $S \subseteq \mathbb{R}[t]$ (one variable) be the semiring generated by t, $1 - t^2$ and $\mathbb{R}_+$. Then $X_S = [0, 1]$ is compact, but $c - t \notin S$ for any $c \in \mathbb{R}$ (see Exercise 5.4.5).

4. Pólya's theorem 5.4.1 can be seen as a particular case of Handelman's theorem 5.4.3: The variety $V = \mathbb{P}^{n-1} \setminus \mathcal{V}(h)$ considered in the proof of 5.4.1 is affine space $\mathbb{A}^{n-1}$ with linear coordinates $y_i = \frac{x_i}{h}$ $(i = 1, \dots, n-1)$. Under the resulting isomorphism $\mathbb{R}[V] \cong \mathbb{R}[y] = \mathbb{R}[y_1, \dots, y_{n-1}]$, the semiring $S \subseteq \mathbb{R}[V]$ defined in the proof of 5.4.1 corresponds to the semiring $S' \subseteq \mathbb{R}[y]$ generated by $\mathbb{R}_+$ together with $y_1, \dots, y_{n-1}$ and $1 - (y_1 + \cdots + y_{n-1})$. Applying Theorem 5.4.3 to S' and translating back to $\mathbb{R}[V]$ gives Theorem 5.4.1.

5. Originally, the theorems of Pólya and Handelman discussed in this section were proved by very different arguments. The proofs presented here are essentially due to Wörmann (1996), see [20]. With the Archimedean positivstellensatz in our pocket, observe how natural and easy the proofs became. A similar remark holds for a classical theorem by Quillen, for which an easy proof was presented in Exercise 5.3.4 using the Archimedean positivstellensatz.

Exercises

5.4.1 The dehomogenized version of Pólya's theorem is false: Find a polynomial $f \in \mathbb{R}[x_1, \dots, x_n]$ (necessarily inhomogeneous) with $f > 0$ on $\mathbb{R}^n$, such that $(1 + x_1 + \cdots + x_n)^N \cdot f$ has a negative coefficient for each $N \geq 0$.

5.4.2 Let $f \in \mathbb{R}[x_1, \dots, x_n]$ be a homogeneous polynomial that is non-negative on the positive orthant $\mathbb{R}^n_+$. If there exist $\xi, \eta \in \mathbb{R}^n_+$ with $f(\xi) = 0$, $f(\eta) \neq 0$ and $\{i \colon \xi_i = 0\} \subseteq \{i \colon \eta_i = 0\}$, show that $(x_1 + \cdots + x_n)^N \cdot f$ has a negative coefficient for every $N \geq 0$.

5.4.3 Let c be a positive real number and let $f(x, y) = (x + y)^2 + c(x - y)^2$. Show that if $n \in \mathbb{N}$ is even with $n < c - 1$, then the form $(x + y)^n \cdot f(x, y)$ has a negative coefficient.

5.4.4 Prove the assertions made in Remark 5.4.6.2.

5.4.5 Let $\mathbb{R}[t]$ be the polynomial ring in one variable. For $a \in \mathbb{R}$ let S_a be the semiring in $\mathbb{R}[t]$ generated by $t - a$, $1 - t^2$ and $\mathbb{R}_+$. The set X_{S_a} is compact and non-empty for $a \leq 1$. Show that S_a fails to be Archimedean for $a \leq 0$, but is Archimedean for $a > 0$.

5.4.6 Let $f = f(x) \in R[x] = R[x_1, \dots, x_n]$ be a form that is strictly positive definite. Following Habicht (1940), we give an explicit proof for the fact that f is a sum of squares of rational functions. Apart from the use of Pólya's theorem, the proof is completely elementary.

(a) Use Pólya's theorem to show: There is a form $g(x) \neq 0$ such that the product $f(x)g(x)$ has non-negative coefficients and contains only even monomials $x^{2\beta}$.
(b) Let $\deg(f) = 2d$ and put $F(x, u) = f(x) + u^{2d}$, where u is a new variable. For every $i \geq 0$, show that there are integers $0 \leq j < 2d$ and $k \geq 0$ with $u^i = F(x, u)P(x, u) \pm f(x)^k u^j$ for some polynomial $P(x, u)$ and some sign $\pm$.
(c) Use (a) to find a form $G(x, u)$ such that $F(x, u)G(x, u) = \sum_{\beta, i} c_{\beta, i} x^{2\beta} u^{2i}$ with $c_{\beta, i} \geq 0$. Use (b) to rewrite the right-hand side modulo $F(x, u)$ as a polynomial of degree $\leq 4d-2$ in u with coefficients in $R[x]$. Now compare the coefficients of suitable powers of u.

Hints: For (a), note that the product $\prod_{\varepsilon \in \{0,1\}^n} f(\varepsilon_1 x_1, \dots, \varepsilon_n x_n)$ has only even monomials. For (b) write $i = 2dk + j$.

5.5 Schmüdgen's Positivstellensatz and Consequences

We now apply the Archimedean positivstellensatz to preorderings and quadratic modules. If $f_1, \dots, f_r \in \mathbb{R}[x] = \mathbb{R}[x_1, \dots, x_n]$, recall that $\mathcal{S}(f_1, \dots, f_r) = \{\xi \in \mathbb{R}^n \colon f_i(\xi) \geq 0 \ (i = 1, \dots, r)\}$ denotes the associated basic closed set in $\mathbb{R}^n$. The central result here is:

5.5.1 Theorem. (Schmüdgen) *Let $f_1, \dots, f_r \in \mathbb{R}[x] = \mathbb{R}[x_1, \dots, x_n]$ be polynomials for which the basic closed set $K = \mathcal{S}(f_1, \dots, f_r) \subseteq \mathbb{R}^n$ is compact. Then the preordering $PO(f_1, \dots, f_r)$ in $\mathbb{R}[x]$ contains every polynomial that is strictly positive on K.*

5.5.2 Corollary. *Let V be an affine $\mathbb{R}$-variety for which $V(\mathbb{R})$ is compact. Every regular function $f \in \mathbb{R}[V]$ that is strictly positive on $V(\mathbb{R})$ is a sum of squares in $\mathbb{R}[V]$.*

Proof. Choose a closed embedding $V \subseteq \mathbb{A}^n$, and let the vanishing ideal $I = \mathcal{I}(V)$ be generated by $f_1, \dots, f_r \in \mathbb{R}[x]$. The preordering $T := PO(\pm f_1, \dots, \pm f_r)$ in $\mathbb{R}[x]$ satisfies $T = \Sigma\mathbb{R}[x]^2 + I$ (easy, see Exercise 5.1.1). Since $\mathcal{S}(\pm f_1, \dots, \pm f_r) = V(\mathbb{R})$, the assertion is a particular case of Theorem 5.5.1. Alternatively, it is also a direct consequence of Theorem 5.5.3 below. □

For K as in Theorem 5.5.1, Schmüdgen derived 5.5.1 from his solution of the K-moment problem, that he proved by combining tools from operator theory with the Krivine–Stengle positivstellensatz. We present a purely algebraic proof that is based on the Archimedean positivstellensatz, and also on Krivine–Stengle. Another algebraic proof, based on Pólya's theorem, is due to Schweighofer [193].

Theorem 5.5.1 follows from the next theorem, combined with the Archimedean positivstellensatz:

5.5.3 Theorem. (Wörmann) *Let A be a finitely generated $\mathbb{R}$-algebra, let $T \subseteq A$ be a finitely generated preordering for which X_T is compact. Then T is Archimedean.*

Proof. Let the $\mathbb{R}$-algebra A be generated by $x_1, \dots, x_n$, and let $V = \mathrm{Spec}(A)$. The set X_T is a (basic closed) semialgebraic set in $V(\mathbb{R})$, since the preordering T is finitely generated. So the subset $X(T)$ of $\mathrm{Sper}(A)$ is the unique constructible set associated with the semialgebraic set X_T (Proposition 4.1.2). Since X_T is compact by hypothesis, there exists a real number c such that $\sum_i x_i^2 < c$ on X_T. Therefore $f := c - \sum_i x_i^2$ satisfies $f > 0$ on X_T, and hence $f > 0$ on the subset $X(T)$ of $\mathrm{Sper}(A)$. To show that T is Archimedean, it suffices to find $b \in \mathbb{R}$ with $b + f \in T$, according to the criterion in 5.2.7(b). By the Krivine–Stengle positivstellensatz 3.2.7, there exists $t \in T$ with $(1+t)f \in T$. This implies

$$(1+t)f + t \cdot \sum_i x_i^2 \;=\; f + ct \,\in T. \tag{5.6}$$

Let $Q = T + fT$, the preordering generated by T and f. Since $(1+t)f \in T$ we have

$$(1+t)Q \,\subseteq T. \tag{5.7}$$

Now Q is Archimedean by 5.2.7(b) since $f \in Q$. So there exists $a \in \mathbb{R}$ with $a - t \in Q$. Therefore (5.7) gives

$$c(1+t)(a-t) \;=\; ca + c(a-1)t - ct^2 \in T. \tag{5.8}$$

Finally

$$c\Big(\frac{a}{2} - t\Big)^2 \;=\; c\frac{a^2}{4} - act + ct^2 \,\in T. \tag{5.9}$$

Adding (5.6), (5.8) and (5.9) we get $f + c\big(a + \frac{a^2}{4}\big) \in T$, and the proof is complete. □

5.5.4 Example. Let us illustrate Theorem 5.5.1 in the simplest possible case, univariate polynomials. Let $h = t(1-t) \in \mathbb{R}[t]$ and $T = PO(h)$, write $K = X_T = [0,1]$. By 5.5.1, every $f \in \mathbb{R}[t]$ with $f|_K > 0$ can be written $f = s + th$ with sums of squares s, t in $\mathbb{R}[t]$.

So far this is not deep. In fact, the conclusion is true and elementary even when only $f|_K \geq 0$ is assumed. It gets interesting after we give it a slight twist. Instead of h and T consider $h' = t^3(1-t)$ and $T' := PO(h')$. We still have $K = X_{T'} = [0,1]$. But now $t \notin T'$, since in an identity $t = s + s'h'$ with sums of squares s, s', the polynomial s would be divisible by t and hence by t^2, giving a contradiction. On the other hand, Schmüdgen's theorem gives for every real number $\varepsilon > 0$ an identity $t + \varepsilon = s + s'h'$ with sums of squares s, $s' \in \mathbb{R}[t]$. If $\varepsilon \to 0$ then the degrees of s and s' necessarily explode to infinity, see Exercise 5.5.5.

This simple example already indicates that, in the general situation of Theorem 5.5.1, it will usually be hard to find explicit preordering representations, for given polynomials that are strictly positive on K. For the concrete example discussed here, reasonable upper degree bounds depending on $\varepsilon > 0$ are in fact available. But in general the situation is rather dismaying. We will return to degree bounds in Section 6.6.

5.5.5 Remarks.

1. As the previous example already suggests, Schmüdgen's theorem 5.5.1 becomes false when $\mathbb{R}$ is replaced by a non-Archimedean real closed field R. See Exercise 5.5.5. Over Archimedean R, the statement remains true.

2. In Theorem 5.5.3, one cannot drop the conditions that A and T are finitely generated. See Exercise 5.5.1 for easy counterexamples in $\mathbb{R}[t]$, and see Marshall [136] for a counterexample in $\mathbb{R}[x_1, x_2, \dots]$, the polynomial ring in infinitely many variables.

5.5.6 We now discuss a series of applications of Schmüdgen's theorem, starting with Hilbert's 17th problem. Let R be a real closed field and let $f \in R[x] = R[x_1, \dots, x_n]$ be a form that is *positive definite*, i.e. $f(\xi) > 0$ for all $0 \neq \xi \in R^n$. By Artin's solution of Hilbert 17 (Theorem 1.5.21), there exists a non-zero form $h \in R[x]$ such that fh^2 is sos. By the Krivine–Stengle positivstellensatz 3.2.7, such h can be chosen to be positive definite itself (Exercise 3.3.3). If $R = \mathbb{R}$ (or if R is Archimedean), we can use Theorem 5.5.1 to prove a much stronger statement:

5.5.7 Theorem. *Let $f, h \in \mathbb{R}[x] = \mathbb{R}[x_1, \dots, x_n]$ be two positive definite forms, and assume that* $\deg(h)$ *divides* $\deg(f)$. *Then for suitable $N \geq 1$, the form $h^N f$ is a sum of squares of forms.*

5.5.8 Corollary. (Reznick) *For every positive definite form $f \in \mathbb{R}[x]$, the product $(x_1^2 + \dots + x_n^2)^N \cdot f$ is a sum of squares of forms for sufficiently large $N \geq 1$.*

Proof of Theorem 5.5.7. The complement V of the hypersurface $h = 0$ in $\mathbb{P}^{n-1}$ is an affine $\mathbb{R}$-variety with coordinate ring

$$\mathbb{R}[V] = \left\{\frac{g}{h^r} : g \in \mathbb{R}[x] \text{ homogeneous, } \deg(g) = r \cdot \deg(h)\right\}.$$

Let $m = \frac{\deg(f)}{\deg(h)}$. Then $\frac{f}{h^m}$ lies in $\mathbb{R}[V]$ and is strictly positive on $V(\mathbb{R})$. Since $V(\mathbb{R}) = \mathbb{P}^{n-1}(\mathbb{R})$ is compact, $\frac{f}{h^m}$ is a sum of squares in $\mathbb{R}[V]$ by Corollary 5.5.2. This means that there exist $k \geq 0$ and forms $g_1, \dots, g_r \in \mathbb{R}[x]$ of degree $k \cdot \deg(h)$, such that

$$\frac{f}{h^m} = \sum_{i=1}^{r} \left(\frac{g_i}{h^k}\right)^2.$$

Multiplying this identity by h^{2N} for sufficiently large N we get the assertion. □

5.5.9 Remarks.

1. Given any psd form $f \in \mathbb{R}[x] = \mathbb{R}[x_1, \dots, x_n]$, there always exists a sum of squares form $h \neq 0$ such that fh is a sum of squares of forms. So far, this is just Artin's solution of Hilbert 17. According to Corollary 5.5.8, h can be chosen to be a power of $x_1^2 + \dots + x_n^2$ whenever f is positive definite. In this sense, $x_1^2 + \dots + x_n^2$ is a *uniform common denominator* for positive definite forms.

2. In fact a stronger result holds, since the condition that $\deg(h)$ divides $\deg(f)$ can be dropped in Theorem 5.5.7. Therefore, any non-constant positive definite form is a uniform denominator for sos representations of arbitrary positive definite forms. The proof needs additional techniques, and we will give it in 6.5.25.

3. For ternary forms ($n = 3$), even more is true: Any non-constant positive definite form is a uniform denominator for all *non-negative* forms. On the other hand, in four or more variables, such a uniform denominator for all non-negative forms does not exist. Both assertions will be proved in the next chapter (Corollary 6.5.28 and Exercise 6.2.4, respectively).

Another important consequence of 5.5.1 is the following characterization of Archimedean quadratic modules in $\mathbb{R}[x_1, \dots, x_n]$:

5.5.10 Theorem. *Let $M \subseteq \mathbb{R}[x] = \mathbb{R}[x_1, \dots, x_n]$ be a quadratic module. The following are equivalent:*

(i) *M is Archimedean;*
(ii) *there is $c \in \mathbb{R}$ with $c - \sum_i x_i^2 \in M$;*
(iii) *there exists $f \in M$ such that the set $\mathcal{S}(f) = \{\xi \in \mathbb{R}^n : f(\xi) \geq 0\}$ is compact;*
(iv) *X_M is compact, and M contains every $f \in \mathbb{R}[x]$ with $f > 0$ on X_M.*

Most interesting is the implication (iii) $\Rightarrow$ (iv), known as *Putinar's positivstellensatz.*

Proof. The implications (i) $\Rightarrow$ (ii) $\Rightarrow$ (iii) are obvious, the equivalence of (i) and (iv) was observed in Corollary 5.3.7. Conversely, if (iii) holds then the preordering $PO(f)$ is Archimedean by Theorem 5.5.3. But $PO(f) \subseteq M$, so a fortiori, M is Archimedean as well. □

Theorem 5.5.3 states that a finitely generated preordering in $\mathbb{R}[x_1, \dots, x_n]$ is Archimedean as soon as its associated basic closed set is compact. For quadratic modules, this conclusion usually fails:

5.5.11 Example. Let $M \subseteq \mathbb{R}[x_1, \dots, x_n]$ be the quadratic module generated by

$$g_i = 2x_i - 1 \ \ (i = 1, \dots, n), \ \ g_{n+1} = 1 - x_1 \cdots x_n.$$

The semialgebraic set $K = X_M = \{\xi \in \mathbb{R}^n : \xi_i \geq \frac{1}{2} \ (i = 1, \dots, n), \ \xi_1 \cdots \xi_n \leq 1\}$ is compact since $\xi_i \leq 2^{n-1}$ $(i = 1, \dots, n)$ for every $\xi \in K$:

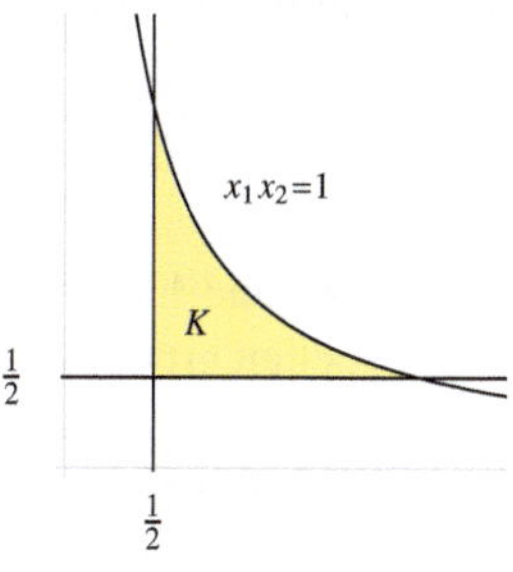

But M is not Archimedean if $n \geq 2$. In other words, $c - \sum_i x_i^2 \notin M$ for any $c \in \mathbb{R}$.

We first show this for $n = 2$, and we relabel the two variables as x, y. Assume that there are sums of squares $s_0, \dots, s_3$ in $\mathbb{R}[x, y]$ with

$$c - (x^2 + y^2) = s_0 + (2x - 1)s_1 + (2y - 1)s_2 + (1 - xy)s_3.$$

For $i = 0, \dots, 3$, the leading (highest degree) form t_i of s_i is itself a sum of squares. The leading forms of the four summands on the right are t_0, $2xt_1$, $2yt_2$ and $-xyt_3$, respectively. Let d be the maximum of their degrees. If $d \leq 2$ then s_1, s_2, s_3 are constant and $\deg(s_0) \leq 2$, which is impossible (look at the coefficient of x^2). Therefore $d \geq 3$, and some of the leading forms add up to zero. More precisely we have $t_0 - xyt_3 = 0$ if $d > 2$ is even, and $xt_1 + yt_2 = 0$ if $d > 2$ is odd. But both are impossible since the rational functions xy and $-\frac{x}{y}$ are indefinite.

Now let $n \geq 3$ and assume $c - \sum_{i=1}^n x_i^2 \in M$. Substituting $x_3 = \cdots = x_n = 1$ we get

$$c' - (x_1^2 + x_2^2) = s_0 + (2x_1 - 1)s_1 + (2x_2 - 1)s_2 + (1 - x_1x_2)s_3$$

for some $c' \in \mathbb{R}$, with sums of squares $s_i \in \mathbb{R}[x_1, x_2]$. This contradicts the case $n = 2$ just discussed.

5.5.12 Remark. Let $g_1, \dots, g_r \in \mathbb{R}[x] = \mathbb{R}[x_1, \dots, x_n]$ be such that the basic closed set $K = \mathcal{S}(g_1, \dots, g_r)$ is compact. The previous example shows that the quadratic module $M = QM(g_1, \dots, g_r)$ in $\mathbb{R}[x]$ need not be Archimedean if $r \geq 2$. In any case, we may choose a polynomial g for which the set $\mathcal{S}(g)$ is compact and contains K. For example, we may take $g = c^2 - \sum_i x_i^2$ if $|\xi| \leq c$ for all $\xi \in K$. The quadratic module $M' = QM(g_1, \dots, g_r, g)$ has $X_{M'} = X_M = K$, and M' is Archimedean by Putinar's theorem 5.5.10. Therefore, every polynomial f with $f|_K > 0$ has a representation

$$f = s_0 + \sum_{i=1}^r s_i g_i + sg$$

with sums of squares $s_0, \dots, s_r, s$ in $\mathbb{R}[x]$. These are $r + 2$ summands. If we just use Schmüdgen's theorem 5.5.1, we get a weighted sos representation with 2^r summands. For this reason, most applications work with Putinar's positivstellensatz 5.5.10.

The example given in 5.5.11 has dimension ≥ 2. Using semiorderings, we prove that there does not exist any such example of dimension one.

5.5.13 Proposition. *Let K be a function field over $\mathbb{R}$ of transcendence degree one. Then every semiordering of K is an ordering.*

Proof. Let M be a semiordering of K, and let $a \leq_M b \Leftrightarrow b - a \in M$ denote the total ordering induced by M on the abelian group $(K, +)$. From Exercise 5.2.8, recall that the subring $O(M) = \{a \in K\colon \exists n \in \mathbb{N}\ -n \leq_M a \leq_M n\}$ of K is a valuation ring of K. If $\mathfrak{m}$ denotes its maximal ideal and $k = O(M)/\mathfrak{m}$ its residue field, then $\overline{M} = \{\overline{a}\colon a \in M\}$ is an Archimedean ordering of k, by the same exercise. Since $O(M)$

clearly contains $\mathbb{R}$, we conclude that $k = \mathbb{R}$. Therefore $O(M)$ is a discrete valuation ring of (the one-dimensional function field) K.

Let t be a prime element of $O(M)$ with $t >_M 0$, and let $g \in \mathfrak{m}$. Then $\frac{t}{g^2} \notin O(M)$, which implies $\frac{t}{g^2} >_M 1$ and hence $t >_M g^2$. The element $u := 1 + g$ therefore satisfies

$$0 <_M u - t <_M \left(1 + \frac{g}{2}\right)^2 <_M u + t.$$

Applying Exercise 5.2.6(c) twice, it follows that $(u - t)^2 <_M (u + t)^2$, and therefore $tu >_M 0$. Since every positive unit of $O(M)$ has the form $v = c^2(1+g)$ with $c \in \mathbb{R}$ and $g \in \mathfrak{m}$, we conclude $vt >_M 0$ for any such v, and hence $vt^n >_M 0$ for every integer n. Therefore M contains the set $Q = \{0\} \cup \{vt^n : n \in \mathbb{Z}, v \in O(M)^*, \overline{v} > 0\}$, which is a positive cone of K by the Baer–Krull construction. It follows that $M = Q$. □

As a consequence of Proposition 5.5.13, one concludes that every quadratic module in a one-dimensional function field over $\mathbb{R}$ is a preordering (use Exercise 5.1.8). We remark that both Proposition 5.5.13 and this consequence remain true when $\mathbb{R}$ is replaced by an arbitrary real closed field.

5.5.14 Proposition. *Let $x = (x_1, \dots, x_n)$, and let M be a finitely generated quadratic module in $\mathbb{R}[x]$. Assume that* $\dim \mathbb{R}[x]/\operatorname{supp}(M) \le 1$, *and that the basic closed set X_M is compact. Then M is Archimedean.*

Proof. Every semiordering N of $\mathbb{R}[x]$ that contains M is (the positive cone of) an ordering. Indeed, the residue field of the prime ideal $\operatorname{supp}(N)$ is a function field over $\mathbb{R}$ of transcendence degree at most one. By Proposition 5.5.13, the semiordering induced by N in this field is an ordering. Therefore N is an ordering as well.

By assumption there is $c \in \mathbb{R}$ such that $\sum_{i=1}^n x_i^2 < c$ on X_M. Put $f = c - \sum_{i=1}^n x_i^2$, then the quadratic module M_1 generated by M and $-f$ satisfies $X_{M_1} = \emptyset$. We claim that M_1 contains -1. Otherwise there exists a semiordering Q of $\mathbb{R}[x]$ with $M_1 \subseteq Q$, by Corollary 5.1.21. But Q is a positive cone as just remarked, and so it would follow that $X(M_1) \neq \emptyset$. Since the quadratic module M_1 is finitely generated, the subset $X(M_1)$ of $\operatorname{Sper} \mathbb{R}[x]$ is (basic closed) constructible. Therefore we get $X_{M_1} \neq \emptyset$, a contradiction.

We have therefore seen that there exist $s \in \Sigma\mathbb{R}[x]^2$ and $g \in M$ with $sf = 1 + g$. This implies $\mathcal{S}(g) \subseteq \mathcal{S}(f)$. In particular, the set $\mathcal{S}(g)$ is compact. So Putinar's criterion 5.5.10 implies that M is Archimedean. □

5.5.15 Remark. Let $M \subseteq \mathbb{R}[x]$ be a finitely generated quadratic module such that X_M is compact. Jacobi and Prestel [101] gave necessary and sufficient conditions for M to be Archimedean. Their approach makes essential use of terminology and results from reduced quadratic forms theory, in particular of the Bröcker–Prestel local-global principle for weak isotropy. The conditions require that certain quadratic equations are solvable at "points at infinity" of M. Although these are conditions of very abstract nature, they allow concrete applications. For full details we refer to [159] Chapter 6 or [138] Chapter 8.

We discuss a result that is much more accessible, where compactness of X_M plus some extra condition implies the Archimedean property for a quadratic module M.

5.5.16 Lemma. *Let $g_1, \dots, g_r \in \mathbb{R}[x] = \mathbb{R}[x_1, \dots, x_n]$ be forms of even degree such that $\mathcal{S}(g_1, \dots, g_r) = \{0\}$. Then there exist sums of squares forms $s_0, \dots, s_r$ and an integer $N \geq 0$ such that*

$$(x_1^2 + \cdots + x_n^2)^N + s_0 + \sum_{i=1}^{r} s_i g_i = 0,$$

and such that every summand is homogeneous of degree $2N$.

Note that, conversely, such an identity implies both $\mathcal{S}(g_1, \dots, g_r) = \{0\}$ and evenness of the $\deg(g_i)$.

Proof. Let $h = x_1^2 + \cdots + x_n^2$ and $V = \mathbb{P}^{n-1} \setminus \mathcal{V}(h)$. As before, V is an affine $\mathbb{R}$-variety with compact set of $\mathbb{R}$-points. By 5.5.3, every quadratic module in $\mathbb{R}[V]$ is Archimedean. Write $\deg(g_i) = 2d_i$, and let $M \subseteq \mathbb{R}[V]$ be the quadratic module generated by $g_i h^{-d_i}$ for $i = 1, \dots, r$. Then $X_M = \varnothing$ by the hypothesis of the lemma. Since M is Archimedean it follows that $-1 \in M$ (Corollary 5.2.13). Hence there is an identity

$$-1 = \frac{s_0}{h^{2e_0}} + \sum_{i=1}^{r} \frac{s_i g_i}{h^{d_i + 2e_i}}$$

of rational functions, where s_i is a sum of squares form of degree $4e_i$ for $i = 0, \dots, r$. Clearing denominators we get the claim. □

If $g \in \mathbb{R}[x]$ is any polynomial, let $\tilde{g}$ denote the leading form (highest degree subform) of g.

5.5.17 Lemma. *Let $g_1, \dots, g_r \in \mathbb{R}[x]$. If $\mathcal{S}(\tilde{g}_1, \dots, \tilde{g}_r) = \{0\}$ then the set $\mathcal{S}(g_1, \dots, g_r)$ is compact.* (See Exercise 5.5.7 for the proof.)

5.5.18 Proposition. *Let $g_1, \dots, g_r \in \mathbb{R}[x] = \mathbb{R}[x_1, \dots, x_n]$ be polynomials such that $\mathcal{S}(\tilde{g}_1, \dots, \tilde{g}_r) = \{0\}$. If $\deg(g_i)$ is even for $i = 1, \dots, r$, the quadratic module $M = QM(g_1, \dots, g_r)$ is Archimedean.*

Proof. By Lemma 5.5.16, there are homogeneous sums of squares polynomials $s_0, \dots, s_r$ with

$$s_0 + \sum_{i=1}^{r} s_i \tilde{g}_i = -(x_1^2 + \cdots + x_n^2)^N,$$

where $N \geq 0$ and all summands have the same degree. The polynomial $g = s_0 + \sum_{i=1}^{r} s_i g_i$ lies in M, and $\tilde{g} = -(\sum_i x_i^2)^N$ which is a negative definite form. Therefore $\mathcal{S}(g)$ is compact by Lemma 5.5.17. Now Putinar's criterion 5.5.10 implies that M is Archimedean. □

The conclusion of Proposition 5.5.18 remains true if all degrees $\deg(g_i)$ are odd instead of even. The proof gets more complicated, see [138] Theorem 7.2.3.

5.5.19 Remark. Usually, applications of the Archimedean positivstellensatz are stated either for modules over Archimedean semirings, or for Archimedean quadratic modules. From the observation made in Exercise 5.2.1, one derives many more applications of "mixed" types. For concrete examples, see Exercises 5.5.2 and 5.5.3. For more background we refer to [192].

5.5.20 Although all aspects of non-commutative real algebraic geometry are otherwise ignored in this course, we make an exception here and prove the following matrix version of the Archimedean positivstellensatz. It will play a role again in Chapter 8. Let A be a (commutative) ring. Given a symmetric matrix S with coefficients in A and a ring homomorphism $\alpha\colon A \to R$ into a real closed field R, we say that $\alpha(S) > 0$ if the symmetric matrix $\alpha(S)$ over R is positive definite. Similarly for $\geq$ instead of $>$.

5.5.21 Theorem. *Let A be a ring, let M be an Archimedean quadratic module in A, and let $S \in \mathrm{Sym}_n(A)$ be such that $\alpha(S) > 0$ for every $\alpha \in X_M$. Then there exist symmetric matrices $S_1, \dots, S_m \in \mathrm{Sym}_n(A)$ and ring elements $a_1, \dots, a_m$ in M with*

$$S = \sum_{i=1}^{m} a_i S_i^2 = \sum_{i=1}^{m} a_i S_i^{\mathsf{T}} S_i.$$

When A is a polynomial ring over $\mathbb{R}$ and the quadratic module M is finitely generated, we'll prove in Theorem 8.6.6 that the degrees of the coefficients of $S_1, \dots, S_m$ can be bounded in terms of S.

Proof. Let $B := A[S]$ denote the subring of the matrix ring $\mathrm{M}_n(A)$ that is generated by A and the matrix S. Note that B is a commutative ring that is contained in $\mathrm{Sym}_n(A)$. The proof will show that the matrices S_i can in fact be chosen to lie in B. The ring extension $A \subseteq B$ is finite by the Hamilton–Cayley theorem. Therefore the quadratic module M^B, generated by M in B, is again Archimedean, see Corollary 5.2.10. Let $s := S$, considered as an element of B. We claim that $s > 0$ on the subset X_{M^B} of $\mathrm{Sper}(B)$. For this let $\beta\colon B \to \mathbb{R}$ be a homomorphism that lies in X_{M^B}. Write $\alpha = \beta|_A$, then $\beta \in X_{M^B}$ means that $\alpha \in X_M$. We claim that $\beta(s)$ is an eigenvalue of the symmetric matrix $\alpha(S) \in \mathrm{Sym}_n(\mathbb{R})$. For this let p_T denote the characteristic polynomial of a matrix T. Taking $T = S$, we have $p_S \in A[t]$ and $p_S(s) = 0$. This implies that $\beta(p_S(s)) = p_{\alpha(S)}(\beta(s)) = 0$, which shows that $\beta(s)$ is an eigenvalue of the symmetric matrix $\alpha(S)$.

Now $\alpha \in X_M$ implies $\alpha(S) > 0$ by the hypothesis. Therefore $\beta(s)$, being an eigenvalue of the matrix $\alpha(S)$, is strictly positive. We have thus shown that $s \in B$ is strictly positive on X_{M^B}. The positivstellensatz 5.3.3 therefore implies $s \in M^B$ since M^B is Archimedean. And $s \in M^B$ means an identity as claimed in the theorem (cf. Remark 5.1.8). □

Exercises

5.5.1 Find a non-Archimedean preordering $T \subseteq \mathbb{R}[t]$ in the univariate polynomial ring for which X_T is compact. (By Theorem 5.5.3, T cannot be finitely generated.)

5.5.2 Let $T_1 \subseteq \mathbb{R}[x]$ and $T_2 \subseteq \mathbb{R}[y]$ be finitely generated preorderings where $x = (x_1, \dots, x_m)$ and $y = (y_1, \dots, y_n)$, and assume that the corresponding basic closed sets $K_1 = \mathcal{S}(T_1) \subseteq \mathbb{R}^m$ and $K_2 = \mathcal{S}(T_2) \subseteq \mathbb{R}^n$ are compact. Show that every polynomial $f = f(x, y) \in \mathbb{R}[x, y]$ with $f > 0$ on $K_1 \times K_2$ can be written in the form

$$f(x, y) = f_1(x)g_1(y) + \cdots + f_r(x)g_r(y)$$

with $r \geq 1$ and with $f_i(x) \in T_1$, $g_i(y) \in T_2$ $(i = 1, \dots, r)$.

5.5.3 Consider the cylinder $K = \{\xi \in \mathbb{R}^3 : \xi_1^2 + \xi_2^2 \leq 1, |\xi_3| \leq 1\}$ in $\mathbb{R}^3$, and let $f \in \mathbb{R}[x_1, x_2, x_3]$ be strictly positive on K. Show that f can be written as a finite sum of products

$$(p_1^2 + (1 - x_1^2 - x_2^2)p_2^2) \cdot (1 + x_3)^m (1 - x_3)^n$$

with $p_1, p_2 \in \mathbb{R}[x_1, x_2]$ and $m, n \geq 0$.

5.5.4 Let $f_1, \dots, f_r \in \mathbb{R}[x] = \mathbb{R}[x_1, \dots, x_n]$ be linear polynomials such that $K = \mathcal{S}(f_1, \dots, f_r)$ is non-empty and compact, i.e. is a polytope. Show that the quadratic module $QM(f_1, \dots, f_r)$ is Archimedean.

5.5.5 Consider the univariate polynomial $f = t^3(1 - t)$ in $\mathbb{R}[t]$.

(a) Show for every fixed $d \geq 1$ that

$$\{p + qf : p, q \in \mathbb{R}[t] \text{ sos with } \deg(p) \leq d, \ 4 + \deg(q) \leq d\}$$

is a closed subset of $\mathbb{R}[t]_{\leq d}$.

(b) Conclude that in any sequence of representations $t + \frac{1}{n} = p_n + q_n f$ with sums of squares $p_n, q_n \in \mathbb{R}[t]$ for $n \geq 1$, one necessarily has $\deg(p_n), \deg(q_n) \to \infty$ as $n \to \infty$.

(c) Let R be a non-Archimedean real closed field, and let $T \subseteq R[t]$ be the preordering generated by f. If $\varepsilon \in R$ is a positive infinitesimal element, the linear polynomial $t + \varepsilon$ is strictly positive on $\mathcal{S}(f) = [0, 1]$ but is not contained in T.

5.5.6 If C is an affine curve over $\mathbb{R}$, every semiordering in $\mathbb{R}[C]$ is (the positive cone of) an ordering. (*Hint*: Proposition 5.5.13)

5.5.7 Prove Lemma 5.5.17.

5.5.8 Let A be a finitely generated $\mathbb{R}$-algebra, let $V = \operatorname{Spec}(A)$ be the associated affine $\mathbb{R}$-variety. Use Schmüdgen's theorem to prove equivalence of the following two conditions:

(i) The topological space $V(\mathbb{R}) = \operatorname{Hom}(A, \mathbb{R})$ is compact;

(ii) for a suitable integer $n \geq 1$, there exists a surjective homomorphism of $\mathbb{R}$-algebras from $\mathbb{R}[x_1, \dots, x_n]/\langle 1 - x_1^2 - \cdots - x_n^2 \rangle$ onto A.

In other words, every affine $\mathbb{R}$-variety V for which $V(\mathbb{R})$ is compact is isomorphic to a closed subvariety of the "sphere variety" $\sum_{i=1}^n x_i^2 = 1$, for some n.

5.5.9 At the cost of becoming more technical, it is possible to formulate versions of Schmüdgen's theorem with an arithmetic flavor. We illustrate this by just one example (statement (b) below).

(a) Let A be a finitely generated algebra over a ring k, and let $T \subseteq A$ be a finitely generated preordering for which $T \cap k$ is Archimedean in k and X_T is compact. Imitate the proof of 5.5.3 to show that T is Archimedean.

(b) Let $V \subseteq \mathbb{A}^n$ be an affine $\mathbb{R}$-variety that is defined by polynomial equations with rational coefficients. Let a polynomial $f \in \mathbb{Z}[x] = \mathbb{Z}[x_1, \dots, x_n]$ be given with $f > 0$ on $V(\mathbb{R})$. Using (a), show that there exist $m \geq 0$ and finitely many polynomials $g_1, \dots, g_r \in \mathbb{Z}[x]$ such that $2^m f = \sum_i g_i^2$ (as elements in $\mathbb{R}[V]$).

5.6 Pure States and the Goodearl–Handelman Theorem

In this section and the next, we'll present an alternative approach to the Archimedean positivstellensatz which is based on (pure) states of partially ordered rings. To prepare for this, we introduce the formal setup here and prove a key result. This requires the notion of locally convex vector space, together with the Hahn–Banach and Krein–Milman theorems. For a brief summary of this background we refer to Appendix B. We also use the Eidelheit–Kakutani separation theorem B.14, that applies to real vector spaces of arbitrary dimension without topology. What is outlined here is just a small part of a far more comprehensive theory, for which we refer to work of Goodearl and Handelman [74], [79].

Let $(G, +)$ be an abelian group, written additively, and let $M \subseteq G$ be a subsemigroup, always containing the neutral element 0. The theory to be discussed is empty unless $G = M - M$.

5.6.1 Definition.

(a) A group homomorphism $\varphi \colon G \to \mathbb{R}$ with $\varphi|_M \geq 0$ is called a *state* of (G, M).
(b) An element $u \in M$ with $G = M + \mathbb{Z}u$ is called an *order unit* of (G, M).
(c) Let u be an order unit of (G, M). A *monic state* of (G, M, u) is a state φ of (G, M) with $\varphi(u) = 1$.

5.6.2 Remark. An element $u \in M$ is an order unit of (G, M) if, and only if, for every $x \in G$ there exists $n \in \mathbb{Z}$ with $nu - x \in M$. We see a direct line to the concept of Archimedean modules (Section 5.2): If A is a ring and $M \subseteq A$ is an additive semigroup, then M is Archimedean if, and only if, $1 \in M$ and $u = 1$ is an order unit of (A, M).

5.6.3 Let $\mathrm{Hom}(G, \mathbb{R})$ denote the set of all group homomorphisms $\varphi \colon G \to \mathbb{R}$. By identifying φ with the tuple $(\varphi(g))_{g \in G}$, we embed $\mathrm{Hom}(G, \mathbb{R})$ into $\mathbb{R}^G = \prod_{g \in G} \mathbb{R}$. Equip $\mathbb{R}^G$ with the product topology, then $\mathbb{R}^G$ is a locally convex vector space (Example B.5) and $\mathrm{Hom}(G, \mathbb{R})$ is a closed vector subspace of $\mathbb{R}^G$. In particular, $\mathrm{Hom}(G, \mathbb{R})$ is a locally convex vector space by itself.

Assume that u is an order unit of (G, M). By $S = S(G, M, u)$ we denote the set of all monic states of (G, M, u), considered as a subset of $\mathrm{Hom}(G, \mathbb{R})$.

5.6.4 Proposition. *The set S of monic states of (G, M, u) is a compact convex subset of* $\mathrm{Hom}(G, \mathbb{R})$.

Proof. It is clear that S is convex, and also that S is a closed subset of $\mathrm{Hom}(G,\mathbb{R})$, and hence of $\mathbb{R}^G$. For any $x \in G$ there exists an integer $n_x \geq 1$ with $n_x u \pm x \in M$, since u is an order unit. Therefore every $\varphi \in S$ satisfies $|\varphi(x)| \leq n_x$, and we see that $S \subseteq \prod_{x\in G}[-n_x, n_x]$. So Tikhonov's theorem implies that S is compact. □

If K is a convex set (in some $\mathbb{R}$-vector space), recall (B.1) that a point $x \in K$ is an *extreme point* of K if $x = (1-t)y + tz$ with $0 < t < 1$ and $y, z \in K$ implies $y = z = x$. In the situation of Proposition 5.6.4, the Krein–Milman theorem (see B.9) implies that the set S of monic states is the closed convex hull of its extreme points. These latter are of particular importance, so we define:

5.6.5 Definition. The extreme points of $S(G, M, u)$ are called the *pure states* of (G, M, u).

5.6.6 We need a little more preparation. For the notions of $\mathbb{Q}$-convex sets[2] or cones in a $\mathbb{Q}$-vector space, or of $\mathbb{Q}$-algebraic interior points, we refer to B.16. As before, let G be an abelian group and $M \subseteq G$ a semigroup. Put $G_\mathbb{Q} := G \otimes \mathbb{Q}$ (tensor product over $\mathbb{Z}$), then the set $M_\mathbb{Q} := \{x \otimes \frac{1}{n} \colon x \in M,\ n \in \mathbb{N}\}$ is a $\mathbb{Q}$-convex cone in $G_\mathbb{Q}$. If $x \in G$ then $x \otimes 1 \in M_\mathbb{Q}$ if and only if $nx \in M$ for some $n \geq 1$. See Exercise 5.6.1 for the easy proofs.

If $\varphi \in \mathrm{Hom}(G,\mathbb{R})$, let $\varphi_\mathbb{Q}$ be the extension of φ to a group homomorphism $G_\mathbb{Q} \to \mathbb{R}$. Then $\varphi \geq 0$ on M holds if and only $\varphi_\mathbb{Q} \geq 0$ on $M_\mathbb{Q}$. So there is a natural bijective correspondence between states of (G, M) and states of $(G_\mathbb{Q}, M_\mathbb{Q})$.

If $u \in M$ is an order unit of (G, M) then $u \otimes 1$ is an order unit of $(G_\mathbb{Q}, M_\mathbb{Q})$ (Exercise 5.6.1). Clearly, the bijection between states of (G, M) and of $(G_\mathbb{Q}, M_\mathbb{Q})$ restricts to bijections between monic states of (G, M, u) and of $(G_\mathbb{Q}, M_\mathbb{Q}, u \otimes 1)$, and similarly for pure states. For these reasons, tensoring with $\mathbb{Q}$ is harmless when we want to study states of (G, M).

5.6.7 Lemma. *Let V be a $\mathbb{Q}$-vector space and let $C \subseteq V$ be a $\mathbb{Q}$-convex cone. A point $u \in C$ is an order unit of (V, C) if, and only if, u is a $\mathbb{Q}$-algebraic interior point of C.*

Proof. Let u be an order unit and let $v \in V$. There is $n \in \mathbb{N}$ with $nu + v \in C$, hence also $u + \frac{1}{n}v \in C$. This already proves one direction. Conversely assume that u is a $\mathbb{Q}$-algebraic interior point of C, and let $v \in V$. By assumption there is $0 < t \in \mathbb{Q}$ with $u + tv \in C$. We may assume $t = \frac{1}{n}$ with $n \in \mathbb{N}$, and get $v \in \mathbb{Z}u + C$. □

Let G be an abelian group and $M \subseteq G$ a semigroup. Using the theorems of Krein–Milman and Eidelheit (Appendix B) we prove the main result of this section:

5.6.8 Theorem. (Goodearl, Handelman) *Let $u \in M$ be an order unit of (G, M) and let $x \in G$. Assume that $\varphi(x) > 0$ holds for every pure state φ of (G, M, u). Then $nx \in M$ for some integer $n \geq 1$.*

[2] The notion of $\mathbb{Q}$-convex sets has no connection with M-convex semigroups or ideals, as considered in Sections 3.6 and 5.1.

Proof. We may replace G, M, u and x by $G\otimes\mathbb{Q}$, $M_\mathbb{Q}$, $u\otimes 1$ and $x\otimes 1$, respectively, see 5.6.6 and Lemma 5.6.7. Assume therefore that G is a $\mathbb{Q}$-vector space, $M \subseteq G$ is a $\mathbb{Q}$-convex cone and $u \in M$ is a $\mathbb{Q}$-algebraic interior point of M. Under the assumption of Theorem 5.6.8, we have to prove $x \in M$.

Let $S = S(G, M, u)$ be the set of monic states, which is a compact convex subset of the locally convex vector space $V := \mathrm{Hom}(G, \mathbb{R})$ (see 5.6.4). We first show $\varphi(x) > 0$ for every $\varphi \in S$. The evaluation map $e_x\colon V \to \mathbb{R}$, $e_x(\varphi) = \varphi(x)$ is a continuous linear form on V. By Krein–Milman (Theorem B.9), the convex hull of the set of pure states is dense in S. Since $\varphi(x) > 0$ for every pure state φ, it follows that $e_x(\varphi) = \varphi(x) \geq 0$ for every $\varphi \in S$. Assume that there is a state $\varphi \in S$ with $\varphi(x) = 0$. Then $\ker(e_x)$ is a closed supporting hyperplane of S, so it contains an extreme point of S (Corollary B.11). This contradicts the hypothesis, and hence $\varphi(x) > 0$ for every $\varphi \in S$.

Now assume that $x \notin M$. Apply Corollary B.19 (to Eidelheit's theorem) to the $\mathbb{Q}$-vector space G, the $\mathbb{Q}$-cone $M \subseteq G$ (with $\mathbb{Q}$-algebraic interior point u) and the element $x \in G$. This gives a group homomorphism $\varphi\colon G \to \mathbb{R}$ with $\varphi|_M \geq 0$, $\varphi(u) = 1$ and $\varphi(x) \leq 0$. But this contradicts the first part of the proof since φ is a monic state. Therefore we must have $x \in M$. □

5.6.9 Remark. The concept of states on partially ordered abelian groups or rings generalizes states on C^*-algebras. If A is a unital C^*-algebra, the set $\{x^*x\colon x \in A\}$ is a closed convex cone in A for which the unit 1 is an order unit. A state of A is a continuous linear functional $f\colon A \to \mathbb{C}$ that satisfies $f(x^*x) \geq 0$ for all $x \in A$. States on C^*-algebras have a natural interpretation in classical quantum mechanics. Over the years, the notion of states was generalized from this classical context to the generality sketched here. Theorem 5.6.8 is essentially due to Effros, Handelman and Shen [61]. The version stated here corresponds to Theorem 4.12 in [74]. We refer to [74] for much more details and background.

Exercises

5.6.1 Let G be an abelian group and $M \subseteq G$ a subsemigroup, and let $M_\mathbb{Q} \subseteq G_\mathbb{Q}$ be defined as in 5.6.6.

(a) For every $x \in G$, show that $x \otimes 1 \in M_\mathbb{Q}$ if and only if there is $n \in \mathbb{N}$ with $nx \in M$.
(b) If $u \in M$ is an order unit of (G, M), show that $u \otimes 1$ is an order unit of $(G_\mathbb{Q}, M_\mathbb{Q})$. Give an example to show that the converse usually fails.

5.6.2 Let (G, M, u) be as in Theorem 5.6.8, and assume that $x \in G$ satisfies $\varphi(x) > 0$ for every pure state φ of (G, M, u). Show that nx is an order unit of (G, M), for some $n \geq 1$.

5.7 Application to Archimedean Stellensätze

The purpose of this section is to present an alternative and more recent approach to Archimedean stellensätze, based on pure states and the Goodearl–Handelman theorem. In this way we will not only get quick and elegant proofs of our previous results from Section 5.3, but we'll arrive at new applications that go substantially further. This section is largely taken from [37].

Let A be a ring (always with $\frac{1}{2} \in A$). We'll work with "modules" M over semirings $S \subseteq A$ for which we don't always require that $1 \in M$. Therefore we define:

5.7.1 Definition. Let A be a ring, let $S \subseteq A$ be a semiring. An S*-pseudomodule* in A is a non-empty set $M \subseteq A$ with $M + M \subseteq M$ and $SM \subseteq M$. If $S = \Sigma A^2$ we also speak of a *quadratic pseudomodule.*

We consider S-pseudomodules M in ideals I of A for which (I, M) has an order unit u. We'll see that it is often possible to characterize the pure states of (I, M, u) quite explicitly. The first result is a step in this direction:

5.7.2 Proposition. *Let A be a ring and $I \subseteq A$ an ideal. Let $S \subseteq A$ be an Archimedean semiring, let $M \subseteq I$ be an S-pseudomodule, and let $u \in M$ be an order unit of (I, M). Then every pure state φ of (I, M, u) satisfies the following multiplicative law:*

$$\forall\, a \in A\ \forall\, b \in I\ \ \varphi(ab) = \varphi(au) \cdot \varphi(b). \tag{5.10}$$

5.7.3 Let $u \in M$ be an order unit of (I, M), let $\varphi\colon I \to \mathbb{R}$ be an additive map. Given any $a \in A$ with $\varphi(au) \neq 0$, define $\varphi_a\colon I \to \mathbb{R}$ by

$$\varphi_a(b) := \frac{\varphi(ab)}{\varphi(au)} \quad (b \in I).$$

Then φ_a is an additive map and $\varphi_a(u) = 1$. We call φ_a the *localization* of φ with respect to a. If φ was a state of (I, M) and if $aM \subseteq M$, then $\varphi(au) > 0$ and $\varphi_a|_M \geq 0$, so then φ_a is a monic state of (I, M, u). If $a_1, a_2 \in A$ satisfy $\varphi(a_i u) > 0$ for $i = 1, 2$, then a direct calculation shows

$$\varphi(a_1 u) \cdot \varphi_{a_1} + \varphi(a_2 u) \cdot \varphi_{a_2} = \varphi((a_1 + a_2)u) \cdot \varphi_{a_1+a_2}. \tag{5.11}$$

Proof of Proposition 5.7.2. Let φ be a pure state of (I, M, u). Both sides of (5.10) are bi-additive in (a, b). Therefore, and since $A = S - S$, it suffices to prove (5.10) for $a \in S$. Let $a \in S$, and note that $\varphi(au) \geq 0$.

Case 1: Assume $\varphi(au) = 0$, we have to show $\varphi(aI) = 0$. Since u is an order unit of (I, M) we have $I = M + \mathbb{Z}u$, and therefore $aI = aM + \mathbb{Z}au$. So it suffices to show $\varphi(aM) = 0$. Let $x \in M$. Then $\varphi(ax) \geq 0$, and there is an integer n such that $y := nu - x \in M$. So

$$0 \leq \varphi(ay) = n\varphi(au) - \varphi(ax) = -\varphi(ax),$$

proving $\varphi(ax) = 0$.

Case 2: Assume $\varphi(au) > 0$. Since S is Archimedean, there is $n \in \mathbb{Z}$ with $n-a \in S$. Take n so large that $\varphi(au) < n$, hence $\varphi((n-a)u) > 0$. We can form the localized (monic) states φ_a and φ_{n-a}. By 5.11,

$$\varphi(au) \cdot \varphi_a + \varphi((n-a)u) \cdot \varphi_{n-a} \;=\; \varphi(nu) \cdot \varphi_n,$$

Now $\varphi(nu) = n$ and $\varphi_n = \varphi$, so

$$\frac{\varphi(au)}{n} \cdot \varphi_a + \left(1 - \frac{\varphi(au)}{n}\right) \cdot \varphi_{n-a} \;=\; \varphi.$$

On the left we have a proper convex combination of the monic states φ_a and φ_{n-a}. Since φ is a pure state, we conclude $\varphi_a = \varphi$. This is identity (5.10). □

If $I = A$ in Proposition 5.7.2 and M is an S-module, we may choose $u = 1$ since M is Archimedean. Thus we get:

5.7.4 Corollary. *Let $S \subseteq A$ be an Archimedean semiring and $M \subseteq A$ an S-module. Every pure state of $(A, M, 1)$ is a ring homomorphism $A \to \mathbb{R}$, and hence an element of X_M.* □

This already implies the positivstellensatz 5.3.1 for modules over Archimedean semirings:

5.7.5 Corollary. (Modules over Archimedean semirings) *Let $S \subseteq A$ be an Archimedean semiring and $M \subseteq A$ an S-module. If $f \in A$ satisfies $f > 0$ on X_M, there exists an integer $n \geq 1$ with $nf \in 1 + M$.*

Proof. Since X_M is compact, there exists $m \in \mathbb{N}$ such that $f > \frac{1}{m}$ on X_M, and so $mf - 1 > 0$ on X_M. Every pure state φ of $(A, M, 1)$ is an element of X_M by 5.7.4, so φ satisfies $\varphi(mf - 1) > 0$. Therefore, by Theorem 5.6.8, there is $n \geq 1$ with $n(mf - 1) \in M$. In particular, $nmf \in 1 + M$. □

Next we are going to work towards a proof of the positivstellensatz for Archimedean quadratic modules. First we establish the analogue of Proposition 5.7.2 for quadratic pseudomodules.

5.7.6 Lemma. *Let $n \geq 1$, and let*

$$t_n(x) \;=\; \sum_{k=0}^{n} \binom{1/2}{k} (-x)^k \;=\; 1 - \frac{x}{2} - \frac{x^2}{8} - \cdots - \frac{1 \cdot 3 \cdots (2n-3)}{2^n\, n!}\, x^n,$$

the n-th Taylor polynomial of $\sqrt{1-x}$. Then the polynomial $t_n(x)^2 - (1-x)$ has non-negative coefficients in $\mathbb{Z}[\frac{1}{2}]$.

Proof. Write $t_n(x)^2 - (1-x) = \sum_{k \geq 0} c_k x^k$. Then $c_k = 0$ for $k \leq n$ and for $k > 2n$. If $n < k \leq 2n$ we have

$$c_k \;=\; (-1)^k \sum_{i=k-n}^{n} \binom{1/2}{i} \binom{1/2}{k-i}.$$

All integers i, $k-i$ in this sum are ≥ 1, and so the i-th summand in the sum has sign $(-1)^{i-1}\cdot(-1)^{k-i-1}$. Hence $c_k > 0$. Since $\binom{1/2}{n}$ lies in $\mathbb{Z}[\frac{1}{2}]$ for all $n \geq 0$, this proves the lemma. □

5.7.7 Lemma. *Let A be a ring, let $I \subseteq A$ be an ideal and $M \subseteq I$ a quadratic pseudomodule with order unit $u \in M$. Let $a \in A$ with $aM \subseteq M$ and $(1-2a)u \in M$. Then we have $\varphi((1-a)M) \geq 0$ for any monic state φ of (I, M, u).*

Proof. If a, $b \in M$, let $a \leq_M b$ stand for $b - a \in M$, as before. We have $au \leq_M \frac{u}{2}$ by hypothesis, and we may multiply the inequality by a since $aM \subseteq M$. Thus we inductively conclude $a^k u \leq_M 2^{-k}u$ for $k \geq 0$.

Let $b \in M$. There is $r \geq 0$ with $2^r u - b \in M$. To show $\varphi((1-a)b) \geq 0$ we may replace b by $2^{-r}b$. So we can assume $u - b \in M$. We'll show $\varphi((1-a)b) > -\varepsilon$ for every real number $\varepsilon > 0$.

For $n \in \mathbb{N}$ let $t_n(x)$ be as in 5.7.6 and write $p_n(x) := t_n(x)^2 - (1-x)$. We have $p_n(\frac{1}{2}) < \varepsilon$ for some $n \in \mathbb{N}$ since the Taylor series converges for $|x| < 1$. Fix such n and write $p(x) := p_n(x)$. By 5.7.6 we have

$$p(x) = \sum_{k=0}^{2n} c_k x^k$$

with non-negative coefficients $c_k \in \mathbb{Z}[\frac{1}{2}]$. Since $aM \subseteq M$, this implies $p(a)M \subseteq M$. Moreover $b \leq_M u$ implies $p(a)b \leq_M p(a)u$, and so $\varphi(p(a)b) \leq \varphi(p(a)u)$. On the other hand we have

$$\varphi(p(a)u) = \sum_k c_k \varphi(a^k u) \leq \sum_k c_k 2^{-k} = p\Big(\frac{1}{2}\Big) < \varepsilon.$$

So

$$\varphi(t_n(a)^2 b) - \varphi((1-a)b) = \varphi(p(a)b) \leq \varphi(p(a)u) < \varepsilon,$$

which implies

$$\varphi((1-a)b) > \varphi(t_n(a)^2 b) - \varepsilon \geq -\varepsilon$$

since M is a quadratic pseudomodule. □

Now we proceed as in the proof of 5.7.2, to show:

5.7.8 Proposition. *Let $I \subseteq A$ be an ideal, and let $M \subseteq I$ be a quadratic pseudomodule with order unit u. Every pure state φ of (I, M, u) satisfies $\varphi(ab) = \varphi(au)\cdot\varphi(b)$ (5.10) for $a \in A$, $b \in I$.*

Proof. Since the semiring ΣA^2 is generating, it suffices to prove (5.10) for $a \in \Sigma A^2$. So let $a \in \Sigma A^2$. If $\varphi(au) = 0$ then $\varphi(aI) = 0$ is shown exactly as in the first case in the proof of 5.7.2. If $\varphi(au) > 0$, choose $k \in \mathbb{N}$ with $2^k u - au \in M$. In order to prove 5.10 we may replace a by $2^{-(k+1)}a$. After doing so, we have $(1-2a)u \in M$, and therefore $\varphi((1-a)M) \geq 0$ by Lemma 5.7.7. Now one argues as in the second case of 5.7.2, to see that φ is a proper convex combination of φ_a and φ_{1-a}. Since both are monic states of (I, M, u), and since φ is a pure state, we get $\varphi = \varphi_a$, which is the claim. □

As before, the case $I = A$ and $u = 1$ gives the positivstellensatz, now for quadratic modules:

5.7.9 Corollary. (Archimedean quadratic modules) *Let $M \subseteq A$ be an Archimedean quadratic module. If $f \in A$ satisfies $f > 0$ on X_M, then $nf \in M$ for some $n \in \mathbb{N}$.*

Proof. By 5.7.8 (applied with $I = A$ and $u = 1$), every pure state φ of $(A, M, 1)$ is a ring homomorphism, so it lies in X_M. Hence $\varphi(f) > 0$ holds by hypothesis, and the claim follows from Theorem 5.6.8. □

As in Corollary 5.7.5, we can in fact conclude $nf \in 1 + M$ for some $n \in \mathbb{N}$.

So far we have used the multiplicative condition (5.10) only for $I = A$. But (5.10) is relevant for proper ideals as well, and we'll now have a closer look at this case.

5.7.10 Lemma. *Let A be a ring and $I \subseteq A$ an ideal, let $u \in I$, and let $\varphi\colon I \to \mathbb{R}$ be an additive map with $\varphi(u) = 1$. The following are equivalent:*

(i) *The multiplicative law* (5.10) *holds, i.e.* $\forall\, a \in A\ \forall\, b \in I\ \ \varphi(ab) = \varphi(au) \cdot \varphi(b)$;
(ii) *there exists a ring homomorphism $\phi\colon A \to \mathbb{R}$ such that φ is ϕ-linear, i.e. $\varphi(ab) = \phi(a) \cdot \varphi(b)$ for all $a \in A$ and $b \in I$.*

Moreover, ϕ in (ii) is uniquely determined by φ and satisfies $\phi(a) = \varphi(au)$ $(a \in A)$.

Proof. (i) ⇒ (ii) Letting $b = u$ in (ii) gives $\phi(a) = \varphi(au)$ for all $a \in A$. So we have to define ϕ in this way. This map ϕ is additive. Moreover it satisfies $\phi(1) = 1$ and

$$\phi(a_1a_2) \;=\; \varphi(a_1 \cdot a_2u) \;=\; \varphi(a_1u) \cdot \varphi(a_2u) \;=\; \phi(a_1) \cdot \phi(a_2)$$

for $a_1, a_2 \in A$. Hence ϕ is a ring homomorphism $A \to \mathbb{R}$. The ϕ-linearity of φ is just condition (5.10) rewritten.

(ii) ⇒ (i): We have $\phi(a) = \varphi(au)$ $(a \in A)$ as above, so (ii) gives $\varphi(ab) = \varphi(au)\varphi(b)$ $(a \in A,\, b \in I)$. □

Summarizing, we find the following dichotomy:

5.7.11 Theorem. *Let $S \subseteq A$ be a preordering or an Archimedean semiring. Let $I \subseteq A$ be an ideal, $M \subseteq I$ an S-pseudomodule and $u \in M$ an order unit of (I, M). If $\varphi\colon I \to \mathbb{R}$ is a pure state of (I, M, u), there exists a unique ring homomorphism $\phi\colon A \to \mathbb{R}$ that makes φ a ϕ-linear map. It is given by $\phi(a) = \varphi(au)$ $(a \in A)$ and satisfies $\phi \in X_T$ where $T = \{t \in A\colon tu \in M\}$ (an S-module). Moreover, exactly one of the following two alternatives holds:*

(1) $\phi(u) \neq 0$ *and* $\varphi(b) = \frac{\phi(b)}{\phi(u)}$ *for all* $b \in I$;
(2) $\phi(I) = 0$ *and* $\varphi(I^2) = 0$.

Proof. The multiplicative law $\varphi(ab) = \varphi(au)\varphi(b)$ (5.10) holds for $a \in A$, $b \in I$, by Proposition 5.7.2 (if S is an Archimedean semiring) and Proposition 5.7.8 (if S is a preordering). In both cases, by Lemma 5.7.10, $\phi(a) = \varphi(au)$ ($a \in A$) defines a ring homomorphism $\phi\colon A \to \mathbb{R}$ that makes φ a ϕ-linear map. It follows that

$$\phi(b) = \varphi(ub) = \phi(u)\varphi(b) \tag{5.12}$$

holds for any $b \in I$. For any $t \in T$ we have $\phi(t) = \varphi(tu) \ge 0$ since $tu \in M$. This shows $\phi \in X_T \subseteq X_S$. There are two possible cases: (1) If $\phi(u) = \varphi(u^2) \neq 0$ then $\varphi(b) = \frac{\phi(b)}{\phi(u)}$ for $b \in I$, by (5.12). (2) If $\phi(u) = \varphi(u^2) = 0$, then $\phi(I) = 0$ by (5.12), and so $\varphi(I^2) = 0$ by the ϕ-linearity of φ. □

5.7.12 Remarks.

1. In general, both $\phi(u) > 0$ and $\phi(u) < 0$ are possible in case (1), and accordingly, both $\phi \in X_M$ and $\phi \in X_{-M}$. In many standard situations however, the second cannot occur. For example, when $M = I \cap N$ for some quadratic module N of A, then necessarily $\phi \in X_M$ since $u^2 \in M$. The same reasoning applies when M is a semiring.

2. In Theorem 5.7.11, monic states of type (2) need not be pure, as can be seen from the discussion in Remark 5.7.13 below. On the other hand, if M is a quadratic pseudomodule, then monic states of type (1) are indeed pure under suitable conditions on M. See Exercise 5.7.5 for one result in this direction.

5.7.13 Remark. For a geometric interpretation of Theorem 5.7.11, consider the following setting. Let $A = \mathbb{R}[V]$ be the coordinate ring of an affine $\mathbb{R}$-variety, and let $W \subseteq V$ be a closed subvariety with vanishing ideal $I \subseteq A$. Let $S \subseteq A$ be an Archimedean semiring, let $M \subseteq I$ be an S-pseudomodule with order unit $u \in M$, and let $\varphi\colon I \to \mathbb{R}$ be a pure state of (I, M, u). According to Theorem 5.7.11, one of the following two cases holds. Either (1) there is a point $\xi \in V(\mathbb{R})$ with $u(\xi) \neq 0$ and $\xi \in X_S$, such that $\varphi(f) = \frac{f(\xi)}{u(\xi)}$ for all $f \in I$. In particular $\xi \notin W(\mathbb{R})$, and φ can be thought of as "normalized" evaluation at ξ. Or else (2), there is $\xi \in W(\mathbb{R})$ such that φ is a linear map $I \to A/\mathfrak{m}_\xi = \mathbb{R}$ of A-modules, where $\mathfrak{m}_\xi$ is the maximal ideal of A at ξ.

Let us consider case (2), and assume in addition that ξ is a non-singular point of both V and W. Then the conormal exact sequence (see A.6.18) implies that φ is an element of the normal space $N_\xi(W, V) = T_\xi(V)/T_\xi(W)$ of W at ξ. More concretely, $\varphi(f) = (\partial_v f)(\xi)$ for $f \in I$, the directional derivative of f at ξ in some direction $v \in T_\xi(V)$, which is well-defined modulo $T_\xi(W)$.

If we make stronger assumptions on M then more can be said. Assume that $M = S \cap I$, where either $S = PO_{\mathbb{R}[V]}(g_1, \dots, g_r)$ is a finitely generated preordering, or S is generated as a semiring in A by $\mathbb{R}_+$ and finitely many elements $g_1, \dots, g_r \in A$. Let $K = \mathcal{S}_V(g_1, \dots, g_r)$ be the associated basic closed set in $V(\mathbb{R})$. Then $\xi \in K$ in both cases (1) and (2), since $\phi \in X_T \subseteq X_S = K$ in Theorem 5.7.11. To say more on pure states of type (2), note that $I = M - M$ since (I, M) has an order unit. In particular, I is generated by M as an ideal, so we may assume that $I = \langle g_1, \dots, g_s \rangle$

with $s \le r$, and that $g_{s+1}, \dots, g_r \notin I$. If we further assume that W is irreducible and $K \cap W(\mathbb{R})$ is Zariski dense in W, then for any point $\xi \in K \cap W(\mathbb{R})$ the following is true: The image $\overline{M}$ of $M = S \cap I$ in $I/I\mathfrak{m}_\xi = N_\xi(W, V)^\vee$ is the polyhedral convex cone generated by $\overline{g}_1, \dots, \overline{g}_s$. See Exercise 5.7.7 for the proof, and see Exercise 5.7.6 for a simple application of Theorem 5.7.11.

5.7.14 Lemma. *Let $S \subseteq A$ be an Archimedean semiring and let $M \subseteq A$ be an S-pseudomodule. Let $g_1, \dots, g_r \in M$, and let $I = \langle g_1, \dots, g_r \rangle$. Then $u = g_1 + \cdots + g_r$ is an order unit of $(I, M \cap I)$.*

Therefore, if A is Noetherian and $I \subseteq A$ is any ideal, then $(I, M \cap I)$ has an order unit if and only if I is generated by $M \cap I$.

Proof. Let $f \in I$, say $f = \sum_{i=1}^r a_i g_i$ with $a_i \in A$. Since S is Archimedean, there is $n \in \mathbb{N}$ with $n - a_i \in S$ for $i = 1, \dots, r$. It follows that $nu - f = \sum_{i=1}^r (n - a_i) g_i \in M$. □

In general, Lemma 5.7.14 does not extend to Archimedean quadratic modules, see Exercise 5.7.3 for an example.

Theorem 5.7.11 has important applications in the case where $I \subseteq A$ is a proper ideal, namely to non-negative polynomials with zeros. For a detailed study we refer to Chapter 6. For now we give just one direct application to polytopes, as an illustration of Remark 5.7.13. The following extends Handelman's theorem 5.4.3:

5.7.15 Proposition. *Let $K \subseteq \mathbb{R}^n$ be a non-empty polytope described by linear inequalities $g_1 \ge 0, \dots, g_r \ge 0$, and let $S \subseteq \mathbb{R}[x] = \mathbb{R}[x_1, \dots, x_n]$ be the semiring generated by $\mathbb{R}_+$ and $g_1, \dots, g_r$. Let F be a face of K. If $f \in \mathbb{R}[x]$ satisfies $f|_F = 0$, $f|_{K \setminus F} > 0$ and $\partial_{\eta - \xi} f(\xi) > 0$ for every $\xi \in F$ and every vertex $\eta \notin F$ of K, then $f \in S$.*

Here ∂_v denotes the directional derivative in direction v, i.e. $\partial_v = \sum_{i=1}^n v_i \frac{\partial}{\partial x_i}$.

Proof. Write $A := \mathbb{R}[x]$. The semiring S is Archimedean by 5.2.7(a), since for every linear polynomial $g \in A$ there exists $c \in \mathbb{R}$ with $g + c \in \mathrm{cone}(1, g_1, \dots, g_r)$ (Proposition 5.4.4). Relabelling, we may assume $g_i|_F \equiv 0$ for $1 \le i \le s$ and $g_j|_F \not\equiv 0$ for $s + 1 \le j \le r$. Let W denote the affine hull of F in $\mathbb{R}^n$, then $I := \langle g_1, \dots, g_s \rangle$ is the full vanishing ideal of W. If f is as in the theorem then $f \in I$, and we want to show $f \in S \cap I$. By 5.7.14, $u := g_1 + \cdots + g_s$ is an order unit of $(I, S \cap I)$. So it suffices to show $\varphi(f) > 0$ for every pure state φ of $(I, S \cap I, u)$.

Let φ be such a pure state. In case (1) (cf. Remark 5.7.13), ϕ is essentially evaluation at some point $\xi \in K \setminus F$, and so $\varphi(f) > 0$ by assumption. Let φ be of type (2). Then there is $\xi \in F$ such that $\varphi = \partial_{\xi, v}$, the partial derivative at ξ in some direction $v \in \mathbb{R}^n$. Translating the coordinate system, we may assume $\xi = 0$. Then W is the linear span of F, and v is determined modulo W. We show that there exist $w \in W$ and a real number $c > 0$ such that $cv + w \in K$. This will complete the proof, since then $cv + w = \sum_i s_i \eta_i$ with real numbers $s_i > 0$, $\sum_i s_i = 1$, where the η_i are vertices of K with $\eta_i \notin F$ for at least one index i; and so $\varphi(f) = \partial_v f(\xi) = \frac{1}{c} \partial_{cv+w} f(\xi) = \frac{1}{c} \sum_i s_i \partial_{\eta_i} f(\xi) > 0$.

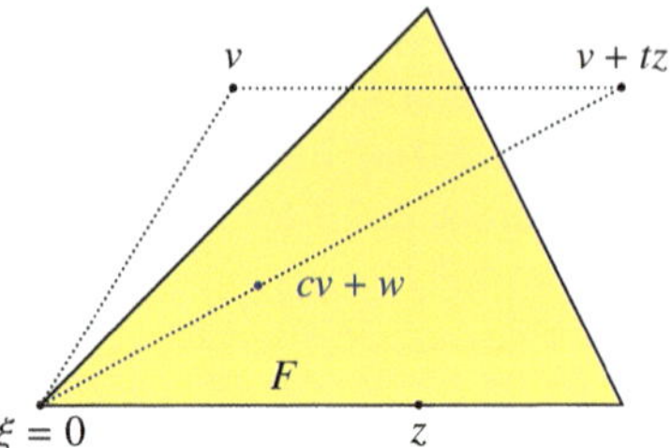

Fix a point $z \in \text{relint}(F)$, then $g_j(z) > 0$ holds for every index $s+1 \le j \le r$. By choosing $t > 0$ sufficiently large, we therefore get $g_j(v+tz) = g_j(v) + tg_j(z) > 0$ for every $j \in \{s+1, \dots, r\}$ with $g_j(0) = 0$ (note that g_j is a linear form for such j). With t being fixed, it follows for sufficiently small $c > 0$ that $g_j(c(v+tz)) \ge 0$ holds for all indices $s+1 \le j \le r$, since $g_j(0) \ne 0$ implies $g_j(0) > 0$. But now $g_j(c(v+tz)) \ge 0$ holds for all indices $j = 1, \dots, r$, since for $1 \le j \le s$ we have $g_j(c(v+tz)) = cg_j(v) = c\partial_v g_j(0) = c\varphi(g_j)$. Therefore $c(v+tz) \in K$ for such t and c. □

5.7.16 Remark. Techniques and arguments as in Remark 5.7.13 or Proposition 5.7.15 can also be applied to rings of "arithmetic" nature. For example, if A is a finitely generated $\mathbb{Q}$-algebra and M is a quadratic pseudomodule in some ideal I of A, with order unit u of (I, M), then the pure states of (I, M, u) are the same as "after tensoring with $\mathbb{R}$". See Exercise 5.7.10 for a precise formulation.

In the next chapter we'll see applications to non-negative polynomials with zeros, that go far beyond Proposition 5.7.15. The key to many of these applications will be the following theorem:

5.7.17 Theorem. (Archimedean local-global principle) *Let A be a ring, let $S \subseteq A$ be an Archimedean semiring and $M \subseteq A$ an S-module. Assume that $f \in A$ is such that, for every maximal ideal $\mathfrak{m}$ of A, there exists $s \in S \setminus \mathfrak{m}$ with $sf \in M$. Then $nf \in M$ for some $n \in \mathbb{N}$.*

Proof. Let $I = \text{supp}(M + Af)$, and let $\tilde{J}$ be the ideal generated by $M \cap I$. For every $\mathfrak{m} \in \text{Max}(A)$ there is $s \in S \setminus \mathfrak{m}$ with $sf \in M$, and $sf \in \tilde{J}$ since $f \in I$. Letting $\mathfrak{m}$ range over all maximal ideals of A, we find elements $s_1, \dots, s_r \in S$ with $\langle s_1, \dots, s_r \rangle = \langle 1 \rangle$ such that $s_i f \in \tilde{J}$ for $i = 1, \dots, r$. Take an identity $\sum_{i=1}^r a_i s_i = 1$ and multiply it by f, to see that $f \in \tilde{J}$. So there exist finitely many elements $x_1, \dots, x_m \in M \cap I$ with $f \in \langle x_1, \dots, x_m \rangle$. Since $I = \text{supp}(M+Af)$, there exist further elements $y_1, \dots, y_m \in M \cap I$ such that $x_i + y_i \in Af$ $(i = 1, \dots, m)$. Consider the ideal $J := \langle x_1, \dots, x_m, y_1, \dots, y_m \rangle$. Then $f \in J$, and Lemma 5.7.14 implies that $u := \sum_{i=1}^m (x_i + y_i)$ is an order unit of $(J, M \cap J)$. There exists $b \in A$ with $u = bf$, since $x_i + y_i \in Af$ for each i.

Let φ be a pure state of $(J, M \cap J, u)$, we show $\varphi(f) > 0$. Let $\phi \in X_S$ be the ring homomorphism associated with φ (Theorem 5.7.11). From $1 = \varphi(u) = \varphi(bf) = \phi(b)\varphi(f)$ we see $\varphi(f) \ne 0$. By the hypothesis in the theorem, there exists $s \in S$ with $\phi(s) \ne 0$ (hence $\phi(s) > 0$) and $sf \in M$. From $\varphi(sf) = \phi(s)\varphi(f) \ge 0$ we therefore get $\varphi(f) > 0$. So Theorem 5.6.8 gives the conclusion of the theorem. □

With Theorem 5.7.17, the membership question for modules over Archimedean semirings gets reduced to local rings. In many situations this makes the question more accessible, as will be seen in Chapter 6. A second and independent proof of Theorem 5.7.17, that does not rely on pure states, will be given in Section 6.2.

Exercises

5.7.1 Let A be a ring and let M be a quadratic module in A. If (A, M) has an order unit u, show that M is Archimedean.

5.7.2 Let A be a ring and $S \subseteq A$ an Archimedean semiring, let $M \subseteq A$ be an S-pseudomodule, and let $f \in M$. Show that $I := \operatorname{supp}(M + Af)$ is an ideal of A and that f is an order unit of $(I, M \cap I)$.

5.7.3 Show that Exercise 5.7.2 does not generalize to quadratic modules, by considering the following example: Let M be the quadratic module generated by x, y and $1-x-y$ in $A = \mathbb{R}[x, y]$. Then M is Archimedean and the ideal $I = Ax$ is generated by an element of M. But $(I, M \cap I)$ does not have an order unit. (*Hint*: Start by showing $M \cap I \subseteq x \cdot \Sigma A^2 + I^2$.)

5.7.4 Let A be a ring containing $\mathbb{Q}$, let $S \subseteq A$ be an Archimedean semiring containing $\mathbb{Q}_+$, and let $M \subseteq A$ be an S-pseudomodule. A given element $f \in A$ lies in M if and only if there exists an ideal I in A with $f \in A$ that has the following two properties:

(1) $(I, M \cap I)$ has an order unit u;
(2) $\varphi(f) > 0$ for every pure state φ of $(I, M \cap I, u)$.

Moreover, when $f \in M$, the ideals I with the above properties are precisely the ideals satisfying $Af \subseteq I \subseteq \operatorname{supp}(M + Af)$.

5.7.5 Let A be a ring with $\mathbb{R} \subseteq A$, let $I \subseteq A$ be an ideal, and let M be a quadratic pseudomodule in I with order unit $u \in M$ and with $x^2 \in M$ for every $x \in I$. The goal of this exercise is to give a proof of the following result (compare Theorem 5.7.11):

Theorem: *Every monic state φ of (I, M, u) satisfying $\varphi(ab) = \varphi(au)\varphi(b)$ for all $a \in A$, $b \in I$ and $\varphi(u^2) \neq 0$ is pure.*

To this end, prove the following steps:

(a) For real numbers $a, b \in [0, 1]$ one has $\sqrt{ab} + \sqrt{(1-a)(1-b)} \leq 1$, with equality only if $a = b$.
(b) Let $S(I, M) = \{\psi \in \operatorname{Hom}(I, \mathbb{R})\colon \psi|_M \geq 0\}$, the vector space of additive maps $I \to \mathbb{R}$ which are non-negative on M. Every $\psi \in S(I, M)$ is $\mathbb{R}$-linear.
(c) Let $\psi \in S(I, M)$. For all $f, g \in I$ show $\psi(fg)^2 \leq \psi(f^2)\psi(g^2)$ (1). For every $f \in M$ show $\psi(f^2)^2 \leq \psi(f)\psi(f^3)$ (2).

Hint for (c): To show (1), consider $\psi((tf-g)^2)$ for $t \in \mathbb{R}$. Use a similar argument to show (2). Let $\varphi \in S(I, M)$ as in the theorem, and let $\varphi = \varphi_1 + \varphi_2$ where $\varphi_1, \varphi_2 \in S(I, M)$. Put $\lambda = \varphi(u^2)$.

(d) Show $\lambda > 0$.
(e) Let $f \in M$ with $\varphi(f) = 1$, put $a := \varphi_1(f)$ and $b := \frac{\varphi_1(f^3)}{\lambda^2}$. Calculate $\varphi(f^2)$, and use (c)(2) and (a) to conclude: $\varphi_1(f^2) = \lambda\varphi_1(f)$.
(f) Let $f, g \in M$ with $\varphi(f) = \varphi(g) = 1$. Use (c)(1) to show $\varphi_1(f) = \varphi_1(g)$.
(g) Conclude that $\varphi_1 = \varphi_1(u) \cdot \varphi$, and use this to prove the theorem.

Hint for (f): Consider $\varphi((f-g)^2)$.

5.7.6 Let $W \subseteq V$ be a closed embedding of affine $\mathbb{R}$-varieties, and let $I = \langle g_1, \ldots, g_r \rangle$ be the vanishing ideal of W in $A = \mathbb{R}[V]$. Let $T \subseteq A$ be a finitely generated preordering with $g_1, \ldots, g_r \in T$, for which $K = \mathcal{S}_V(T)$ is compact. Let $f \in I$ be an element with $f > 0$ on $K \setminus W(\mathbb{R})$, and assume for every $\xi \in K \cap W(\mathbb{R})$ that there exists an identity $f = g + \sum_{i=1}^r g_i h_i$ with $g \in I^2$, $h_i \in A$ and $h_i(\xi) > 0$ for $i = 1, \ldots, r$. Then $f \in T$.

5.7.7 As in Remark 5.7.13, let $W \subseteq V$ be a closed embedding of irreducible affine $\mathbb{R}$-varieties. Let $A = \mathbb{R}[V]$, let $I \subseteq A$ be the vanishing ideal of W, and let $M = PO_A(g_1, \ldots, g_r)$ and $K = \mathcal{S}_V(M)$ with elements $g_i \in A$. Assume that $I = \langle g_1, \ldots, g_s \rangle$ with $1 \le s \le r$, that $g_i \notin I$ for $i = s+1, \ldots, r$, and that $K \cap W(\mathbb{R})$ is Zariski dense in W. Then for any point $\xi \in K \cap W(\mathbb{R})$, show that the image of $M \cap I$ in $I/I\mathfrak{m}_\xi$ is the convex cone generated by $\overline{g}_1, \ldots, \overline{g}_s$.

5.7.8 Let $x = (x_1, \ldots, x_n)$, let $A = \mathbb{R}[\![x]\!]$ be the ring of formal power series, with maximal ideal $\mathfrak{m}$, and let $f \in A$ be a power series with initial form f_{2d} of degree $2d$. If f_{2d} lies in the interior of the sum of squares cone $\Sigma_{n,2d}$ (cf. Section 2.4), prove that f is a sum of squares in A. (*Hint*: $(\mathfrak{m}^{2d}, \Sigma A^2 \cap \mathfrak{m}^{2d})$ has an order unit.)

5.7.9 This exercise is aiming at a generalization of Pólya's theorem 5.4.1 that allows for non-linear multipliers. Let $f, g \in \mathbb{R}[x] = \mathbb{R}[x_1, \ldots, x_n]$ be non-constant forms such that $g > 0$ on $C = \mathbb{R}^n_+ \setminus \{0\}$, and such that all coefficients of f are strictly positive. Prove that $f^m g$ has strictly positive coefficients for some $m \ge 0$, as follows.

(a) Reduce to proving that $f^m g$ has non-negative coefficients for some m.
(b) Try to imitate the proof of Pólya's theorem, defining a ring A of rational functions and a suitable semiring $S \subseteq A$. The proof should go through as long as $\deg(g)$ divides $\deg(f)$.
(c) In the general case, choose integers $k, r \ge 0$ such that $q := x_1^k g/f^r$ has total degree zero. Now use the Archimedean local-global principle 5.7.17 to show that $q \in S$.

5.7.10 Let A be a $\mathbb{Q}$-algebra, let I be an ideal in A, let $g_1, \ldots, g_r \in I$ and put $M = (\Sigma A^2)g_1 + \cdots + (\Sigma A^2)g_r$, a quadratic pseudomodule in M. Assume that $u \in M$ is an order unit of (I, M). Let $B = A \otimes \mathbb{R}$ and $J = IB = I \otimes \mathbb{R}$, and put $N = (\Sigma B^2)g_1 + \cdots + (\Sigma B^2)g_r$, a quadratic pseudomodule in N.

(a) Show that u is an order unit of (J, N).
(b) If an $\mathbb{R}$-linear map $\psi \colon J \to \mathbb{R}$ satisfies $\psi|_M \ge 0$, show that $\psi|_N \ge 0$.
(c) If $p \in I$ is such that $\psi(p) > 0$ for every pure state ψ of (J, N, u), conclude that $p \in M$.

5.8 Notes

A brief summary of the historical genesis of the Archimedean positivstellensatz has already been given in 5.3.8. Further instances of the positivstellensatz were proved by Jacobi [100], namely for Archimedean modules over the semiring ΣA^{2n}. This has applications to the representation of positive polynomials as sums of higher (even) powers, see chapter 7 of [159].

Semiorderings were originally introduced by Prestel [158] for fields, and extended to rings by Bröcker [32]. Using the local-global principle for weakly isotropic quadratic forms, proved by Prestel and Bröcker around 1973, one can give characterizations of Archimedean quadratic modules in terms of certain residue quadratic forms. This has a number of attractive consequences, see [159] chapter 6 and [138] chapter 8. Our Proposition 5.5.18 is just one simple application of this sort, that can be proved without using this theory.

Pólya's theorem 5.4.1 is from 1928 and was published in [155]. The elementary and explicit answer to Hilbert 17 for positive definite polynomials from Exercise 5.4.6 is due to Habicht [78] from 1940. Handelman's theorem is in [80]. The proof of Pólya's and Handelman's theorems from the Archimedean positivstellensatz is in [20]. Schmüdgen's positivstellensatz was proved in [190]. In his paper, Schmüdgen first solved the K-moment problem for compact basic closed sets K in $\mathbb{R}^n$, using the Krivine-Stengle positivstellensatz together with operator-theoretic techniques. The K-moment problem was then used to prove Theorem 5.5.1. The purely algebraic proof given here is again in Berr–Wörmann [20]. Putinar's positivstellensatz is in [161]. The example in 5.5.11 is taken from Jacobi–Prestel [101]. Lemma 5.2.8 is due to Brumfiel [35].

Reznick's theorem 5.5.8 on uniform denominators is taken from [165]. The stronger version 5.5.7, and its further improvement Theorem 6.5.25, are in Scheiderer [184]. The basic ideas for proving the multiplicative properties of pure states in Section 5.7 go back to Segal [195] in 1947, and are essentially all contained in Bonsall, Lindenstrauss and Phelps [30] from 1966. The Archimedean local-global principle was first proved for Archimedean preorderings in Scheiderer [183]. The version stated in Theorem 5.7.17 is from [37].

Chapter 6
Positive Polynomials with Zeros

In the previous chapter we discussed polynomials that are strictly positive on a given basic closed set. Under Archimedean hypotheses, it was shown that "denominator-free" certificates for strict positivity do exist, most of them relying on sums of squares. Now we allow the non-negative polynomials to have zeros, and explore the extent to which certificates still exist. The question may be seen as a broad generalization of the problem originally considered by Hilbert in 1888. A central tool in our analysis will be the Archimedean local-global principle, already established in Theorem 5.7.17. This principle often enables us to reduce the question to the case of local rings.

We start with elementary cases in Section 6.1, and then prove a general negative result in dimension greater than or equal to 3. Section 6.2 introduces saturated preorderings together their basic properties. Then a second and independent proof of the Archimedean local-global principle is presented that is based on real spectrum techniques. The next two sections focus on quadratic modules in local and semilocal rings. Section 6.5 shifts back to a global perspective and presents a variety of consequences. To mention just one example, it is shown that psd = sos holds on all non-singular real surfaces with a compact set of $\mathbb{R}$-points (Theorem 6.5.19). Section 6.6 addresses the existence question for degree bounds in weighted sos representations, and discusses basic results in this direction.

6.1 First Examples, and a General Negative Result

6.1.1 Let V be an affine variety over a real closed field R. Given polynomials $g_1, \dots, g_r \in R[V]$ on V, we may consider the basic closed set

$$K = \mathcal{S}_V(g_1, \dots, g_r) = \{\xi \in V(R) \colon g_1(\xi) \ge 0, \dots, g_r(\xi) \ge 0\}$$

on V on the one hand, and the preordering

C. Scheiderer, *A Course in Real Algebraic Geometry*, Graduate Texts in Mathematics 303,
https://doi.org/10.1007/978-3-031-69213-0_6

$$T = PO_V(g_1, \ldots, g_r) = \sum_{e \in \{0,1\}^r} \Sigma R[V]^2 \cdot g_1^{e_1} \cdots g_r^{e_r}$$

in $R[V]$, generated by the g_i, on the other. If $R = \mathbb{R}$ and K is compact, then T contains every $f \in \mathbb{R}[V]$ with $f|_K > 0$ (Schmüdgen's theorem 5.5.1). To what extent can one expect that such a result extends to polynomials that are just non-negative on K, possibly with zeros? This is the main question studied in this chapter.

In general, for V and K as above, the preordering $\mathcal{P}_V(K) = \mathcal{P}(K) := \{f \in R[V] \colon f|_K \geq 0\}$ is called the *saturated preordering* associated with K (terminology will be generalized in Section 6.2). If $K = \mathcal{S}_V(g_1, \ldots, g_r)$, and if $T = PO_V(g_1, \ldots, g_r)$ contains $\mathcal{P}(K)$ (hence $T = \mathcal{P}(K)$), this fact may be considered a *denominator-free nichtnegativstellensatz* on K. We have seen two situations where such a result holds, namely the affine line $V = \mathbb{A}^1$ and the plane curve $V = \mathcal{V}(1 - x^2 - y^2)$ (the circle). In either case, every non-negative polynomial on V is a sum of two squares in $R[V]$ (Lemma 2.3.2 and Proposition 2.3.1, respectively).

We start this section by discussing yet another case where such a result is true.

6.1.2 Let R be a real closed field, let $K \subsetneq R$ be a proper and non-empty closed semialgebraic set. Then K has the form

$$K = \,]-\infty, b_0] \cup [a_1, b_1] \cup \cdots \cup [a_m, b_m] \cup [a_{m+1}, \infty[$$

where

$$-\infty \leq b_0 < a_1 \leq b_1 < \cdots < a_m \leq b_m < a_{m+1} \leq \infty$$

and $m \geq 0$ (we adopt the convention that $]-\infty, -\infty] := [\infty, \infty[:= \emptyset$). The polynomials

$$p_i := (t - b_i)(t - a_{i+1}) \quad (i = 0, \ldots, m)$$

(with $t - b_0 := 1$ if K is bounded below (case $b_0 = -\infty$) and $t - a_{m+1} := -1$ if K is bounded above (case $a_{m+1} = \infty$)) are called the *natural generators* for the set K. Clearly the description $K = \mathcal{S}(p_0, \ldots, p_m)$ holds.

For an example let $K = [-2, -1] \cup \{0\} \cup [1, \infty[$. Then the natural generators are $t + 2$, $(t + 1)t = t^2 + t$ and $t(t - 1) = t^2 - t$. For $K = [-2, -1] \cup [1, 2]$ they are $t + 2$, $(t + 1)(t - 1) = t^2 - 1$ and $2 - t$.

6.1.3 Proposition. *Let $K \subsetneq R$ be as before ($K \neq R$, $K \neq \emptyset$), with natural generators $p_0, \ldots, p_m$.*

(a) *$\mathcal{P}(K) = PO(p_0, \ldots, p_m)$.*
(b) *Conversely, if K is unbounded, and if $G \subseteq R[t]$ is any set of polynomials that generates the preordering $\mathcal{P}(K)$, there exist constants $c_0, \ldots, c_m > 0$ in R such that $c_i p_i \in G$ for $i = 0, \ldots, m$.*

Assertion (a) is a denominator-free nichtnegativstellensatz for the set K. When K is unbounded, the combination of (a) and (b) says that the saturated preordering $\mathcal{P}(K)$ has a unique (up to positive scaling) minimal system of generators, which is given by the natural generators.

Proof. We prove (a) under the assumption that K has no isolated points. See Remark 6.1.4 and Exercise 6.1.4 for the general case. So assume $a_i < b_i$ for $i = 1, \dots, m$. Given $f \neq 0$ in $R[t]$ with $f|_K \geq 0$, we have to prove $f \in PO(p_0, \dots, p_m)$. We may factor f as $f = f_1 g$ with polynomials $f_1, g \in R[t]$, in such a way that $g \geq 0$ on R, and all roots of f_1 are real and simple. Then g is a sum of (two) squares in $R[t]$, and $f_1|_K \geq 0$ since K has no isolated point. Replacing f by f_1, we may therefore assume that f is real-rooted with only simple roots. Then in each of the closed intervals $[b_i, a_{i+1}]$ $(1 \leq i \leq m-1)$, the number of roots of f is even. To prove (a) it therefore suffices to show:

(1) If $b \leq \beta < \alpha \leq a$ then $(t-\alpha)(t-\beta) \in PO((t-a)(t-b))$;
(2) $t - \alpha \in PO(t-a)$ for $\alpha \leq a$, and $\beta - t \in PO(b-t)$ for $\beta \geq b$.

Statements (2) are obvious, and (1) is not hard either, see Exercise 6.1.2. This proves (a) when K has no isolated point.

For the proof of (b) let K be unbounded. Let $p = p_i$ be one of the natural generators for K. We start by reasoning that p generates an extreme ray (see 8.1.14 or B.1) in the R-convex cone $\mathcal{P}(K)$. So assume $p = q_1 + q_2$ with $q_1, q_2 \in \mathcal{P}(K)$. Then $\deg(q_j) \leq \deg(p)$ for $j = 1, 2$ since K is unbounded. Since p is real-rooted and q_1, q_2 vanish in the roots of p, it follows that q_1, q_2 are non-negative scalar multiples of p. Now assume that $g_1, \dots, g_r \in \mathcal{P}(K)$ are such that $p \in PO(g_1, \dots, g_r)$, say

$$p = \sum_{e \in \{0,1\}^r} s_e \cdot g_1^{e_1} \cdots g_r^{e_r}$$

with sums of squares s_e in $R[t]$. By the previous argument, every summand is a non-negative scalar multiple of p. Since p is not a product of two non-constant members of $\mathcal{P}(K)$, this implies $g_j \in R_+ p$ for some $j \in \{1, \dots, r\}$. So we have shown that every generating system for the preordering $\mathcal{P}(K)$ contains p, up to positive scaling. □

6.1.4 In the situation of Proposition 6.1.3, let $S \subseteq K$ be the set of isolated points of K. We briefly sketch how to prove 6.1.3(a) when $S \neq \emptyset$. Let $T = PO(p_0, \dots, p_m)$, the preordering in $R[t]$ generated by the canonical generators for K, and let $f \in R[t]$ with $f|_K \geq 0$. One can prove $f \in T$ by induction on $\sigma(f) := \sum_{c \in S} \mathrm{ord}_c(f)$. Note that f lies in T if $\sigma(f) = 0$, by the argument used before. So assume that $f(c) = 0$ for some $c \in S$. If $\mathrm{ord}_c(f) \geq 3$ then f may be replaced by $(t-c)^{-2} f$, while keeping the hypothesis $f|_K \geq 0$. Thus we can assume $\mathrm{ord}_c(f) \in \{1, 2\}$. Through a case-by-case discussion one shows that f has a factorization $f = f_1 f_2$ with $f_1, f_2 \in \mathcal{P}(K)$, and with $f_1(c) = 0$ and $\deg(f_1) \leq 2$. It follows that $f_1 \in T$, by one of (1) or (2) in the proof of 6.1.3. Since $\sigma(f_2) < \sigma(f)$, this suffices for the induction. We refer to Exercise 6.1.4 for full details.

6.1.5 Remarks.

1. Let $K \subseteq R$ be a closed semialgebraic set ($K \neq \emptyset$, $K \neq R$) and let $g_1, \dots, g_r \in R[t]$ be polynomials with $K = \mathcal{S}(g_1, \dots, g_r)$. The inclusion

$$PO(g_1, \dots, g_r) \subseteq \mathcal{P}(K) \qquad (6.1)$$

holds by definition. Whether or not (6.1) is an equality depends strongly on the choice of the g_i, as we see from Proposition 6.1.3. The set K can always be described by one single inequality $g \geq 0$, for example by letting g be the product of the natural generators. But as long as K is unbounded, the inclusion (6.1) is proper unless all natural generators are among $g_1, \dots, g_r$ (up to scaling).

2. The uniqueness property in 6.1.3(b) breaks down completely when K is bounded. In fact, we'll later see for $R = \mathbb{R}$ and for compact K, that the saturated preordering $\mathcal{P}(K)$ can always be generated by two polynomials, and even by a single polynomial if K has no isolated point (Proposition 6.5.16).

3. Let $K \subseteq R$ be a closed semialgebraic set (with $K \neq \emptyset$ and $K \neq R$). We have seen that the saturated preordering $\mathcal{P}(K)$ is generated by the natural generators $p_0, \dots, p_m$ for K. Easy examples (e.g. Exercise 5.1.3(b)) show, on the other hand, that usually the p_i do not generate $\mathcal{P}(K)$ as a quadratic module. When K is unbounded, it can be shown that $QM(p_0, \dots, p_m) = \mathcal{P}(K)$ holds if, and only if, K is either a half-line, or a half-line together with an isolated point. See [115] Theorem 2.5, and see also Exercise 6.1.3. On the other hand, when $R = \mathbb{R}$ and K is compact, we will later see (Section 6.5) that $p_0, \dots, p_m$ generate $\mathcal{P}(K)$ even in the sense of quadratic modules.

For arbitrary (closed) semialgebraic sets K in R, the preordering $\mathcal{P}(K)$ is finitely generated, as we have seen. But for sets K of higher dimension, this usually fails. In fact, the following is true:

6.1.6 Theorem. *Let V be an affine R-variety, let $g_1, \dots, g_r \in R[V]$, and let $K = \mathcal{S}_V(g_1, \dots, g_r)$, a basic closed set in $V(R)$. If $\dim(K) \geq 3$, there exists a polynomial $f \in R[V]$ with $f \geq 0$ on $V(R)$ but $f \notin PO(g_1, \dots, g_r)$.*

6.1.7 Corollary. *If $K \subseteq V(R)$ is a closed semialgebraic set of dimension $\dim(K) \geq 3$, the saturated preordering $\mathcal{P}(K)$ is not finitely generated.* □

6.1.8 To prove Theorem 6.1.6 we need to work in suitable local rings, so we start with recalling a few basic facts (see A.4.7). If $(A, \mathfrak{m}, k)$ is a regular local ring, recall that $\mathrm{gr}(A) = \bigoplus_{\nu \geq 0} \mathrm{gr}_\nu(A)$ is the associated graded ring, where $\mathrm{gr}_\nu(A) = \mathfrak{m}^\nu/\mathfrak{m}^{\nu+1}$. The order of an element $f \neq 0$ in A is $\omega(f) = \max\{\nu \geq 0 \colon f \in \mathfrak{m}^\nu\}$. The leading form of f is the residue class $L(f)$ of f in $\mathrm{gr}_n(A) = \mathfrak{m}^n/\mathfrak{m}^{n+1}$ where $n = \omega(f)$. If $a_1, \dots, a_d$ is a regular system of parameters in A, the graded ring $\mathrm{gr}(A)$ is the polynomial ring over k in the variables $y_i := L(a_i)$ $(i = 1, \dots, d)$. In particular, $L(f)$ is a homogeneous k-polynomial in $y_1, \dots, y_d$ of degree $\omega(f)$, for every $f \neq 0$.

6.1.9 Lemma. *Let $(A, \mathfrak{m})$ be a regular local ring whose residue field k is real. If $0 \neq f \in A$ is a sum of m squares in A, then $\omega(f) = 2s$ is even, and $L(f) \in \mathrm{gr}_{2s}(A)$ is a sum of m squares of elements in $\mathrm{gr}_s(A)$.*

Proof. Let $f = \sum_{i=1}^m f_i^2$ with $f_i \in A$, and let $s = \min_i \omega(f_i)$. Then $f_i^2 \in \mathfrak{m}^{2s}$ for all i, and there is at least one index i with $\omega(f_i^2) = 2s$. Since $\mathrm{gr}(A)$ is a polynomial ring over the real field k, we see that $\omega(f) = 2s$, and that $L(f)$ is the sum of the $L(f_j^2) = L(f_j)^2$ for those indices j for which $\omega(f_j) = s$. □

6.1.10 Corollary. *Let $(A, \mathfrak{m}, k)$ be a regular local ring with real residue field k, and let $a_1, \dots, a_n$ be a regular system of parameters in A. Let $g \in A[x_1, \dots, x_n]$ be a homogeneous polynomial of degree d whose coefficient-wise reduction $\overline{g}$ modulo $\mathfrak{m}$ is not sos in $k[x_1, \dots, x_n]$. Then, for arbitrary $h \in \mathfrak{m}^{d+1}$, the element $f := g(a_1, \dots, a_n) + h$ is not a sum of squares in A.*

Proof. The element $f \in A$ has order $\omega(f) = d$ and has leading form $L(f) = \overline{g}(y_1, \dots, y_n) \in \mathrm{gr}_d(A) = \mathfrak{m}^d/\mathfrak{m}^{d+1}$, with $y_i = L(a_i) \in \mathrm{gr}_1(A)$. If f were a sum of squares in A then $L(f)$ would be a sum of squares in $\mathrm{gr}(A)$ (Lemma 6.1.9). By the hypothesis of the corollary, this is not the case. □

6.1.11 Example. Let k be a real field, and let $f = \sum_\alpha c_\alpha x^\alpha$ be a formal power series over k, where $x = (x_1, \dots, x_n)$. If $\omega(f) = d$, and if the leading form $\sum_{|\alpha|=d} c_\alpha x^\alpha$ of f is not sos as a polynomial in $k[x]$, then f is not a sum of squares in $k[\![x]\!]$.

Proof of Theorem 6.1.6. The Zariski closure Z of K in V has dimension $\dim(Z) = \dim(K)$. So there exists an irreducible component Y of Z with $\dim(Y) = d \ge 3$. After relabelling the g_i we may assume that $g_{s+1}, \dots, g_r$ vanish identically on Y, while $g_1, \dots, g_s$ don't. Since $K \cap Y(R) = \mathcal{S}_Y(g_1, \dots, g_s)$ is Zariski dense in Y, the same is true for the basic open set $\mathcal{U}_Y(g_1, \dots, g_s)$ in $Y(R)$. The non-singular locus Y_{reg} of Y is open and dense in Y, so there exists a regular R-point ξ of Y that satisfies $g_i(\xi) > 0$ for $i = 1, \dots, s$. Since the local ring $A := \mathcal{O}_{Y,\xi}$ is regular of dimension d, there exist a sequence $a_1, \dots, a_d$ in $R[V]$ that generates the maximal ideal of A.

Let $p \in R[x_1, x_2, x_3]$ be a psd form that is not sos in $R[x_1, x_2, x_3]$, for example the Motzkin form. The element $f := p(a_1, a_2, a_3)$ of $R[V]$ is non-negative on $V(R)$, since f is the pullback of p under the polynomial map $(a_1, a_2, a_3)\colon V \to \mathbb{A}^3$. We show that f is not contained in the preordering $T := PO_{R[V]}(g_1, \dots, g_r)$.

To see this, let $\widehat{A}$ be the completion of A (A.4.4) and assume $f \in T$. Then $f \in \widehat{T} = PO_{\widehat{A}}(g_1, \dots, g_r)$ as well, using the natural homomorphisms $R[V] \to R[Y] \to A \to \widehat{A}$. For $i = 1, \dots, s$, the element g_i is a square in $\widehat{A}$ since $g_i(\xi) > 0$ and A has residue field R (compare A.4.8). For $i = s + 1, \dots, r$, on the other hand, we have $g_i = 0$ in A, and hence in $\widehat{A}$ as well. So $f \in \widehat{T}$ means that f is actually a sum of squares in $\widehat{A}$. On the other hand, f fails to be sos in $\widehat{A}$, according to Corollary 6.1.10. This contradiction completes the proof of Theorem 6.1.6. □

6.1.12 Remarks.

1. The conclusion of Corollary 6.1.7 can be extended to sets $K \subseteq \mathbb{R}^2$ that contain an open convex cone. This will be shown in 6.6.24.

2. In view of the previous results, it is natural to ask for a characterization of those (basic closed) semialgebraic sets K whose saturated preordering $\mathcal{P}(K)$ is finitely generated. Later in this chapter we'll see answers in typical cases. To get there we need to work in an abstract real spectrum setting, and we start preparing for this in the next section.

Exercises

6.1.1 Let $K \subseteq R^n$ be a closed semialgebraic set. If the preordering $\mathcal{P}(K) = \{f \in R[x] \colon f|_K \geq 0\}$ is finitely generated, show that K is basic closed.

6.1.2 Let R be a real closed field and let $a, b \in R$ with $|a|, |b| \leq 1$. In the polynomial ring $R[t]$, show that $(t-a)(t-b) \in PO(t^2-1)$.

6.1.3 Let $K = \{0\} \cup [1, \infty[\subseteq R$, and let $M \subseteq R[t]$ be the quadratic module generated by the natural generators for K. Show that M is a preordering in $R[t]$.

6.1.4 Fill in the missing details in the proof of Proposition 6.1.3(a), see 6.1.4.

6.1.5 Coordinate rings of nonrational affine curves over $\mathbb{R}$ do not in general satisfy psd = sos, unlike $\mathbb{R}[t]$. Prove this for the following example. Let $g \in \mathbb{R}[x]$ be a univariate polynomial of odd degree ≥ 3, and let C be the plane affine curve $y^2 = g(x)$. Show that there exists $p \in \mathbb{R}[C] = \mathbb{R}[x, y]/\langle y^2 - g(x)\rangle$ with $p > 0$ on $C(\mathbb{R})$, such that p fails to be sos in $\mathbb{R}[C]$. (*Hint*: Show that $g(x)$ can be assumed to be monic, and consider the element $p = x + c \in \mathbb{R}[C]$ for suitable $c \in \mathbb{R}$.)

6.1.6 Let A be a ring. In 3.2.9 it was remarked that a psd element f of A need not satisfy an identity $(1+s)f = t$ with $s, t \in \Sigma A^2$. Give an example for such A and f. (*Hint*: You may use Example 6.1.11.)

6.2 Saturated Preorderings, and the Archimedean Local-Global Principle Revisited

We prepare the setup to study nichtnegativstellensätze systematically, using the real spectrum. A key role will eventually be played by the Archimedean local-global principle, proved in the previous chapter using the notion of pure states (Theorem 5.7.17). We'll present a second, independent proof at the end of this section (Theorem 6.2.19). It avoids pure states and is based on real spectrum techniques. Always let A be a ring that contains $\frac{1}{2}$.

6.2.1 Notation. If $M \subseteq A$ is any subset, recall from 3.1.6 the notation $X(M) = \{\alpha \in \mathrm{Sper}(A) \colon \forall f \in M\ f(\alpha) \geq 0\}$. Subsets of $\mathrm{Sper}(A)$ of this form are called *pro-basic closed*, since they are the intersections of families of basic closed constructible sets. If $Y \subseteq \mathrm{Sper}(A)$ is any subset, we write $\mathcal{P}(Y) := \{f \in A \colon f|_Y \geq 0\}$ for the preordering of all ring elements that are non-negative on Y. Note that this generalizes the notation $\mathcal{P}(K)$ introduced for semialgebraic sets in 6.1.1. If we consider elements of Y as positive cones in A, then simply $\mathcal{P}(Y) = \bigcap_{P \in Y} P$. With this notation, note that a subset Y of $\mathrm{Sper}(A)$ is pro-basic closed if, and only if, $Y = X(\mathcal{P}(Y))$.

6.2.2 Remark. The operators X and $\mathcal{P}$ reverse inclusions. For any subsets $M \subseteq A$ and $Y \subseteq \mathrm{Sper}(A)$, note the equivalence

$$Y \subseteq X(M) \quad \Leftrightarrow \quad M \subseteq \mathcal{P}(Y). \tag{6.2}$$

Indeed, either condition is equivalent to $M \subseteq P$ for every positive cone $P \in Y$. Hence the tautological inclusions $M \subseteq \mathcal{P} \circ X(M)$ and $Y \subseteq X \circ \mathcal{P}(Y)$ hold, and we conclude

$$\mathcal{P} \circ X \circ \mathcal{P} = \mathcal{P}, \quad X \circ \mathcal{P} \circ X = X.$$

Technically speaking, (6.2) means that the pair $(X, \mathcal{P})$ of operators forms a Galois connection, aka adjunction pair. We will not make use of this terminology.

6.2.3 Lemma and Definition. *For any preordering $T \subseteq A$, the following conditions are equivalent:*

(i) *T is an intersection of positive cones of A;*
(ii) *$T = \mathcal{P}(Y)$ for some subset Y of* Sper(A)*;*
(iii) *$T = \mathcal{P}(X(T))$;*
(iv) *if $f \in A$ satisfies an identity $sf = f^{2m} + t$ with $m \geq 0$ and $s, t \in T$, then $f \in T$.*

The preordering T is said to be saturated *if (i)–(iv) hold.*

Proof. (i) ⇒ (iii): If $T = \bigcap_i P_i$ with positive cones P_i, then $P_i \in X(T)$ for all i, and so $\bigcap_{P \in X(T)} P \subseteq T$ is clear. The reverse inclusion holds anyway. The implications (iii) ⇒ (ii) ⇒ (i) are obvious, and the equivalence between (iii) and (iv) is a consequence of the nichtnegativstellensatz 3.2.8. □

Remark 6.2.2 therefore implies that the operators X and $\mathcal{P}$ induce a bijective correspondence between pro-basic closed sets in Sper(A) and saturated preorderings in A.

6.2.4 Corollary. *For any subset $T \subseteq A$ there exists a unique smallest saturated preordering S in A with $T \subseteq S$. We call S the saturation of T and write $S =$* Sat(T)*. The saturation satisfies $X(\mathrm{Sat}(T)) = X(T)$ and $\mathrm{Sat}(T) = \mathcal{P}(X(T))$. When T is a preordering, it is given by*

$$\mathrm{Sat}(T) = \{f \in A \colon \exists m \in \mathbb{N}\ \exists s, t \in T\ sf = f^{2m} + t\}.$$

Proof. S is the intersection of all saturated preorderings that contain T, and is therefore saturated by 6.2.3(i). Since $T \subseteq \mathcal{P}(X(T))$ and $\mathcal{P}(X(T))$ is saturated, we have $S \subseteq \mathcal{P}(X(T))$, hence $X(T) \subseteq X(S)$. The reverse inclusion is clear from $T \subseteq S$. Hence $X(S) = X(T)$, and therefore $S = \mathcal{P}(X(T))$ by 6.2.3(ii). The description of Sat(T) for T a preordering is the nichtnegativstellensatz 3.2.8. □

6.2.5 Remarks. We illustrate the concept of saturation with several remarks and examples.

1. The concept of saturation is irrelevant in non-real rings (see 3.2.16). In such a ring, $T = A$ is the only preordering, and is in particular saturated.

2. In a field, every preordering is saturated (Proposition 1.1.28). In most other real rings this statement is false.

3. The smallest saturated preordering in A is

$$\mathrm{Sat}(\Sigma A^2) = A_+ = \{f \in A \colon \forall \alpha \in \mathrm{Sper}(A)\ f(\alpha) \geq 0\}$$

(3.2.18) and consists of the psd (positive semidefinite) elements of A. By definition, the preordering ΣA^2 is saturated if, and only if, every psd element is a sum of squares in A, i.e. $A_+ = \Sigma A^2$. This is a key property of a ring, therefore we introduce a special phrase for it: Given a ring A, we say that *psd* = *sos holds in A* if $A_+ = \Sigma A^2$ holds. Otherwise we say that *psd* $\neq$ *sos* holds in A.

As remarked in 3.2.19, examples of rings with psd = sos are fields, the polynomial ring $R[t]$ in one variable, or the coordinate ring of the circle, over a real closed field R. Typical non-examples are polynomial rings in more than one variable over R (Hilbert 2.4.9). Other non-examples are coordinate rings of affine R-varieties V with $\dim V(R) \geq 3$ (Theorem 6.1.6), or also regular local rings of dimension ≥ 3 with real residue field (Corollary 6.1.10).

4. Every ring homomorphism $\varphi\colon A \to B$ satisfies $\varphi(A_+) \subseteq B_+$.

5. If V is an affine R-variety and $K = \mathcal{S}_V(g_1, \dots, g_r)$ is a basic closed set in $V(R)$, with $g_i \in R[V]$, the saturation of the preordering $T = PO_V(g_1, \dots, g_r)$ is $\mathcal{P}(K) = \{f \in R[V]\colon f|_K \geq 0\}$. Note that this is in agreement with terminology 6.1.1 for $\mathcal{P}(K)$.

6. From a general view point, a *nichtnegativstellensatz* in a ring A is nothing else but the statement that a certain preordering T in A is saturated. Usually such a result will only be of interest when the preordering T is finitely generated, like $T = \Sigma A^2$, or like in Proposition 6.1.3.

We discuss a few technical results related to saturation. They will be useful later in this chapter.

6.2.6 Proposition. *Let A be a ring and let T be a saturated preordering in A. For any multiplicative set $S \subseteq A$, the preordering T_S in A_S is again saturated.*

Here $T_S = \{\frac{t}{s^2}\colon t \in T,\ s \in S\}$, the preordering in A_S generated by T (5.1.11).

Proof. Identify $\mathrm{Sper}(A_S)$ with a subset of $\mathrm{Sper}(A)$ in the usual way (3.1.9), and similarly for the Zariski spectra. Let $X = X_A(T)$, and write X_S for $X_{A_S}(T_S) = X \cap \mathrm{Sper}(A_S)$. Given $f \in A_S$ with $f \geq 0$ on X_S, we have to show $f \in T_S$, and we may assume $f \in A$. Let $W := \{\alpha \in X\colon f(\alpha) < 0\}$. For every $\alpha \in W$ there is $s \in S$ with $s(\alpha) = 0$, since $\alpha \notin \mathrm{Sper}(A_S)$. Therefore $W \subseteq \bigcup_{s \in S} Z_A(s)$. Now W is pro-constructible in $\mathrm{Sper}(A)$, and the sets $Z_A(s)$ are constructible in $\mathrm{Sper}(A)$. So there exist finitely many elements $s_1, \dots, s_n \in S$ such that $W \subseteq \bigcup_{i=1}^n Z_A(s_i)$ (Proposition 3.4.13(b)). Putting $s := s_1 \cdots s_n$, the element s lies in S and $W \subseteq Z_A(s)$. Therefore $g := s^2 f \in A$ satisfies $g \geq 0$ on X. Hence $g \in T$ since T is saturated, and so $f = \frac{g}{s^2}$ lies in T_S. □

6.2.7 Corollary. *If* psd = sos *holds in the ring A, and if $S \subseteq A$ is any multiplicative subset,* psd = sos *holds in A_S as well.*

Proof. Enough to apply 6.2.6 with $T = \Sigma A^2$, since $T_S = \Sigma A_S^2$. □

6.2.8 Corollary. *Let A be a ring. If A has a prime ideal $\mathfrak{p}$ with real residue field for which the local ring $A_\mathfrak{p}$ is regular of dimension ≥ 3, then* psd $\neq$ sos *in A.*

Proof. By 6.2.7 it suffices to show psd $\neq$ sos for $A_{\mathfrak{p}}$. This follows from Corollary 6.1.10: If $a_1, a_2, a_3 \in A$ form a regular sequence in $A_{\mathfrak{p}}$, and if $p(x_1, x_2, x_3)$ denotes the Motzkin form, the element $p(a_1, a_2, a_3) \in A$ is psd in A but not sos. □

6.2.9 Lemma. *Let A be a ring. For every quadratic module M in A we have*

$$\sqrt{\operatorname{supp}(M)} \subseteq \bigcap_{\alpha \in X(M)} \operatorname{supp}(\alpha) = \operatorname{supp}(\operatorname{Sat}(M)).$$

When M is a preordering, the inclusion is an equality.

Proof. Recall that $\operatorname{supp}(M) = M \cap (-M)$. The second equality follows from $\operatorname{Sat}(M) = \bigcap_{P \in X(M)} P$ by applying the support to both sides. It follows that the support of $\operatorname{Sat}(M)$ is an intersection of (real) prime ideals, and therefore is a (real) radical ideal. From $M \subseteq \operatorname{Sat}(M)$ we get $\sqrt{\operatorname{supp}(M)} \subseteq \operatorname{supp}(\operatorname{Sat}(M))$. Conversely let M be a preordering, and let $f \in \operatorname{supp}(\operatorname{Sat}(M))$. Then f vanishes identically on $X(\operatorname{Sat}(M)) = X(M)$, and so $f \in \sqrt{\operatorname{supp}(M)}$ by the abstract real nullstellensatz 3.2.10. □

6.2.10 Proposition. *Let $Y \subseteq \operatorname{Sper}(A)$ be a pro-constructible set and consider the ideal $I = \bigcap_{\alpha \in Y} \operatorname{supp}(\alpha)$ of A. For every prime ideal $\mathfrak{p}$ of A with $I \subseteq \mathfrak{p}$, there exists $\alpha \in Y$ with $\operatorname{supp}(\alpha) \subseteq \mathfrak{p}$. In particular,*

$$\dim(A/I) = \sup\{\dim(A/\operatorname{supp}(\alpha)) \colon \alpha \in Y\}.$$

Proof. Let $\mathfrak{p} \supseteq I$, and assume $\operatorname{supp}(\alpha) \not\subseteq \mathfrak{p}$ for every $\alpha \in Y$. Then, for every $\alpha \in Y$, there exists $f_\alpha \in \operatorname{supp}(\alpha)$ with $f_\alpha \notin \mathfrak{p}$, which implies $Y \subseteq \bigcup_{\alpha \in Y} Z(f_\alpha)$. Since the sets $Z(f_\alpha)$ are constructible and Y is pro-constructible, there exists a finite subcovering. Hence there are finitely many elements $\alpha_1, \dots, \alpha_n \in Y$ such that $Y \subseteq \bigcup_{i=1}^n Z(f_{\alpha_i})$. Let $f := f_{\alpha_1} \cdots f_{\alpha_n}$, then $Y \subseteq Z(f)$, and so $f \in \bigcap_{\alpha \in Y} \operatorname{supp}(\alpha) = I$. Since $f \notin \mathfrak{p}$, this contradicts the hypothesis $I \subseteq \mathfrak{p}$. □

6.2.11 Remarks.

1. Let us re-interpret the preceding results in a geometrical setting. Let R be a real closed field and let $T \subseteq R[x] = R[x_1, \dots, x_n]$ be a finitely generated preordering. We put $K = \mathcal{S}(T) \subseteq R^n$, a basic closed set in R^n. Let V be the Zariski closure of K in $\mathbb{A}^n$ and let $\mathcal{I}(V) \subseteq R[x]$ be the vanishing ideal of V. Then $X(T) = \widetilde{K}$ (the constructible set in $\operatorname{Sper} R[x]$ associated with K) and $\bigcap_{\alpha \in \widetilde{K}} \operatorname{supp}(\alpha) = \bigcap_{\xi \in K} \mathfrak{m}_\xi = \mathcal{I}(V)$ (the second equality holds by definition of V). So Lemma 6.2.9 implies $\sqrt{\operatorname{supp}(T)} = \mathcal{I}(V)$. In particular we see $\dim R[x]/\operatorname{supp}(T) = \dim(K)$.

2. For (finitely generated) quadratic modules M the situation is quite different in general, since the dimension of the semialgebraic set $\mathcal{S}(M)$ can be strictly smaller than the Krull dimension of the ring $R[x]/\operatorname{supp}(M)$. See Exercise 6.2.6 for an example, and Exercise 6.2.7 for another example that in addition is Archimedean.

6.2.12 Remark. Every ring homomorphism $\varphi\colon A \to B$ satisfies $\varphi(A_+) \subseteq B_+$. Conversely, assume that φ is surjective and that $g \in B_+$ is given. When does there exist $f \in A_+$ with $\varphi(f) = g$? Geometrically this means that we are given a closed subvariety W of an affine R-variety V, together with a psd polynomial g on W. When can g be extended to a psd polynomial f on all of V?

Clearly, every sum of squares in B lifts to a sum of squares in A under φ. On the other hand there are easy examples where the above question has a negative answer (Exercise 6.2.2). We discuss a condition that is sufficient for a positive answer. It will be used in the proof of Theorem 6.2.19 below.

6.2.13 Lemma. *Let A be a ring and $I \subseteq A$ an ideal, let $Y \subseteq \mathrm{Sper}(A)$ be a closed set, and let $f \in A$ satisfy $f \geq 0$ on $Y \cap Z(I)$. For every $\alpha \in Y \cap Z(I) \cap Z(f)$, assume that there exists $h_\alpha \in I$ with $h_\alpha \geq 0$ on Y and $f + h_\alpha \geq 0$ on a neighborhood of α in Y. Then there is $h \in I$ with $f + h \geq 0$ on Y.*

Proof. It suffices to prove for every $\alpha \in Y$ that there exists $h_\alpha \in I$ with $h_\alpha \geq 0$ on Y and $(f + h_\alpha)(\alpha) \geq 0$. By compactness of the constructible topology, this will imply the existence of finitely many $\alpha_1, \dots, \alpha_n \in Y$ for which $Y \subseteq \bigcup_{i=1}^n X(f + h_{\alpha_i})$. Then the element $h := \sum_{i=1}^n h_{\alpha_i}$ lies in I and satisfies $f + h \geq 0$ on Y.

So let $\alpha \in Y$. First assume that α has a specialization β in $Z(I)$, so $f(\beta) \geq 0$ holds by the hypothesis. If $f(\beta) > 0$ then $f(\alpha) > 0$, and we may take $h = 0$. If $f(\beta) = 0$ then, by assumption, there exists $h \in I \cap \mathcal{P}(Y)$ with $f + h \geq 0$ near β on Y, and in particular $(f + h)(\alpha) \geq 0$. There remains the case where $\{\alpha\} \cap Z(I) = \varnothing$. This means $-1 \in P_\alpha + I$, hence there is $g \in I$ with $g(\alpha) \geq 1$. Therefore the element $h := (1 + f^2)g^2 \in I$ satisfies $h(\alpha) > |f(\alpha)|$, which is enough to conclude. □

6.2.14 Corollary. *If $f \in A$ is psd on a neighborhood of $Y \cap Z(I)$ in Y, there is $h \in I$ with $f + h \geq 0$ on Y. When $f > 0$ on $Y \cap Z(I)$, there is $h \in I$ with $f + h > 0$ on Y.*

Proof. Let $f \geq 0$ on a neighborhood of $Y \cap Z(I)$ in Y. Then the condition in the previous lemma is satisfied with $h_\alpha = 0$ for every α, implying the assertion. When $f > 0$ on $Y \cap Z(I)$ we find, for every $\alpha \in Y$, an element $h_\alpha \in I$ with $h_\alpha \geq 0$ on Y and $(f + h_\alpha)(\alpha) > 0$. (For $\alpha \in Z(I)$ take $h_\alpha = 0$, for $\alpha \notin Z(I)$ take $h_\alpha = g^2$ with $g \in I$ satisfying $g(\alpha) \neq 0$.) Using such h_α, the argument is completed as in the proof of Lemma 6.2.13. □

6.2.15 Remark. The condition of Lemma 6.2.13 is sufficient for the desired conclusion, but it is by no means necessary. For example, let $A = R[x, y]$ and $I = Ax$, and take $Y = \mathrm{Sper}(A)$ in 6.2.13. The image of $A_+ \to (A/I)_+$ obviously contains $f = x + y^2$, but there is no psd polynomial h in I that would make $f + h$ psd locally around the origin.

We are now heading for a second proof of the Archimedean local-global principle. From 5.2.17, recall the notion of absolute boundedness for subsets of the real spectrum.

6.2.16 Lemma. *Let Y be a subset of $\mathrm{Sper}(A)$ that is pro-basic closed (6.2.1) and absolutely bounded (5.2.17). Given elements $f, g \in A$ with $f \geq 0$ on Y and $g < 0$ on $Y \cap Z(f)$, there exists a positive integer N such that $Nf > g$ on Y.*

Proof. Let $Y_1 = Y \cap \{g \geq 0\}$ and put $T = \mathcal{P}(Y_1)$. Then $Y_1 = X(T)$ since Y_1 is pro-basic closed, and so $f > 0$ on $X(T)$ holds by hypothesis. By the positivstellensatz 3.2.7 there are $s, t \in T$ with $sf = 1 + t$. On the other hand there exist integers $m, n \geq 1$ with $m > g$ and $n > s$ on Y_1, since Y_1 is absolutely bounded. It follows that $mnf > msf \geq m > g$ on Y_1, and therefore $mnf > g$ on Y as well. □

The following lemma is the key for our second proof of the Archimedean local-global principle:

6.2.17 Lemma. *Let $Y \subseteq \operatorname{Sper}(A)$ be a set that is pro-basic closed and absolutely bounded, and let $f, g \in A$ be non-negative on Y. Then any $h \in Af + Ag$ with $h > 0$ on Y can be written $h = sf + tg$ with elements $s, t \in A$ that are strictly positive on Y.*

Proof. Start with arbitrary elements $a, b \in A$ for which $af + bg = h$. We have $a > 0$ on $Y \cap Z(g)$ and $b > 0$ on $Y \cap Z(f)$. Hence by 6.2.16 there exist positive integers N_1, N_2 such that $N_1 g > -a$ and $N_2 f > -b$ on Y.

The topological space $Y^{\max} = Y^{\text{arch}}$ is compact (5.2.19). Recall that every $s \in A$ defines a continuous $\mathbb{R}$-valued function $\hat{s}$ on $Y^{\max}$ (5.2.20). Let $\varphi \colon Y^{\max} \to \mathbb{R}$ be the function defined by $\varphi(\xi) = \max\left\{-N_1, -\frac{\hat{b}(\xi)}{\hat{f}(\xi)}\right\}$ if $\hat{f}(\xi) \neq 0$, and by $\varphi(\xi) = -N_1$ if $\hat{f}(\xi) = 0$. This function φ is continuous. To see this we only need to consider φ in a neighborhood of $\xi \in Y^{\max}$ with $\hat{f}(\xi) = 0$, i.e. $f(\xi) = 0$. Now $U := U(b - N_1 f)$ is an open neighborhood of $Y \cap Z(f)$, and for $\eta \in U \cap Y^{\max}$ with $f(\eta) \neq 0$ we have $\frac{\hat{b}(\eta)}{\hat{f}(\eta)} > N_1$, and hence $\varphi(\eta) = -N_1$. This means that φ is constant in a neighborhood of ξ. In a similar way let $\psi \colon Y^{\max} \to \mathbb{R}$ be defined by $\psi(\xi) := \min\left\{N_2, \frac{\hat{a}(\xi)}{\hat{g}(\xi)}\right\}$ for $\hat{g}(\xi) \neq 0$ and $\psi(\xi) = N_2$ for $\hat{g}(\xi) = 0$. Then ψ is continuous by an analogous reasoning.

We claim that $\varphi < \psi$ holds (pointwise) on $Y^{\max}$. By the choice of N_1 and N_2, the inequalities $-N_1 < \hat{a}/\hat{g}$ and $-\hat{b}/\hat{f} < N_2$ hold on $Y^{\max}$ whenever the denominators do not vanish, and $-\hat{b}/\hat{f} < \hat{a}/\hat{g}$ holds since $h > 0$. On $Y^{\max} \cap Z(f)$ we have $\varphi = -N_1$, while on $Y^{\max} \cap Z(g)$ we have $\psi = N_2$, hence the inequality $\varphi < \psi$ holds there as well.

Therefore the Stone–Weierstrass theorem (Corollary 5.2.26) implies the existence of an element $c \in A$ satisfying $\varphi < c < \psi$ on $Y^{\max}$. This implies that the inequalities

$$-b < cf \quad \text{and} \quad cg < a$$

hold everywhere on $Y^{\max}$, and hence on Y as well. Hence the elements $s := a - cg$ and $t := b + cf$ in A satisfy $s, t > 0$ on Y and $sf + tg = h$. □

We generalize the lemma to the case of more than two generators:

6.2.18 Proposition. *Let $Y \subseteq \operatorname{Sper}(A)$ be a set that is pro-basic closed and absolutely bounded. Let $f_1, \dots, f_r$ in A be non-negative on Y, and let $h \in Af_1 + \cdots + Af_r$ with $h > 0$ on Y. Then there exist $a_1, \dots, a_r \in A$ with $a_1 f_1 + \cdots + a_r f_r = h$ and with $a_i > 0$ on Y $(i = 1, \dots, r)$.*

Proof. The case $r = 1$ is trivial, and $r = 2$ is Lemma 6.2.17. Let $r > 2$ and assume that the claim has already been proved for $r-1$. Put $A' = A/\langle f_r\rangle$ and write $f_i' = f_i + \langle f_r\rangle$ for $i = 1, \dots, r-1$. The pro-basic closed set $Y' := Y \cap Z(f_r)$ in $\mathrm{Sper}(A') = Z(f_r)$ is absolutely bounded. By the inductive hypothesis there are $b_1, \dots, b_{r-1} \in A$ with $b_i > 0$ on $Y \cap Z(f_r)$ and with

$$b_1 f_1 + \cdots + b_{r-1} f_{r-1} \equiv h \pmod{f_r}.$$

By Corollary 6.2.14 there exist $c_1, \dots, c_{r-1} \in A$ with $c_i \equiv b_i \pmod{f_r}$ and with $c_i > 0$ on Y ($i = 1, \dots, r-1$). Put $f := \sum_{i=1}^{r-1} c_i f_i$ and $g := f_r$. Then $f, g \geq 0$ on Y and $h \in Af + Ag$. Apply Lemma 6.2.17 to f and g, this gives $s, t \in A$ with $s > 0$, $t > 0$ on Y and with $h = sf + tg$. This implies the assertion. □

We can now give a second proof of the Archimedean local-global principle. For convenience, here is the statement again:

6.2.19 Theorem. (Archimedean local-global principle) *Let A be a ring, let $S \subseteq A$ be an Archimedean semiring and $M \subseteq A$ an S-module. Assume that $f \in A$ is such that, for every maximal ideal $\mathfrak{m}$ of A, there exists $s \in S \setminus \mathfrak{m}$ with $sf \in M$. Then $nf \in M$ for some integer $n \geq 1$.*

Proof. The set $\{s \in S : sf \in M\}$ is not contained in any maximal ideal of A. Hence there exist finitely many elements $s_1, \dots, s_r$ in S with $s_i f \in M$ for every i and with $\langle s_1, \dots, s_r\rangle = \langle 1\rangle$. The pro-basic closed set $X(S) \subseteq \mathrm{Sper}(A)$ is absolutely bounded since S is Archimedean (5.2.18). From Proposition 6.2.18 we therefore get elements $a_1, \dots, a_r \in A$ with $\sum_{i=1}^r a_i s_i = 1$ such that $a_i > 0$ on $X(S)$. By the Archimedean positivstellensatz 5.3.1, there is an integer $n \geq 1$ with $na_i \in S$ for $i = 1, \dots, r$. It follows that $nf = \sum_{i=1}^r (na_i)(s_i f) \in M$. □

6.2.20 Remark. It seems not to be known whether an analogue of Theorem 6.2.19 exists for Archimedean quadratic modules. Let M be an Archimedean quadratic module in a ring A, and let $f \in A$. If for every maximal ideal $\mathfrak{m}$ of A there exists $s \in A \setminus \mathfrak{m}$ with $s^2 f \in M$, does it follow that $f \in M$? At least, it is true that f lies in the quadratic module generated by all products pq with $p, q \in M$ (Exercise 6.2.8).

We'll return to the local-global principle in Section 6.5 below.

Exercises

6.2.1 Let k be a field of characteristic $\neq 2$. Prove that psd = sos holds in the ring $k[[t]]$ of formal power series.

6.2.2 Let $A = \mathbb{R}[x, y]$ and $I = \langle y^2 - x^3\rangle \subseteq A$. Prove that the map $A_+ \to (A/I)_+$ fails to be surjective.

6.2.3 Let A be a connected Noetherian ring that is real but not real reduced (3.2.16, 3.2.17), i.e. $\mathrm{Sper}(A) \neq \varnothing$ and $\sqrt[\mathrm{re}]{\langle 0\rangle} \neq \{0\}$.

(a) Show that psd $\neq$ sos in A.

(b) Conclude that psd $\neq$ sos holds in every connected algebra A of finite type over a field with $\dim(A) \geq 3$ and $\mathrm{Sper}(A) \neq \varnothing$.

Hint: The hypothesis A connected means that the only elements $a \in A$ with $a^2 = a$ are $a = 0$ and $a = 1$. Use Nakayama's lemma in (a) to show $I^2 \neq I$ for the real nilradical $I = \sqrt[r]{\langle 0\rangle}$ of A.

6.2.4 Let $n \geq 4$. This exercise shows that there is no uniform denominator for all psd forms in $\mathbb{R}[x] = \mathbb{R}[x_1, \dots, x_n]$ (see Remark 5.5.9.3). Given a non-zero form $h \in \mathbb{R}[x]$, show that there exists a psd form $f \in \mathbb{R}[x]$ such that fh^N is not a sum of squares for any $N \geq 0$. (*Hint*: Fix a point $\xi \in \mathbb{R}^n$ with $h(\xi) \neq 0$ and work in the local ring of $\mathbb{P}^{n-1}$ at $[\xi]$.)

6.2.5 Let $\mathfrak{m}$ be a maximal ideal of the polynomial ring $R[x] = R[x_1, \dots, x_n]$ whose residue field is the field $R(\sqrt{-1})$. When $n \geq 4$, show that psd $\neq$ sos holds in the localization $R[x]_{\mathfrak{m}}$. (Use the fact that any localization of a regular local ring is again regular, A.4.5.) *Remark*: Benoist [18] has shown that the statement is true for $n = 3$ as well.

6.2.6 Consider the quadratic module $M = QM(x, y, -xy)$ in $\mathbb{R}[x, y]$. Show that the semialgebraic set $K = \mathcal{S}(M)$ is one-dimensional, but that $\mathrm{supp}(M) = \{0\}$. (*Hint*: Exercise 5.1.9)

6.2.7 With a bit more effort we can also construct a quadratic module as in Exercise 6.2.6, which in addition is Archimedean. Let $M \subseteq \mathbb{R}[x, y]$ be the quadratic module that is generated by $1 - x^2 - y^2$, $-xy$, $x - y$ and $y - x^2$. Show that M is Archimedean and $\mathcal{S}(M)$ is a point, but that $\mathrm{supp}(M) = \{0\}$.

Remark: Some valuation theory is needed to identify the support. The quotient field $\mathbb{R}(x, y)$ admits a valuation v with residue field $\mathbb{R}$ and value group Γ, such that $0 < v(x) < v(y) < 2v(x)$ and $v(x)$, $v(y)$ are $\mathbb{Z}$-linearly independent in Γ. Assuming the existence of such v, you should be able to show $\mathrm{supp}(M) = \{0\}$.

6.2.8 Let M be an Archimedean quadratic module in a ring A, and let $f \in A$. For every maximal ideal $\mathfrak{m}$ of A, assume that there exists $s \in A \setminus \mathfrak{m}$ with $s^2 f \in M$. Prove that f lies in the quadratic module generated by all products pq with $p, q \in M$.

6.3 Sums of Squares in Local Rings

We study preorderings and their saturations in local rings A. Given a preordering T in A and an element f in the saturation of T, we try to identify conditions that imply $f \in T$. As a first step we show that $f > 0$ on $X(T)$ always implies $f \in T$. In general, the question becomes easier when we replace A and T by their completions (the completion $\widehat{T}$ of T will be defined below). The main results say that, under suitable conditions, $f \in \widehat{T}$ implies $f \in T$.

We continue to assume $\frac{1}{2} \in A$ for every ring A. As before, an element $a \in A$ is said to be psd if $a \in A_+$ (6.2.5).

6.3.1 Theorem. *Let $(A, \mathfrak{m}, k)$ be a local ring, and let $u \in A^*$ be a psd unit of A. Then u is a sum of squares in A.*

Proof. Consider the ring $B = A[t]/\langle t^2 + u\rangle = A[\tau]$, where τ denotes the coset of t in B. Since $\mathrm{Sper}(B)$ is empty we have $-1 \in \Sigma B^2$ (Corollary 3.2.16). So there are $a_i, b_i \in A$ with $-1 = \sum_{i=1}^r (a_i + b_i\tau)^2$ in B. In particular,

$$-1 = \sum_{i=1}^r a_i^2 - u \sum_{i=1}^r b_i^2 \tag{6.3}$$

holds in A, which says $u\sum_i b_i^2 = 1+\sum_i a_i^2$. Therefore, if $\sum_{i=1}^r b_i^2$ is a unit in A, then

$$u = \Big(\sum_{i=1}^r b_i^2\Big)^{-1}\Big(1+\sum_{i=1}^r a_i^2\Big)$$

is a sum of squares in A and we are done. When the field $k = A/\mathfrak{m}$ is real, this will hold automatically, since by (6.3) we cannot have $b_i \in \mathfrak{m}$ for all i. For the rest of the proof we assume $\sum_i b_i^2 \in \mathfrak{m}$, and we'll show how to find another identity (6.3) which has the desired property.

For $x, y \in A^r$ write $\langle x,y\rangle = \sum_{i=1}^r x_i y_i$. By enlarging r if necessary, we may assume that $b_i \in \mathfrak{m}$ holds for at least two indices i, say $b_1, b_2 \in \mathfrak{m}$. There exist $w_1, w_2 \in A$ such that both $w_1^2+w_2^2$ and $1+u(w_1^2+w_2^2)$ are units; for example, we may take $w_1 = 1$ and $w_2 = 0$ or 1. The tuples $w = (w_1, w_2, 0, \dots, 0)$, $a = (a_1, \dots, a_r)$ and $b = (b_1, \dots, b_r)$ in A^r satisfy $\langle b, w\rangle \in \mathfrak{m}$. Since $\langle b,b\rangle \in \mathfrak{m}$ by assumption, we have $\langle a,a\rangle \equiv -1 \pmod{\mathfrak{m}}$ from (6.3). Let

$$\gamma = \frac{\langle a,a\rangle - u\langle b,w\rangle}{\langle a,a\rangle - u\langle w,w\rangle} \in A^*$$

(note that the denominator is a unit by the choice of w). Further let

$$(a',\, b') := (a,\, b) - 2\gamma(a,\, w) \in A^r \oplus A^r.$$

Now we calculate:

$$\begin{aligned}
\langle a',a'\rangle - u\cdot\langle b',b'\rangle &= (1-2\gamma)^2\langle a,a\rangle - u\cdot\langle b-2\gamma w, b-2\gamma w\rangle \\
&= \big(\langle a,a\rangle - u\langle b,b\rangle\big) + 4\gamma\big((\gamma-1)\langle a,a\rangle + u(\langle b,w\rangle - \gamma\langle w,w\rangle)\big) \\
&= -1 + 4\gamma\big(\gamma(\langle a,a\rangle - u\langle w,w\rangle) - (\langle a,a\rangle - u\langle b,w\rangle)\big) \\
&= -1. \qquad (6.4)
\end{aligned}$$

On the other hand, $\langle b',b'\rangle$ is a unit in A since $\langle b',b'\rangle \equiv \langle b-2\gamma w,\ b-2\gamma w\rangle \equiv 4\gamma^2\langle w,w\rangle$ modulo $\mathfrak{m}$. Using identity (6.4) instead of (6.3), we conclude $u \in \Sigma A^2$ by the argument at the beginning of the proof. □

What is behind this proof is a transversality argument, which is a standard technique in quadratic forms theory. Recall that a ring $A \neq \{0\}$ is semilocal if it has only finitely many maximal ideals. To generalize the previous theorem to semilocal rings we could use a similar technique. But it is easier to deduce the semilocal case directly from the local case:

6.3.2 Proposition. *Let A be a semilocal ring and let M be a quadratic module in A. If $f \in A$ is such that $f \in M_{\mathfrak{m}}$ for every maximal ideal $\mathfrak{m}$ of A, then $f \in M$.*

Here $M_{\mathfrak{m}}$ denotes the extension of M to the localization $A_{\mathfrak{m}}$ of A (see 5.1.8), which is

$$M_{\mathfrak{m}} = \Big\{\frac{x}{s^2} : x \in M,\ s \in A \setminus \mathfrak{m}\Big\}.$$

Proof. Let $\mathfrak{m}_1, \dots, \mathfrak{m}_r$ be the maximal ideals of A. For every index $i = 1, \dots, r$ there exists, by assumption, an element $s_i \in A$ with $s_i \notin \mathfrak{m}_i$ and $s_i^2 f \in M$. By the Chinese remainder theorem there are elements $a_1, \dots, a_r \in A$ with $a_i \equiv 1 \pmod{\mathfrak{m}_i}$ and $a_i \equiv 0 \pmod{\mathfrak{m}_j}$, for all $i \neq j$ in $\{1, \dots, r\}$. The element $u = \sum_{i=1}^r (a_i s_i)^2$ is a unit in A, and $uf = \sum_{i=1}^r a_i^2 s_i^2 f \in M$. Since $u^{-1} = \sum_i (a_i s_i u^{-1})^2$ is a sum of squares in A, we conclude $f \in M$. □

6.3.3 Corollary. *In a semilocal ring, every psd unit is a sum of squares.*

Proof. Let $f \in A^*$ be psd. Then f is a psd unit in $A_\mathfrak{m}$ for every maximal ideal $\mathfrak{m}$ of A, and is therefore a sum of squares in $A_\mathfrak{m}$ by Theorem 6.3.1. Apply Proposition 6.3.2 with $M = \Sigma A^2$ to conclude that f is sos in A. □

In fact, the statement can be generalized further. For this the following elementary observation is useful. Let A be a ring, let g_i $(i \in I)$ be a family of elements of A and let $T = PO_A(g_i\colon i \in I)$ be the preordering generated by the g_i. We formally adjoin square roots of the g_i and write

$$B = A[x_i\colon i \in I]/\langle x_i^2 - g_i\colon i \in I\rangle.$$

6.3.4 Lemma. *Given $f \in A$, we have $f \in T$ if and only if f is sos in B, and also $f \in \mathrm{Sat}(T)$ if and only if f is psd in B.*

Proof. See Exercise 6.3.1. □

6.3.5 Corollary. *Let A be a semilocal ring and let T be a preordering in A. Then $\mathrm{Sat}(T) \cap A^* \subseteq T$.*

Note that this generalizes Corollary 6.3.3 (which in turn generalizes Theorem 6.3.1).

Proof. Let $f \in \mathrm{Sat}(T) \cap A^*$. Then $X(-f) \cap X(T) = X(-f) \cap \bigcap_{t \in T} X(t) = \varnothing$ since $f > 0$ on $X(T)$. By compactness of the constructible topology there exist finitely many $t_1, \dots, t_r \in T$ such that $f > 0$ on $X(t_1, \dots, t_r)$. Consider the ring

$$B = A[x_1, \dots, x_r] \,/\, \langle x_i^2 - t_i,\ i = 1, \dots, r\rangle.$$

Since A is semilocal and the ring extension $A \subseteq B$ is finite, B is a semilocal ring as well. Moreover, f is a psd unit in B. By Corollary 6.3.3, f is sos in B, which implies $f \in T$ by 6.3.4. □

After these generalities, here is a key lemma:

6.3.6 Lemma. *Let A be a semilocal ring, let T be a preordering in A and let $f \in \mathrm{Sat}(T)$. If $f \in T + Af^2$ then $f \in T$.*

Proof. By Proposition 6.3.2 we may assume that A is local, with maximal ideal $\mathfrak{m}$. By hypothesis there is an identity $f = t + f^2 g$ with $t \in T$ and $g \in A$. If f is a unit in A then $f \in T$ by Corollary 6.3.5. So assume that $f \in \mathfrak{m}$, which implies $1 - fg \in A^*$ and

$f(1-fg) = t$. We claim that $1-fg \in \mathrm{Sat}(T)$. Indeed, $f \ge 0$ on $X(T)$ and $1-fg \equiv 1$ on $Z(f)$, so the claim follows from $f(1-fg) \in T$. So $1-fg \in T$ by Corollary 6.3.5, and hence $f = t(1-fg)^{-1}$ lies in T as well. □

From this lemma we get a further generalization of Corollary 6.3.5:

6.3.7 Corollary. *If T is a preordering in a semilocal ring A, then T contains every $f \in A$ that is strictly positive on $X(T)$.*

When A has a maximal ideal $\mathfrak{m}$ whose residue field $A/\mathfrak{m}$ is non-real, the hypothesis is strictly more general than $f \in A^* \cap \mathrm{Sat}(T)$.

Proof. Let $f > 0$ on $X(T)$. Since the preordering $T' = T + Af^2$ in A satisfies $X(T') = X(T) \cap Z(f) = \varnothing$, we have $T' = A$ by Theorem 3.2.3. So Lemma 6.3.6 implies $f \in T$. □

The next result is much stronger in general than Lemma 6.3.6:

6.3.8 Proposition. *Let A be a semilocal ring, let $T \subseteq A$ be a preordering and let $f \in \mathrm{Sat}(T)$. If $f \notin T$, there is an ideal J of A with $f \notin T + J$ and with $\sqrt{J} = \sqrt{\mathrm{supp}(T + Af)}$.*

Proof. We start with a simple observation: If $f \in \mathrm{Sat}(T)$ and $g \in \mathrm{supp}(T)$, and if $f \in T + A(f+g)^2$, then $f \in T$. Indeed, we have $f + g \in T + A(f+g)^2$, and so $f+g \in \mathrm{Sat}(T)$ implies $f+g \in T$ by Lemma 6.3.6. This implies $f \in T$ since $-g \in T$.

Now let $f \in \mathrm{Sat}(T)$. We have $X(T + Af) = X(T) \cap Z(f) = X(T + Af^2)$, and so Lemma 6.2.9 implies

$$\sqrt{\mathrm{supp}(T + Af)} = \sqrt{\mathrm{supp}(T + Af^2)}.$$

Choose a family $(g_\lambda)_{\lambda\in\Lambda}$ of elements that generates the ideal $\mathrm{supp}(T + Af^2)$ and put

$$J := Af^2 + \sum_{\lambda\in\Lambda} A(f+g_\lambda)^2.$$

It is easy to see that $\sqrt{J} = \sqrt{\mathrm{supp}(T + Af^2)}$. Let $f \in T + J$ be given. We'll show that $f \in T$, which will complete the proof. There exist finitely many elements among the g_λ, say $g_1, \dots, g_r$, with

$$f \in T + \langle f^2, (f+g_1)^2, \dots, (f+g_r)^2\rangle.$$

Writing $T' = T + \langle f^2, (f+g_1)^2, \dots, (f+g_{r-1})^2\rangle$ we have $f \in T' + A(f+g_r)^2$. Since $f \in \mathrm{Sat}(T')$ and $g_r \in \mathrm{supp}(T')$, the observation from the beginning of the proof implies $f \in T'$. By iterating this step we get $f \in T + \langle f^2\rangle$, and applying Lemma 6.3.6 a last time (now we use $f \in \mathrm{Sat}(T)$) gives $f \in T$, as desired. □

6.3.9 Definition. Let $(A, \mathfrak{m})$ be a local Noetherian ring, and let $(\widehat{A}, \widehat{\mathfrak{m}})$ denote its completion (see A.4.4). Given a preordering T in A, the *completion* $\widehat{T}$ of T is the preordering in $\widehat{A}$ that is generated by $i(T)$, where $i \colon A \to \widehat{A}$ denotes the natural map.

6.3.10 Lemma. *Let* $(A, \mathfrak{m})$ *be a local Noetherian ring. For any preordering* T *in* A *and any* $n \geq 0$, *one has* $i^{-1}(\widehat{T} + \widehat{\mathfrak{m}}^n) = T + \mathfrak{m}^n$.

Proof. Only "$\subseteq$" needs a proof. Given $f \in A$ with $i(f) \in \widehat{T} + \widehat{\mathfrak{m}}^n$, we have

$$i(f) \equiv g_1^2 i(t_1) + \cdots + g_r^2 i(t_r) \pmod{\widehat{\mathfrak{m}}^n}$$

with suitable $t_j \in T$ and $g_j \in \widehat{A}$ $(j = 1, \ldots, r)$. The natural map $A/\mathfrak{m}^n \to \widehat{A}/\widehat{\mathfrak{m}}^n$ is bijective. In particular, there exist elements $h_j \in A$ with $g_j \equiv i(h_j) \pmod{\widehat{\mathfrak{m}}^n}$, for $j = 1, \ldots, r$. So $t := \sum_j h_j^2 t_j$ lies in T and satisfies $i(t) \equiv i(f) \pmod{\widehat{\mathfrak{m}}^n}$. Therefore $f - t$ lies in $i^{-1}(\widehat{\mathfrak{m}}^n) = \mathfrak{m}^n$, proving $f \in T + \mathfrak{m}^n$. □

In the sequel we'll use $i\colon A \to \widehat{A}$ for identifying a local Noetherian ring A with a subring of its completion.

6.3.11 Theorem. *Let* $(A, \mathfrak{m})$ *be a local Noetherian ring, let* T *be a preordering in* A *and let* $f \in \mathrm{Sat}(T)$. *Assume that* $\mathrm{supp}(\alpha) = \mathfrak{m}$ *holds for every* $\alpha \in Z(f) \cap X(T)$. *Then the following conditions are equivalent:*

(i) $f \in T$,
(ii) $f \in \widehat{T}$,
(iii) $f \in T + \mathfrak{m}^n$ *for every* $n \geq 0$,

and they hold when $\widehat{T}$ *is saturated (in* $\widehat{A}$*).*

Proof. Clearly, $f \in \mathrm{Sat}_A(T)$ implies $f \in \mathrm{Sat}_{\widehat{A}}(\widehat{T})$. Therefore (ii) holds when $\widehat{T}$ is saturated. The implication (i) $\Rightarrow$ (ii) is trivial, and (ii) $\Rightarrow$ (iii) follows from Lemma 6.3.10. To prove (iii) $\Rightarrow$ (i) let $T' = T + Af$, so $X(T') = Z(f) \cap X(T)$. By Lemma 6.2.9, the hypothesis in the theorem says $\mathfrak{m} \subseteq \sqrt{\mathrm{supp}(T')}$. Assume that $f \in T + \mathfrak{m}^n$ for every $n \geq 0$, but $f \notin T$. Then Proposition 6.3.8 gives an ideal $J \subseteq A$ for which $f \notin T + J$ and $\sqrt{J} = \sqrt{\mathrm{supp}(T')}$, so $\mathfrak{m} \subseteq \sqrt{J}$. Since A is Noetherian, there exists $n \geq 0$ with $\mathfrak{m}^n \subseteq J$. This implies $f \notin T + \mathfrak{m}^n$, which contradicts the assumption. □

6.3.12 Remark. To consider Theorem 6.3.11 in a geometric setting, let $K = \mathcal{S}(g_1, \ldots, g_r)$ be a basic closed set in R^n, where $g_1, \ldots, g_r \in R[x] = R[x_1, \ldots, x_n]$, and let $f \in R[x]$ with $f \geq 0$ on K. The question whether f lies in the preordering $T = PO_{R[x]}(g_1, \ldots, g_r)$ is usually hard, and we may relax it by studying it locally at a given point $\xi \in K$. This means to work in the local ring $\mathcal{O}_\xi = R[x]_{\mathfrak{m}_\xi}$ (with $\mathfrak{m}_\xi \subseteq R[x]$ the maximal ideal of ξ), and to discuss whether f lies in $T_\xi = PO_{\mathcal{O}_\xi}(g_1, \ldots, g_r)$. The assumption $f \in \mathcal{P}(K)$ implies $f \in \mathrm{Sat}_{\mathcal{O}_\xi}(T_\xi)$. Let $Z_f = \mathcal{Z}(f) \cap K$ be the zero set of f in K. Then the subset $Z_{\mathcal{O}_\xi}(f) \cap X_{\mathcal{O}_\xi}(T_\xi)$ of $\mathrm{Sper}(\mathcal{O}_\xi)$ is identified with the intersection of $\widetilde{Z_f}$ with $\mathrm{Sper}(\mathcal{O}_\xi)$ (inside $\mathrm{Sper}\, R[x] = \widetilde{R^n}$). So the condition

$$\mathrm{supp}(\alpha) = \mathfrak{m}_{\mathcal{O}_\xi} \quad \text{for every } \alpha \in Z_{\mathcal{O}_\xi}(f) \cap X_{\mathcal{O}_\xi}(T_\xi)$$

from Theorem 6.3.11 holds trivially if $f(\xi) \neq 0$. More interestingly, it is satisfied if f has only finitely many zeros in K. It may however fail when ξ is just an isolated point of Z_f.

6.3.13 Remark. Let $(A, \mathfrak{m})$ be a local Noetherian ring and let $T \subseteq A$ be a preordering. Theorem 6.3.11 states that the obvious inclusion $T \subseteq \mathrm{Sat}(T) \cap \widehat{T}$ (of preorderings in A) is an equality, provided that $\mathrm{supp}(\alpha) = \mathfrak{m}$ for every $\alpha \in Z(f) \cap X(T)$. Without such a condition, the inclusion can be strict. We give two examples:

1. In Exercise 6.3.3, A is the local ring of a reducible plane curve in a non-real singular point, such that psd $\neq$ sos in A. Since the non-real residue field implies psd = sos in $\widehat{A}$, the conclusion "f sos in $\widehat{A} \Rightarrow f$ sos in A" for psd elements f in A fails.

2. There also exist examples with regular local rings. Consider the localization A of $\mathbb{R}[x_1, \dots, x_n]$ in a non-real maximal ideal. Again psd = sos holds in $\widehat{A}$, but it fails in A when $n \geq 4$ (Exercise 6.2.5). In fact, psd $\neq$ sos is even true for $n = 3$, according to [18], but this is much harder to prove.

We conclude with a series of first applications of Theorem 6.3.11.

6.3.14 Corollary. *Let A be a local Noetherian ring of dimension one, let $T \subseteq A$ be a preordering, and let $f \in \mathrm{Sat}(T)$ not be a zero divisor in A. Then $f \in T \Leftrightarrow f \in \widehat{T}$.*

For zero divisors $f \in \mathrm{Sat}(T)$, it may happen that $f \in \widehat{T}$ but $f \notin T$, see the example in Exercise 6.3.3.

Proof. By Lemma 6.2.9 we have $\bigcap_{\alpha \in X(T) \cap Z(f)} \mathrm{supp}(\alpha) = \sqrt{\mathrm{supp}(T + Af)}$. Since f is not a zero divisor, f is not contained in any minimal prime ideal of A. Therefore $\mathfrak{m} \subseteq \sqrt{Af}$ and hence $\mathfrak{m} \subseteq \sqrt{\mathrm{supp}(T + Af)}$, which shows that the hypothesis of Theorem 6.3.11 is satisfied. □

6.3.15 Corollary. psd = sos *holds in every discrete valuation ring.*

Proof. Let $(A, \mathfrak{m}, k)$ be a discrete valuation ring and let $f \neq 0$ be a psd element of A. If k is non-real then $f > 0$ on $\mathrm{Sper}(A)$, and so f is sos by Corollary 6.3.7. When k is real, the assertion is immediate from Corollary 6.3.14 (plus Exercise 6.2.1), since $\widehat{A} \cong k[[t]]$. Alternatively, here is a more direct argument: Let k be real and write $f = ut^n$ where t is a prime element and $u \in A^*$. Then n is even by the Baer–Krull theorem (3.5.11). Assume that $u(\alpha) < 0$ for some $\alpha \in \mathrm{Sper}(A)$. Then $t(\alpha) = 0$ since f is psd, so $\mathrm{supp}(\alpha) = \mathfrak{m}$. Again by Baer–Krull, there is a proper generalization β of α in $\mathrm{Sper}(A)$. Then $\mathrm{supp}(\beta) = \{0\}$ and $u(\beta) < 0$, implying $f(\beta) < 0$, a contradiction. Therefore u is psd in A, and so u and f are sos in A by Theorem 6.3.1. □

6.3.16 (Plane curve singularities) Let R be a real closed field, let C be a plane affine curve over R. So $C = \mathcal{V}(f) \subseteq \mathbb{A}^2$ for some non-constant polynomial $f \in R[x, y]$ without multiple factors. The coordinate ring of C is $R[C] = R[x, y]/\langle f \rangle$, and the curve C is irreducible over R if and only if the polynomial f is irreducible in $R[x, y]$. An R-point ξ of C is singular if both partial derivatives $f_x = \frac{\partial f}{\partial x}$, $f_y = \frac{\partial f}{\partial y}$ vanish at ξ. The simplest type of singularities are nodes (or A_1-singularities), which are singular points ξ for which the Hessian matrix $D^2 f = \begin{pmatrix} f_{xx} & f_{xy} \\ f_{yx} & f_{yy} \end{pmatrix}$ is invertible at ξ. Nodes come in two types over R, since R is not algebraically closed: The singular R-point ξ of C

is an *ordinary node* if the symmetric matrix $D^2 f(\xi)$ is indefinite, and is an *acnode*[1] if $D^2 f(\xi)$ is (positive or negative) definite:

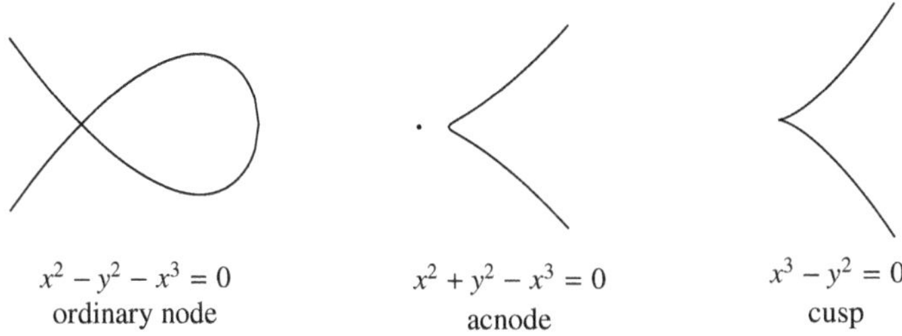

$x^2 - y^2 - x^3 = 0$
ordinary node

$x^2 + y^2 - x^3 = 0$
acnode

$x^3 - y^2 = 0$
cusp

In other words, if affine coordinates (x, y) are chosen such that $\xi = (0, 0)$ is a node of C, then after a suitable linear coordinate change, f can be brought in one of the two forms $f(x, y) = x^2 - y^2 + g(x, y)$ or $f(x, y) = x^2 + y^2 + g(x, y)$, with $g \in \langle x, y\rangle^3$. The first corresponds to an ordinary node, the second to an acnode. Note that an acnode is an isolated point of $C(R)$, while an ordinary node isn't. It is easy to see that the completed local ring $\widehat{\mathcal{O}}_{C,\xi} = R[\![x, y]\!]/\langle f\rangle$ is isomorphic to $R[\![x, y]\!]/\langle x^2 - y^2\rangle \cong R[\![x, y]\!]/\langle xy\rangle$ in the first case and to $R[\![x, y]\!]/\langle x^2 + y^2\rangle$ in the second. See Exercise 6.3.6.

The next simple type of plane curve singularities are cusps (alias A_2-singularities). The origin is an *(ordinary) cusp* of C if, after suitable linear coordinate change, f has the form $f(x, y) = y^2 + f_3(x, y) + g(x, y)$ where $g \in \langle x, y\rangle^4$ and f_3 is a cubic form with $f_3(x, 0) \neq 0$. The completed local ring $\widehat{\mathcal{O}}_{C,\xi}$ is easily shown to be isomorphic to $R[\![x, y]\!]/\langle y^2 - x^3\rangle$ in this case (Exercise 6.3.6).

6.3.17 Remarks.

1. Let C be a plane curve over R as above, let $\xi \in C(R)$. When ξ is either a regular point or an ordinary node of C, the property psd = sos holds in the local ring $\mathcal{O}_{C,\xi}$ (Corollary 6.3.15 for ξ regular and Exercise 6.3.7 for ξ an ordinary node). On the other hand, psd = sos does not hold in the local ring (and hence not in $R[C]$ either) when ξ is an acnode or an ordinary cusp (Exercise 6.3.8).

2. All curve singularities (plane or not) with the psd = sos property have been determined in [179]. They are precisely the ordinary multiple points with independent tangents, i.e. the local rings whose completion is isomorphic to

$$R[\![x_1, \ldots, x_n]\!]/\langle x_i x_j \colon 1 \le i < j \le n\rangle$$

for some $n \ge 1$. Informally speaking, these are those curve singularities in real n-space that locally look like a union of coordinate axes.

Applications of the techniques of this section to varieties of higher dimension will be given in the next sections.

---

[1] terminology is not uniform in the literature

Exercises

6.3.1 Prove Lemma 6.3.4.

6.3.2 Let $(A, \mathfrak{m}, k)$ be a local Henselian ring with $\mathrm{char}(k) \neq 2$. We denote the residue map $A \to k$ by $a \mapsto \overline{a}$.

(a) If $f \in A^*$ is such that $\overline{f}$ is a sum of n squares in k, then f is a sum of n squares in A.
(b) Let $T \subseteq A$ be a preordering. If $f \in T \cap A^*$, show that $f + g \in T$ for every $g \in \mathfrak{m}$.

6.3.3 Let $B = \mathbb{R}[x, y]$ and $f = x$, $g = y^2 + 1 - x \in B$, and let $\mathfrak{m} = \langle x, y^2 + 1\rangle \subseteq B$. The local ring $A := B_\mathfrak{m}/\langle fg\rangle$ has Krull dimension one. Show that f is psd in A and sos in $\widehat{A}$, but that f is not sos in A.

6.3.4 Let k be a field (with $\mathrm{char}(k) \neq 2$).

(a) Show that $k[\![x, y]\!]/\langle xy\rangle$ is isomorphic to the subring of $k[\![t]\!]\times k[\![t]\!]$ consisting of all pairs (f, g) with $f(0) = g(0)$.
(b) Show that $k[\![x, y]\!]/\langle y^2 - x^3\rangle$ is isomorphic to the subring of $k[\![t]\!]$ consisting of all power series with no linear term.

6.3.5 Let k be a field (with $\mathrm{char}(k) \neq 2$).

(a) Show that psd = sos holds in $k[\![x, y]\!]/\langle xy\rangle$.
(b) If the field k is real, show that psd = sos holds neither in $k[\![x, y]\!]/\langle x^2 + y^2\rangle$ nor in $k[\![x, y]\!]/\langle y^2 - x^3\rangle$.

(This should be easy if you use Exercise 6.3.4.)

6.3.6 Let C be a plane curve over the real closed field R, let $\xi \in C(R)$ be a singular point of C, and let $\mathcal{O}_\xi = \mathcal{O}_{C,\xi}$ be the local ring of C at ξ (see 6.3.16. Show that $\widehat{\mathcal{O}}_\xi$ is isomorphic to

(a) $R[\![x, y]\!]/\langle x^2 - y^2\rangle \cong R[\![x, y]\!]/\langle xy\rangle$ if ξ is an ordinary node,
(b) $R[\![x, y]\!]/\langle x^2 + y^2\rangle$ if ξ is an acnode,
(c) $R[\![x, y]\!]/\langle y^2 - x^3\rangle$ if ξ is a cusp.

6.3.7 Let C be a plane affine curve over R and let $\xi \in C(R)$ be an ordinary node. Show that psd = sos holds in the local ring $\mathcal{O}_{C,\xi}$. (Use Exercise 6.3.5 and a result from the text, but note that the case of zero divisors needs an extra argument.)

6.3.8 Let C be a plane affine curve over R, and let $\xi \in C(R)$ be a singular point of C.

(a) If ξ is an acnode (Remark 6.3.16), show that psd $\neq$ sos in $R[C]$.
(b) Try to prove the same conclusion if ξ is a cusp.

Hint: Assuming that ξ is the origin, show in case (a) that there is $g \in \langle x, y\rangle^2$ for which $f := x + g$ is psd on $C(R)$. For (b), try to modify the argument suitably.

6.3.9 Let A be a discrete valuation ring and let $t \in A$ be a prime element.

(a) The ring $A[x]/\langle x^2 - t\rangle$ is a discrete valuation ring.
(b) Which of the preorderings $PO(t)$, $PO(t, -t)$ or $PO(t, -t^2)$ in A are saturated?

6.3.10 Let $g_1, \ldots, g_r \in A = \mathbb{R}[\![t]\!]$ (formal power series in one variable), let $T = PO(g_1, \ldots, g_r)$.

(a) Assume $X(T) = X(t)$. Then T is saturated iff $\omega(g_i) = 1$ for some index i.
(b) Assume $X(T) = X(t, -t)$. Then T is saturated iff there are indices i, j with $\omega(g_i) = \omega(g_j) = 1$ and with $g_i g_j \leq 0$ on $\mathrm{Sper}(A)$.

6.4 Two-Dimensional Local Rings

The main result of this section (Theorem 6.4.7) says that psd = sos holds in every regular local ring of dimension two. Not surprisingly, the proof becomes harder than for the analogous result in dimension one (6.3.15). Using techniques from the previous sections, we reduce to the case of a formal power series ring. Then we employ the Weierstrass division and preparation theorems to deal with this case.

We start by recalling two classical theorems from the theory of quadratic forms. As usual, all rings are assumed to contain $\frac{1}{2}$, and in particular, all fields have characteristic different from two.

6.4.1 Theorem. (Artin) *Let k be a field, let $f \in k[t]$ be a monic and irreducible polynomial. If the residue field $k[t]/\langle f\rangle$ is non-real, f is a sum of squares in $k[t]$.*

It is obvious that the converse is true as well.

Proof. We may assume that the field k is real. By assumption there is an identity

$$fh = \sum_{i=1}^{n} g_i^2 \tag{6.5}$$

with polynomials g_i, $h \in k[t]$ such that $h \neq 0$ and f divides none of the g_i. We may assume $\deg(g_i) < \deg(f)$ for all i, which implies $\deg(h) < \deg(f)$. It suffices to find such an identity for which h is a constant. Indeed, this constant will then be the leading coefficient of the right-hand sum of squares, so it is sos in k, and we will be done.

Assume that $\deg(h) \geq 1$ in (6.5). We'll find a new identity of the same type in which the degree of h has become smaller. Write $g_i = q_i h + r_i$ for $i = 1, \dots, n$ where q_i, $r_i \in k[t]$ and $\deg(r_i) < \deg(h)$. Then $r_i \neq 0$ for at least one index i, since otherwise we would have $fh = h^2 \sum_i q_i^2$ and hence $f = h \sum_i q_i^2$, contradicting that f is irreducible. We use the identity

$$\Big(\sum_{i=1}^{n} g_i^2\Big)\Big(\sum_{i=1}^{n} r_i^2\Big) = \Big(\sum_{i=1}^{n} r_i g_i\Big)^2 + \sum_{1\leq i<j\leq n} \big(r_i g_j - r_j g_i\big)^2. \tag{6.6}$$

From $r_i \equiv g_i \pmod h$ we get $\sum_i r_i^2 \equiv \sum_i r_i g_i \equiv \sum_i g_i^2 \equiv 0 \pmod h$, and also $r_i g_j - r_j g_i \equiv 0 \pmod h$ for all $i < j$. So the right-hand side of (6.6) has the form $h^2 q$ where $q \in k[t]$ is sos. On the left we get a product decomposition $\sum_i r_i^2 = hh'$ with $0 \neq h' \in k[t]$, where $\deg(h') < \deg(h)$. Re-writing (6.6) gives $fh^2h' = h^2 q$, and after cancelling we get $fh' = q$. Since $\deg(h') < \deg(h)$, the proof is complete. □

6.4.2 Proposition. *Let k be a field. If $f \in k[t]$ is a sum of squares in $k(t)$, it is a sum of squares in $k[t]$.*

Proof. We may assume that f has no multiple factors. By assumption there is an identity $fh^2 = \sum_{i=1}^{n} g_i^2$ with polynomials h, $g_i \neq 0$. In particular, the leading coefficient of f is sos in k. Let $f_1 \in k[t]$ be irreducible such that f_1 divides f as well as each g_i, say $f = f_1 \tilde{f}$ and $g_i = f_1 \tilde{g}_i$. Then f_1 doesn't divide $\tilde{f}$, and the above identity implies $h = f_1 \tilde{h}$ for some polynomial $\tilde{h}$. So we get

$$f(f_1\tilde{h})^2 = f_1^2 \sum_i \tilde{g}_i^2,$$

and cancelling gives $f\tilde{h}^2 = \sum_i \tilde{g}_i^2$. Repeating the argument if necessary, we eventually arrive at an identity $fh^2 = \sum_{i=1}^n g_i^2$ with $\gcd(f, g_1, \dots, g_n) = 1$. Now every monic irreducible factor f' of f has a non-real residue field, and so each such factor f' is a sum of squares in $k[t]$ by Theorem 6.4.1. Since the leading coefficient of f is sos, f itself is sos in $k[t]$. □

We remark that Cassels [38] proved a stronger theorem in 1964: If a polynomial $f \in k[t]$ is a sum of m squares in $k(t)$, f is a sum of m squares in $k[t]$.

6.4.3 Lemma. *Let B be a discrete valuation ring with real residue field and with quotient field K. Then $\Sigma B[t]^2 = B[t] \cap \Sigma K(t)^2$. In particular,* psd = sos *holds in the polynomial ring $B[t]$.*

Proof. Let π be a prime element of B and let $f \in B[t]$ be a sum of squares in $K(t)$. By Proposition 6.4.2, f is a sum of squares in $K[t]$. Clearing denominators of elements in K, this means an identity $\pi^{2n} f = f_1^2 + \cdots + f_r^2$ with $f_1, \dots, f_r \in B[t]$ and $n \geq 0$. If $n \geq 1$ then reduction modulo π shows that each f_i has all coefficients divisible by π, using that the residue field of B is real. Therefore, we inductively see that f is sos in $B[t]$. Every psd element of $B[t]$ is psd in $K(t)$ and hence lies in $\Sigma K(t)^2$, which shows the last statement. □

The next lemma is an easy consequence of Theorem 3.6.8 and the fact that every localization of a regular local ring is again regular (A.4.5):

6.4.4 Lemma. *Let A be a regular local ring, let $K = \mathrm{qf}(A)$. Then $A_+ = A \cap \Sigma K^2$.*

Proof. $A_+ \subseteq K_+ = \Sigma K^2$ is clear. For the converse let $f \in A \cap \Sigma K^2$. By Corollary 3.6.9, any point $\beta \in \mathrm{Sper}(A)$ has a generalization α in $\mathrm{Sper}(K)$, and $f(\alpha) \geq 0$ implies $f(\beta) \geq 0$. So f lies in A_+. □

After these preparations we are going to prove the main result for rings of formal power series:

6.4.5 Theorem. *For every field k (with* $\mathrm{char}(k) \neq 2$*),* psd = sos *holds in the ring $k[[x, y]]$ of formal power series in two variables.*

Proof. Our proof uses the Weierstrass division and preparation theorems, see A.4.9 for more details. Write $A = k[[x, y]]$, which is a regular local ring. Given a psd element $f \neq 0$ in A, we need to show that f is sos in A. By Theorem 6.3.1 we may assume that f is not a unit in A. Since A is a unique factorization domain (*ufd*) we may write $f = f_1 g^2$ with $f_1, g \in A$, where f_1 doesn't contain any repeated factor. By Lemma 6.4.4, the element f_1 is psd in A as well, and it suffices to show that f_1 is sos in A.

So assume that f doesn't contain any repeated factor in A. After a suitable linear coordinate change we have $f = ug$, where u is a unit in A and $g \in k[\![x]\!][y]$ is a Weierstrass polynomial in y (Weierstrass preparation theorem, see A.4.9). This means that g has the form

$$g = y^m + \sum_{i=0}^{m-1} a_i(x)y^i$$

where $a_i(x) \in k[\![x]\!]$ are power series that satisfy $a_i(0) = 0$ $(i = 0, \dots, m-1)$. Write $B := k[\![x]\!]$. Since the polynomial ring $B[y]$ is a ufd, we can write $g = p_1 \cdots p_r$ where each p_i is irreducible in $B[y]$ and monic as a polynomial in y. Then each p_i is a Weierstrass polynomial in y by itself. For each index $i = 1, \dots, r$, the ring inclusion $B[y] \subseteq A$ induces an isomorphism $B[y]/p_iB[y] \to A/p_iA$ of the residue rings. This is a direct consequence of the Weierstrass division theorem, see A.4.9. Therefore, each p_i is irreducible also as an element of A.

For each index i, the quotient field of $A/\langle p_i \rangle$ is non-real. Indeed, say $i = 1$ and assume that there is $\alpha \in \mathrm{Sper}(A)$ with $\mathrm{supp}(\alpha) = \langle p_1 \rangle$. Since $A_{\langle p_1 \rangle}$ is a discrete valuation ring, there exist $\alpha_1, \alpha_2 \in \mathrm{Sper}(A)$ with $\mathrm{supp}(\alpha_j) = \langle 0 \rangle$, with $\alpha_j \rightsquigarrow \alpha$ $(j = 1, 2)$ and with $p_1(\alpha_1) > 0$ and $p_1(\alpha_2) < 0$. Since $f = up_1 \cdots p_r$ is psd in A, one of the remaining factors $u, p_2, \dots, p_r$ has to have opposite signs in α_1 and α_2 as well. But this implies that this factor is divisible by p_1, a contradiction.

On the other hand, let $K = \mathrm{qf}(B) = k(\!(x)\!)$. The quotient field of $A/\langle p_i \rangle \cong B[y]/\langle p_i \rangle$ is isomorphic to $K[y]/\langle p_i \rangle$. Indeed, Gauss's lemma shows that p_i is irreducible as a polynomial over K, and that the natural homomorphism $B[y]/\langle p_i \rangle \to K[y]/\langle p_i \rangle$ is injective. So this map induces an isomorphism of $\mathrm{qf}(B[y]/\langle p_i \rangle) \cong \mathrm{qf}(A/\langle p_i \rangle)$ with $K[y]/\langle p_i \rangle$.

It follows that the field $K[y]/\langle p_i \rangle$ is non-real. Since p_i is monic in $K[y]$, Theorem 6.4.1 implies that p_i is a sum of squares in $K[y]$. By Lemma 6.4.3, therefore, p_i is a sum of squares in $B[y]$, and hence in A as well. Since $f = up_1 \dots p_r$ is psd in A, the unit u is psd in A as well by Lemma 6.4.4. Therefore u is sos in A (Theorem 6.3.1). Altogether we have shown that f is a sum of squares in A. □

When the ground field is real closed, we can say more:

6.4.6 Corollary. *Let R be a real closed field. Then every psd element of $R[\![x, y]\!]$ is a sum of two squares.*

Proof. Let again $A = R[\![x, y]\!]$, $B = R[\![x]\!]$ and $K = R(\!(x)\!)$. Every psd unit in A is a square (Exercise 6.3.2). Following the proof of Theorem 6.4.5, and since a product of two sums of two squares is a sum of two squares, we may assume that $f \in B[y]$ is irreducible and monic in y, and that f is sos in $B[y]$. We'll show that f is a sum of two squares in $B[y]$. The field $K[y]/\langle f \rangle = \mathrm{qf}(B[y]/\langle f \rangle)$ is a finite non-real extension of K, see the previous proof. Therefore it contains $i = \sqrt{-1}$, see Exercise 6.4.2 for a proof. This means there exist polynomials $p, q \in B[y]$ which are not divisible by f, and such that f divides $p^2 + q^2$. We claim that f becomes reducible in the ring $B[i, y] = C[\![x]\!][y]$, where $C = R(\sqrt{-1})$. Indeed, since f divides $(p + iq)(p - iq)$ (and $B[i, y]$ is a ufd), f would otherwise divide one of the factors.

This would mean that f divides both p and q in $B[y]$, a contradiction. It follows that $f = u(g + ig')(g - ig') = u(g^2 + g'^2)$ with $u \in B^*$ and suitable $g, g' \in B[y]$. Since u is a square in B, we see that f is a sum of two squares in $B[y]$. □

As a second step, we show that Theorem 6.4.5 extends to arbitrary regular local rings of dimension two:

6.4.7 Theorem. *Let $(A, \mathfrak{m})$ be a regular local ring of dimension at most two (with $\frac{1}{2} \in A$). Then* psd = sos *holds in A.*

Proof. If $\dim(A) = 1$ then A is a discrete valuation ring, and this case was already settled in Corollary 6.3.15. So let $\dim(A) = 2$. By the Auslander–Buchsbaum theorem, A is a unique factorization domain, see A.4.5. Let $f \in A$ be psd, say $f = p_1 \cdots p_r$ with irreducible factors p_i in A. As in the proof of Theorem 6.4.5 we may assume that f has no multiple factors. By the same argument as there, the field $\mathrm{qf}(A/\langle p_i \rangle)$ is non-real for each i. Therefore $\mathrm{supp}(\alpha) = \mathfrak{m}$ holds for every zero $\alpha \in \mathrm{Sper}(A)$ of f. By Theorem 6.3.11 it therefore suffices to show that f is sos in the completion $\widehat{A}$ of A. If $k = A/\mathfrak{m}$ is non-real then $\mathrm{Sper}(\widehat{A}) = \varnothing$ (Exercise 3.2.5), hence $\Sigma \widehat{A}^2 = \widehat{A}$ (3.2.16). If k is real then $\widehat{A} \cong k[[x, y]]$, see A.4.6, and so f is sos in $\widehat{A}$ by Theorem 6.4.5. □

For greater flexibility in geometric applications, we extend Theorem 6.4.7 to certain preorderings:

6.4.8 Proposition. *Let $(A, \mathfrak{m})$ be a regular local ring with $\dim(A) = 2$, and let $f, g \in A$ with $\langle f, g \rangle = \mathfrak{m}$. Then both preorderings $PO(f)$ and $PO(f, g)$ in A are saturated.*

Proof. Let $B_1 = A[x]/\langle x^2 - f \rangle$ and $B_2 = A[x, y]/\langle x^2 - f,\ y^2 - g \rangle$. Both are regular local rings of dimension two (Exercise 6.4.1). Hence they satisfy psd = sos (Theorem 6.4.7), and the claim follows using Lemma 6.3.4. □

The one-dimensional analogue to 6.4.8 was proved in Exercise 6.3.9.

6.4.9 Remark. Concerning regular local rings A, we have seen that psd = sos holds in A whenever $\dim(A) \le 2$ (Theorem 6.4.7). Another case where psd = sos holds is when the field of fractions $K = \mathrm{qf}(A)$ is non-real. Indeed, this implies that $\mathrm{Sper}(A)$ is empty (Corollary 3.6.9), and so $\Sigma A^2 = A$ by 3.2.16. On the other hand, psd = sos fails whenever $\dim(A) \ge 3$ and the residue field of A is real (Corollary 6.2.8). What is left is the case where A has dimension ≥ 3, the residue field is non-real and the quotient field K is real. When $\dim(A) \ge 4$ and A is the local ring of a point on an R-variety, Exercise 6.2.5 shows that again psd = sos fails in A. The same is true when the variety is over an arbitrary field, as long as $K = \mathrm{qf}(A)$ is real.

The case $\dim(A) = 3$ is more subtle and was settled only recently by Benoist. He proved that the result is the same as in dimensions ≥ 4 ([18] Theorem 0.5). So altogether, when A is the local ring of an algebraic variety over a field at some non-singular point, the property psd = sos holds if and only if $\dim(A) \le 2$ or $K = \mathrm{qf}(A)$ is non-real.

The negative results for $\dim(A) \geq 3$ hold for local rings of algebraic varieties, but they do not extend to general regular local rings. In fact Benoist constructs, for every $d \geq 3$, a regular local $\mathbb{R}$-algebra A with $\dim(A) = d$, for which $K = \mathrm{qf}(A)$ is real and psd = sos holds ([18] Theorem 4.2). In this example, $\mathrm{Sper}(A)$ consists of a single point only, whose support is the zero ideal.

We would like to mention the notion of bad points on a real variety, since it is closely related to the "psd = sos" property of local rings. For simplicity we restrict to varieties over $\mathbb{R}$.

6.4.10 Definition. Let $f \in \mathbb{R}[x] = \mathbb{R}[x_1, \dots, x_n]$ be a polynomial. A point $\xi \in \mathbb{C}^n$ is a *bad point* of f if $h(\xi) = 0$ for every $h \in \mathbb{R}[x]$ for which fh^2 is a sum of squares in $\mathbb{R}[x]$. By $B(f) \subseteq \mathbb{C}^n$ we denote the *bad locus* of f, namely the set of all bad points of f.

By definition, $B(f)$ is always a Zariski closed $\mathbb{R}$-subvariety of $\mathbb{A}^n$. When f is psd on $\mathbb{R}^n$, the bad locus $B(f)$ is a proper subvariety of $\mathbb{A}^n$, by Artin's theorem 1.5.21. Otherwise $B(f) = \mathbb{A}^n$. Directly from the definition we see:

6.4.11 Lemma. *A point $\xi \in \mathbb{C}^n$ is a bad point of f if and only if f is not a sum of squares in the local ring $\mathcal{O}_\xi = \mathbb{R}[x]_{\mathfrak{m}_\xi}$ at ξ.* □

Theorem 6.4.7 therefore implies that psd polynomials in two variables have no bad points. In other words, for a psd polynomial $f \in \mathbb{R}[x_1, x_2]$ there are no local obstructions against f being a sum of squares in $\mathbb{R}[x_1, x_2]$. More generally, if $f \in \mathbb{R}[x_1, \dots, x_n]$ is psd then $B(f)$ (is empty or) has codimension at least three. Indeed, let $V \subseteq \mathbb{A}^n$ be an irreducible component of $B(f)$, and let $\mathfrak{p} = \mathcal{I}(V) \subseteq \mathbb{R}[x]$ be the prime ideal corresponding to V. Since f is not a sum of squares in the local ring $\mathbb{R}[x]_\mathfrak{p}$, and since this ring is regular, it must have dimension ≥ 3 by Theorem 6.4.7.

6.4.12 Remark. Benoist's paper [18] mentioned before contains several other remarkable results closely related to bad points. To mention two of them, he constructs a psd polynomial $f \in \mathbb{R}[x_1, x_2, x_3]$ whose bad locus consists of just two complex-conjugate points ([18] Theorem 3.6). This is the first example where the $\mathbb{R}$-points are not dense in $B(f)$. He also constructs a psd polynomial $f \in \mathbb{R}[x_1, x_2, x_3]$ with a bad point $\xi \in \mathbb{R}^3$ such that f is a sum of squares in the completion $\widehat{\mathcal{O}}_\xi$. Whether such f exists (in any number of variables) had been a question of Brumfiel ([51] p. 62).

Exercises

6.4.1 Let $(A, \mathfrak{m})$ be a regular local ring, and let $a_1, \dots, a_r \in \mathfrak{m}$ be linearly independent modulo $\mathfrak{m}^2$. Let positive integers $n_1, \dots, n_r$ be given. Then

$$B = A[x_1, \dots, x_r]/\langle x_i^{n_i} - a_i \colon i = 1, \dots, r\rangle$$

is again a regular local ring, and $\dim(B) = \dim(A)$. (*Hint*: Use going-up and consider the Artinian ring $B/\mathfrak{m}_A B$.)

6.4.2 Let k be a field and let $K = k((x))$, the field of formal Laurent series over k. Find all finite field extensions of K up to K-isomorphism when k is (a) algebraically closed of characteristic zero, (b) real closed. In particular, show in case (b) that every non-real finite extension of K contains $\sqrt{-1}$. *Hint*: When k is algebraically closed, recall that the field $k((x^{1/\infty}))$ of Puiseux series is algebraically closed (Puiseux' theorem A.4.10)

6.4.3 Let $f \in \mathbb{R}[x_1, \dots, x_n]$ be homogeneous, let $d = \deg(f)$.

(a) Show that f is a sum of squares if and only if the origin in $\mathbb{C}^n$ is not a bad point of f.
(b) Let $0 \neq \xi \in \mathbb{C}^n$, let k be an index with $\xi_k \neq 0$. Show that $\xi \in B(f) \Leftrightarrow fx_k^{-d}$ is not a sum of squares in the local ring of $\mathbb{P}^{n-1}$ at $[\xi] = (\xi_1 : \cdots : \xi_n)$.

6.4.4 Let $f \in \mathbb{R}[x_1, \dots, x_n]$ be psd. Show that f vanishes in each of its bad points. When $f > 0$ on $\mathbb{R}^n$, show that f has no (real or complex) bad points.

6.4.5 Let $f \in \mathbb{R}[x] = \mathbb{R}[x_1, \dots, x_n]$ be psd. Show that there exists $h \in \mathbb{R}[x]$ such that $h^2 f$ is sos, and such that the only real zeros of h are the real bad points of f.

6.5 Global Results

We now combine local saturatedness results, as obtained in the previous sections, with the Archimedean local-global principle (Theorem 6.2.19), to obtain global results. We start with versions of Theorem 6.2.19 that are more convenient in the case of preorderings or quadratic modules. Let A always be a ring. We assume $\mathbb{Q} \subseteq A$ for simplicity.

6.5.1 Theorem. *Assume $\mathbb{Q} \subseteq A$, let $T \subseteq A$ be an Archimedean preordering and let $f \in \mathrm{Sat}(T)$. If $f \in T_{\mathfrak{m}}$ holds for every maximal ideal $\mathfrak{m}$ of A with* $\mathrm{supp}(T + Af) \subseteq \mathfrak{m}$, *then $f \in T$.*

Proof. Let $f \in \mathrm{Sat}(T)$ be as in the theorem. The condition $f \in T_{\mathfrak{m}}$ means that there exists $s \in A$, $s \notin \mathfrak{m}$, with $s^2 f \in T$. In view of Theorem 6.2.19, it suffices therefore to show $f \in T_{\mathfrak{m}}$ for every maximal ideal $\mathfrak{m}$ with $\mathrm{supp}(T + Af) \not\subseteq \mathfrak{m}$. So let $\mathfrak{m}$ be such a maximal ideal. Clearly $f \in \mathrm{Sat}_{A_{\mathfrak{m}}}(T_{\mathfrak{m}})$. If $f \notin \mathfrak{m}$ then f is a unit in $A_{\mathfrak{m}}$, and so $f \in T_{\mathfrak{m}}$ by Corollary 6.3.5. So assume $f \in \mathfrak{m}$. By assumption on $\mathfrak{m}$, there exist $t_i \in T$, $a_i \in A$ $(i = 1, 2)$ with $t_1 + a_1 f = -t_2 + a_2 f \notin \mathfrak{m}$. Hence t_1, t_2 are units in $A_{\mathfrak{m}}$, and so t_1, $t_2 > 0$ on $X_{A_{\mathfrak{m}}}(T_{\mathfrak{m}})$. Since $t_1 + t_2 = (a_2 - a_1)f$ and $f \geq 0$ on $X_{A_{\mathfrak{m}}}(T_{\mathfrak{m}})$, we conclude that $f > 0$ on $X_{A_{\mathfrak{m}}}(T_{\mathfrak{m}})$ as well. So Corollary 6.3.7 implies $f \in T_{\mathfrak{m}}$. □

6.5.2 Remark. In a geometric setting, Theorem 6.5.1 becomes more intuitive. Let V be an affine $\mathbb{R}$-variety and let $K = \mathcal{S}_V(g_1, \dots, g_r)$ be a basic closed set in $V(\mathbb{R})$, with $g_i \in \mathbb{R}[V]$. We assume that K is compact, which means that the preordering $T = PO(g_1, \dots, g_r)$ in $\mathbb{R}[V]$ is Archimedean (Theorem 5.5.3). Given a polynomial $f \in \mathbb{R}[V]$ with $f|_K \geq 0$, Theorem 6.5.1 provides a necessary and sufficient criterion, in terms of "local" conditions, for f to be contained in T. Namely, for each maximal ideal $\mathfrak{m}$ of $\mathbb{R}[V]$ with $\mathrm{supp}(T + Af) \subseteq \mathfrak{m}$ we have the condition $f \in T_{\mathfrak{m}}$. These maximal ideals correspond to the (complex) points in the Zariski closure Z of $\mathcal{Z}_V(f) \cap K$, the zero set of f in K, as we see from applying Remark 6.2.11 to the preordering

$T + Af$. Beware that, in general, it is *not* enough to check $f \in T_{\mathfrak{m}}$ for the real points in Z, i.e. for the maximal ideals $\mathfrak{m}$ with residue field $\mathbb{R}$. An example illustrating this point is given in Exercise 6.5.6.

For quadratic modules we only have a weaker statement:

6.5.3 Proposition. *Assume $\mathbb{Q} \subseteq A$, and let $M \subseteq A$ be a module over an Archimedean preordering T of M. If $f \in A$ satisfies $f \in M_{\mathfrak{m}}$ for every maximal ideal $\mathfrak{m}$ of A with* $\operatorname{supp}(M) \subseteq \mathfrak{m}$, *then $f \in M$.*

Proof. By Theorem 6.2.19 it suffices to show $f \in M_{\mathfrak{m}}$ for every maximal ideal $\mathfrak{m}$ with $\operatorname{supp}(M) \not\subseteq \mathfrak{m}$. For such $\mathfrak{m}$ there is $s \in \operatorname{supp}(M)$ with $s \notin \mathfrak{m}$. Since $\operatorname{supp}(M)$ is an ideal we have $s^2 f \in \operatorname{supp}(M)$, and so $f \in M_{\mathfrak{m}}$ (compare Lemma 5.1.11). □

In any case, the question of being saturated can be decided locally:

6.5.4 Corollary. *($\mathbb{Q} \subseteq A$) Let $M \subseteq A$ be a module over an Archimedean preordering. If $M_{\mathfrak{m}}$ is saturated (in $A_{\mathfrak{m}}$) for every maximal ideal $\mathfrak{m}$ of A with* $\operatorname{supp}(M) \subseteq \mathfrak{m}$, *then M is saturated (in A).*

Of course, the converse is true anyway, without any Archimedean hypothesis (Proposition 6.2.6).

Proof. If $f \in \operatorname{Sat}(M)$ then $f \in \operatorname{Sat}(M_{\mathfrak{m}})$ for any $\mathfrak{m}$. So the claim follows from 6.5.3. □

We now discuss selected geometric applications. In all of them we are given an affine $\mathbb{R}$-variety V and a compact basic closed set $K = \mathcal{S}_V(g_1, \dots, g_r)$ in $V(\mathbb{R})$, and we consider the preordering T and the quadratic module M generated by $g_1, \dots, g_r$ in $\mathbb{R}[V]$. Given a point $\xi \in V(\mathbb{C})$, we denote by T_ξ the preordering generated by T in the local ring $\mathcal{O}_{V,\xi}$, and by $\widehat{T}_\xi$ the preordering generated by T in the completed local ring $\widehat{\mathcal{O}}_{V,\xi}$. Similarly for M_ξ and $\widehat{M}_\xi$.

Let $f \in \mathbb{R}[V]$ be a polynomial with $f|_K \geq 0$. Assuming that f has only finitely many zeros in K, we have:

6.5.5 Proposition. *Let V be an affine $\mathbb{R}$-variety, let $T \subseteq \mathbb{R}[V]$ be a finitely generated preordering with associated basic closed set $K = \mathcal{S}_V(T)$, and let $f \in \mathbb{R}[V]$ satisfy $f \geq 0$ on K. If K is compact and f has only finitely many zeros $\xi_1, \dots, \xi_r$ in K, and if $f \in \widehat{T}_{\xi_i}$ holds for each i, then $f \in T$.*

Proof. The preordering T is Archimedean by Theorem 5.5.3. By Theorem 6.5.1 it therefore suffices to show $f \in T_{\mathfrak{m}}$ for every maximal ideal $\mathfrak{m}$ of $\mathbb{R}[V]$ that contains $\operatorname{supp}(T + \langle f \rangle)$. These maximal ideals are the maximal ideals of the $\mathbb{C}$-points in the Zariski closure of $\mathcal{Z}_V(f) \cap K$ (Lemma 6.2.9). Since $\mathcal{Z}_V(f) \cap K = \{\xi_1, \dots, \xi_r\}$ is a finite set of $\mathbb{R}$-points, it means we have to show $f \in T_{\xi_i}$ for each i. By assumption we know $f \in \widehat{T}_{\xi_i}$, so the assertion follows from Theorem 6.3.11. □

There is also a version for quadratic modules that gives the same conclusion. The hypotheses K compact and $|\mathcal{Z}_V(f) \cap K| < \infty$ have to be replaced by assumptions which are equivalent in the preorderings case, but are stronger for quadratic modules:

6.5.6 Proposition. *Let V be an affine $\mathbb{R}$-variety, let $M \subseteq \mathbb{R}[V]$ be a finitely generated Archimedean quadratic module, let $K = \mathcal{S}_V(M)$, and let $f \in \mathbb{R}[V]$ satisfy $f|_K \geq 0$. Assume that the ideal $J := \operatorname{supp}(M+\langle f^2\rangle)$ of $\mathbb{R}[V]$ satisfies $\dim \mathbb{R}[V]/J = 0$. If $f \in \widehat{M}_\xi$ for every $\xi \in \mathcal{Z}_V(f) \cap K$, then $f \in M$.*

The proof is based on the following general lemma, which is an Archimedean analog of Lemma 6.3.6:

6.5.7 Lemma. *Let A be a ring, let $M \subseteq A$ be an Archimedean quadratic module and let $f \in \operatorname{Sat}(M)$. If $f \in M + \langle f^2\rangle$ then $f \in M$.*

Proof. Since $A = (\Sigma A^2) - (\Sigma A^2)$, we have $M + \langle f^2\rangle = M - f^2 \Sigma A^2$. So the assumption gives an identity $f = x - sf^2$ with $x \in M$ and $s \in \Sigma A^2$. Therefore $x = f(1 + sf)$, and both f and $1 + sf$ are non-negative on $Y := X(M)$. By Lemma 6.2.17 there exist $a, b \in A$ with

$$1 = af + b(1 + sf) \tag{6.7}$$

and with $a > 0$, $b > 0$ on Y. Hence also $ab > 0$ on Y, and so $a, b, ab \in M$ by the Archimedean positivstellensatz 5.3.3. Multiply (6.7) by bf to see that $bf = abf^2 + b^2 f(1 + sf) = abf^2 + b^2 x$ lies in M. Multiply (6.7) by f to see $f = af^2 + bf + bsf^2 \in M$. □

Proof of Proposition 6.5.6. Write $A = \mathbb{R}[V]$. By Lemma 6.5.7 it suffices to show $f \in M + J$, since $M + J \subseteq M + \langle f^2\rangle$. Being 0-dimensional and Noetherian, the ring A/J is semilocal. Every minimal = maximal ideal of A/J is real by Proposition 5.1.17, so it is the maximal ideal $\mathfrak{m}_\xi/J$ corresponding to some point $\xi \in \mathcal{Z}_V(f) \cap K$. Applying Proposition 6.3.2 to the quadratic module $(M + J)/J$ in A/J, it suffices to show, for every such maximal ideal $\mathfrak{m}/J = \mathfrak{m}_\xi/J$, that the quadratic module generated by M in $(A/J)_{\mathfrak{m}/J}$ contains f. Consider the commutative diagram

$$\begin{array}{ccccc} A & \longrightarrow & A_{\mathfrak{m}} & \longrightarrow & \widehat{A_{\mathfrak{m}}} \\ \downarrow & & \downarrow & & \downarrow \\ A/J & \longrightarrow & A_{\mathfrak{m}}/JA_{\mathfrak{m}} & \xrightarrow{\sim} & \widehat{A_{\mathfrak{m}}}/J\widehat{A_{\mathfrak{m}}} \end{array}$$

of ring homomorphisms. The map labelled $\sim$ is an isomorphism since $\dim(A/J) = 0$. In view of the natural isomorphism $(A/J)_{\mathfrak{m}/J} \cong A_{\mathfrak{m}}/JA_{\mathfrak{m}}$, this completes the proof. □

Using Proposition 6.5.6, we can decide the question left open in Remark 6.1.5.3:

6.5.8 Proposition. *Let C be a non-singular affine curve over $\mathbb{R}$, and let $M \subseteq \mathbb{R}[C]$ be a quadratic module that is finitely generated. If the subset $\mathcal{S}(M)$ of $C(\mathbb{R})$ is compact then M is a preordering.*

Proof. The quadratic module M is Archimedean by Proposition 5.5.14. Given $0 \neq f, g \in M$, we have to show $fg \in M$. For this we may assume that C is irreducible. Now Proposition 6.5.6 applies since the ideal $\operatorname{supp}(M + \langle (fg)^2 \rangle)$ is 0-dimensional. Therefore, the question may be decided in the completed local rings $\widehat{\mathcal{O}}_{C,\xi}$ ($\xi \in C(\mathbb{R})$). These rings are isomorphic to the formal power series ring $\mathbb{R}[[t]]$. By an easy argument (Exercise 6.5.9), every quadratic module in $\mathbb{R}[[t]]$ is a preordering. So we are done. □

A prominent case where Propositions 6.5.5 and 6.5.6 apply is when suitable assumptions are made on the Hessian of f at each of its zeros:

6.5.9 Definition. Let V be an affine $\mathbb{R}$-variety, let $\xi \in V(\mathbb{R})$ be a non-singular $\mathbb{R}$-point of V with maximal ideal $\mathfrak{m}_\xi$, and let $\boldsymbol{g} = (g_1, \dots, g_n)$ with $g_i \in \mathbb{R}[V]$ be a regular system of parameters at ξ. Let $f \in \mathbb{R}[V]$ with $f(\xi) = 0$, and write $f = \sum_{i=1}^n a_i g_i + \sum_{i,j=1}^n b_{ij} g_i g_j + h$ with $a_i, b_{ij} = b_{ji} \in \mathbb{R}$ and $h \in \mathfrak{m}_\xi^3$.

(a) f satisfies the *Hessian conditions* at ξ if $a_1 = \cdots = a_n = 0$ (i.e. $f \in \mathfrak{m}_\xi^2$) and if the Hessian matrix $(b_{ij})_{1 \le i,j \le n}$ (of f at ξ, with respect to $\boldsymbol{g}$) is positive definite.
(b) Let $0 \le r \le n$, then f satisfies the *boundary Hessian conditions at ξ with respect to $g_1, \dots, g_r$* if $a_1 > 0, \dots, a_r > 0$ and $a_{r+1} = \cdots = a_n = 0$, and if the symmetric matrix $(b_{ij})_{r+1 \le i,j \le n}$ (of size $n - r$) is positive definite.

6.5.10 Remarks.

1. The Hessian conditions (a) are independent of the chosen regular parameter system $\boldsymbol{g}$. Similarly, the boundary Hessian conditions (b) depend only on the first r parameters $g_1, \dots, g_r$. Note that (a) is the particular case $r = 0$ of (b).

2. Let $V = \mathbb{A}^n$ be affine n-space and let $g = x = (x_1, \dots, x_n)$ be the cartesian coordinates. For $0 \le r \le n$ let $K_r = \{\xi \in \mathbb{R}^n : \xi_1 \ge 0, \dots, \xi_r \ge 0\}$. If $f \in \mathbb{R}[x]$ is a polynomial with $f(0) = 0$, the following hold by elementary calculus (Exercise 6.5.1): If f satisfies the boundary Hessian conditions at $\xi = 0$ with respect to $x_1, \dots, x_r$, then $f > 0$ on $(U \cap K_r) \setminus \{0\}$ for some neighborhood U of the origin. Conversely, if $f|_{K_r}$ has a local minimum at the origin, the "non-strict" boundary Hessian conditions hold, i.e. $\frac{\partial f}{\partial x_i}(0) \ge 0$ ($i = 1, \dots, r$) and $\left(\frac{\partial^2 f}{\partial x_i \partial x_j}(0)\right)_{r+1 \le i,j \le n} \succeq 0$. The same holds, *mutatis mutandis*, around any non-singular $\mathbb{R}$-point on any affine $\mathbb{R}$-variety, and with respect to any regular system of parameters. Therefore, the boundary Hessian conditions (b) at ξ are a natural sufficient condition, for the restriction of f to $\mathcal{S}_V(g_1, \dots, g_r)$ to have a strict local minimum at ξ.

6.5.11 Theorem. *Let V be an affine $\mathbb{R}$-variety, let $T \subseteq \mathbb{R}[V]$ be a finitely generated preordering for which $K = \mathcal{S}_V(T)$ is compact, and let $f \in \mathbb{R}[V]$ with $f|_K \ge 0$. For every $\xi \in \mathcal{Z}(f) \cap K$, assume that ξ is a non-singular point of V, and that there is a sequence $g_1, \dots, g_r$ in T which is part of a regular system of parameters at ξ, such that f satisfies the boundary Hessian conditions at ξ with respect to $g_1, \dots, g_r$. Then $f \in T$.*

For the proof we need the following lemma:

6.5.12 Lemma. *If k is a field and $f \in k[[x_1, \dots, x_n]]$ satisfies $\omega(f - \sum_{i=1}^n x_i^2) \geq 3$, then f is a sum of squares in $k[[x_1, \dots, x_n]]$.*

Proof. The lemma is a particular case of Exercise 5.7.8, where a proof is given that uses pure states. The following direct proof is taken from Marshall [138]. If $n = 1$ then $f = x^2 f_1$ where $f_1 \in 1 + \mathfrak{m}$, so f_1 and hence f is a square. Let $n > 1$ and let $g = f - \sum_{i=1}^n x_i^2$. Since $\omega(g) \geq 3$ we may write $g = x_1^2 g_1 + x_1 g_2 + g_3$, where the g_i are power series such that x_1 occurs neither in g_2 nor in g_3, and $\omega(g_i) \geq i$ for $i = 1, 2, 3$. So we have

$$f = \frac{1}{2} x_1^2 (1 + 2g_1) + \frac{1}{2}(x_1 + g_2)^2 + h \tag{6.8}$$

with

$$h = x_2^2 + \cdots + x_n^2 + g_3 - \frac{1}{2} g_2^2.$$

Both the first two summands in (6.8) are squares. Since g_2, g_3 involve only the variables $x_2, \dots, x_n$, the series h is sos by the inductive hypothesis. Therefore f is sos as well. □

The following is an easy generalization (see Exercise 6.5.5 for the proof):

6.5.13 Corollary. *Let R be a real closed field, let $x = (x_1, \dots, x_n)$ and $1 \leq i \leq r$. Let $f, g \in R[[x]]$ be such that $f = \sum_{i=r+1}^n a_i x_i + \sum_{i,j=1}^n a_{ij} x_i x_j + g$ where a_i, $a_{ij} = a_{ji} \in R$ and $\omega(g) \geq 3$. Assume that $a_{r+1}, \dots, a_n > 0$ and that the symmetric matrix $(a_{ij})_{1 \leq i,j \leq r}$ is positive definite. Then f lies in the preordering of $R[[x]]$ that is generated by $x_{r+1}, \dots, x_n$.*

Proof of Theorem 6.5.11. By Corollary 6.5.13, f lies in $\widehat{T_\xi}$ for every $\xi \in \mathcal{Z}_V(f) \cap K$. Since the boundary Hessian conditions imply that each zero of f in K is isolated in K, the zero set $\mathcal{Z}_V(f) \cap K$ is finite. So the theorem follows from Proposition 6.5.5. □

6.5.14 Remark. When we replace T in 6.5.11 by an Archimedean quadratic module M, the same proof works if we add the assumption that $\mathbb{R}[V]/\operatorname{supp}(M + \langle f \rangle)$ is 0-dimensional, see 6.5.6. In fact, this additional assumption may be shown to hold automatically, therefore it is not needed. We refer to Marshall's paper [137] for the details.

6.5.15 Example. Let $\xi \in \mathbb{R}^n$, and let $f \in \mathbb{R}[x] = \mathbb{R}[x_1, \dots, x_n]$ be such that $f(\xi) = 0$ and $u := \nabla f(\xi) \neq 0$. Assume that the Hessian matrix $H := D^2 f(\xi)$ satisfies $v^\top H v < 0$ for every $v \neq 0$ with $\langle u, v \rangle = 0$. (Using terminology that will be introduced in Section 8.6, the hypothesis says that f is strictly quasi-concave at ξ.) Then the tangent $t_\xi(x) = \langle u, x - \xi \rangle$ to the hypersurface $f = 0$ at $x = \xi$ satisfies a boundary Hessian condition at ξ with respect to f. This follows from the Taylor expansion $f(\xi + x) = \langle u, x \rangle + \frac{1}{2} x^\top H x + \cdots$ of f around ξ (where $\cdots$ stands for terms of degree ≥ 3), which shows $t_\xi(x) = f(x) - \frac{1}{2}(x - \xi)^\top H (x - \xi) - \cdots$

We now consider preorderings on curves.

6.5.16 Proposition. *Let C be an irreducible affine curve over $\mathbb{R}$ without real singular points, and let $g_1, \dots, g_r \in \mathbb{R}[C]$ be such that $K = \mathcal{S}_C(g_1, \dots, g_r)$ is compact. The preordering $T = PO(g_1, \dots, g_r)$ in $\mathbb{R}[C]$ is saturated if, and only if, the following two conditions hold:*

(1) *For every boundary point ξ of K (relative to $C(\mathbb{R})$) there is an index i with* $\mathrm{ord}_\xi(g_i) = 1$;
(2) *for every isolated point ξ of K there are indices i, j with* $\mathrm{ord}_\xi(g_i) = \mathrm{ord}_\xi(g_j) = 1$ *such that $g_i g_j \le 0$ in a neighborhood of ξ on $C(\mathbb{R})$.*

Proof. Assume that $C \subseteq \mathbb{A}^n$. Let ξ be a boundary point of K that is not an isolated point of K. Then there exists $f \in \mathbb{R}[C]$ with $f \ge 0$ on K and with $\mathrm{ord}_\xi(f) = 1$. An easy way to see this is as follows: Choose a point $\eta \in C(\mathbb{R})$ that is close to ξ and not contained in K, and consider the polynomial $f_\eta(x) = |x - \eta|^2 - |\xi - \eta|^2$ ($x \in \mathbb{R}^n$). If η was chosen close enough to ξ, then $f_\eta \in \mathbb{R}[C]$ has the desired property. When ξ is an isolated point of K, a variation of the argument shows that there exist f_1, $f_2 \in \mathbb{R}[C]$ with $\mathrm{ord}_\xi(f_1) = \mathrm{ord}_\xi(f_2) = 1$ and with $f_1 f_2 \le 0$ on $C(\mathbb{R})$ in a neighborhood of ξ.

Now assume that T is saturated. For every boundary point ξ of K, the completed localized preordering $\widehat{T_\xi}$ in $\widehat{\mathcal{O}}_{C,\xi}$ is saturated as well, by the previous argument combined with Exercise 6.3.10. The same exercise shows that (1) and (2) are satisfied.

Conversely assume that (1) and (2) are true. Then for every $\xi \in K$, the completed localized preordering $\widehat{T_\xi}$ is saturated by Exercise 6.3.10. Hence the uncompleted preordering T_ξ is saturated (Corollary 6.3.14), and so Theorem 6.5.1 implies that T itself is saturated. □

6.5.17 Examples.

1. Without going into details we remark, for C and K as above, that there always exist two polynomials h_1, $h_2 \in \mathbb{R}[C]$ with $K = \mathcal{S}_C(h_1, h_2)$ that satisfy the conditions in Proposition 6.5.16. When K has no isolated points, one can even find a single polynomial h_1 with these properties. When K has at least one isolated point, this is not possible by 6.5.16. In any case, the saturated preordering $\mathcal{P}(K)$ of any compact semialgebraic set $K \subseteq C(\mathbb{R})$ is finitely generated (for C a non-singular affine curve).

2. For an example, consider the set $K = \{0\} \cup [2, 3] \subseteq \mathbb{R}$. The natural generators (6.1.2) for K are $p_0 = t$, $p_1 = t(t-2)$ and $p_2 = 3 - t$. Their product $g = p_0 p_1 p_2$ satisfies $K = \{\xi \in \mathbb{R} \colon g(\xi) \ge 0\}$, but $PO(g)$ fails to be saturated according to 6.5.16. An explicit case of a polynomial h with $h|_K \ge 0$ but $h \notin PO(g)$ is $h = t$, as witnessed in the (completed) local ring of the origin.

3. Let us re-consider the saturated preordering $\mathcal{P}(K) \subseteq \mathbb{R}[t]$ of a closed semialgebraic set $K \subseteq \mathbb{R}$. It was shown in 6.1.3 that $\mathcal{P}(K)$ is always finitely generated by the natural generators for K, and that these are the essentially unique minimal system of generators when K is unbounded. From Proposition 6.5.16 we see that the bounded (compact) case is entirely different: $\mathcal{P}(K)$ can always be generated by two polynomials, and there is no uniqueness of the generators whatsoever. When K has no isolated points, $\mathcal{P}(K)$ is even generated by the product $p_0 \cdots p_m$ of the natural generators alone.

Let us show that this last result doesn't extend to non-Archimedean real closed base fields R. Let $\varepsilon > \delta > 0$ be two positive infinitesimals in R, and consider $K = [-1, 0] \cup [\delta, \varepsilon]$. The natural generators for K are $p_0 = t + 1$, $p_1 = t(t - \delta)$ and $p_2 = \varepsilon - t$, their product is $p = p_0p_1p_2$. We show that $PO(p)$ doesn't contain p_1. Otherwise there would be an identity $p_1 = \sigma_0 + \sigma_1 p_0p_1p_2$ with σ_0, σ_1 sos in $R[t]$. It follows that σ_0 is divisible by p_1 and hence by p_1^2, so we get

$$1 = s_0p_1 + s_1p_0p_2 \tag{6.9}$$

with sums of squares s_0, s_1 in $R[t]$. Let $\mathcal{O} \subseteq R$ be the convex hull of $\mathbb{Z}$ in R. We cannot have $s_0, s_1 \in \mathcal{O}[t]$, since reduction modulo $\mathfrak{m}_{\mathcal{O}}$ would then give $1 = \overline{s}_0t^2 - \overline{s}_1t(t+1)$ in $\mathbb{R}[t]$, which is absurd. Let $c > 0$ be the maximum of the absolute values of the coefficients of s_0 and s_1. So $c \notin \mathcal{O}$, and dividing (6.9) by c gives $0 = s_0'p_1 + s_1'p_0p_2$ with s_0', s_1' sums of squares in $\mathcal{O}[t]$. Reduction modulo $\mathfrak{m}_{\mathcal{O}}$ gives $0 = \overline{s}_0't^2 - \overline{s}_1't(t+1)$ in $\mathbb{R}[t]$ where $\overline{s}_0'$, $\overline{s}_1'$ are sums of squares in $\mathbb{R}[t]$ and at least one of them is non-zero. Again this is impossible.

6.5.18 Remarks.

1. Let C be an affine curve over $\mathbb{R}$ that is non-singular and irreducible. When $C(\mathbb{R})$ is compact, we have seen (6.5.16) that psd = sos holds for $\mathbb{R}[C]$. There are in fact many more cases when psd = sos holds, as illustrated by the following example. Consider the plane affine curve $C = \mathcal{V}(x^3 + y^3 + xy + 1) \subseteq \mathbb{A}^2$, with non-singular projective model $\overline{C} = \mathcal{V}(x^3 + y^3 + xyz + z^3) \subseteq \mathbb{P}^2$. Obviously $C(\mathbb{R})$ is not compact. There are three points of $\overline{C}$ on the line $z = 0$ (the "points at infinity" of the affine curve C), namely $P = (1 : -1 : 0)$ and the pair Q, Q' of complex conjugate points $(\omega : 1 : 0)$ and $(1 : \omega : 0)$ with ω a primitive sixth root of unity. The curve $C_0 := \overline{C} \setminus \{Q, Q'\}$ is again affine (by Riemann–Roch), and contains the original curve C as an open subcurve, namely $C = C_0 \setminus \{P\}$. Algebraically, this means that $\mathbb{R}[C]$ is the localization of $\mathbb{R}[C_0]$ in a suitable multiplicative set. Since $C_0(\mathbb{R})$ is compact, psd = sos holds on C_0, and therefore it holds on C as well (6.2.6).

2. The previous example can be generalized in a straightforward way. For every irreducible non-singular affine curve C over $\mathbb{R}$, there is a non-singular projective curve $\overline{C}$, unique up to isomorphism, that contains C as a Zariski dense open subset. If at least one of the finitely many points in $\overline{C} \setminus C$ has complex residue field, then psd = sos holds on C, exactly as in the example before. On the other hand, if $\overline{C}$ has genus ≥ 1 and all points in $\overline{C} \setminus C$ are $\mathbb{R}$-points, it can be shown that the psd = sos property fails for C. An example are the hyperelliptic curves of odd degree ≥ 3 considered in Exercise 6.1.5. These results can be extended further in various directions, see [180] or [153] for more details. Altogether, saturatedness questions on affine curves over $\mathbb{R}$ are essentially well understood.

We now turn to algebraic surfaces. From the Archimedean local-global principle and the discussion in Section 6.4, we immediately see:

6.5.19 Theorem. *Let V be a non-singular affine $\mathbb{R}$-variety of dimension two for which $V(\mathbb{R})$ is compact. Then* psd = sos *holds in* $\mathbb{R}[V]$.

Proof. The preordering of sums of squares in $\mathbb{R}[V]$ is Archimedean (Theorem 5.5.3). For every maximal ideal $\mathfrak{m}$ in $\mathbb{R}[V]$, the local ring $\mathbb{R}[V]_\mathfrak{m}$ is regular of dimension two, so psd = sos holds in this ring (Theorem 6.4.7). Hence the theorem follows from Corollary 6.5.4. □

6.5.20 Remarks.

1. Theorem 6.5.19 contrasts remarkably with the fact that the psd = sos property fails for the two-dimensional polynomial ring (Theorem 2.4.9).

2. Certain singularities may be allowed in Theorem 6.5.19 without affecting the conclusion. Indeed, the proof goes through as long as psd = sos holds in every local ring of V. For example, this property can be shown to hold when the singularities of V are of type E_8, corresponding to an equation $x^2 + y^3 + z^5 = 0$ in local coordinates.

3. The hypothesis that V is non-singular in Theorem 6.5.19 cannot be relaxed to $V_{\mathrm{sing}}(\mathbb{R}) = \varnothing$. A concrete example illustrating this point is the surface V in affine 3-space with equation $z^2 + g_1(x,y)g_2(x,y) = 0$, where $g_1(x,y) = 1 - x^2 - y^2$ and $g_2(x,y) = 1 - 2x^2 - 3y^2$. See Exercise 6.5.7.

6.5.21 Remark. A prominent case to which Theorem 6.5.19 can be applied is trigonometric polynomials. Generally, by an n-variate trigonometric polynomial we mean a function $f\colon \mathbb{R}^n \to \mathbb{R}$ of the form

$$f(t_1, \ldots, t_n) = p(\cos(t_1), \sin(t_1), \ldots, \cos(t_n), \sin(t_n))$$

where $p \in \mathbb{R}[x_1, y_1, \ldots, x_n, y_n]$ is a polynomial in the ordinary sense. The $\mathbb{R}$-algebra of n-variate trigonometric polynomials $\mathbb{R}[\cos(t_j), \sin(t_j)\colon j = 1, \ldots, n]$ is naturally isomorphic to

$$A_n := \mathbb{R}[x_1, y_1, \ldots, x_n, y_n]/\langle 1 - x_j^2 - y_j^2\colon j = 1, \ldots, n\rangle \cong A_1 \otimes_\mathbb{R} \cdots \otimes_\mathbb{R} A_1,$$

and can be considered as the ring of polynomial functions on the n-dimensional torus. Similar as for ordinary polynomials, one may ask whether every non-negative trigonometric polynomial can be written as a sum of squares of trigonometric polynomials. This question can now be decided completely:

1. ($n = 1$) Every univariate non-negative trigonometric polynomial is a sum of (two) squares of trigonometric polynomials. This is essentially the Fejér–Riesz theorem (Proposition 2.3.1).
2. ($n = 2$) Every bivariate non-negative trigonometric polynomial is a sum of squares of trigonometric polynomials. This follows from Theorem 6.5.19, applied to the donut surface $V = \mathcal{V}(1 - x_1^2 - x_2^2, 1 - x_3^2 - x_4^2) \subseteq \mathbb{A}^4$.
3. If $n \geq 3$, there exist n-variate trigonometric polynomials f that are non-negative but cannot be written as a sum of squares of trigonometric polynomials. This follows from Theorem 6.1.6. To see explicit examples, one may start with any psd homogeneous (ordinary) polynomial $p(x_1, \ldots, x_n)$ that is not a sum of squares, such as the Motzkin form, and take $f(t_1, \ldots, t_n) = p(\cos(t_1), \ldots, \cos(t_n))$, cf. the proof of 6.1.6.

4. n-variate trigonometric polynomials that are strictly positive everywhere are always sums of squares of trigonometric polynomials, regardless of the value of n. This is a particular case of Schmüdgen's theorem 5.5.1.

We also get saturatedness results for 2-dimensional compact semialgebraic sets. Here is an example:

6.5.22 Theorem. *Let $g_1, \dots, g_r \in \mathbb{R}[x_1, x_2]$ be irreducible polynomials such that $K = S(g_1, \dots, g_r) \subseteq \mathbb{R}^2$ is compact. Assume the following:*

(1) *For $i = 1, \dots, r$ and every $\xi \in K \cap \mathcal{Z}(g_i)$ we have $\nabla g_i(\xi) \neq 0$;*
(2) *if $i \neq j$ and $\xi \in K \cap \mathcal{Z}(g_i, g_j)$, then $\nabla g_i(\xi)$ and $\nabla g_j(\xi)$ are linearly independent;*
(3) *$K \cap \mathcal{Z}(g_i, g_j, g_k) = \varnothing$ for any triple i, j, k of pairwise distinct indices.*

Then the preordering $PO(g_1, \dots, g_r)$ in $\mathbb{R}[x_1, x_2]$ is saturated.

Here, of course, $\nabla g = (\frac{\partial g}{\partial x_1}, \frac{\partial g}{\partial x_2})$ denotes the gradient of g.

Proof. Put $T = PO(g_1, \dots, g_r)$, so $\mathrm{Sat}(T) = \mathcal{P}(K)$. For any boundary point ξ of K, it follows from conditions (1)–(3) that the local dimension of K at ξ is two. Therefore, if f, $g \in \mathbb{R}[x, y]$ are non-zero polynomials with $g|_K \geq 0$ and $(fg)|_K \geq 0$, then $f|_K \geq 0$ as well.

Let $f \in \mathbb{R}[x, y]$ be given with $f|_K \geq 0$, we have to show $f \in T$. By decomposing f into irreducible factors, we may assume that f is not divisible by any of the g_i, and neither by the square of any non-constant polynomial. It follows that the set $\mathcal{Z}(f) \cap K$ is finite, and it suffices to show $f \in \widehat{T_\xi}$ for every zero ξ of f in K (Theorem 6.5.5). Let $\xi \in \mathcal{Z}(f) \cap K$. After relabelling the g_i there is $0 \leq t \leq 2$ with $g_i(\xi) = 0$ for $1 \leq i \leq t$ and $g_i(\xi) > 0$ for $t + 1 \leq i \leq r$. Moreover the gradients $\nabla g_i(\xi)$ $(1 \leq i \leq t)$ are linearly independent. The preordering $\widehat{T_\xi}$ is generated by $g_1, \dots, g_t$, and it is saturated by Proposition 6.4.8. Therefore T itself is saturated by 6.5.5. □

Examples where Theorem 6.5.22 applies are (compact convex) polygons, or more generally compact basic closed sets $K \subseteq \mathbb{R}^2$ whose boundary curves are smooth and whose real pairwise intersection points are transversal (with no three of them intersecting in a boundary point of K).

6.5.23 Remark. The aforementioned techniques can be extended to obtain saturatedness results for certain two-dimensional non-compact sets K, similar to what was remarked in 6.5.18 for curves. An example is given in Exercise 6.5.8.

Although these sets K are not compact, they are close to being compact, in the sense that the ring $B_V(K) \subseteq \mathbb{R}[V]$ of K-bounded polynomials has transcendence degree two. For several years it had been an open question whether there exists a two-dimensional set $K \subseteq \mathbb{R}^2$ whose ring of K-bounded polynomials has transcendence degree one, and whose saturated preordering $\mathcal{P}(K)$ is finitely generated. In particular, the *strip conjecture* was discussed, according to which $K = [-1, 1] \times \mathbb{R} \subseteq \mathbb{R}^2$ should be such an example. In 2010, Marshall [139] succeeded in proving this conjecture:

6.5.24 Theorem. (Marshall) *Every polynomial* $f \in \mathbb{R}[x, y]$ *that is non-negative on the strip* $[-1, 1] \times \mathbb{R}$ *can be written in the form*

$$f(x, y) = \sum_{i=1}^{r} g_i(x, y)^2 + (1 - x^2) \sum_{j=1}^{s} h_j(x, y)^2$$

with polynomials $g_i, h_j \in \mathbb{R}[x, y]$.

In other words, the preordering $T = PO(1 - x^2)$ in $\mathbb{R}[x, y]$ is saturated. Unfortunately we do not have room here to include the beautiful proof.

Finally we take up the question of uniform denominators for positive (semi-)definite forms. Using the local-global principle, we are now in a position to lift the degree restriction that was needed in Theorem 5.5.7. To get there we start with the following general result:

6.5.25 Theorem. *Let* $x = (x_1, \ldots, x_n)$, *let* $h_1, \ldots, h_r \in \mathbb{R}[x]$ *be homogeneous of even degrees, and let*

$$K = \{\xi \in \mathbb{R}^n : h_1(\xi) \geq 0, \ldots, h_r(\xi) \geq 0\}.$$

Let $f, g \in \mathbb{R}[x]$ *be forms of even degree, with* $\deg(g) > 0$, *and assume that* f, g *are strictly positive on* $K \setminus \{(0, \ldots, 0)\}$. *If one of the following conditions*

(1) $\deg(g)$ *divides* $\deg(f)$,
(2) *the interior of* K *is non-empty*

holds, there exists $N \geq 0$ *and homogeneous sums of squares* $s_e \in \mathbb{R}[x]$ $(e \in \{0, 1\}^r)$ *such that*

$$fg^N = \sum_e s_e \cdot h_1^{e_1} \cdots h_r^{e_r} \tag{6.10}$$

and such that every summand has the same degree.

6.5.26 Corollary. *Given any two positive definite forms* $f, g \in \mathbb{R}[x_1, \ldots, x_n]$, *where* $\deg(g) > 0$, *there is* $N \geq 0$ *such that* fg^N *is a sum of squares of forms.*

6.5.27 Corollary. *For any positive definite form* f *in* $\mathbb{R}[x]$, *there is some odd power* f^{2k+1} *that is a sum of squares of forms.*

Proof. Apply the previous corollary with $g = f^2$. □

Proof of Theorem 6.5.25. As in 5.5.7, the complement V of the hypersurface $g = 0$ in $\mathbb{P}^{n-1}$ is an affine $\mathbb{R}$-variety, whose coordinate ring $\mathbb{R}[V]$ consists of all fractions $\frac{q}{g^m}$ where $q \in \mathbb{R}[x]$ is homogeneous of degree $\deg(g^m)$. Put $p = x_1^2 + \cdots + x_n^2$ and choose integers $d_i, e_i \geq 0$ with $\deg(h_i p^{d_i}) = \deg(g^{2e_i})$ $(i = 1, \ldots, r)$. The fractions $H_i := \frac{h_i p^{d_i}}{g^{2e_i}}$ are regular functions on V, and

$$K' := \Big\{\eta \in V(\mathbb{R}) : H_1(\eta) \geq 0, \ldots, H_r(\eta) \geq 0\Big\}$$

is a basic closed (compact) subset of $V(\mathbb{R})$. Note that the points of K' are identified with the points $[\xi] \in \mathbb{P}^{n-1}(\mathbb{R})$ where $\xi \in K$, $\xi \neq 0$. Let T be the preordering in $\mathbb{R}[V]$ that is generated by $H_1, \ldots, H_r$, and note that T is Archimedean since K' is compact.

First assume (1). The proof is a straightforward generalization of the proof of Theorem 5.5.7: Let $m \geq 1$ be the integer with $\deg(f) = \deg(g^m)$, then $\varphi := \frac{f}{g^m}$ lies in $\mathbb{R}[V]$ and is strictly positive on K'. So $\varphi \in T$ by Schmüdgen's theorem. After multiplication by a sufficiently high even power of g, this gives an identity of the form (6.10).

Now assume (2), so $\mathrm{int}(K) \neq \varnothing$. We choose a linear form l in $\mathbb{R}[x]$ with $l(\xi) = 0$ for some interior point ξ of K, and such that l does not divide $h_1 \cdots h_r$. Moreover choose integers $m, k \geq 0$ with $\deg(l^{2k} f) = \deg(g^m)$. Then $\varphi := \frac{l^{2k} f}{g^m} \in \mathbb{R}[V]$ satisfies $\varphi \geq 0$ on K'. Assume that we know $\varphi \in T$ (it will be shown below). After clearing denominators, this gives an identity

$$l^{2k} f g^N = \sum_e \sigma_e \cdot h_1^{e_1} \cdots h_r^{e_r}$$

in $\mathbb{R}[x]$, where the σ_e are homogeneous sums of squares and each summand on the right has the same degree. Each of these summands is non-negative on a neighborhood of ξ in $\mathbb{R}^n$, which implies that each of them is divisible by l^{2k} (see Exercise 4.6.3). Thus l^{2k} divides σ_e for every e, so we may cancel l^{2k} and get the desired conclusion.

To prove $\varphi \in T$ we use the Archimedean local-global principle, viz. Theorem 6.5.1. For every point $\eta \in V(\mathbb{C})$ we show that φ lies in T_η, the preordering generated by T in the local ring $\mathcal{O}_\eta := \mathcal{O}_{V,\eta}$. Let $\nu \in \{1, \ldots, n\}$ be an index with $\eta_\nu \neq 0$. All three fractions

$$\tilde{f} = \frac{f}{x_\nu^{\deg(f)}}, \quad \tilde{g} = \frac{g}{x_\nu^{\deg(g)}}, \quad \tilde{l} = \frac{l}{x_\nu}$$

lie in $\mathcal{O}_\eta$, and $\tilde{g}$ is a unit in $\mathcal{O}_\eta$. By definition we have $\varphi = \tilde{l}^{2k} \tilde{f} \tilde{g}^{-m}$. To prove $\varphi \in T_\eta$ it therefore suffices to show that $\tilde{f}$ and $\tilde{g}$ lie in T_η. Both are strictly positive on the basic closed constructible subset of $\mathrm{Sper}(\mathcal{O}_\eta)$ that is associated with T_η, since f and g are strictly positive on $K \setminus \{0\}$. Therefore $\tilde{f}, \tilde{g} \in T_\eta$ by Corollary 6.3.7, and the proof is complete. □

For three homogeneous variables, Corollary 6.5.26 can be improved even further:

6.5.28 Theorem. *Let $g \in \mathbb{R}[x_1, x_2, x_3]$ be an arbitrary positive definite form which is not constant. Then for any non-negative form $f \in \mathbb{R}[x_1, x_2, x_3]$, there exists an integer $N \geq 1$ such that fg^N is a sum of squares of forms.*

Proof. Let $V = \mathbb{P}^2 \setminus \mathcal{V}(g)$, a non-singular affine surface over $\mathbb{R}$. Going back to the proof of Theorem 6.5.25, case (2), for $n = 3$ and $K = \mathbb{R}^3$, we choose l, k, m as was done there. The fraction $\varphi = \frac{l^{2k} f}{g^m} \in \mathbb{R}[V]$ is psd on $V(\mathbb{R})$, and it suffices to show that φ is sos in $\mathbb{R}[V]$. But this is true by Theorem 6.5.19, since V is a non-singular affine surface and $V(\mathbb{R})$ is compact. □

6.5.29 Remarks.

1. Let $n \geq 1$. For every psd polynomial $f \in \mathbb{R}[x_1, \ldots, x_n]$ there exist sos polynomials p, q in $\mathbb{R}[x_1, \ldots, x_n]$ with $(1 + p)f = q$ if, and only if, $n \leq 2$. Indeed, the

negative answer for $n \geq 3$ was remarked in Exercise 6.1.6, whereas the positive answer for $n = 2$ follows from Theorem 6.5.28.

2. In fact, even the singular conic $g = x_1^2 + x_2^2$ is a uniform denominator in the sense of Theorem 6.5.28. That is, for every psd ternary form $f \in \mathbb{R}[x_1, x_2, x_3]$ there exists $N \geq 1$ such that the form fg^N is a sum of squares of forms [107]. This seems surprising at first, since g has (projectively) a real zero. The proof rests on the cylinder theorem [189], according to which the psd = sos property holds for the ring $\mathbb{R}[x, y, z]/\langle 1 - x^2 - y^2 \rangle$. In turn, this theorem is a strengthening of Marshall's strip theorem (Theorem 6.5.24).

Exercises

6.5.1 Let $n \geq 1$ and $0 \leq r \leq n$. Put $K = \{\xi \in \mathbb{R}^n \colon \xi_1 \geq 0, \dots, \xi_r \geq 0\}$ and let $f \in \mathbb{R}[x] = \mathbb{R}[x_1, \dots, x_n]$ with $f(0) = 0$ satisfy the boundary Hessian conditions at the origin with respect to $x_1, \dots, x_r$. Show that $f > 0$ on $(U \cap K) \setminus \{0\}$ for some neighborhood U of the origin.

6.5.2 Consider the plane affine curve C over $\mathbb{R}$ with equation

$$x^4 + y^4 = x^2 - y^2.$$

Either prove that psd = sos holds in $\mathbb{R}[C]$, or find an explicit element of $\mathbb{R}[C]$ that is psd but not sos.

6.5.3 Same as Exercise 6.5.2, but for the curve

$$x^4 + y^4 = x^2 + y^2.$$

6.5.4 Consider the plane affine curve $C = \mathcal{V}(x^4 + y^2 - x^3)$. The curve C has a cusp singularity at the origin, and $C(\mathbb{R})$ is compact. We prove that the preordering of all psd polynomials on C is not finitely generated. Write $P = \{f \in \mathbb{R}[x, y] \colon f \geq 0 \text{ on } C(\mathbb{R})\}$. Let $\mathfrak{m} = \langle x, y \rangle \subseteq \mathbb{R}[x, y]$, the maximal ideal at the origin $O = (0, 0)$. For $f \in \mathbb{R}[x, y]$ let $\omega(f) = \sup\{n \geq 0 \colon f \in \mathfrak{m}^n\}$, the vanishing order of f at O.

(a) If $f \in P$ has $\omega(f) = 1$, show that there are $a, b \in \mathbb{R}$ with $f \in ax + by + \mathfrak{m}^2$ and $a > 0$.
(b) Conversely, show for every $b \in \mathbb{R}$ that there exists $f \in P$ with $f \in x + by + \mathfrak{m}^2$.
(c) Use (a) and (b) to prove that the preordering P is not finitely generated.

6.5.5 Prove Corollary 6.5.13.

6.5.6 Let $p_1, p_2 \in \mathbb{R}[x, y]$ be irreducible polynomials for which the plane affine curves $C_i = \mathcal{V}(p_i)$ $(i = 1, 2)$ are non-singular, let $T = PO(p_1 p_2)$ and $K = \mathcal{S}(p_1 p_2)$. We assume that K is compact and $\mathcal{S}(-p_1, -p_2) = \varnothing$, and that $C_1(\mathbb{C}) \cap C_2(\mathbb{C}) \neq \varnothing$. Find examples of such pairs p_1, p_2, then prove for any such pair:

(a) T is Archimedean and $p_1, p_2 \in \mathrm{Sat}(T)$;
(b) $p_1, p_2 \in T_{\mathfrak{m}}$ for every maximal ideal $\mathfrak{m}$ of $\mathbb{R}[x, y]$ with residue field $\mathbb{R}$;
(c) $p_1, p_2 \notin T$.

6.5.7 Let $p_1, p_2 \in \mathbb{R}[x, y]$ be as in Exercise 6.5.6. Show that the surface $V \subseteq \mathbb{A}^3$ with equation $z^2 = p_1(x, y) p_2(x, y)$ is irreducible and has no real singular points, and that moreover $V(\mathbb{R})$ is compact. Yet psd = sos does not hold on V.

6.5.8 Consider the preordering $T = PO(x, 1-x, y, 1-xy)$ in $\mathbb{R}[x, y]$ with associated set $K = \mathcal{S}(T) \subseteq \mathbb{R}^2$. Show that T is saturated, although K has dimension two and is not compact. *Hint*: The preordering $PO(u-u^2, v-v^2) \subseteq \mathbb{R}[u, v]$ is saturated. Use this to prove the claim via a suitable substitution of variables.

6.5.9 Let R be a real closed field. Show that every quadratic module in the power series ring $R[[t]]$ is a preordering.

6.6 Stability

An important question for applications is the study of degree bounds in weighted sums of squares representations of polynomials. Quite a bit is known, and we will only scratch on the surface.

6.6.1 Let $n \geq 1$ and $x = (x_1, \dots, x_n)$, and write $\Sigma = \Sigma\mathbb{R}[x]^2$ in the following. Let $M = QM(g_1, \dots, g_r)$ be a finitely generated quadratic module in $\mathbb{R}[x]$, and put $g_0 = 1$. For $d \geq 0$ let us write (temporarily)

$$M_d(g_1, \dots, g_r) := \left\{\sum_{i=0}^{r} s_i g_i \colon s_i \in \Sigma,\ \deg(s_i g_i) \leq d \text{ for } i = 0, \dots, r\right\}. \tag{6.11}$$

Note that the inclusion $M_d(g_1, \dots, g_r) \subseteq M \cap \mathbb{R}[x]_{\leq d}$ of convex cones will usually be strict, and that $M = \bigcup_{d \geq 0} M_d(g_1, \dots, g_r)$. For every d, the cone $M_d(g_1, \dots, g_r)$ is semialgebraic as a subset of the finite-dimensional vector space $\mathbb{R}[x]_{\leq d}$. Indeed, it follows from Corollary 2.1.17 that every sum of squares s_i in (6.11) can be written as a sum of at most $\binom{n+k_i}{n}$ many squares, where $k_i = \lfloor \frac{1}{2}(d - \deg(g_i)) \rfloor$. So $M_d(g_1, \dots, g_r)$ is the image of a polynomial map $\mathbb{R}^N \to \mathbb{R}[x]_{\leq d}$, for suitable N.

6.6.2 Definition. Let $M = QM(g_1, \dots, g_r)$ as before. A *stability bound* for M (with respect to the system $g_1, \dots, g_r$ of generators) is a map $\varphi \colon \mathbb{Z}_+ \to \mathbb{Z}_+$ with the property that $M \cap \mathbb{R}[x]_{\leq d} \subseteq M_{\varphi(d)}(g_1, \dots, g_r)$ holds for every $d \geq 0$.

6.6.3 Remark. Assume that a stability bound φ as in Definition 6.6.2 exists and is known. Then it is possible, for a given polynomial f, to test membership of f in M effectively. If $\deg(f) = d$, this means that one can—in principle—check whether there exist sums of squares s_i with $f = \sum_{i=0}^{r} s_i g_i$ and $\deg(s_i) \leq \varphi(d) - \deg(g_i)$ for $i = 0, \dots, r$. As will be discussed in Chapter 8 in more detail (Section 8.4 and Examples 8.3.4), this is the feasibility question for an explicit semidefinite program. Under mild assumptions, this question can be decided effectively, at least if the degrees are not too big.

The existence of a stability bound 6.6.2 depends only on the quadratic module M, and not on the generators chosen. This is a consequence of the next lemma:

6.6.4 Lemma. *Let $g_1, \dots, g_r \in \mathbb{R}[x]$ and $M = QM(g_1, \dots, g_r)$, and let $g \in M$. There exists a stability bound for M with respect to $g_1, \dots, g_r$ if, and only if, there exists one with respect to $g_1, \dots, g_r, g$.*

Proof. The "only if" is clear anyway. Let $\psi\colon \mathbb{Z}_+ \to \mathbb{Z}_+$ be a stability bound with respect to $g_1, \dots, g_r, g$. Choose a representation $g = \sigma_0 + \sigma_1 g_1 + \cdots + \sigma_r g_r$ of g, with sums of squares σ_i, and let $e = \max\{\deg(\sigma_i g_i)\colon i = 0, \dots, r\}$. Note that $e \ge \deg(g)$. We claim that $\varphi(d) := \psi(d) + e - \deg(g)$ is a stability bound with respect to $g_1, \dots, g_r$. Indeed, given $f \in M$ with $\deg(f) \le d$, there is an identity $f = s_0 + s_1 g_1 + \cdots + s_r g_r + sg$ with sums of squares s and s_i, and with $\deg(sg), \deg(s_i g_i) \le \psi(d)$. So

$$f = \sum_{i=0}^{r} (s_i + s\sigma_i) g_i. \tag{6.12}$$

Since $\deg(s_i g_i) \le \psi(d)$, $\deg(sg) \le \psi(d)$ and $\deg(s\sigma_i g_i) \le \deg(s) + e$, each summand in (6.12) has degree at most $\max\{\psi(d), \deg(s) + e\}$. From $\deg(s) \le \psi(d) - \deg(g)$ it follows that this number is at most $\psi(d) + e - \deg(g)$. □

6.6.5 Definition. A finitely generated quadratic module M in $\mathbb{R}[x]$ is *stable* if a stability bound exists for M with respect to some (equivalently, any) finite system of generators.

6.6.6 Examples.

1. The quadratic module $QM(1-t^2)$ in $\mathbb{R}[t]$ is stable, with stability bound $\varphi(d) = d + 1$ (Exercise 6.6.1).

2. Let $M \subseteq \mathbb{R}[x_1, \dots, x_n]$ be a finitely generated quadratic module such that $\mathcal{S}(M)$ contains a non-empty open cone in $\mathbb{R}^n$. Then M is stable with stability bound $\varphi(d) = d$ (Exercise 6.6.2).

3. Stengle [203] analyzed the following example. Consider the polynomial $f = 1 - t^2$ in $\mathbb{R}[t]$. For every $\varepsilon > 0$, the quadratic module $M = PO(f^3)$ in $\mathbb{R}[t]$ contains $f + \varepsilon$, according to Theorem 5.5.1. But when we represent $f + \varepsilon$ as an element of M, the degrees of the summands necessarily explode as $\varepsilon \to 0$. Stengle proved that there exists a constant $c > 0$ such that, in every identity $f + \varepsilon = p + qf^3$ with p, q sos, one has $\deg(p) > c\varepsilon^{-1/2}$. In particular, this shows that the quadratic module M is not stable.

4. Looking for an analogue of the stability property in commutative algebra, one might ask: Given an ideal $I = \langle p_1, \dots, p_r \rangle$ in the polynomial ring $k[x] = k[x_1, \dots, x_n]$ over a field, when does there exist a function $\varphi\colon \mathbb{Z}_+ \to \mathbb{Z}_+$ with

$$I \cap k[x]_{\le d} \subseteq p_1 k[x]_{\le\varphi(d)} + \cdots + p_r k[x]_{\le\varphi(d)}$$

for all d? The answer is, always, and this is an easy consequence of the theory of Gröbner bases.

6.6.7 It is easy to generalize the notion of stability from polynomial rings to finitely generated $\mathbb{R}$-algebras A, without the need to fix a system of algebra generators. A quadratic module $M = QM(g_1, \dots, g_r)$ in A is called *stable* if for every finite-dimensional linear subspace U of A there exists a finite-dimensional linear subspace W of A with

$$M \cap U \subseteq (\Sigma W^2) + (\Sigma W^2)g_1 + \cdots + (\Sigma W^2)g_r.$$

Here ΣW^2 denotes the set of sums of squares of elements in W. For the polynomial ring $A = \mathbb{R}[x]$, this definition is clearly equivalent to the one given in 6.6.5. The proof of Lemma 6.6.4 carries over to this more general setup without difficulty, so the definition just given is independent of the system $g_1, \ldots, g_r$ of generators of M.

6.6.8 Lemma. *Let A be a finitely generated $\mathbb{R}$-algebra and M a finitely generated quadratic module in A. If I is an ideal of A with $I \subseteq \operatorname{supp}(M)$, then M is stable (in A) if and only if M/I is stable (in A/I).*

Proof. The "only if" part is directly clear. For the converse let $I = \langle h_1, \ldots, h_s \rangle$. Let $g_1, \ldots, g_r$ be a generating system for M such that the $\pm h_j$ are among the g_i. For any linear subspace V of A let

$$\Sigma_g(V) := (\Sigma V^2) + g_1(\Sigma V^2) + \cdots + g_r(\Sigma V^2).$$

Let $U \subseteq A$ be a subspace with $\dim(U) < \infty$. Since M/I is stable by assumption, there is a subspace $W \subseteq A$ with $\dim(W) < \infty$ such that $M \cap U \subseteq \Sigma_g(W) + I$. Let $L \subseteq A$ be a subspace with $\dim(L) < \infty$ and with $U + \Sigma_g(W) \subseteq L$. Then $M \cap U \subseteq (I \cap L) + \Sigma_g(W)$. Let $V \subseteq A$ be a subspace with $1 \in V$, $\dim(V) < \infty$ and with $I \cap L \subseteq h_1 V + \cdots + h_s V$. Since

$$h_j v = \Big(\frac{1+v}{2}\Big)^2 h_j - \Big(\frac{1-v}{2}\Big)^2 h_j,$$

and since $\{\pm h_1, \ldots, \pm h_s\} \subseteq \{g_1, \ldots, g_r\}$, we get $I \cap L \subseteq \Sigma_g(V)$. Hence $M \cap U \subseteq \Sigma_g(V) + \Sigma_g(W) \subseteq \Sigma_g(V + W)$. □

By the lemma, there is no loss of generality if we restrict our discussion of stable quadratic module to the polynomial ring $\mathbb{R}[x]$ (instead of coordinate rings $\mathbb{R}[V]$ of affine varieties V).

6.6.9 There is an alternative characterization of stable quadratic modules that is both useful and instructive. We need a few notational preparations. If R is a real closed field, a subset M of $R[x] = R[x_1, \ldots, x_n]$ will be called *locally semialgebraic* if, for every finite-dimensional linear subspace U of $R[x]$, the subset $M \cap U$ of U is semialgebraic. If in addition M is contained in some finite-dimensional subspace U, we say that M is a *semialgebraic* subset of $R[x]$. For example, if $M \subseteq \mathbb{R}[x]$ is a stable quadratic module, then M is locally semialgebraic as a subset of $\mathbb{R}[x]$.

We generalize base field extension (Definition 4.1.7) from semialgebraic sets to locally semialgebraic sets, as follows. If M is a locally semialgebraic subset of $\mathbb{R}[x]$ and $R \supseteq \mathbb{R}$ is a real closed overfield, let the set $M_R \subseteq R[x]$ be defined by $M_R \cap U_R = (M \cap U)_R$ for every finite-dimensional $\mathbb{R}$-subspace $U \subseteq \mathbb{R}[x]$. Here $U_R = U \otimes_{\mathbb{R}} R$ denotes the R-subspace of $R[x]$ spanned by U. Clearly, M_R is a locally semialgebraic subset of $R[x]$.

Now let $M = QM(g_1, \ldots, g_r)$ be a quadratic module in $\mathbb{R}[x]$, and put $g_0 = 1$ as usual. For $k \geq 0$ let M_k be the set of all sums $\sum_{i=0}^{r} s_i g_i$ with s_i sos in $\mathbb{R}[x]$ and $\deg(s_i) \leq k$. Then M_k is a semialgebraic subset of $\mathbb{R}[x]$, so we may consider its extension $(M_k)_R$ to $R \supseteq \mathbb{R}$. The union of these extended sets $(M_k)_R$, over $k \geq 0$, coincides with the quadratic module $M^{R[x]} = QM_{R[x]}(g_1, \ldots, g_r)$ that is generated by M in $R[x]$. In other words, we have

$$M^{R[x]} = \bigcup_{k \geq 0} (M_k)_R.$$

6.6.10 Proposition. *For every finitely generated quadratic module M in $\mathbb{R}[x]$, the following are equivalent:*

(i) *M is stable;*
(ii) *for every real closed field extension $R \supseteq \mathbb{R}$, the quadratic module $M^{R[x]}$ generated by M in $R[x]$ is a locally semialgebraic subset of $R[x]$;*
(iii) *M is a locally semialgebraic subset of $\mathbb{R}[x]$, and $M_R = M^{R[x]}$ holds for every real closed extension $R \supseteq \mathbb{R}$.*

To prevent confusion we remark (again) that $M_R \subseteq R[x]$ is the base field extension of the locally semialgebraic set $M \subseteq \mathbb{R}[x]$ from $\mathbb{R}$ to R, while $M^{R[x]}$ is the quadratic module generated by M in $R[x]$.

Proof. We fix a generating system $g_1, \ldots, g_r$ of M and use the notation M_k ($k \geq 0$) introduced in 6.6.9 above.

(i) $\Rightarrow$ (iii): Given a subspace $U \subseteq \mathbb{R}[x]$ with $\dim(U) < \infty$, there is $m \geq 0$ with $M \cap U = M_m \cap U$, since M is stable. In particular, $M \cap U$ is a semialgebraic set. For $R \supseteq \mathbb{R}$ we have $M^{R[x]} = M_R$ since

$$M^{R[x]} \cap U_R = \bigcup_{k \geq 0} (M_k)_R \cap U_R = \bigcup_{k \geq 0} (M_k \cap U)_R = (M_m \cap U)_R = M_R \cap U_R$$

for every U as before. The implication (iii) $\Rightarrow$ (ii) is obvious. To see (ii) $\Rightarrow$ (i) let $U \subseteq \mathbb{R}[x]$, $\dim(U) < \infty$. For every real closed field $R \supseteq \mathbb{R}$ we have

$$M^{R[x]} \cap U_R = \bigcup_{k \geq 0} (M_k)_R \cap U_R = \bigcup_{k \geq 0} (M_k \cap U)_R, \tag{6.13}$$

and this subset of U_R is R-semialgebraic since $M^{R[x]}$ is a locally semialgebraic set by hypothesis (ii). Consider the countable ascending union

$$M \cap U = \bigcup_{k \geq 0} (M_k \cap U)$$

of semialgebraic sets in U. We have to show that $M \cap U = M_k \cap U$ for some $k \geq 0$. From Exercise 1.6.6, recall that $\mathbb{R}$ has a real closed field extension S that is $\aleph_1$-saturated. Since the countable union $\bigcup_{k \geq 0} (M_k \cap U)_S$ is an S-semialgebraic set, the saturatedness property implies that it this set is covered by finitely many sets $(M_k \cap U)_S$. In other words, there is $k \geq 0$ with $(M_i \cap U)_S = (M_k \cap U)_S$ for all $i \geq k$. This means $M_i \cap U = M_k \cap U$ for $i \geq k$, and so $M \cap U = M_k \cap U$. □

6.6.11 Examples.

1. For every real closed field R and every semialgebraic set $K \subseteq R^n$, the saturated preordering $\mathcal{P}(K) \subseteq R[x]$ is a locally semialgebraic set in $R[x]$. Indeed, if we fix a degree d, the intersection $\mathcal{P}(K) \cap R[x]_{\leq d}$ can be described by an R-formula in the coefficients of the polynomials, therefore it is a semialgebraic set by Tarski's theorem. If $n = 1$, we see from this remark and from Proposition 6.6.10 that, for every closed semialgebraic set $K \subseteq \mathbb{R}$, the saturated preordering $P = \mathcal{P}(K)$ in $\mathbb{R}[t]$ is stable. Indeed, P is generated (as a preordering) by the natural generators for K, and the same is true over any real closed field $R \supseteq \mathbb{R}$ (Proposition 6.1.3(a)).

2. If $C \subseteq \mathbb{A}^n$ is any affine non-singular curve over $\mathbb{R}$ and $K \subseteq C(\mathbb{R})$ is any compact semialgebraic set, it can be shown [186] that the saturated preordering $\mathcal{P}(K)$ is stable. This result is much harder to prove, at least when the curve is not rational.

3. Let $K \subseteq \mathbb{R}^n$ be a compact semialgebraic set of dimension two whose saturated preordering $\mathcal{P}(K)$ is finitely generated. For example, K may be a polygone or a disk in the plane, or the 2-sphere in $\mathbb{R}^3$, see Section 6.5. Then it is known that $\mathcal{P}(K)$ fails to be stable. This will be proved below in the case where $K \subseteq \mathbb{R}^2$ (Corollary 6.6.23).

More generally, the main goal for this section is to prove that Archimedean quadratic modules are never stable, as long as their associated basic closed sets have dimension at least two. A full proof will only be given for dimension ≥ 3. In this context it is useful to introduce an auxiliary topology on the polynomial ring (see also Example B.5.3):

6.6.12 Definition. If V is an $\mathbb{R}$-vector space of at most countable dimension, define a topology τ on V as follows: A subset $M \subseteq V$ is τ-open if and only if $M \cap U$ is open in U for every finite-dimensional linear subspace U of V (with respect to the Euclidean topology on U). We will refer to the topology τ as the *canonical topology* on V.

6.6.13 Remarks.

1. From the definition it is clear that, with respect to τ, every linear form $V \to \mathbb{R}$ is continuous, and every linear subspace of V is closed.

2. In particular, we may consider the canonical topology τ on the polynomial ring $\mathbb{R}[x] = \mathbb{R}[x_1, \dots, x_n]$ (or more generally, on any finitely generated $\mathbb{R}$-algebra). For every set $S \subseteq \mathbb{R}^n$, the saturated preordering $\mathcal{P}(S)$ in $\mathbb{R}[x]$ is τ-closed, since for every $\xi \in \mathbb{R}^n$ the evaluation map $\varphi_\xi \colon \mathbb{R}[x] \to \mathbb{R}$, $p \mapsto p(\xi)$ is continuous.

Given a quadratic module M in $\mathbb{R}[x]$, say finitely generated, it is usually a difficult task to describe the closure $\overline{M}$ of M with respect to the canonical topology. We are going to show that closures of stable quadratic modules are better behaved in this respect, and we start with two useful lemmas of a general nature.

6.6.14 Lemma. *For any ring A and any ideal $I \subseteq A$, we have* $1 + \sqrt[\mathrm{re}]{I} \subseteq I + \Sigma A^2$.

Proof. Recall that $\frac{1}{2} \in A$ is a general hypothesis. For every $a \in A$ and every $n \geq 1$, one sees by induction that

$$n - a + \frac{4a^{2^n}}{2^{2^{n+1}}} \in \Sigma A^2. \tag{6.14}$$

Indeed, for $n = 1$ this is the identity $1 - a + \frac{a^2}{4} = (1 - \frac{a}{2})^2$, and the inductive step follows from

$$n + 1 - a + \frac{4a^{2^{n+1}}}{2^{2^{n+2}}} = \Big(n - a + \frac{4a^{2^n}}{2^{2^{n+1}}}\Big) + \Big(1 - \frac{2a^{2^n}}{2^{2^{n+1}}}\Big)^2.$$

If $a \in \sqrt[\mathrm{re}]{I}$ then $-a^{2m} \in I + \Sigma A^2$ for some $m \geq 0$, by the abstract real nullstellensatz 3.2.15. Let $n = 2^k$ be a power of 2 with $2^n \geq 2m$, so $-a^{2^n} \in I + \Sigma A^2$. Applying (6.14) to na instead of a shows that $n(1 - a) \in I + \Sigma A^2$, and hence $1 - a \in I + \Sigma A^2$ since $\frac{1}{2} \in A$. □

Explicitly, the identity used to prove (6.14) is

$$n - a = \Big(1 - \frac{a}{2}\Big)^2 + \Big(1 - \frac{a^2}{8}\Big)^2 + \Big(1 - \frac{a^4}{128}\Big)^2 + \cdots + \Big(1 - \frac{a^{2^{n-1}}}{2^{2^n - 1}}\Big)^2 - \frac{a^{2^n}}{2^{2^{n+1}-2}}$$

for $n \geq 1$.

6.6.15 Lemma. *Let M be a quadratic module in a ring A, let $I :=$ supp(M) be its support ideal. Then $\sqrt{I}$ is an M-convex ideal, i.e.* $\mathrm{supp}(M + \sqrt{I}) = \sqrt{I}$.

Proof. In Proposition 5.1.17, it was proved that every minimal prime divisor $\mathfrak{p}$ of I is M-convex. Hence $\sqrt{I}$ is M-convex, being an intersection of M-convex (prime) ideals. This means $\mathrm{supp}(M + \sqrt{I}) = \sqrt{I}$, see Remark 5.1.12. □

6.6.16 Proposition. *Let A be a finitely generated $\mathbb{R}$-algebra and let M be a quadratic module in A. Then* $\sqrt{\mathrm{supp}(M)} \subseteq \mathrm{supp}(\overline{M})$, *and equality holds if M is a finitely generated preordering.*

Here, of course, $\overline{M}$ denotes the closure of M in the canonical topology of A. It is easy to see that $\overline{M}$ is again a quadratic module in A (Exercise 6.6.4).

Proof. Lemma 6.6.14 implies $\varepsilon + \sqrt{\mathrm{supp}(M)} \subseteq M$ for every real number $\varepsilon > 0$. Hence the closure of M contains $\sqrt{\mathrm{supp}(M)}$, which proves the first assertion. When M is finitely generated we have $\overline{M} \subseteq \mathrm{Sat}(M)$, since $\mathrm{Sat}(M)$ is closed (Remark 6.6.13). In particular, then, $\mathrm{supp}(\overline{M}) \subseteq \mathrm{supp}(\mathrm{Sat}(M))$. If in addition M is a preordering, $\mathrm{supp}(\mathrm{Sat}(M)) = \sqrt{\mathrm{supp}(M)}$ holds by Lemma 6.2.9, which gives the reverse inclusion in this case. □

The next theorem ensures that the closure of a finitely generated quadratic module can be controlled, if the module is stable:

6.6.17 Theorem. *Let $M \subseteq \mathbb{R}[x]$ be a finitely generated quadratic module which is stable.*

(a) $\overline{M} = M + \sqrt{\mathrm{supp}(M)}$.
(b) *The quadratic module $\overline{M}$ is again finitely generated and stable.*

Since the essential step in the proof of (a) will be used again in Section 8.5, we isolate it as a separate technical lemma:

6.6.18 Lemma. *Let $M = QM(g_0, \dots, g_r) \subseteq \mathbb{R}[x]$ be a quadratic module and let $I \subseteq \mathbb{R}[x]$ be an M-convex radical ideal. Moreover let $W_0, \dots, W_r$ and V be finite-dimensional linear subspaces of $\mathbb{R}[x]$ with $g_i W_i W_i \subseteq V$ for all i. Then the subset*

$$P = (\Sigma W_0^2)g_0 + \cdots + (\Sigma W_r^2)g_r + (V \cap I)$$

of V is closed.

Proof. Here, of course, ΣW_i^2 denotes the set of sums of squares of elements of W_i. Note that every element in ΣW_i^2 is a sum of $m_i = \dim(W_i)$ many squares from W_i (Corollary 2.1.16). For $i = 0, \dots, r$ let $J_i := (I : g_i) = \{p \in \mathbb{R}[x] \colon pg_i \in I\}$, an ideal in $\mathbb{R}[x]$. Consider the map

$$\phi\colon \bigoplus_{i=0}^{r} W_i^{m_i} \to V, \quad (p_{ij})_{\substack{0\le i\le r \\ 1\le j\le m_i}} \mapsto \sum_{i=0}^{r}\sum_{j=1}^{m_i} p_{ij}^2 g_i.$$

The map ϕ induces a map

$$\overline{\phi}\colon \bigoplus_{i=0}^{r} (W_i/W_i \cap J_i)^{m_i} \to V/V \cap I, \tag{6.15}$$

and $\mathrm{im}(\overline{\phi})$ is the image of $P \subseteq V$ in $V/V \cap I$. We show that $\overline{\phi}$ is "anisotropic", meaning that only the zero tuple is mapped to zero. So let $\overline{p} = (\overline{p}_{ij})$ be a tuple in the left-hand direct sum of (6.15) for which $\overline{\phi}(\overline{p}) = 0$, i.e. with $\sum_{i,j} p_{ij}^2 g_i \in I$. Since I is M-convex we have $p_{ij}^2 g_i \in I$ for all i, j, and $I = \sqrt{I}$ implies $p_{ij} g_i \in I$, hence $p_{ij} \in J_i$ for all i, j. We may therefore apply Lemma 2.4.7: Since the map $\overline{\phi}$ is homogeneous of degree 2, this lemma implies that the image set $\mathrm{im}(\overline{\phi})$ is closed. Since $V \cap I \subseteq P$, this means that P is closed in V. □

Proof of Theorem 6.6.17. Put $I := \sqrt{\mathrm{supp}(M)}$. To prove (a) it suffices, in view of Lemma 6.6.16, to show that $M + I$ is closed. Let $1 = g_0, g_1, \dots, g_r$ be a generating system for the quadratic module M, and let $U \subseteq \mathbb{R}[x]$ be a given linear subspace, with $1 \in U$ and $\dim(U) < \infty$. Since M is stable, there exists a linear subspace $W \subseteq \mathbb{R}[x]$ with $\dim(W) < \infty$ such that $M \cap U \subseteq (\Sigma W^2)g_0 + \cdots + (\Sigma W^2)g_r$, cf. 6.6.7. The ideal I is M-convex by Lemma 6.6.15. Let $V \subseteq \mathbb{R}[x]$ be a finite-dimensional subspace that contains U together with $g_i WW$ for all i. Then Lemma 6.6.18 implies that the cone $P = (V \cap I) + \sum_i (\Sigma W^2)g_i$ is closed in V. If $f = g + h$ lies in $(M + I) \cap U$, with $g \in M$ and $h \in I$, then $h + \varepsilon \in M$ for all $\varepsilon > 0$, by Lemma 6.6.14. Therefore $f + \varepsilon = g + (h + \varepsilon) \in M \cap U$ for $\varepsilon > 0$, and so $f \in \overline{M \cap U}$, which implies $f \in P$ since $M \cap U \subseteq P$ and P is closed. Altogether this shows that $(M + I) \cap U = P \cap U$ is closed. Since U was arbitrary, this proves that $M + I$ is closed, and so (a) has been shown. Since the quadratic module $(M + I)/I$ in $\mathbb{R}[x]/I$ is generated by the residue classes of $g_0, \ldots, g_r$, the argument just given implies that $(M+I)/I$ is stable. Therefore Lemma 6.6.8 implies that $M+I$ itself is stable, which completes the proof of the theorem. □

6.6.19 Corollary. *A finitely generated and stable quadratic module is closed if, and only if, its support is a radical ideal.*

Proof. If M is closed then $\mathrm{supp}(M)$ is radical by Corollary 6.6.16, even without the assumption that M is stable. If M is stable and $\mathrm{supp}(M)$ is radical, then $\overline{M} = M$ by Theorem 6.6.17. □

Here is one application:

6.6.20 Corollary. *If M is a finitely generated quadratic module in $\mathbb{R}[x]$ for which $\mathcal{S}(M)$ contains a non-empty open cone in $\mathbb{R}^n$, then M is closed (and stable).*

Proof. M is stable, see Exercise 6.6.2. So Corollary 6.6.19 implies that M is closed, since clearly $\mathrm{supp}(M) = \{0\}$. □

Note that the corollary generalizes the closedness of the sos cone $\Sigma_{n,\le 2d}$ in $\mathbb{R}[x]_{\le 2d}$ (Proposition 2.4.6). On the other hand, Theorem 6.6.17 gives many examples of quadratic modules for which stability fails:

6.6.21 Corollary. *Let $M \subseteq \mathbb{R}[x]$ be a finitely generated quadratic module that is Archimedean. If the associated basic closed set $K = \mathcal{S}(M)$ has* $\dim(K) \ge 3$, *then M is not stable.*

Proof. By the Archimedean positivstellensatz 5.3.1, M contains every polynomial that is strictly positive on K. Hence the closure of M is saturated, $\overline{M} = \mathcal{P}(K)$. Since $\dim(K) \ge 3$, the saturated preordering $\mathcal{P}(K)$ cannot be finitely generated (Corollary 6.1.7). Therefore M cannot be stable, according to Theorem 6.6.17. □

Corollary 6.6.21 remains true for $\dim(K) = 2$, but the proof becomes considerably more technical [182]. We can however give an easily accessible proof in the case where K is a subset of $\mathbb{R}^2$:

6.6.22 Theorem. *Let $M \subseteq \mathbb{R}[x] = \mathbb{R}[x_1, \ldots, x_n]$ be a stable quadratic module. If $n \ge 2$, and if the basic closed set $K = \mathcal{S}(M)$ has non-empty interior in $\mathbb{R}^n$, there exists a polynomial $p \in \mathbb{R}[x]$ that is strictly positive on $\mathbb{R}^n$ and that is not contained in M.*

Proof. Fix a polynomial $f \in \mathbb{R}[x]$ with $f > 0$ on $\mathbb{R}^n$ such that f is not sos. For instance, we may take $f = 1 + g$ where g is the (inhomogeneous) Motzkin polynomial (Example 2.2.10.1). Let $M = QM(g_1, \ldots, g_r)$. Since K has non-empty interior by assumption, we may translate the coordinate system suitably to get $g_i(0) > 0$ for $i = 1, \dots, r$. Let $f_c(x) := f(cx)$ for $c \in \mathbb{R}$, we'll prove that $f_c \notin M$ for sufficiently large $c > 0$. Assume to the contrary that for arbitrarily large values $c > 0$ there is an identity

$$f(cx) = \sum_{i=0}^{r} g_i(x) \sum_{j=1}^{N_c} p_{ij}^{(c)}(x)^2 \tag{6.16}$$

(where $g_0 = 1$). Since M is stable, there is an integer $d \geq 0$ such that an identity (6.16) exists with $\deg(p_{ij}^{(c)}) \leq d$ for all i, j and c. We may assume $N_c = N < \infty$ for all c (viz., N may be taken to be the number of monomials of degree $\leq d$). Replacing x by $\frac{x}{c}$ we get

$$f(x) = \sum_{i=0}^{r} g_i\Big(\frac{x}{c}\Big) \sum_{j=1}^{N} p_{ij}^{(c)}\Big(\frac{x}{c}\Big)^2. \tag{6.17}$$

Since $g_i(0) > 0$ for all i, there are real numbers ρ, $\alpha > 0$ such that $g_i(u) \geq \alpha > 0$ for all $u \in \mathbb{R}^n$ with $|u| < \rho$. Hence, for any $v \in \mathbb{R}^n$ there exists $c_0 > 0$ in $\mathbb{R}$ such that

$$\sup_{c>c_0} \sup_{i,j} \Big| p_{ij}^{(c)}\Big(\frac{v}{c}\Big)\Big| < \infty. \tag{6.18}$$

Indeed, for $c > c_0 := \frac{|v|}{\rho}$ we have $\left|\frac{v}{c}\right| < \rho$, so

$$f(v) = \sum_{i,j} g_i\Big(\frac{v}{c}\Big) p_{ij}^{(c)}\Big(\frac{v}{c}\Big)^2 \geq \alpha \sum_{i,j} p_{ij}^{(c)}\Big(\frac{v}{c}\Big)^2,$$

and hence the supremum (6.18) is at most $\sqrt{f(v)/\alpha}$.

There exist finitely many points $\xi_1, \dots, \xi_N \in \mathbb{R}^n$ such that the evaluation map $\mathbb{R}[x]_{\leq d} \to \mathbb{R}^N$, $p \mapsto (p(\xi_1), \dots, p(\xi_N))$ is bijective. Therefore, the previous argument shows that, for some $\gamma > 0$, the family of all polynomials $p_{ij}^{(c)}(\frac{x}{c})$, $c > \gamma$, is bounded in the vector space $\mathbb{R}[x]_{\leq d}$. So there exists a sequence $c_\nu \to \infty$ with the property that, for any pair i, j of indices, the sequence of polynomials $p_{ij}^{(c_\nu)}(\frac{x}{c_\nu})$ converges (coefficientwise) against some polynomial $p_{ij}(x) \in \mathbb{R}[x]_{\leq d}$, for $\nu \to \infty$. Now consider identity (6.17) for $c = c_\nu$ and pass to the limit $\nu \to \infty$. It follows that

$$f(x) = \sum_{i=0}^{r} g_i(0) \sum_{j=1}^{N} p_{ij}(x)^2.$$

This contradicts our choice of f, which was supposed not to be a sum of squares. □

6.6.23 Corollary. *Let $n \geq 2$. If $M \subseteq \mathbb{R}[x_1, \dots, x_n] = \mathbb{R}[x]$ is a finitely generated Archimedean quadratic module such that $K = \mathcal{S}(M)$ has non-empty interior in $\mathbb{R}^n$, then M is not stable.*

Proof. Immediate from Theorem 6.6.22 and the Archimedean positivstellensatz 5.3.1. □

6.6.24 Corollary. *Let $n \geq 2$, and let $K \subseteq \mathbb{R}^n$ be a closed semialgebraic set that contains a non-empty open convex cone. Then the saturated preordering $\mathcal{P}(K)$ does not contain any quadratic submodule that is finitely generated and dense in $\mathcal{P}(K)$ (with respect to the canonical topology).*

Proof. Let $g_1, \dots, g_r \in \mathcal{P}(K)$ and let $M = QM(g_1, \dots, g_r)$. The quadratic module M is closed by Corollary 6.6.20, but $M \neq \mathcal{P}(K)$ by Theorem 6.6.22. □

Hilbert's 1888 results (Theorem 2.4.9) imply that psd = sos fails in $\mathbb{R}[x]$ when $n \geq 2$. Corollary 6.6.24 can be seen both as a strengthening and a generalization of this result. In the case $K = \mathbb{R}^n$, it says that the cone $P = \mathcal{P}(\mathbb{R}^n)$ of psd polynomials does not contain any preordering that is dense in P and finitely generated. For an even stronger version that applies to specific degrees, see Exercise 6.6.7.

Exercises

6.6.1 Let M be the quadratic module generated by $g = 1-t^2$ in the univariate polynomial ring $\mathbb{R}[t]$. Show that M is stable and has stability bound $\varphi(d) = d + 1$ with respect to the generator g.

6.6.2 Let $M \subseteq \mathbb{R}[x] = \mathbb{R}[x_1, \dots, x_n]$ be a finitely generated quadratic module such that $\mathcal{S}(M)$ contains a non-empty open cone in $\mathbb{R}^n$. Then M is stable with stability bound $\varphi(d) = d$.

6.6.3 This exercise contains a partial converse to Remark 6.6.11.1.

(a) Let $S \subseteq \mathbb{R}$ be an infinite compact semialgebraic set. Show that the saturated preordering $\mathcal{P}(S)$ is the only quadratic module in $\mathbb{R}[t]$ that is finitely generated and stable and satisfies $\mathcal{S}(M) = S$.
(b) Show that (a) fails if S is a finite set, e.g. $S = \{0\}$.

Hint for (a): Proposition 5.5.14.

6.6.4 Let A be a finitely generated $\mathbb{R}$-algebra, let M be a quadratic module in A and let $\overline{M}$ denote the closure of M with respect to the canonical topology of A (Definition 6.6.12).

(a) Show that $\overline{M}$ is a quadratic module in A, and that $\overline{M}$ is a preordering if M is one.
(b) If M is a finitely generated preordering that is proper, show that the preordering $\overline{M}$ is proper as well.
(c) Give an example of a proper preordering in $\mathbb{R}[t]$ that is dense in $\mathbb{R}[t]$.

6.6.5 Let V be an $\mathbb{R}$-vector space of countable infinite dimension, and let τ be the canonical topology on V. Given a subset $M \subseteq V$, let $M^{\ddagger}$ denote the sequential closure of M in V. That is, $M^{\ddagger}$ consists of all limits $\lim_{n\to\infty} v_n$ in V, where $(v_n)_{\geq 1}$ is an arbitrary convergent sequence in V with $v_n \in M$ for all n.

(a) Let $(v_n)_{n\geq 1}$ be a sequence in V that converges with respect to τ. Show that $\{v_n : n \geq 1\}$ is contained in some finite-dimensional linear subspace of V.
(b) Conclude that $M^{\ddagger} = \bigcup_U \overline{M \cap U}$, union over the finite-dimensional linear subspaces U of V.

6.6.6 Consider the univariate polynomial ring $\mathbb{R}[t]$ with the canonical topology. Let $S \subseteq \mathbb{R}[t]$ be the semiring of all polynomials with non-negative coefficients, and let

$$M = \{p \in S : p \neq 0,\ p(0) \cdot \deg(p) \geq 1\}.$$

Determine the sequential closure $M^\ddagger$ of M and conclude that $M^\ddagger \neq (M^\ddagger)^\ddagger = \overline{M} = S$. In particular, $M^\ddagger$ is not closed with respect to τ.

6.6.7 Let $n \geq 2$, let $P = \mathcal{P}(\mathbb{R}^n)$ be the cone of all psd polynomials in $\mathbb{R}[x] = \mathbb{R}[x_1, \dots, x_n]$, and let $2d = 6$ (if $n = 2$) or $2d = 4$ (if $n \geq 3$), respectively. Show that P is "very non-finitely generated" as a preordering, in the following sense: Given any finite number of psd polynomials $g_1, \dots, g_r \in P$, let $T = PO(g_1, \dots, g_r)$. Then show for any $k \geq 2d$ that $T \cap \mathbb{R}[x]_{\leq k}$ is not dense in $P \cap \mathbb{R}[x]_{\leq k}$. (*Hint*: It suffices to prove this for $k = 2d$.)

6.7 Notes

Proposition 6.1.3 is due to Kuhlmann, Marshall and Schwartz [115], [116]. Theorem 6.1.6 is proved in [178]. Lemma 6.2.17 is essentially the "basic lemma" from [116]. The proofs of Theorem 6.4.1 and Proposition 6.4.2 are taken from Artin's original paper [5]. Otherwise, the results from Sections 6.3 and 6.4 are mostly taken from Scheiderer [179], some with simplified proofs. Theorem 6.4.5 about sums of squares in $k[\![x, y]\!]$ was proved before for $k = \mathbb{R}$ by Bochnak and Risler [28] (for the ring of convergent power series, and using analytic arguments).

According to Delzell [51], the existence of bad points was first noted by Straus in a 1956 letter to Kreisel. The fact that the bad locus $B(f)$ of a polynomial has codimension ≥ 3 was proved by Delzell [51].

Theorem 6.5.11 is due to Marshall [137], improving on a previous weaker version of Scheiderer [181]. Theorems 6.5.19 and 6.5.22 are from [183], as well as Theorem 6.5.28. Theorem 6.5.25 is from [184], and the results on stability in Section 6.6 are mostly taken from [182]. Concrete upper degree bounds were proved by Schweighofer [194] and Nie–Schweighofer [147]. Recently there has been considerable progress on improving these bounds, and also explicit lower bounds have been obtained. We refer to Slot [199] and Baldi–Slot [10] for results in this direction and for an overview.

Chapter 7
Sums of Squares on Projective Varieties

The question whether non-negative polynomials can be expressed as sums of squares of polynomials will now be examined in the context of projective real varieties. Every form $f = f(x_0, \dots, x_n)$ of even degree takes a well-defined sign at any real point of projective space $\mathbb{P}^n$. For any projective $\mathbb{R}$-variety $X \subseteq \mathbb{P}^n$ and any even number $2d$, we can therefore consider the convex cone $P_{X,2d}$ of non-negative forms of degree $2d$ on X. It is evident that this cone contains the cone $\Sigma_{X,2d}$ of sums of squares of forms of degree d. The central result in this chapter presents a complete classification of all irreducible projective $\mathbb{R}$-varieties X with Zariski dense $\mathbb{R}$-points, and all even degrees $2d$, such that $P_{X,2d} = \Sigma_{X,2d}$ holds. Assuming that X is not contained in a hyperplane is not a serious restriction, and neither is the assumption $2d = 2$. Under these conditions, every non-negative quadratic form on X is a sum of squares of linear forms on X if, and only if, X is a variety of minimal degree.

7.1 Varieties of Minimal Degree

7.1.1 Let $X \subseteq \mathbb{P}^n$ be a projective $\mathbb{R}$-variety, with homogeneous vanishing ideal $I = \mathfrak{I}(X) \subseteq \mathbb{R}[x] = \mathbb{R}[x_0, \dots, x_n]$ and homogeneous coordinate ring $\mathbb{R}[X] = \mathbb{R}[x]/I$. Recall that the ring $\mathbb{R}[X]$ is graded, its d-th graded piece being equal to

$$\mathbb{R}[X]_d = (\mathbb{R}[x]_d + I)/I \cong \mathbb{R}[x]_d/(I \cap \mathbb{R}[x]_d), \quad d \geq 0.$$

Let $f \in \mathbb{R}[X]$ be homogeneous of degree d and let $\xi \in X(\mathbb{R})$, say $\xi = [u]$ where $0 \neq u \in \mathbb{R}^{n+1}$ is an affine representative of ξ. The value $f(u)$ depends on u, and not just on ξ, unless $f(u) = 0$. But when $d = \deg(f)$ is even, the sign of $f(u)$ is independent of u, and we'll write $f(\xi) > 0$, $f(\xi) = 0$ or $f(\xi) < 0$ in this case, depending on whether sign $f(u)$ is 1, 0 or -1, respectively. The notation $f(\xi) \geq 0$ or $f(\xi) \leq 0$ is defined similarly. For all $d \geq 0$ it is clear that

$$P_{X,2d} := \{f \in \mathbb{R}[X]_{2d} \colon f \geq 0 \text{ on } X(\mathbb{R})\}$$

C. Scheiderer, *A Course in Real Algebraic Geometry*, Graduate Texts in Mathematics 303,
https://doi.org/10.1007/978-3-031-69213-0_7

is a convex cone in $\mathbb{R}[X]_{2d}$. This cone is closed since $\{f \in P_{X,2d} \colon f(\xi) \ge 0\}$ is a closed halfspace in $\mathbb{R}[X]_{2d}$ for every $\xi \in X(\mathbb{R})$. On the other hand we may consider the cone $\Sigma_{X,2d}$ in $\mathbb{R}[X]_{2d}$ that consists of all sums of squares of elements of $\mathbb{R}[X]_d$. The inclusion $\Sigma_{X,2d} \subseteq P_{X,2d}$ is obvious, and so it is natural to ask: *When is it true that $\Sigma_{X,2d} = P_{X,2d}$?*

Of course, this is a natural projective analogue of the question we have been pursuing for affine $\mathbb{R}$-varieties in the previous chapters. In Section 7.2, a fairly complete answer will be given, and we'll see that it features some characteristic differences to the case of affine varieties. Other than for those, there exist projective varieties of arbitrary dimension on which $\Sigma_{X,2} = P_{X,2}$ holds. In the projective setting, there always exist degree bounds for sums of squares representations, unlike in the affine case (cf. Section 6.6). Therefore Archimedean effects don't play a role, and the answer will be the same over any real closed field. It is just for simplicity that we are going to work over $\mathbb{R}$.

In this first section we are are going to discuss general preparations from algebraic geometry. Harris' book [83], in particular chapters 18 and 19, is an excellent (though sometimes demanding) background reading for the following. In Section 7.2 we'll return to the question outlined above, and prove the main results.

7.1.2 k always denotes an algebraically closed field. Let $X \subseteq \mathbb{P}^n$ be a non-empty projective k-variety, with vanishing ideal $I = \mathfrak{I}(X) \subseteq k[x] = k[x_0, \dots, x_n]$ and homogeneous coordinate ring $k[X] = k[x]/I = \bigoplus_d k[X]_d$. Recall (A.6.19) that the Hilbert polynomial $P_X(t)$ of X is the unique polynomial with rational coefficients that satisfies $P_X(i) = \dim k[X]_i$ for all sufficiently large integers i. Let $m = \deg P_X(t)$ and let $c \in \mathbb{Q}^*$ be the leading (highest) coefficient of $P_X(t)$. Then $m = \dim(X)$, the dimension of X. By definition, the *degree* of X is $\deg(X) := c \cdot m!$. This is a positive integer.

7.1.3 Alternatively, dimension and degree can be characterized in terms of linear sections of X. The dimension $\dim(X)$ is the largest number m with the property that $L \cap X \neq \varnothing$ for every linear subspace $L \subseteq \mathbb{P}^n$ of codimension m. The cardinality of $L \cap X$ is the same for all sufficiently general $(n-m)$-planes L. This cardinality is the degree $\deg(X)$ of X ([83] Lecture 18).

7.1.4 Examples. Here is a short list of examples that are relevant for what follows.

1. If $f \in k[x]_d$ has no muliple factors and $X = \mathcal{V}(f) \subseteq \mathbb{P}^n$ is the associated hypersurface, the Hilbert polynomial of X is $P_X(t) = \binom{t+n}{n} - \binom{t+n-d}{n} = \frac{d}{(n-1)!}\, t^{n-1} + \cdots$ Therefore $\deg(X) = d$. Intersecting X with a sufficiently general line $L \subseteq \mathbb{P}^n$ gives the same result, $|L \cap X| = d$. Indeed, when $u, v \in k^{n+1}$ are sufficiently general, the polynomial $f(tu + v) \in k[t]$ has degree d and has d different roots.

2. Any linear subspace $L \subseteq \mathbb{P}^n$ has degree one, since $P_L(t) = \binom{t+m}{m} = \frac{t^m}{m!} + \cdots$ where $m = \dim(L)$. A general linear subspace of dimension $n - m$ intersects L in a single point, by linear algebra.

3. If $\dim(X) = m$, and if $X_1, \dots, X_r$ are the m-dimensional irreducible components of X, then $\deg(X) = \sum_{i=1}^r \deg(X_i)$.

4. Let n, $d \geq 1$ be integers and let $v_d = v_{n,d} \colon \mathbb{P}^n \to \mathbb{P}^N$ be the degree d Veronese map, with $N = \binom{n+d}{n} - 1$, see A.6.11. So $v_d(\xi) = (\xi_0^{a_0} \cdots \xi_n^{a_n})_{|a|=d}$. The image variety $V = V_{n,d} = v_d(\mathbb{P}^n)$ is an irreducible subvariety of $\mathbb{P}^N$ of dimension n. We calculate its degree: If $L \subseteq \mathbb{P}^N$ is the intersection of n general hyperplanes, then $L \cap V = v_d(Y)$ where Y is the intersection of n general hypersurfaces of degree d in $\mathbb{P}^n$. Therefore $\deg(V) = d^n$ by Bézout's theorem ([84] Thm. I.7.7, [83] Lecture 18). Alternatively, we get the degree through the Hilbert polynomial: Via v_d we have $k[V]_j = k[x_0, \ldots, x_n]_{dj}$ for all $j \geq 0$, so

$$\dim k[V]_j = \binom{dj+n}{n} = \frac{(dj+n)\cdots(dj+1)}{n!} = \frac{d^n}{n!} j^n + \cdots$$

and again we see $\deg(V) = d^n$.

7.1.5 *Cones* over projective varieties: Let $L' \cong \mathbb{P}^m$, $L \cong \mathbb{P}^{n-m-1}$ be two complementary linear subspaces of $\mathbb{P}^n$ (meaning that $L \cap L' = \varnothing$), let $Y \subseteq L'$ be a subvariety. The *cone* $X = C_L(Y)$ *over* Y (with vertex space L) is the union of all $(n-m)$-planes $L \vee y$ in $\mathbb{P}^n$, where $y \in Y$. Alternatively, $X = \overline{\pi^{-1}(Y)}$ where $\pi \colon \mathbb{P}^n \smallsetminus L \to L'$ is the linear projection from L. We may choose linear coordinates in such a way that $L' = \mathcal{V}(x_{m+1}, \ldots, x_n)$ and $L = \mathcal{V}(x_0, \ldots, x_m)$. Then

$$X = \{(x_0 : \cdots : x_n) \colon x_0 = \cdots = x_m = 0 \text{ or } (x_0 : \cdots : x_m) \in Y\}$$

If $J = \mathcal{I}(Y) \subseteq k[x_0, \ldots, x_m]$ is the vanishing ideal of Y, then $\mathcal{I}(X)$ is the ideal generated by J in $k[x_0, \ldots, x_n]$. It is easy to see (Exercise 7.1.8) that $\dim(X) = \dim(Y) + (n-m)$, hence $\operatorname{codim}_{\mathbb{P}^n}(X) = \operatorname{codim}_{\mathbb{P}^m}(Y)$, and that $\deg(X) = \deg(Y)$.

7.1.6 Definition. A projective variety $X \subseteq \mathbb{P}^n$ is *non-degenerate* if X is not contained in any hyperplane, or equivalently, if $\dim k[X]_1 = n+1$.

7.1.7 Proposition. *If $X \subseteq \mathbb{P}^n$ is an irreducible and non-degenerate variety, then*

$$\deg(X) \geq 1 + \operatorname{codim}(X).$$

Proof. We give a sketch of the proof. For a different argument see [83] 18.9–18.12. The proof is by induction on $\operatorname{codim}(X) = n - \dim(X)$, the cases $\operatorname{codim}(X) = 0, 1$ being obvious. So let $\operatorname{codim}(X) > 1$. For $x \in X$ we consider the linear projection $\pi \colon X \dashrightarrow \mathbb{P}^{n-1}$ with centre x. The closed image $Y = \overline{\pi(X \smallsetminus \{x\})}$ of X is irreducible and is non-degenerate in $\mathbb{P}^{n-1}$. If $x \in X$ is chosen general enough, the variety X is easily seen not to be a union of lines through x. For such x, the general fibre of $\pi \colon X \smallsetminus \{x\} \to Y$ is finite and non-empty, which implies $\dim(X) = \dim(Y)$ (see e.g. [83] Theorem 11.12).

To complete the argument let $d = \dim(X)$ and let $x \in X$ be a general point as before, so $\operatorname{codim}(Y) = \operatorname{codim}(X) - 1$. If $H_1, \ldots, H_d \subseteq \mathbb{P}^{n-1}$ are general hyperplanes then

$$Y \cap \bigcap_{i=1}^{d} H_i = \pi\Big(X \cap \bigcap_{i=1}^{d} (H_i \vee x)\Big).$$

If $x \in X$ was chosen general enough, the intersection $X \cap \bigcap_{i=1}^d (H_i \vee x)$ has exactly $\deg(X)$ many points, with x being one of them. For such x, therefore, $Y \cap \bigcap_{i=1}^d H_i$ has at most $\deg(X) - 1$ points, which shows $\deg(Y) \le \deg(X) - 1$. We conclude

$$\deg(X) - \operatorname{codim}(X) \ \ge \ (1 + \deg(Y)) - (1 + \operatorname{codim}(Y)) \ = \ \deg(Y) - \operatorname{codim}(Y),$$

and the right-hand side is ≥ 1 by the inductive hypothesis. □

In view of Proposition 7.1.7, it is natural to make the following definition:

7.1.8 Definition. An irreducible and non-degenerate variety $X \subseteq \mathbb{P}^n$ has *minimal degree* if $\deg(X) = 1 + \operatorname{codim}(X)$.

7.1.9 Examples.

1. $\mathbb{P}^n$ is a variety of minimal degree. An irreducible hypersurface in $\mathbb{P}^n$ is non-degenerate and of minimal degree if and only if it is a quadric, i.e. has degree 2.

2. If X is a cone over an irreducible and non-degenerate variety $Y \subseteq \mathbb{P}^m$, then X is non-degenerate as well, and X is of minimal degree if and only if Y is of minimal degree (see 7.1.5).

3. Veronese embeddings give rise to more examples. Let $V = v_d(\mathbb{P}^n) \subseteq \mathbb{P}^N$ where $n \ge 1$, $d > 1$ and $N = \binom{n+d}{n} - 1$. Clearly, V is irreducible and non-degenerate, and $\deg(V) = d^n$ (Example 7.1.4.4). Since $\dim(V) = n$, we have

$$1 + \operatorname{codim}(V) \ = \ \binom{d+n}{n} - n.$$

If $n = 1$ then $1 + \operatorname{codim}(V) = d = \deg(V)$. So the rational normal curve $v_d(\mathbb{P}^1) \subseteq \mathbb{P}^d$ is a variety of minimal degree, for every $d \ge 1$. If $n = 2$ then

$$1 + \operatorname{codim}(V) \ = \ \binom{d+2}{2} - 2 \ = \ d^2 - \binom{d-1}{2} \ \le \ d^2 \ = \ \deg(V),$$

with equality if and only if $d \le 2$. So the Veronese surface $v_2(\mathbb{P}^2) \subseteq \mathbb{P}^5$ is of minimal degree as well. An easy argument shows that no other Veronese varieties are of minimal degree (Exercise 7.1.1).

7.1.10 We discuss another class of varieties. Let $r \ge 0$ and $d_0, \dots, d_r \ge 0$ be integers, and let $n + 1 = \sum_{i=0}^r (d_i + 1)$. Fix linear subspaces $U_0, \dots, U_r$ in $\mathbb{P}^n$ with $\dim(U_i) = d_i$ that are projectively independent (i.e. none intersects the linear span of the others). For each index $i = 0, \dots, r$ fix a parametrized rational normal curve $\phi_i \colon \mathbb{P}^1 \to U_i$ of degree d_i, i.e. let

$$\phi_i(s:t) \ = \ (s^{d_i} : s^{d_i - 1} t : \cdots : t^{d_i})$$

in suitable linear coordinates on U_i. Let $X = X(d_0, \dots, d_r) \subseteq \mathbb{P}^n$ be the union of the r-dimensional linear subspaces $\phi_0(y) \vee \cdots \vee \phi_r(y)$, for $y \in \mathbb{P}^1$. This union is a closed irreducible subvariety of $\mathbb{P}^n$, see below, and is called a *rational normal scroll*. We have $\dim(X) = r + 1$, except when $d_0 = \cdots = d_r = 0$, in which case $X = \mathbb{P}^r$.

We sketch the proof of the claims just made. Let $\varphi_i\colon \mathbb{A}^2 \to \mathbb{A}^{d_i+1}$, $\varphi_i(s,t) = (s^{d_i}, s^{d_i-1}t, \ldots, t^{d_i})$ for $i = 0, \ldots, r$. If coordinates in $\mathbb{P}^n$ are chosen suitably, the points in X are the $\phi(u_0, \ldots, u_r; s, t) := [u_0\varphi_0(s,t), \ldots, u_r\varphi_r(s,t)]$ in $\mathbb{P}^n$, where $(0,0) \neq (s,t) \in \mathbb{A}^2$ and $(0,\ldots,0) \neq (u_0,\ldots,u_r) \in \mathbb{A}^{r+1}$. When $(d_0,\ldots,d_r) \neq (0,\ldots,0)$, we see that X admits an open dense embedding $\mathbb{A}^{r+1} \to X$, given by sending $(u_1,\ldots,u_r,t)$ to

$$\left(1 : t : \cdots : t^{d_0} : u_1 : u_1 t : \cdots : u_1 t^{d_1} : \cdots : u_r : u_r t : \cdots : u_r t^{d_r}\right). \tag{7.1}$$

In particular, X is an irreducible and rational variety. If $d_0, \ldots, d_m \geq 1$ and $d_{m+1} = \cdots = d_{m+s} = 0$ then $X(d_0,\ldots,d_{m+s})$ is the cone over $X(d_0,\ldots,d_m)$ with vertex space $\mathbb{P}^{s-1}$. When $d_i \geq 1$ for every i, it is not hard to see that X is identified with the projective variety of all matrices of size $2 \times (n - r)$ of the form

$$\begin{pmatrix} x_{0,0} & \cdots & x_{0,d_0-1} & x_{1,0} & \cdots & x_{1,d_1-1} & \cdots\cdots & x_{r,0} & \cdots & x_{r,d_r-1} \\ x_{0,1} & \cdots & x_{0,d_0} & x_{1,1} & \cdots & x_{1,d_1} & \cdots\cdots & x_{r,1} & \cdots & x_{r,d_r} \end{pmatrix} \tag{7.2}$$

that have rank one (Exercise 7.1.3). In previous notation, the point $\phi(u_0,\ldots,u_r;s,t)$ in X corresponding to this matrix is given by $u_i\varphi_i(s,t) = (x_{i,0},\ldots,x_{i,d_i})$ for $i = 0,\ldots,r$. It can be shown [65] that the vanishing ideal of X is generated by all 2×2 minors of the matrix (7.2), but we won't use this fact.

For special values of $d_0,\ldots,d_r$ we get familiar varieties. If $r = 0$ then $X(d_0)$ is the rational normal curve of degree d_0 in $\mathbb{P}^{d_0}$. If $d_0 = \cdots = d_r = 0$ then $X = \mathbb{P}^r$. For $d_0 = \cdots = d_r = 1$ we get the Segre variety $(\mathbb{P}^1)^{r+1} \subseteq \mathbb{P}^{2r+1}$ of rank one matrices of size $2 \times (r+1)$.

If $d_i \geq 1$ for all i then the scroll variety $X(d_0,\ldots,d_r)$ is non-singular, see Exercise 7.1.7.

7.1.11 This remark is for readers with a background in toric varieties: From the open embedding (7.1), we see that X is the toric variety associated with the lattice polytope

$$P = \operatorname{conv}\Big(\{0\} \times [0,d_0] \cup \{e_1\} \times [0,d_1] \cup \cdots \cup \{e_r\} \times [0,d_r]\Big)$$

in $\mathbb{Z}^r \oplus \mathbb{Z} = \mathbb{Z}^{r+1}$.

7.1.12 Proposition. *All rational normal scrolls are non-degenerate as projective varieties, and are varieties of minimal degree. In fact, if $(d_0,\ldots,d_r) \neq (0,\ldots,0)$ then $X(d_0,\ldots,d_r)$ has degree $d_0 + \cdots + d_r$.*

Proof. (Sketch) Let $(d_0,\ldots,d_r) \neq (0,\ldots,0)$, let $X = X(d_0,\ldots,d_r) \subseteq \mathbb{P}^n$ and write $d = \sum_{i=0}^r d_i$. Then $\dim(X) = r + 1$ and $n = r + d$, so $\operatorname{codim}(X) = d - 1$, and we have to show $\deg(X) = d$. A hyperplane $H \subseteq \mathbb{P}^n$ corresponds to a sequence $p_0,\ldots,p_r$ of binary forms $p_j = p_j(s,t)$, not all of them zero and with $\deg(p_j) = d_j$ $(j = 0,\ldots,r)$. Using the notation from 7.1.10, the intersection $H \cap X$ consists of those $\phi(u_0, \dots, u_r; s, t)$ for which $\sum_{j=0}^r u_j p_j(s,t) = 0$. This already shows that X is non-degenerate. Moreover, if $H_0, \dots, H_r$ are $r+1$ general hyperplanes in $\mathbb{P}^n$, their intersection with X corresponds to the solutions of a system of $r+1$ equations

$$\sum_{j=0}^{r} u_j p_{ij}(s,t) = 0 \quad (i = 0, \dots, r), \tag{7.3}$$

where $p_{ij}(s,t)$ is a general binary form of degree d_j for all $0 \le i,\ j \le r$. Now (7.3) can be considered as a system of $r+1$ linear equations in the unknowns $u_0, \dots, u_r$, that depends on the parameter $(s:t) \in \mathbb{P}^1$. Therefore, the determinant of this system is a form of degree $d = \sum_{i=0}^r d_i$ in (s,t). For a sufficiently general choice of the p_{ij}, there are exactly d values of $(s:t)$ for which this determinant vanishes, and for these values the matrix has corank one. This means that $X \cap \bigcap_{i=0}^r H_i$ has exactly d points, for sufficiently general $H_0, \dots, H_r$. □

So far we have seen several examples of varieties of minimal degree (7.1.9, 7.1.12). A classical theorem due to del Pezzo and Bertini says that the list is complete:

7.1.13 Theorem. *An irreducible and non-degenerate variety $X \subseteq \mathbb{P}^n$ of minimal degree is one of the following:*

(1) *A hypersurface of degree two,*
(2) *a rational normal scroll, or*
(3) *a cone over the Veronese surface $v_2(\mathbb{P}^2)$ in $\mathbb{P}^5$.*

Unfortunately, a proof of this theorem is beyond the scope of this course. It can be found in [65].

7.1.14 Definition. Let $X \subseteq \mathbb{P}^n$ be a projective variety with $\operatorname{codim}(X) = e$ and with vanishing ideal $I = \mathcal{I}(X)$. If X is non-degenerate, the number

$$\varepsilon(X) := \binom{e+1}{2} - \dim(I_2)$$

is called the *quadratic deficiency* of X.

7.1.15 Proposition. *Let $X \subseteq \mathbb{P}^n$ be an irreducible non-degenerate variety of minimal degree. Then $\varepsilon(X) = 0$.*

Proof. (Sketch) Let $e = \operatorname{codim}(X)$ and $I = \mathcal{I}(X)$, we have to show $\dim(I_2) = \binom{e+1}{2}$. For this we'll present an *ad hoc* argument that uses the classification 7.1.13. If X is a cone over a non-degenerate variety Y, it is very easy to see $\varepsilon(X) = \varepsilon(Y)$ (Exercise 7.1.8).

In each of the cases (1)–(3), there is an obvious homogeneous ideal I' with zero set $\mathcal{V}(I') = X$, so we have $I = \sqrt{I'}$. We work with I'_2 instead of I_2, so effectively we only prove $\varepsilon(X) \le 0$. That is however all that will be used (in the proof of Theorem 7.2.7 below). It is in fact not hard to show that $I' = I$ in each case.

If $X = \mathcal{V}(f)$ is a hypersurface of degree two then $I' = I$ is the principal ideal generated by f, so $\dim(I_2) = 1$ as asserted. For the Veronese surface $X = v_2(\mathbb{P}^2)$ we have $e = 3$. Let I' be the ideal generated by the six 2×2 minors of the matrix in Exercise 7.1.2. These minors are linearly independent, and so $\dim(I'_2) = 6 = \binom{e+1}{2}$. Now let $X = X(d_0, \dots, d_r)$ be a rational normal scroll with $d_0, \dots, d_r \geq 1$. The matrix (7.2) in 7.1.10 has $d_0 + \cdots + d_r = (n+1) - (r+1) = e + 1$ many columns, so its number of 2×2 minors is $\binom{e+1}{2}$. Again, these minors are linearly independent, and so the ideal I' they generate satisfies the assertion. □

7.1.16 Remark. It can be shown [130] that $\varepsilon(X) \geq 0$ holds for every irreducible and non-degenerate variety $X \subseteq \mathbb{P}^n$, and that moreover $\varepsilon(X) = 0$ if and only if X is of minimal degree. The combination of both statements can be phrased by saying that a variety of minimal degree, and of given codimension e, is contained in the maximal possible number of linearly independent quadrics, namely $\binom{e+1}{2}$ many. And that varieties of minimal degree are characterized by this condition. In other words, $\varepsilon(X)$ measures how many independent quadrics are "missing" that contain X. We will only use the weaker statement 7.1.15.

The following easy lemma will be used in the next section:

7.1.17 Lemma. *Let $X \subseteq \mathbb{P}^n$ be a non-degenerate variety and let $m = \dim(X)$. Then $\varepsilon(X) = \dim k[X]_2 - (m+1)(n+1) + \binom{m+1}{2}$.*

Proof. Let $e = \operatorname{codim}(X) = n - m$. We have

$$\dim k[X]_2 = \binom{n+2}{2} - \dim(I_2) = \binom{n+2}{2} + \varepsilon(X) - \binom{e+1}{2}$$

(second equality by the definition of $\varepsilon(X)$), hence $\varepsilon(X) = \dim k[X]_2 + \binom{e+1}{2} - \binom{n+2}{2}$. The lemma follows from rewriting this expression:

$$\begin{aligned} \binom{e+1}{2} - \binom{n+2}{2} &= \frac{1}{2}\big((n-m+1)(n-m) - (n+2)(n+1)\big) \\ &= \binom{m+1}{2} - (m+1)(n+1). \end{aligned}$$ □

Exercises

7.1.1 Let $n \geq 1$, $d \geq 2$ be integers. Show that the degree d Veronese embedding $v_d(\mathbb{P}^n) \subseteq \mathbb{P}^{\binom{n+d}{n}-1}$ of $\mathbb{P}^n$ is a variety of minimal degree (if and) only if $n = 1$ or $(n, d) = (2, 2)$ (cf. Example 7.1.9.3).

7.1.2 Verify the following description of the Veronese surface V in $\mathbb{P}^5$ as a determinantal variety: $V = v_2(\mathbb{P}^2)$ consists of all points $y = (y_0 : \cdots : y_5)$ for which

$$\operatorname{rk}\begin{pmatrix} y_0 & y_1 & y_2 \\ y_1 & y_3 & y_4 \\ y_2 & y_4 & y_5 \end{pmatrix} = 1.$$

7.1.3 Let $r \ge 0$ and $d_0, \dots, d_r \ge 1$, let $n + 1 = \sum_{i=0}^{r}(d_i + 1)$, and let $X = X(d_0, \dots, d_r) \subseteq \mathbb{P}^n$ be the rational normal scroll as defined in 7.1.10. Show that in suitable linear coordinates x_{ij} on $\mathbb{P}^n$ $(0 \le j \le d_i, 0 \le i \le r)$, X is the (projective) set of all matrices of size $2 \times (n - r)$ of the form

$$\begin{pmatrix} x_{0,0} & \cdots & x_{0,d_0-1} & x_{1,0} & \cdots & x_{1,d_1-1} & \cdots\cdots & x_{r,0} & \cdots & x_{r,d_r-1} \\ x_{0,1} & \cdots & x_{0,d_0} & x_{1,1} & \cdots & x_{1,d_1} & \cdots\cdots & x_{r,1} & \cdots & x_{r,d_r} \end{pmatrix}$$

that have rank 1.

7.1.4 Let $X \subseteq \mathbb{P}^4$ be the linear projection of the Veronese surface $V = v_2(\mathbb{P}^2) \subseteq \mathbb{P}^5$ from a point $p \in V$. Show that $X = X(1, 2)$, the scroll surface in $\mathbb{P}^4$. Moreover, show that X is the blowing-up of $\mathbb{P}^2$ in a point.

7.1.5 Which lines are contained in a scroll surface $X(a, b)$?

7.1.6 Prove that a rational normal scroll of dimension > 1 is never a Veronese variety $v_d(\mathbb{P}^n)$ with $d > 1$. (*Hint*: First show that $v_d(\mathbb{P}^n)$ does not contain any line if $d > 1$.)

7.1.7 Let $r \ge 0$ and let $d_0, \dots, d_r \ge 1$ be positive integers. Show that the rational normal scroll $X(d_0, \dots, d_r)$ is a non-singular variety.

7.1.8 Let $H \subseteq \mathbb{P}^n$ be a hyperplane, let $Y \subseteq H$ be a closed subvariety that is non-degenerate, and let $X \subseteq \mathbb{P}^n$ be the cone over Y (with vertex some point in $\mathbb{P}^n \setminus H$). Compute the Hilbert function of X in terms of the Hilbert function of Y. Use the result to express dimension, degree and quadratic deficiency of X in terms of the same data for Y.

7.1.9 Let $X \subseteq \mathbb{P}^n$ be a non-degenerate variety of dimension m. By the Hilbert–Serre theorem (see A.6.19), the Hilbert series $H_X(t) = \sum_{i\ge0} \dim(k[X]_i)\, t^i$ is a rational function $H_X(t) = p(t)/(1 - t)^{m+1}$ where $p \in \mathbb{Z}[t]$ is a polynomial with $p(1) \ne 0$. Show that

$$p(t) = 1 + \operatorname{codim}(X)\, t + \varepsilon(X)\, t^2 + \text{(higher order terms)}$$

In particular, the quadratic deficiency of X is the quadratic coefficient of $p(t)$.

7.2 Sums of Squares and Varieties of Minimal Degree

After the review of algebraic geometry background in the previous section, we now return to varieties defined over the field $\mathbb{R}$ of real numbers.

7.2.1 We start by introducing a tool that is also important otherwise, the *apolarity pairing*. For $i = 0, \dots, n$ let $\partial_i = \frac{\partial}{\partial x_i}$, a linear differential operator on the polynomial ring $A = \mathbb{R}[x] = \mathbb{R}[x_0, \dots, x_n]$. Write $\partial = (\partial_0, \dots, \partial_n)$, and write $\partial^\alpha = \partial_0^{\alpha_0} \cdots \partial_n^{\alpha_n}$ for every multi-index $\alpha = (\alpha_0, \dots, \alpha_n) \in \mathbb{Z}_+^{n+1}$. Given a polynomial $f = \sum_\alpha c_\alpha x^\alpha \in \mathbb{R}[x]$, let $f(\partial)$ be the differential operator on $A = \mathbb{R}[x]$ defined by

$$f(\partial) = \sum_\alpha c_\alpha \partial^\alpha.$$

As usual, let A_m denote the space of forms of degree m in A. If f is homogeneous of degree d, then $f(\partial)(A_m) \subseteq A_{m-d}$ for $m \ge d$, and $f(\partial)(A_m) = 0$ for $m < d$.

Let $\alpha, \beta \in \mathbb{Z}_+^{n+1}$ with $|\alpha| = |\beta| = m$. One directly checks that $\partial^\alpha x^\beta = 0$ if $\alpha \ne \beta$, and that $\partial^\alpha x^\alpha = \alpha_0! \cdots \alpha_n! = \alpha!$. For $m \ge 0$ and $f, g \in A_m$, put

$$\langle f, g\rangle := \frac{1}{m!} f(\partial)(g) = \frac{1}{m!} g(\partial)(f) = \langle g, f\rangle \in \mathbb{R}.$$

Then

$$A_m \times A_m \to \mathbb{R}, \quad (f, g) \mapsto \langle f, g\rangle = \langle g, f\rangle, \tag{7.4}$$

is an inner product (positive definite and bilinear) on A_m, the *apolarity pairing*. Division by $m!$ is just a convenient normalization. Under this pairing, the monomials of degree m are pairwise orthogonal, and

$$\langle x^\alpha, x^\alpha\rangle = \frac{1}{m!}\partial^\alpha(x^\alpha) = \frac{\alpha!}{m!} = \frac{\alpha_0! \cdots \alpha_n!}{m!}$$

holds if $|\alpha| = m$. We observe an elementary but crucial property:

7.2.2 Proposition. *Let $m \geq 0$. Given a point $u = (u_0, \dots, u_n)$ in $\mathbb{R}^{n+1}$, let $l_u = \sum_{i=0}^n u_i x_i \in A_1$ be the corresponding linear form, and let $\varphi_u \in A_m^\vee$ be evaluation in u, defined by $\varphi_u(f) = f(u)$ for $f \in A_m$.*

(a) *$\langle f, l_u^m\rangle = f(u)$ holds for all $m \geq 0$ and all $f \in A_m$.*
(b) *The linear isomorphism $\phi\colon A_m \to A_m^\vee$ induced by the inner product* (7.4) *satisfies $\phi(l_u^m) = \varphi_u$ for every $u \in \mathbb{R}^{n+1}$.*

Proof. It is enough to prove (a) when $f = x_{i_1} \cdots x_{i_m}$ is a monomial. Since $\partial_i(l_u) = u_i$ $(i = 0, \dots, n)$, the product rule gives $\partial_{i_1} \cdots \partial_{i_m}(l_u^m) = m u_{i_m} \cdot \partial_{i_1} \cdots \partial_{i_{m-1}}(l_u^{m-1})$. By induction, this implies $\partial_{i_1} \cdots \partial_{i_m}(l_u^m) = m!\, u_{i_1} \cdots u_{i_m}$, which is assertion (a). (b) is a direct consequence of (a). □

Now let $X \subseteq \mathbb{P}^n$ be a projective $\mathbb{R}$-variety, with homogeneous coordinate ring $S_X = \mathbb{R}[X]$, and let $\widehat{X} \subseteq \mathbb{A}^{n+1}$ be the affine cone over X (A.6.10). From 7.1.1, recall the definition of the convex cones $\Sigma_{X,2d} \subseteq P_{X,2d}$ in $S_{X,2d}$ (the sos and the psd cone of forms of degree $2d$).

7.2.3 Lemma. *Assume that the projective variety X has a Zariski dense set $X(\mathbb{R})$ of real points. Then $\Sigma_{X,2d}$ is closed in $S_{X,2d}$ for every $d \geq 0$.*

Proof. The homogeneous coordinate ring S_X agrees with the affine coordinate ring of the affine cone $\widehat{X}$. This ring is real reduced since $\widehat{X}(\mathbb{R})$ is Zariski dense in $\widehat{X}$, see Corollary 3.3.7. So the lemma is a particular case of Exercise 3.2.6. □

By arguments similar to 2.4.6, we see that both cones $\Sigma_{X,2d}$ and $P_{X,2d}$ are semialgebraic and full-dimensional, as subsets of $S_{X,2d}$.

If C is a convex cone in a finite-dimensional $\mathbb{R}$-vector space V, recall that the dual cone of C is $C^* = \{\lambda \in V^\vee : \forall\, x \in C\ \lambda(x) \geq 0\}$. With the usual identification $(V^\vee)^\vee = V$, one has $C^{**} = \overline{C}$, i.e., the bi-dual $C^{**} = (C^*)^*$ of C is the closure of C (see 8.1.21).

7.2.4 Proposition. *Assume that $X(\mathbb{R})$ is Zariski dense in X, let $d \geq 0$. The dual cone $(P_{X,2d})^* \subseteq S_{X,2d}^\vee$ is the conic hull of all point evaluations in points $u \in \widehat{X}(\mathbb{R})$.*

Proof. For the proof write $S = S_X$ and $P_{2d} = P_{X,2d}$. Let $I = \mathfrak{I}(X)$, so $S = \mathbb{R}[x]/I$. For $u \in \widehat{X}(\mathbb{R})$ and $p \in I_{2d}$ we have $p(u) = 0$, therefore point evaluation in u induces a linear form $\psi_u \in S_{2d}^\vee$. The convex cone $C := \operatorname{cone}\{\psi_u \colon u \in \widehat{X}(\mathbb{R})\}$ in $S_{2d}^\vee$ satisfies $C^* = P_{2d}$, by the definition of P_{2d}. Hence $\overline{C} = P_{2d}^*$ holds by cone duality, and it remains to see that C is closed in $S_{2d}^\vee$.

We continue to write $A = \mathbb{R}[x]$. Let $Q \subseteq A_{2d}$ be the convex cone of all finite sums $f = \sum_{i=1}^r (l_{u_i})^{2d}$ with $u_1, \dots, u_r \in \widehat{X}(\mathbb{R})$. Each $f \in Q$ is a sum of $\dim(A_{2d}) = \binom{n+2d}{n}$ many powers $(l_{u_i})^{2d}$, by the cone version of Carathéodory's theorem (8.1.14). Since $\sum_i (l_{u_i})^{2d} = 0$ in A_{2d} implies $u_i = 0$ for all i, it follows from Lemma 2.4.7 that the cone Q is closed in A_{2d}. If $\phi\colon A_{2d} \to A_{2d}^\vee$ is the apolarity isomorphism, Proposition 7.2.2 implies that the cone $C' := \phi(Q)$ is generated by the φ_u, for $u \in \widehat{X}(\mathbb{R})$. So this cone is closed in $A_{2d}^\vee$. Since $C' \subseteq (I_{2d})^\perp = (A_{2d}/I_{2d})^\vee = S_{2d}^\vee$, we may consider C' as a closed cone in $S_{2d}^\vee$. This identifies C' with C above, thereby completing the proof. □

We'll return to the cone Q and the apolarity pairing in Section 8.1. Here is the main result of this section:

7.2.5 Theorem. (Blekherman-Smith-Velasco) *Let $X \subseteq \mathbb{P}^n$ be an irreducible and non-degenerate $\mathbb{R}$-variety, and assume that $X(\mathbb{R})$ is Zariski dense in X. Then $\Sigma_{X,2} = P_{X,2}$ holds if, and only if, X is a variety of minimal degree.*

The theorem identifies those projective varieties X (irreducible and with Zariski dense $\mathbb{R}$-points) that satisfy psd = sos in degree two. From this one can derive the answer for all even degrees, see 7.2.16 below.

7.2.6 We first give a sketch of proof for the forward direction. So let $X \subseteq \mathbb{P}^n$ be irreducible and non-degenerate, with $X(\mathbb{R})$ Zariski dense in X. For a leaner notation we write Σ_2, P_2 and S_2 instead of $\Sigma_{X,2}$, $P_{X,2}$ and $S_{X,2}$, respectively. Assuming that X is not of minimal degree, we want to show $P_2 \not\subseteq \Sigma_2$. Since both convex cones are closed, it is equivalent to show $\Sigma_2^* \not\subseteq P_2^*$. By Proposition 7.2.4, the convex cone P_2^* is generated by all point evaluations $\psi_u \in S_2^\vee$ with $u \in \widehat{X}(\mathbb{R})$. We are going to construct a sum of squares form $f \in \Sigma_2$ and a non-zero element $\lambda \in \Sigma_2^*$, in such a way that $f > 0$ on $X(\mathbb{R})$ and $\lambda(f) = 0$ hold. This will prove the claim, since assuming $\lambda = \sum_i a_i \psi_{u_i}$ (with $u_i \neq 0$ in $\widehat{X}(\mathbb{R})$ and $a_i > 0$ in $\mathbb{R}$) would imply $\lambda(f) = \sum_i a_i f(u_i) > 0$, contradiction. Let $m = \dim(X)$ and $e = n - m = \operatorname{codim}(X)$. By assumption we have $\deg(X) \geq e + 2$.

The first step is to intersect X with a suitable linear subspace $L \subseteq \mathbb{P}^n$ of codimension m, in order to reduce to a zero-dimensional situation. More precisely, we show that there exist linear forms $h_1, \dots, h_m \in \mathbb{R}[x]$ such that, writing $L := \mathcal{V}(h_1, \dots, h_m) \subseteq \mathbb{P}^n$, the intersection $Z := X \cap L$ has the following properties: Z is a finite set of $\deg(X) \geq e + 2$ many different points, which are in linearly general position (meaning that any $e + 1$ of them span L projectively), and such that at least $e + 1$ of them are $\mathbb{R}$-rational.

To achieve this, one first shows that if $H \subseteq \mathbb{P}^n$ is a sufficiently general hyperplane, the intersection $H \cap X$ is again non-degenerate (in H). For a proof we refer to [83]

Proposition 18.10. Moreover, by Bertini's classical theorem ([83] Thm. 17.16 or [84] Thm. II.8.18), the singularities of $H \cap X$ are contained in X_{sing}, when H is chosen general enough. As long as $m = \dim(X) \geq 2$, the intersection $H \cap X$ is again irreducible for general H, by another facet of Bertini. Since $X(\mathbb{R})$ is Zariski dense in X by assumption, there is a non-empty open set (with respect to the Euclidean topology) of hyperplanes H for which $H \cap X$ contains a real non-singular point of X. Assume $m = \dim(X) \geq 2$ for a moment. Then, by combining the statements above, we conclude that there is a non-empty open set of hyperplanes $H \subseteq \mathbb{P}^n$ for which $H \cap X$ is non-degenerate and irreducible and has Zariski dense $\mathbb{R}$-points. (For the last claim, recall that an irreducible $\mathbb{R}$-variety with one non-singular $\mathbb{R}$-point has Zariski dense $\mathbb{R}$-points, Corollary 1.7.9.)

By inductively applying the step just described, we find linear forms $h_1, \ldots, h_{m-1}$ in $\mathbb{R}[x]$ such that, writing $L' := \mathcal{V}(h_1, \ldots, h_{m-1}) \cong \mathbb{P}^{e+1}$, the intersection $C := X \cap L'$ is an irreducible non-degenerate curve in L' with Zariski dense $\mathbb{R}$-points. Now choose $e+1$ real points on C that are projectively independent and hence span a hyperplane H' in L'. For a sufficiently general choice of these points, the intersection $Z := C \cap H'$ is a finite set of $\deg(X) \geq e+2$ many points in linearly general position in $L := L' \cap H'$, of which at least $e+1$ are real.

So far we have found linear forms $h_1, \ldots, h_m$ in $\mathbb{R}[x]$ such that, with $L := \mathcal{V}(h_1, \ldots, h_m) \cong \mathbb{P}^e$, the intersection $Z := X \cap L$ consists of $\deg(X) \geq e+2$ different points which are in linearly general position in L, and such that $|Z(\mathbb{R})| \geq e+1$. We'll use the $\mathbb{R}$-points in Z to find a linear form $\lambda \in \Sigma_2^* \setminus P_2^*$. For this we have to distinguish two cases.

Case 1: $|Z(\mathbb{R})| \geq e+2$.

Let $0 \neq u_0, \ldots, u_{e+1} \in \mathbb{R}^{n+1}$ be such that $[u_0], \ldots, [u_{e+1}] \in Z(\mathbb{R})$ are in linearly general position in $L \cong \mathbb{P}^e$. So $u_0, \ldots, u_{e+1}$ are linearly dependent, but any $e+1$ of them are linearly independent. Hence there exist non-zero real numbers $a_0, \ldots, a_e$ with $u_{e+1} = \sum_{\nu=0}^{e} a_\nu u_\nu$, and so

$$p(u_{e+1}) = \sum_{\nu=0}^{e} a_\nu p(u_\nu) \tag{7.5}$$

holds for every linear form $p \in S_1$.

For $\nu = 0, \ldots, e+1$ let $\psi_\nu = \psi_{u_\nu} \in S_2^\vee$ be point evaluation in u_ν. Given a tuple $c = (c_0, \ldots, c_e)$ of positive real numbers, consider

$$\lambda_c := \Big(\sum_{\nu=0}^{e} \frac{a_\nu^2}{c_\nu}\Big) \cdot \Big(\sum_{\nu=0}^{e} c_\nu \psi_\nu\Big) - \psi_{e+1} \in S_2^\vee. \tag{7.6}$$

Using (7.5) we get for any $p \in S_1$:

$$\lambda_c(p^2) = \Big(\sum_{\nu=0}^{e} \frac{a_\nu^2}{c_\nu}\Big) \cdot \Big(\sum_{\nu=0}^{e} c_\nu p(u_\nu)^2\Big) - \Big(\sum_{\nu=0}^{e} a_\nu p(u_\nu)\Big)^2. \tag{7.7}$$

So $\lambda_c(p^2) \geq 0$ for any $p \in S_1$, by Cauchy–Schwarz, and hence $\lambda_c \in \Sigma_2^*$. On the other hand, there exists a linear form $p_c \in S_1$ with $p_c(u_\nu) = \frac{a_\nu}{c_\nu}$ for $0 \leq \nu \leq e$, since $u_0, \dots, u_e$ are linearly independent. For this particular form, (7.7) gives $\lambda_c(p_c^2) = 0$.

Now fix a tuple $c = (c_0, \dots, c_e)$ with $c_\nu > 0$, in such a way that p_c does not vanish in any point of $Z(\mathbb{R})$. This is possible since $a_\nu \neq 0$ for every ν. For such a choice of c, the quadratic form $f := p_c^2 + h_1^2 + \cdots + h_m^2$ is strictly positive on $X(\mathbb{R})$. On the other hand, f satisfies $\lambda_c(f) = 0$ since $\psi_\nu(h_j^2) = h_j(u_\nu)^2 = 0$ for $0 \leq \nu \leq e+1$ and $1 \leq j \leq m$. So λ_c cannot be a sum of evaluations in points of $\widehat{X}(\mathbb{R})$, which shows $\lambda_c \in \Sigma_2^* \setminus P_2^*$. This settles the proof in Case 1, as sketched in the outline at the beginning of 7.2.6.

Case 2: $|Z(\mathbb{R})| = e+1$.

Since $|Z(\mathbb{C})| > e+1$, Z contains a pair of complex-conjugate points. So there exist $0 \neq u_1, \dots, u_e \in \mathbb{R}^{n+1}$ and $0 \neq v, \overline{v} \in \mathbb{C}^{n+1}$ such that any $e+1$ of $u_1, \dots, u_e, v, \overline{v}$ are ($\mathbb{C}$-) linearly independent, and such that $[u_1], \dots, [u_e]$, $[v]$, $[\overline{v}]$ lie in Z. Again using that the points of Z are in linearly general position, there exist non-zero complex numbers $a_1, \dots, a_e, b, b'$ with $a_1u_1 + \cdots + a_eu_e + bv + b'\overline{v} = 0$. An easy argument from linear algebra shows that, after suitable (complex) scaling of the tuple $(a_1, \dots, a_e, b, b')$, we have $a_1, \dots, a_e \in \mathbb{R}$ and $b' = \overline{b} \in \mathbb{C}$. Putting $w := -(bv + \overline{bv}) = -2\,\mathrm{Re}(bv)$, we have $0 \neq w \in \mathbb{R}^{n+1}$ and

$$a_1u_1 + \cdots + a_eu_e = w. \tag{7.8}$$

From here on, the argument is very much parallel to Case 1. For $\nu = 1, \dots, e$ let $\psi_\nu = \psi_{u_\nu} \in S_2^\vee$, and define, for any tuple $c = (c_1, \dots, c_e)$ of positive real numbers c_ν,

$$\lambda_c := \Big(\sum_{\nu=1}^{e} \frac{a_\nu^2}{c_\nu}\Big) \cdot \Big(\sum_{\nu=1}^{e} c_\nu\psi_\nu\Big) - \psi_w \in S_2^\vee. \tag{7.9}$$

For any $p \in S_1$ we get, using (7.8):

$$\lambda_c(p^2) = \Big(\sum_{\nu=0}^{e} \frac{a_\nu^2}{c_\nu}\Big) \cdot \Big(\sum_{\nu=0}^{e} c_\nu p(u_\nu)^2\Big) - \Big(\sum_{\nu=0}^{e} a_\nu p(u_\nu)\Big)^2. \tag{7.10}$$

We conclude $\lambda_c \in \Sigma_2^*$ as before, and also the rest of the argument works as in Case 1. With this, the proof of the forward direction in Theorem 7.2.5 is complete.

Now let us look at the reverse direction. In fact we prove a stronger result here:

7.2.7 Theorem. (Blekherman-Plaumann-Sinn-Vinzant) *Let X be an irreducible non-degenerate $\mathbb{R}$-variety of minimal degree, with $X(\mathbb{R})$ Zariski dense in X. Then every psd form in $\mathbb{R}[X]_2$ is a sum of s squares of linear forms, where* $s := \dim(X) + 1$.

Proof. If the assertion has been proved for X, it also follows for $X' = C_p(X)$, the cone over X with vertex p a point (cf. 7.1.5). Indeed, let $S = \mathbb{R}[X]$ and $S' = \mathbb{R}[X']$, then $S' = S[y]$ with a new variable y of degree one. Any quadratic form $f \in S_2'$ on X' can be written $f = ay^2 + by + c$ with $a \in \mathbb{R}$, $b \in S_1$ and $c \in S_2$. If f is psd on X', then clearly either $a = b = 0$ and $c \in P_2$, or $a > 0$ and $4ac - b^2 \in P_2$. In the first case we are done, in the second we see from

$$f = a \cdot \left((x + \frac{b}{2a})^2 + \frac{4ac - b^2}{4a^2} \right)$$

that f is a sum of $\dim(X) + 2 = \dim(X') + 1$ squares of elements in S', by the assumption on X.

So we can assume that X is not a cone, therefore X is non-singular (Exercise 7.1.7). Let $m = \dim(X)$ and $S = \mathbb{R}[X]$. We study the sum of squares map

$$\sigma\colon (S_1)^{m+1} = S_1 \times \cdots \times S_1 \to S_2, \quad (g_0, \ldots, g_m) \mapsto \sum_{i=0}^{m} g_i^2$$

and want to show $\operatorname{im}(\sigma) = P_2$. The image set of σ is closed in S_2, by Lemma 2.4.7 and since $\sum_{i=0}^{m} g_i^2 = 0$ implies $g_1 = \cdots = g_m = 0$ (recall that $X(\mathbb{R})$ is Zariski dense in X). The (total) derivative of σ at a tuple $g = (g_0, \ldots, g_m)$ is the linear map $(D\sigma)_g\colon (S_1)^{m+1} \to S_2$, $(h_0, \ldots, h_m) \mapsto 2\sum_i g_i h_i$. There exist linear forms $g_0, \ldots, g_m \in S_1$ with $\mathcal{V}_X(g_0, \ldots, g_m) = \varnothing$, see 7.1.3. Now the proof of the theorem proceeds in three steps:

(1) *For every tuple* $g = (g_0, \ldots, g_m) \in (S_1)^{m+1}$ *with* $\mathcal{V}_X(g_0, \ldots, g_m) = \varnothing$, *the derivative* $(D\sigma)_g$ *is surjective.*

We can only sketch the proof. It can be shown (see [64] Theorem 4.2 for a proof) that the variety X is arithmetically Cohen–Macaulay, which means that the graded ring S is Cohen–Macaulay. In particular, this property implies that every sequence $g_0, \ldots, g_m$ of non-constant homogeneous elements in S with $\mathcal{V}_X(g_0, \ldots, g_m) = \varnothing$ is S-regular. This latter property means that the graded ring $\operatorname{gr}^J(S) = \bigoplus_{i \geq 0} J^i/J^{i+1}$ associated with the ideal $J = \langle g_0, \ldots, g_m \rangle$ is isomorphic to the polynomial ring $(S/J)[t_0, \ldots, t_m]$. More precisely, the natural homomorphism from $(S/J)[t_0, \ldots, t_m]$ to $\operatorname{gr}^J(S)$ that maps t_i to the class of g_i in $\operatorname{gr}_1^J(S)$ is an isomorphism of graded rings. This in turn implies that the kernel of $(D\sigma)_g\colon (S_1)^{m+1} \to S_2$, $(h_0, \ldots, h_m) \mapsto 2\sum_i g_i h_i$ is generated by the obvious relations $(0, \ldots, -g_j, \ldots, g_i, \ldots, 0)$ $(0 \leq i < j \leq m)$, and so it has dimension $\binom{m+1}{2}$. It follows that

$$\operatorname{rk}(D\sigma)_g = (m+1)\dim(S_1) - \binom{m+1}{2} = (m+1)(n+1) - \binom{m+1}{2}.$$

This number is equal to $\dim(S_2)$ by Lemma 7.1.17, since $\varepsilon(X) = 0$ (Proposition 7.1.15).

As is well-known from calculus, assertion (1) means that the map σ is locally submersive at the tuple g. This implies, in particular, that the image set of σ contains a neighborhood of $\sigma(g)$ in S_2.

Let S_2^{sm} be the set of all $q \in S_2$ whose zero variety $\mathcal{V}_X(q)$ is non-singular. By Bertini's theorem ([83] Thm. 17.16, [84] Thm. II.8.18), S_2^{sm} contains a dense open subset of S_2. Let $P_2^+ \subseteq P_2$ be the open set of all quadratic forms that are strictly positive on $X(\mathbb{R})$. Then $P_2^{sm} := P_2 \cap S_2^{sm}$ is dense in P_2 since P_2 is the closure of its interior. Moreover $P_2^{sm} \subseteq P_2^+$ since every real zero of a psd form q is a singularity of $\mathcal{V}_X(q)$ (the form vanishes of order ≥ 2 there). Altogether, P_2^{sm} is a dense subset of P_2^+.

(2) *The set P_2^{sm} is connected.*

The crucial point is that the difference set $P_2^{+,sing} := P_2^+ \setminus P_2^{sm}$ (of all strictly positive quadratic forms q on X for which $\mathcal{V}_X(q)$ is singular) has codimension ≥ 2 in S_2. To see this, note that every such form q has at least two complex-conjugate singular points in X. On the other hand it can be shown that the forms $q \in S_2$ for which $\mathcal{V}_X(q)$ is singular are a hypersurface in S_2, and that a Zariski dense subset of this hypersurface consists of forms q for which $\mathcal{V}_X(q)$ has only a single (complex) singular point. So the codimension of $P_2^{+,sing}$ is at least two. Since P_2^+ is connected, being an open cone, the set P_2^{sm} is itself connected since it is the complement of a set of codimension ≥ 2. We actually proved this last conclusion (Theorem 4.6.9).

(3) *The set $P_2^{sm} \cap \mathrm{im}(\sigma)$ is open and closed in P_2^{sm}, and is non-empty.*

The set is relatively closed since $\mathrm{im}(\sigma)$ is closed. As mentioned before, there exist tuples $g = (g_0, \dots, g_m) \in S_1^{m+1}$ with $\mathcal{V}_X(g_0, \dots, g_m) = \emptyset$. For such g, the map σ is a submersion locally at g, by (1), and so $q := \sigma(g)$ lies in the interior of $\mathrm{im}(\sigma)$. So the set (3) is non-empty since S_2^{sm} is dense in S_2. On the other hand, if g is a tuple for which $q = \sigma(g) \in P_2^{sm}$, then necessarily $\mathcal{V}_X(g_0, \dots, g_m) = \emptyset$ since any point in this set would be a singularity of $V_X(q)$. Locally at g, the map σ is therefore a submersion by (1), and so $q = \sigma(g)$ is an interior point of P_2^{sm}. So the set $P_2^{sm} \cap \mathrm{im}(\sigma)$ is relatively open in P_2^{sm}. This completes the proof of (3).

From (2) and (3) it follows that P_2^{sm} is contained in $\mathrm{im}(\sigma)$. Since P_2^{sm} is dense in P_2 and $\mathrm{im}(\sigma)$ is closed, we conclude $\mathrm{im}(\sigma) = P_2$. Theorem 7.2.7 is proved. □

In the following we are going to make explicit what Theorem 7.2.5, together with its quantitative refinement Theorem 7.2.7, means concretely for the three classes of varieties X of minimal degree (Theorem 7.1.13).

7.2.8 Remark. We start with the question that Hilbert considered, and settled, in 1888: For which pairs of integers n, $d \geq 1$ is it true that the inclusion $\Sigma_{n,2d} \subseteq P_{n,2d}$ of cones (2.4.5) is an equality? In terms of Veronese embeddings, this means to ask when $\Sigma_{X,2} = P_{X,2}$ holds for $X = V_{n-1,d} = v_d(\mathbb{P}^{n-1})$. The answer found by Hilbert (Theorem 2.4.9) is contained in Theorem 7.2.5: Since the Veronese variety $V_{n-1,d}$ is always non-degenerate, we have equality if and only if $V_{n-1,d}$ is of minimal degree. As discussed in Example 7.1.9.3, this is the case if and only if $n \leq 2$ or $d = 1$ or $(n, 2d) = (3, 4)$.

The quantitative side of the question, again settled by Hilbert, is contained in Theorem 7.2.7. Leaving away the elementary cases $n \leq 2$ or $2d = 2$, the remaining hard case corresponds to the Veronese surface $X = V_{2,2} \subseteq \mathbb{P}^5$. Hilbert's three-squares theorem 2.4.10 is recovered in 7.2.7: Every psd ternary form $f = f(x_0, x_1, x_2)$ of degree four is a sum of three squares of quadratic forms. While in Section 2.4 we had only proved the weaker statement that f is a sum of four squares (Proposition 2.4.11), this gap has now been closed.

We see how Hilbert's theorem turns out to be part of a beautiful general geometric picture. In fact, the proof of Theorem 7.2.7 that was sketched above proceeds by mimicking (in modern language) Hilbert's original proof for ternary quartics. The modern viewpoint notwithstanding, it is not surprising that Hilbert's proof was hard for his contemporaries to accept.

Let us try and see what Theorems 7.2.5 and 7.2.7 mean for the remaining varieties of minimal degree.

7.2.9 Remark. Consider quadric hypersurfaces, so let $q = q(x_0, \dots, x_n)$ be a quadratic form over $\mathbb{R}$ and $X = \mathcal{V}(q) \subseteq \mathbb{P}^n$. Then X irreducible means $\mathrm{rk}(q) \ge 3$, and $X(\mathbb{R})$ Zariski dense means that q is indefinite (by elementary reasoning, or also as a particular instance of the sign-changing criterion 1.7.14). Under these assumptions, the conclusion $P_{X,2} = \Sigma_{X,2}$ of 7.2.5 is contained in the following classical result. It is used in optimization and control theory:

7.2.10 Proposition. (S-Lemma) *Let A, B be real symmetric $n \times n$ matrices with B indefinite. If $x^\top Ax \ge 0$ holds for every $x \in \mathbb{R}^n$ with $x^\top Bx = 0$, there exists a real number t with $A + tB \succeq 0$.*

Of course we proved this (for $\mathrm{rk}(B) \ge 3$) in Theorem 7.2.7. To give a proof along traditional lines, we need some concepts from convexity theory (faces of convex sets and their relative interior) that will be discussed in full detail in Section 8.1. Given symmetric matrices A, $B \in \mathrm{Sym}_n(\mathbb{R})$, let $\langle A, B\rangle = \mathrm{tr}(AB)$ (the trace inner product between A and B), and write $q_A(x) = x^\top Ax$ ($x \in \mathbb{R}^n$) for the quadratic form on $\mathbb{R}^n$ that is associated with A.

7.2.11 Lemma. *Given symmetric matrices A, $B \in \mathrm{Sym}_n(\mathbb{R})$ and real numbers a, b, the following are equivalent:*

(i) *There is $x \in \mathbb{R}^n$ with $q_A(x) = a$ and $q_B(x) = b$;*
(ii) *there is $S \in \mathrm{Sym}_n(\mathbb{R})$ with $\langle S, A\rangle = a$, $\langle S, B\rangle = b$ and $S \succeq 0$.*

Proof. (i) $\Rightarrow$ (ii) is easy: The matrix $S = xx^\top$ satisfies $S \succeq 0$ and $\langle S, A\rangle = q_A(x)$, $\langle S, B\rangle = q_B(x)$. For the converse note that $q_S(x) = \langle S, xx^\top\rangle$. Therefore, assuming that there exists a matrix S as in (ii), we have to prove that there exists such a matrix of rank one. Write $\mathsf{S}^n = \mathrm{Sym}_n(\mathbb{R})$, let $L := \{S \in \mathsf{S}^n \colon \langle S, A\rangle = a,\ \langle S, B\rangle = b\}$ and $K = L \cap \mathsf{S}^n_+$. The set K is closed, convex and non-empty, and it doesn't contain a line. Therefore K has an extreme point T, see Exercise 8.1.13. Let $\mathrm{rk}(T) = r$, we show $r \le 1$. Consider $U := \{S \in \mathsf{S}^n \colon \ker(T) \subseteq \ker(S)\}$, a linear subspace of S^n of dimension $\binom{r+1}{2}$. Then $F := U \cap \mathsf{S}^n_+$ is a face of S^n_+ with $T \in \mathrm{relint}(F)$ (8.2.3), and so $F \cap L$ is a face of $\mathsf{S}^n_+ \cap L = K$ whose relative interior contains T. Since T is an extreme point of K we conclude $F \cap L = \{T\}$. Now L is an affine-linear subspace of S^n of codimension ≤ 2. Since L intersects a neighborhood of T inside U in a single point, we must have $\dim(U) \le 2$. This means $r \le 1$. □

7.2.12 Corollary. *For arbitrary quadratic forms $f, g \in \mathbb{R}[x_1, \dots, x_n]$, the set*

$$\{(f(x), g(x)) \colon x \in \mathbb{R}^n\}$$

is a convex cone in $\mathbb{R}^2$.

Proof. Let $A, B \in \mathsf{S}^n$ be the matrices with $f = q_A$ and $g = q_B$. Then the set in question is the image of the cone S^n_+ under the linear map $S \mapsto (\langle S, A\rangle, \langle S, B\rangle)$, by Lemma 7.2.11. □

7.2.13 *Proof of Proposition 7.2.10*: Write $f = q_A$, $g = q_B$, and consider the open quadrants $Q = \{(a, b) \colon a < 0,\ b > 0\}$ and $Q' = \{(a, b) \colon a < 0,\ b < 0\}$ in $\mathbb{R}^2$. The set $M = \{(f(x), g(x)) \colon x \in \mathbb{R}^n\}$ is a convex cone in $\mathbb{R}^2$, by Corollary 7.2.12. By assumption we have $(t, 0) \notin M$ for $t < 0$. Since M is convex, it can meet at most one of Q and Q'. Replacing g by $-g$ if necessary we may assume $M \cap Q = \varnothing$. From hyperplane separation (Theorem 8.1.5(a)) it follows that there is a non-zero linear form $l(u, v) = au + bv$ with $l \geq 0$ on M and $l \leq 0$ on Q. This means that $aA + bB \succeq 0$. From $l|_Q \leq 0$ we see $a \geq 0$ and $b \leq 0$. Moreover $a \neq 0$ since otherwise M would be contained in the lower half-plane, contradicting that g is indefinite. So $a > 0$, and hence $A + \frac{b}{a}B$ is positive semidefinite. □

7.2.14 Remark. We also see the correct number of squares on quadric hypersurfaces, as in Theorem 7.2.7: In the S-Lemma 7.2.10, the set $J = \{t \in \mathbb{R} \colon A + tB \succeq 0\}$ is a compact interval since B is indefinite. If t is a boundary point of J then $A + tB$ is singular. Hence the quadratic form $q_A + tq_B$ is a sum of $n - 1 = \dim(X_B) + 1$ squares (note that X_B is a hypersurface in $\mathbb{P}^{n-1}$).

7.2.15 Finally, consider rational normal scrolls $X = X(d_0, \dots, d_r)$, with $m := r+1 = \dim(X)$. In this case, Theorem 7.2.7 says the following. If $Q(s, t) = (q_{ij}(s, t))$ is a symmetric $m \times m$ matrix such that q_{ij} is a binary form of degree $d_i + d_j$ for all $1 \leq i, j \leq m$, and if Q is psd (i.e. $Q(s, t) \succeq 0$ for all $s, t \in \mathbb{R}$), then the quadratic form $x^\top Qx$ is a sum of $m + 1$ squares of linear forms. In other words, there is an $m \times (m + 1)$ matrix $P = (p_{ij}(s, t))$ with p_{ij} a binary form of degree d_i for all i, j, and with $Q = PP^\top$.

Or, in a dehomogenized version: Let $q(t, x) = q(t, x_1, \dots, x_m)$ be a polynomial in $\mathbb{R}[t, x]$ that has degree ≤ 2 in the x-variables. If q is psd (on $\mathbb{R}^{m+1}$) then q is a sum of squares in $\mathbb{R}[t, x]$, and in fact, of $m + 2$ squares. The qualitative part is a classical result (Jakubović [102], Rosenblum–Rovnyak [171]), while the quantitative part is more recent (Leep 2006, unpublished). The traditional proofs use techniques from quadratic forms theory and are by no way easy.

7.2.16 Remark. We complement the discussion with some further remarks. Consider irreducible $\mathbb{R}$-varieties $X \subseteq \mathbb{P}^n$ with Zariski dense $\mathbb{R}$-points. Theorem 7.2.5 identifies those varieties X on which every psd quadratic form is a sum of squares of forms. In fact, as already mentioned, the theorem implies the answer for all (even) degrees. Indeed, let $V = v_d(X)$ be the image of X under the degree d Veronese map $v_d\colon \mathbb{P}^n \to \mathbb{P}^N$, and let $\mathbb{P}^r \subseteq \mathbb{P}^N$ be the linear subspace spanned by V. Then $\mathbb{R}[V]_i = \mathbb{R}[X]_{di}$ holds for any $i \ge 0$, and in particular, for $i \le 2$. Therefore $P_{X,2d} = \Sigma_{X,2d}$ is equivalent to $P_{V,2} = \Sigma_{V,2}$. By Theorem 7.2.5, this holds if and only if $V = v_d(X)$ is a variety of minimal degree in $\mathbb{P}^r$.

It can be shown ([24] Remark 4.6) that, except for $X = \mathbb{P}^2$ and $d = 2$ (the Hilbert case of ternary quartics), this can happen with $d \ge 2$ only when $v_d(X)$ is a rational normal curve. There do indeed exist examples where this case occurs, see Exercises 7.2.1 and 7.2.2.

7.2.17 Remark. Let $X \subseteq \mathbb{P}^n$ be a non-degenerate irreducible projective $\mathbb{R}$-variety of minimal degree with Zariski dense $\mathbb{R}$-points, and let $m = \dim(X)$. According to Theorem 7.2.7, every psd quadratic form p on X is a sum of $m + 1$ squares of linear forms on X. As in Section 2.1, every such sos representation has a Gram matrix, which is a symmetric matrix over $\mathbb{R}$ whose rows and columns are indexed by a basis of $\mathbb{R}[X]_1$. Similar as in 2.1.12, we consider two such representations as (orthogonally) equivalent if they have the same Gram matrix. When the form p is sufficiently general, one can show that the number of inequivalent representations of p as a sum of $m + 1$ squares is finite, and is independent of p:

(1) When $X = v_d(\mathbb{P}^1)$ is a rational normal curve, p corresponds to a binary form $f(t_1, t_2)$ of degree $2d$. Clearly, such f is a sum $f = g_1^2 + g_2^2 = (g_1 + ig_2)(g_1 - ig_2)$ of two squares. If f is positive definite with simple complex roots, it is elementary to see that there exist precisely 2^{d-1} inequivalent such representations.
(2) When $X = v_2(\mathbb{P}^2)$ is the Veronese surface in $\mathbb{P}^5$, p corresponds to a psd ternary quartic $f(t_1, t_2, t_3)$, which is a sum $f = q_1^2 + q_2^2 + q_3^2$ of three squares of quadratic forms. If the plane curve $f = 0$ is non-singular, it is known [157] that there exist precisely eight inequivalent such representations.

These examples, together with others, motivated the authors of [23] to conjecture for arbitrary X (irreducible of minimal degree with dense $\mathbb{R}$-points): If $f \in P_{X,2}$ is sufficiently general, there are exactly $2^{n-m} = 2^{\operatorname{codim}(X)}$ inequivalent representations of X as a sum of $m + 1$ squares. This conjecture was later proved by Hanselka–Sinn [82].

7.2.18 Remark. An irreducible non-degenerate variety X in $\mathbb{P}^n$ has *almost minimal degree* if $\deg(X) = 2 + \operatorname{codim}(X)$, that is, if the degree of X is one more than the minimal possible degree. It can be shown that such a variety X satisfies $\varepsilon(X) = 1$ if it is arithmetically Cohen–Macaulay. Under this condition, and if $m = \dim(X)$ and $X(\mathbb{R})$ is Zariski dense, Chua et al [45] showed that every sum of squares in $\mathbb{R}[X]_2$ is a sum of $m + 2$ squares. Examples to which this result applies are the Veronese embeddings $v_3(\mathbb{P}^2) \subseteq \mathbb{P}^9$ (of degree 9) and $v_2(\mathbb{P}^3) \subseteq \mathbb{P}^9$ (of degree 8). Consequently, every sos sextic form in three variables is a sum of four squares of cubic forms, and every sos quartic form in four variables is a sum of five squares of quadratic forms. These particular cases had been proved before in a direct way [185].

7.2.19 Remark. We return to the question raised in Remark 2.4.15. The proper context for this question is the theory of toric varieties. According to [24] (Section 6) and [45] (Theorem 2.1), the following classification can be deduced from the main results of this section:

Let $P \subseteq \mathbb{R}^n$ be a lattice polytope with the property that every non-negative polynomial $f \in \mathbb{R}[x_1, \dots, x_n]$ with Newton polytope $\text{New}(f) \subseteq 2P$ is a sum of squares of polynomials. Assume further that P is normal, meaning that for every integer $k \geq 1$ and every lattice point $v \in kP$, there exist lattice points $v_1, \dots, v_k \in P$ with $v = v_1 + \cdots + v_k$. Then, up to a lattice automorphism of $\mathbb{Z}^n$ and a translation, P is contained in one of the following polytopes:

(1) The m-dimensional unit simplex $S_m = \text{conv}(0, e_1, \dots, e_m) \subseteq \mathbb{R}^m$;
(2) the Cayley polytope of m line segments $[0, d_i]$ (with $d_i \geq 0$), which is the convex hull of $\bigcup_{i=1}^m \{e_i\} \times [0, d_i]$ in $\mathbb{R}^{m+1}$;
(3) the scaled 2-simplex $2S_2 = \text{conv}(0, 2e_1, 2e_2) \subseteq \mathbb{R}^2$;
(4) the free sum $\text{conv}(Q \times \{0\} \cup \{0\} \times \Delta_{k-1}) \subseteq \mathbb{R}^m \times \mathbb{R}^k$, where $Q \subseteq \mathbb{R}^m$ is one of (1)–(3) and $\Delta_{k-1} = \text{conv}(e_1, \dots, e_k) \subseteq \mathbb{R}^k$.

Conversely, for each polytope P in this list it is true that every non-negative polynomial f with $\text{New}(f) \subseteq 2P$ is a sum of squares.

The projective toric varieties associated with (1)–(3) are $\mathbb{P}^m$, smooth rational normal scrolls and the Veronese surface in $\mathbb{P}^5$, respectively. The polytopes (4) correspond to cones over these smooth varieties.

Exercises

7.2.1 Let $X \subseteq \mathbb{P}^3$ be the image of the morphism $\mathbb{P}^1 \to \mathbb{P}^3$, $(s : t) \mapsto (s^4 : s^3t : st^3 : t^4)$.

(a) Show that X is not of minimal degree, and find a quadratic form on X that is psd but not a sum of squares of linear forms.
(b) Prove that $v_2(X)$ is contained in a hyperplane of $\mathbb{P}^9$ and is of minimal degree in $\mathbb{P}^8$. Conclude that $v_2(X) \cong v_8(\mathbb{P}^1)$ and that $P_4(X) = \Sigma_4(X)$.

7.2.2 Same (with different degrees) as Exercise 7.2.1, but for the image X of $\mathbb{P}^1 \to \mathbb{P}^4$, $(s : t) \mapsto (s^6 : s^5t : s^3t^3 : st^5 : t^6)$.

7.2.3 Let $X \subseteq \mathbb{P}^n$ be an arbitrary hypersurface over $\mathbb{R}$ of degree at least three. Show that $\Sigma_{X,2} \neq P_{X,2}$ in the homogeneous coordinate ring $\mathbb{R}[X]$ of X.

7.2.4 Let $X \subseteq \mathbb{P}^n$ be a union of two hyperplanes (over $\mathbb{R}$). Does $P_2(X) = \Sigma_2(X)$ hold?

7.3 Notes

The classification of varieties of minimal degree was achieved by del Pezzo (1886) for surfaces, and by Bertini (1907) in general. Theorem 7.2.5 was proved by Blekherman, Smith and Velasco in 2016 and is the main result of [24]. Our proof of the forward direction follows Proposition 3.2 in this paper. We took the converse from Blekherman, Plaumann, Sinn and Vinzant [23], who proved the quantitative refinement in Theorem 7.2.7. A weaker version of the S-Lemma (Proposition 7.2.10) was proved by Finsler [68] in 1936. The version that is known today is from 1971 and is due to Jakubović [103]. As in our presentation, his proof was based on Corollary 7.2.12, which was proved by Dines [54] in 1941. A detailed survey on the S-Lemma can be found in [154].

Chapter 8
Sums of Squares and Optimization

Since the turn of the millennium, the use of sums of squares techniques has become an indispensable tool in polynomial optimization. This final chapter contains an introduction to some of the most important concepts. We commence with an overview of important general notions related to convexity (Section 8.1). Then spectrahedra are introduced, together with their general properties (Sections 8.2 and 8.3). After a quick primer on conic programming, and in particular on semidefinite programming (Section 8.4), the key technique of moment relaxation is presented in Section 8.5. It will become evident that fundamental results from Chapter 5 are crucial for the iteration to converge. In particular, this holds for the positivstellensätze of Schmüdgen and Putinar. The latter part of the chapter addresses the characterization of spectrahedral shadows, which are the feasible sets of semidefinite programming. On the one hand we present results by Helton and Nie, establishing the existence of a semidefinite representation for convex sets of a very general nature. On the other, we show that prominent convex sets fail to be spectrahedral shadows. Once more, sums of squares lie at the core of the arguments.

The recent book by Netzer and Plaumann [144] on the geometry of linear matrix inequalities has a considerable overlap with this chapter.

8.1 Convex Sets: Basic Concepts and Facts

There exist several excellent and comprehensive introductions to the geometry of convex sets in general, and polyhedra in particular. We only mention the monographs by Grünbaum [77] and Ziegler [214] on polytopes, and by Webster [212] and Barvinok [11] on general convex geometry.

In the following let V be a vector space over $\mathbb{R}$ of finite dimension. Recall that $V^\vee$ denotes the dual space of linear functionals $V \to \mathbb{R}$.

8.1.1 For $x, y \in V$ we put $[x,y] = \{(1-t)x + ty \colon 0 \le t \le 1\}$, and similarly $[x,y[\; = [x,y] \smallsetminus \{y\}$, $]x,y] = [x,y] \smallsetminus \{y\}$ and $]x,y[\; = [x,y] \smallsetminus \{x,y\}$. For $n \ge 0$, the *n-dimensional standard simplex* is $\Delta_n = \{(t_0,\dots,t_n) \colon t_i \ge 0, \sum_{i=0}^n t_i = 1\} \subseteq \mathbb{R}^{n+1}$.

C. Scheiderer, *A Course in Real Algebraic Geometry*, Graduate Texts in Mathematics 303,
https://doi.org/10.1007/978-3-031-69213-0_8

A set $K \subseteq V$ is *convex* if $[x, y] \subseteq K$ for any $x, y \in K$. Arbitrary intersections of convex sets are convex. If $M \subseteq V$ is any subset, the elements $\sum_{i=0}^{n} t_i x_i$ with $n \geq 0$, $t = (t_0, \dots, t_n) \in \Delta_n$ and $x_0, \dots, x_n \in M$ are the *convex combinations* of M. The set of all convex combinations of M is $\mathrm{conv}(M)$, the *convex hull* of M. This is the smallest convex set in V that contains M. A *polytope* is the convex hull of finitely many points. A polytope is an *n-simplex* if it is the convex hull of $n + 1$ affinely independent points.

A *closed halfspace* of V is a set of the form $H = \{x \in V : \lambda(x) \geq c\}$ where $\lambda : V \to \mathbb{R}$ is a non-zero linear functional and $c \in \mathbb{R}$. An *open halfspace* is the complement of a closed halfspace. Any (affine) hyperplane in V defines two closed (and also two open) halfspaces of V. A *polyhedron* in V is an intersection of finitely many closed halfspaces of V.

The *affine hull* $\mathrm{aff}(M)$ of a set $M \subseteq V$ is the smallest affine-linear subspace of V that contains M. So $\mathrm{aff}(M)$ consists of all finite sums $\sum_i a_i x_i$ with $x_i \in M$, $a_i \in \mathbb{R}$ and $\sum_i a_i = 1$ (the *affine combinations* of elements of M). The *dimension* of a non-empty convex set $K \subseteq V$ is $\dim(K) := \dim \mathrm{aff}(K)$, the linear dimension of $\mathrm{span}(K - x)$ for (any) $x \in K$. The empty set has dimension -1. Convex sets of dimension zero are points, convex sets of dimension one are non-degenerate intervals on lines (including half-lines and full lines).

If K_1, K_2 are convex sets in V, their *Minkowski sum* $K_1 + K_2 = \{x_1 + x_2 : x_1 \in K_1, x_2 \in K_2\}$ is again a convex set.

8.1.2 Proposition. (Carathéodory) *Let* $\dim(V) = n < \infty$ *and let* $M \subseteq V$ *be any subset. Then every element of* $\mathrm{conv}(M)$ *is a convex combination of* $n + 1$ *elements in* M.

Proof. It suffices to show that any convex combination $v = \sum_{i=0}^{n+1} a_i v_i$ of $n+2$ points $v_0, \dots, v_{n+1} \in V$ is a convex combination of $n + 1$ of these points. We may assume $a_i > 0$ for all i. Since $\dim(V) = n$, there is a relation $\sum_{i=0}^{n+1} b_i v_i = 0$ such that $\sum_{i=0}^{n+1} b_i v_i = 0$ and $b_i > 0$ for at least one index i. Put $t = \min\{\frac{a_i}{b_i} : b_i > 0,\ i = 0, \dots, n+1\}$. Then $t > 0$ and $v = \sum_{i=0}^{n+1} (a_i - t b_i) v_i$. Moreover, the coefficients $a_i - t b_i$ are non-negative, they sum up to 1, and at least one of them is zero. □

Carathéodory's bound is sharp, as one sees from taking M to be any set of $n + 1$ affinely independent points. As an easy consequence of 8.1.2 we record (Exercise 8.1.1):

8.1.3 Corollary. *The convex hull of any compact set in* V *is compact.* □

8.1.4 Let $K \subseteq V$ be a convex set. The *relative interior* $\mathrm{relint}(K)$ of K is the topological interior of K relative to the affine hull $\mathrm{aff}(K)$ of K. Since the convex hull of any $n + 1$ affinely independent points has non-empty interior relative to their affine hull, the relative interior of any non-empty convex set is non-empty. If $x \in \mathrm{relint}(K)$ then $[x, y[\subseteq \mathrm{relint}(K)$ holds for any $y \in \overline{K}$. As a consequence, both $\mathrm{relint}(K)$ and $\overline{K}$ are convex sets, and $\overline{K}$ is the closure of $\mathrm{relint}(K)$. See Exercise 8.1.3 for the proofs.

A fundamental tool in convex geometry is separation of convex sets by hyperplanes. We discuss two basic versions that hold in finite dimension. For extensions to infinite dimension (Hahn–Banach theorem, Eidelheit–Kakutani theorem) we refer to Appendix B and to the literature.

8.1.5 Theorem. (Hyperplane separation) *Let K, K' be convex subsets of V.*

(a) *Assume that* $\operatorname{relint}(K) \cap \operatorname{relint}(K') = \emptyset$. *Then there exist* $0 \neq f \in V^\vee$ *and* $c \in \mathbb{R}$ *with* $f \leq c$ *on* K *and* $f \geq c$ *on* K'.
(b) *If K is compact, K' is closed and $K \cap K' = \emptyset$, there exist $f \in V^\vee$ and $c, c' \in \mathbb{R}$ with $c < c'$ such that $f \leq c$ on K and $f \geq c'$ on K'.*

Proof. We may assume $V = \mathbb{R}^n$ and $K, K' \neq \emptyset$, and we start by proving (b). Since K' is closed there exists, for any $u \in \mathbb{R}^n$, a unique point $v \in K'$ that is nearest to u, by an elementary geometric argument. If moreover $u \notin K'$, then

$$\langle v - u, u \rangle < \langle v - u, v \rangle \leq \langle v - u, y \rangle \tag{8.1}$$

holds for every $y \in K'$. Let $d_{K'}(x) = \inf_{y \in K'} |y - x|$ denote the distance function to K' (Remark 4.3.6.2). Since this function is continuous, it takes its minimum on the compact set K, say in $u \in K$. Let v be the point in K' that is nearest to u, and let $f(x) = \langle v - u, x \rangle$. Then (8.1) says $f(y) \geq f(v) > f(u)$ for $y \in K'$. On the other hand, u is the point in K that is nearest to v, so reversing the roles of u and v in (8.1) we get $f(x) \leq f(u)$ for all $x \in K$. Combining both inequalities gives (b).

To prove (a), one starts by showing that there exists a nested sequence $K_1 \subseteq K_2 \subseteq \cdots$ of compact convex subsets of K whose union is $\operatorname{relint}(K)$ (Exercise 8.1.7). Similarly, there is a nested sequence $K_1' \subseteq K_2' \subseteq \cdots$ of compact convex sets whose union is $\operatorname{relint}(K')$. By (b), there is a sequence $(v_i)_{i \geq 1}$ of unit vectors in $\mathbb{R}^n$, together with real numbers c_i, such that

$$\langle v_i, x \rangle < c_i < \langle v_i, y \rangle \tag{8.2}$$

for all $x_i \in K_i$, $y_i \in K_i'$ and $i \geq 1$. By compactness of the sphere there is a convergent subsequence of (v_i), so we may assume that $\lim_{i \to \infty} v_i = v$ exists. Put $c = \sup\{\langle v, x \rangle \colon x \in K\}$ and $c' = \inf\{\langle v, y \rangle \colon y \in K'\}$. Assuming $c > c'$ would mean that there exist $x \in K$ and $y \in K'$ with $\langle v, x \rangle > \langle v, y \rangle$, giving a contradiction to (8.2) for sufficiently large i. Therefore $c \leq c'$, which proves (a). □

8.1.6 Corollary. *For any set $M \subseteq V$, the closed convex hull of M is the intersection of all closed halfspaces that contain M.* □

A *supporting hyperplane* of a set $M \subseteq V$ is an affine hyperplane $H \subseteq V$ with $H \cap M \neq \emptyset$, such that M is contained in one of the two closed halfspaces defined by H.

8.1.7 Corollary. *Let $K \subseteq V$ be convex. For any point $x \in K \setminus \operatorname{relint}(K)$, there is a supporting hyperplane H of K with $x \in H$ and $K \not\subseteq H$.*

Proof. Replacing V by $\mathrm{aff}(K)$ we may assume that K has non-empty interior. Apply Theorem 8.1.5(b) to the convex sets $\{x\}$ and K, to find $0 \neq f \in V^\vee$ with $f(y) \geq f(x)$ for every $y \in K$. Clearly, this implies $f(y) > f(x)$ for $y \in \mathrm{int}(K)$, so $H = \{z \in \mathbb{R}^n : f(z) = f(x)\}$ is a hyperplane as required. □

8.1.8 Let $K \subseteq V$ be a closed convex set. A non-empty convex subset F of K is a *face* of K if $x, y \in K$ and $]x, y[\cap F \neq \emptyset$ imply $x, y \in F$. The faces of K that are different from K are called *proper*. Any non-empty intersection of faces of K is a face of K. For every supporting hyperplane H of K, the intersection $F = K \cap H$ is a face of K. Faces of K that arise in this way are said to be *exposed*. By definition, the improper face $F = K$ of K is also considered to be exposed. Since any face is a convex set, it is clear what is meant by the dimension of a face. Faces of dimension zero are called *extreme points*, and we use $\mathrm{Ex}(K)$ to denote the set of all extreme points of K. The set $\mathrm{Ex}(K)$ need not be closed, not even when K is compact. See Exercise 8.1.10 for an example. Faces of dimension one are usually called *edges*. Note that any face of a face of K is a face of K, but that an exposed face of an exposed face of K need not be an exposed face of K (Exercise 8.1.12). An important fact is:

8.1.9 Proposition. *If $K \subseteq V$ is closed and convex, then $F = \mathrm{aff}(F) \cap K$ holds for any face F of K.*

For the proof see Exercise 8.1.11. As an immediate consequence, note that every face of K is closed. Moreover:

8.1.10 Corollary. *For every proper face $F \neq K$ of K we have* $\dim(F) < \dim(K)$. □

8.1.11 Corollary. *Let $K \subseteq V$ be closed and convex. For every $x \in K$ there is a unique face F of K with $x \in \mathrm{relint}(F)$. This F is the smallest face of K that contains x. It is called the* supporting face *of x.*

Proof. If F is a face of K and $x \in \mathrm{relint}(F)$, it is immediate that F is the smallest face that contains x, i.e. that $F \subseteq F'$ for every other face F' of K with $x \in F'$. Given a point $x \in K \setminus \mathrm{relint}(K)$, it therefore remains to find a face F of K with $x \in \mathrm{relint}(F)$. By Corollary 8.1.7 there exists a supporting hyperplane $H \subseteq V$ of K with $x \in K$. Hence $K \cap H$ is a proper (exposed) face of K that contains x, and $\dim(K \cap H) < \dim(K)$. If $x \in \mathrm{relint}(K \cap H)$ we are finished, otherwise continue with $K \cap H$. After finitely many steps we have found a face F of K with $x \in \mathrm{relint}(F)$. □

8.1.12 Corollary. *If $K \subseteq V$ is closed and convex, the union of all proper faces $F \neq K$ of K is the relative boundary $K \setminus \mathrm{relint}(K)$ of K.* □

8.1.13 Theorem. *Let $K \subseteq V$ be a compact convex set. Then K is the convex hull of its extreme points:* $K = \mathrm{conv}(\mathrm{Ex}(K))$.

Proof. The convex set $K' = \mathrm{conv}(\mathrm{Ex}(K))$ is contained in K, and we use induction on $\dim(K)$ to show $K' = K$. Clearly we may assume $\mathrm{int}(K) \neq \varnothing$. Given $x \in \partial K$, there is a proper face F of K with $x \in F$ (8.1.12). Since F is compact convex with $\dim(F) < \dim(K)$, we have $F = \mathrm{conv}(\mathrm{Ex}(F))$ by the inductive hypothesis. This implies $F \subseteq K'$ since $\mathrm{Ex}(F) \subseteq \mathrm{Ex}(K)$, and so K' contains the boundary of K. But it is obvious that $\mathrm{conv}(\partial K) = K$, and so $K' = K$. □

Theorem 8.1.13 generalizes to compact convex sets K in arbitrary locally convex vector spaces, where however it becomes necessary to take the closure of the convex hull, so $K = \overline{\mathrm{conv}(\mathrm{Ex}(K))}$. This is the famous Krein–Milman theorem (Appendix B, Theorem B.9).

8.1.14 By a *convex cone* in V we mean a convex set $C \subseteq V$ with $0 \in C$ such that $tx \in C$ for every $x \in C$ and $t \geq 0$. In view of this property, the condition that C is convex may be replaced by $C + C \subseteq C$. We often simply speak of cones when we mean convex cones. The *conic hull* of a set $M \subseteq V$, denoted $\mathrm{cone}(M)$, is the smallest convex cone that contains M. It consists of all *conic combinations* $\sum_{i=1}^{n} a_i x_i$ (with $a_i \geq 0$, $n \geq 1$) of elements $x_i \in M$. If $\dim(V) = n < \infty$, Carathéodory's theorem for cones says that every element of $\mathrm{cone}(M)$ is a conic combination of n (instead of $n + 1$) elements of M. The affine hull of a convex cone is its linear hull, so $\mathrm{aff}(C) = \mathrm{span}(C) = C - C$. The cone C is *pointed* if $C \cap (-C) = \{0\}$. Note that the closure $\overline{C}$ of a convex cone C is again a convex cone. The relative interior of C is usually not a convex cone by our conventions, since $0 \notin \mathrm{relint}(C)$ unless C is a linear subspace of V. The Minkowski sum $C_1 + C_2$ of two convex cones C_1, C_2 is again a convex cone.

If the convex cone C is closed, every face of C is again a convex cone. The smallest face of C is $C \cap (-C)$, sometimes called the *support* of C. Obviously, extreme points of C are not of interest. If $x \in C$ is such that $x \neq 0$ and the half-line $\mathbb{R}_+ x$ is a face of C, then $\mathbb{R}_+ x$ is called an *extreme ray* of C. The cone analogue of Theorem 8.1.13 states that every closed and pointed convex cone $C \subseteq V$ is the Minkowski sum of its extreme rays. This can either be proved in a similar way as 8.1.13, or can be deduced from 8.1.13.

We continue to assume that the $\mathbb{R}$-vector space V has finite dimension. The next proposition, together with the subsequent lemma, is often useful:

8.1.15 Proposition. *Let $C \subseteq V$ be a closed convex cone and let $f\colon V \to W$ be a linear map of finite-dimensional vector spaces. If $C \cap \ker(f) = \{0\}$, the image cone $f(C)$ is closed in W.*

Equivalently, we have to show that the convex cone $C + \ker(f)$ is closed in V. This follows from the next lemma:

8.1.16 Lemma. *Let $C_1, \ldots, C_r \subseteq V$ be closed convex cones in V, and assume that $x_1 + \cdots + x_r = 0$ with $x_i \in C_i$ $(i = 1, \ldots, r)$ implies $x_1 = \cdots = x_r = 0$. Then the convex cone $C_1 + \cdots + C_r$ is closed as well.*

Proof. We may assume $V = \mathbb{R}^n$. It suffices to prove the case $r = 2$. Assuming that $C_1 + C_2$ fails to be closed, we show that $C_1 \cap (-C_2) \neq \{0\}$. By assumption there are sequences $(x_\nu)_{\nu \geq 1}$ in C_1 and $(y_\nu)_{\nu \geq 1}$ in C_2, as well as a point $z \in \mathbb{R}^n$, such that $x_\nu + y_\nu \to z$ and $z \notin C_1 + C_2$. Both sequences (x_ν) and (y_ν) are unbounded since otherwise, after passing to a subsequence, both sequences would converge, giving $z \in C_1 + C_2$, a contradiction. Again after passing to a suitable subsequence, we have $x_\nu/|x_\nu| \to x$ where $|x| = 1$, and $x \in C_1$ since C_1 is closed. Moreover $x_\nu + y_\nu \to z$ implies that $\frac{x_\nu}{|x_\nu|} + \frac{y_\nu}{|x_\nu|}$ converges to 0. Therefore $\frac{y_\nu}{|x_\nu|} \to -x$, and hence $-x \in C_2$. □

8.1.17 Proposition. *Every finitely generated convex cone is closed.*

Proof. Let $C = \operatorname{cone}(v_1, \dots, v_r) \subseteq V$, we argue by induction on r. When $r = 1$ it is obvious that C is closed. Assuming that the assertion has been proved for cones generated by r elements, let $v \in V$ and $C' = C + \mathbb{R}_+ v$. If $-v \notin C$ then C' is closed by Lemma 8.1.16. So assume that $-v \in C$, and so $C' = C + \mathbb{R}v$. Let $\pi\colon V \to V/\mathbb{R}v$ denote the quotient map. The cone $\pi(C') = \pi(C)$ in $V/\mathbb{R}v$ is generated by r elements, so it is closed by the inductive hypothesis. This means that C' is closed in V. □

To a large extent, the study of convex sets K in $\mathbb{R}^n$ is equivalent to the study of convex cones C in $\mathbb{R}^{n+1}$. The step from K to C is formalized by the concept of homogenization of convex sets, that we now introduce.

8.1.18 Let $K \subseteq V$ be a non-empty convex set. The *recession cone* of K, denoted $\operatorname{rc}(K)$, is the set of all $u \in V$ for which $K + \mathbb{R}_+ u \subseteq K$. This is a convex cone, which is closed if K is closed (Exercise 8.1.15). Let $K^c = \operatorname{cone}(\{1\} \times K) = \{(t, tx) \colon t \geq 0, x \in K\}$, a convex cone in $\mathbb{R} \times V$. We define the *homogenization* of K to be the convex cone

$$K^h := K^c + (\{0\} \times \operatorname{rc}(K)) = K^c \cup (\{0\} \times \operatorname{rc}(K))$$

in $\mathbb{R} \times V$ (check the right-hand equality!). When $K = \emptyset$ we put $K^c = K^h = \{(0, 0)\} \subseteq \mathbb{R} \times V$. It is an easy exercise to show $\dim(K^h) = 1 + \dim(K)$ for any convex set K (Exercise 8.1.16). Homogenization of convex sets commutes with taking closures:

8.1.19 Proposition. *Let $K \subseteq V$ be a convex set. Then K^c is dense in $(\overline{K})^h$, and therefore in K^h as well. Moreover we have $(\overline{K})^h = \overline{K^h}$. In particular, K^h is closed if K is closed.*

Proof. We start by proving the last claim, so let K be closed. Since $\operatorname{rc}(K)$ is closed (Exercise 8.1.15), it suffices to consider a convergent sequence $t_\nu(1, x_\nu) \to (t, u)$ in $\mathbb{R} \times V$ with $t_\nu > 0$ and $x_\nu \in K$ for all ν. Either $t = \lim_{\nu \to \infty} t_\nu > 0$, then $\frac{u}{t} = \lim_{\nu \to \infty} \frac{t_\nu x_\nu}{t_\nu} = \lim_{\nu \to \infty} x_\nu$, so $\frac{u}{t} \in K$ since K is closed, and hence $(u, t) \in K^c \subseteq K^h$. Otherwise $t = 0$, and then $u \in \operatorname{rc}(K)$ holds by Exercise 8.1.15(iii).

Obviously, K^c is dense in $(\overline{K})^c$. To prove that K^c is dense in $(\overline{K})^h$, it therefore suffices to show that $\{0\} \times \operatorname{rc}(\overline{K})$ is contained in the closure of K^c. Let $u \in \operatorname{rc}(\overline{K})$. By Exercise 8.1.15 there exist sequences $(x_\nu)_\nu$ in $\overline{K}$ and $t_\nu \to 0$ in $\mathbb{R}_+$, such that $t_\nu x_\nu \to u$. For every index ν choose $x'_\nu \in K$ with $|x'_\nu - x_\nu| < 1$. Then $t_\nu x'_\nu \to u$ as well, so we may assume that $x_\nu \in K$ for all ν. Now the sequence $t_\nu(1, x_\nu)$ lies in K^c and converges against $(0, u)$.

Since $K^c \subseteq K^h$, the density assertion just proven implies $(\overline{K})^h \subseteq \overline{K^h}$. The reverse inclusion is obvious since $(\overline{K})^h$ is closed. □

8.1.20 Corollary. *The homogenization operator $K \mapsto K^h$ induces a bijective correspondence between the closed convex sets $K \neq \varnothing$ in V and the closed convex cones C in $\mathbb{R}_+ \times V$ with $C \not\subseteq \{0\} \times V$.*

The inverse map sends C to its "dehomogenization" $C^d = \{x \in V\colon (1, x) \in C\}$. Note how the correspondence is formally analogous to the correspondence in algebraic geometry between closed subvarieties of affine and projective space.

Proof. It only remains to show that $(C^d)^h = C$, when C is a closed convex cone in $\mathbb{R}_+ \times V$ with $C \not\subseteq \{0\} \times V$. This means to show, for any $u \in V$, that $u \in \mathrm{rc}(C^d)$ if and only if $(0, u) \in C$. The "if" direction being obvious, let $u \in \mathrm{rc}(C^d)$ and choose $x \in C^d$. Then $(1, x + tu) \in C$ for all $t > 0$. Dividing by t and letting $t \to \infty$ shows $(0, u) \in C$ since C is closed. □

8.1.21 We need to discuss the concepts of duality and polarity for convex sets. Let $M \subseteq V$ be a set. The convex cone that is *dual* to M is

$$M^* = \{f \in V^\vee\colon f(x) \geq 0 \text{ for all } x \in M\}.$$

This is a closed convex cone in $V^\vee$, the dual linear space of V. By canonically identifying $V^{\vee\vee} = (V^\vee)^\vee$ with V, the bi-dual convex cone $M^{**} := (M^*)^*$ is a closed convex cone in V that contains M. From hyperplane separation 8.1.5, it follows immediately that $M^{**} = \overline{\mathrm{cone}(M)}$, the closure of the conic hull of M. In other words, the closed conic hull of M is the intersection of all closed linear halfspaces of V that contain M. In particular, $C^{**} = C$ holds for every closed convex cone C.

We give several examples that illustrate the importance of dual cones.

8.1.22 Example. Let linear functions $f_1, \ldots, f_r \in V^\vee$ on V be given. The dual of the cone $C = \{v \in V\colon f_i(v) \geq 0\}$ in V is $C^* = \mathrm{cone}(f_1, \ldots, f_r)$, the convex cone generated by $f_1, \ldots, f_r$ in $V^\vee$. Indeed, the convex cone $D := \mathrm{cone}(f_1, \ldots, f_r) \subseteq V^\vee$ satisfies $D^* = C$, by the definition of C. Since D is closed by Proposition 8.1.17, we get $D = C^*$ by cone duality.

In a different but equivalent formulation, this statement is known as *Farkas' Lemma*, see Exercise 8.1.25.

8.1.23 Example. For $M \subseteq \mathbb{R}^n$ a non-empty set, let $P_M = \{f \in \mathbb{R}[x]_{\leq 1}\colon f|_M \geq 0\}$ denote the cone of linear polynomials that are non-negative on M. Identifying a linear polynomial $a_0 + \sum_{i=1}^n a_i x_i$ with its vector of coefficients $(a_0, \ldots, a_n)$, we see that P_M is the dual cone of $M^c = \mathrm{cone}(1 \times M) \subseteq \mathbb{R} \times \mathbb{R}^n$. Let $K = \mathrm{conv}(M)$, then the cone $M^c = K^c$ is dense in the homogenization K^h of K (Proposition 8.1.19). Therefore both cones have the same dual cone. In other words, the closed convex cones $P_M = P_K$ and $(K^h)^*$ are naturally (linearly) isomorphic. This also gives a second characterization of K^h, in the case when $K \subseteq \mathbb{R}^n$ is non-empty, closed and convex:

$$K^h = \bigcap_{f \in P_K} \{u = (u_0, \ldots, u_n) \in \mathbb{R} \times \mathbb{R}^n \colon f^h(u) \geq 0\}.$$

Here f^h denotes the homogenization of the linear polynomial f, as in Definition 2.4.3.

8.1.24 Corollary. *Let $f, f_1, \ldots, f_r \in \mathbb{R}[x] = \mathbb{R}[x_1, \ldots, x_n]$ be linear polynomials such that the polyhedron $K := \{u \in \mathbb{R}^n \colon f_1(u) \geq 0, \ldots, f_r(u) \geq 0\}$ is non-empty. If $f \geq 0$ on K, there exist $a_0, \ldots, a_r \geq 0$ in $\mathbb{R}$ with $f = a_0 + a_1 f_1 + \cdots + a_r f_r$.*

Proof. By Example 8.1.23, the homogenization of K is given by

$$K^h = \{u = (u_0, \ldots, u_n) \in \mathbb{R}^{n+1} \colon u_0 \geq 0,\ f_i^h(u) \geq 0 \text{ for } i = 1, \ldots, r\}.$$

Now apply Example 8.1.22 to K^h, and dehomogenize again. □

8.1.25 Example. For another natural occurrence of dual cones fix integers n, $d \geq 0$. Recall that $P_{n,2d}$ is the convex cone of all positive semidefinite (psd) forms of degree $2d$ in $\mathbb{R}[x] = \mathbb{R}[x_1, \ldots, x_n]$. Let $V(n, 2d)$ denote the image set of the affine Veronese map

$$\psi_{n,2d} \colon \mathbb{R}^n \to \mathbb{R}^N, \quad u = (u_1, \ldots, u_n) \mapsto (u^\alpha)_{|\alpha|=2d} \tag{8.3}$$

where $N = \binom{n+2d-1}{n-1}$ is the number of monomials of degree $2d$ in x. An element in the dual cone of the conic hull of $V(n, 2d)$ is a tuple $(c_\alpha)_{|\alpha|=2d}$ of real numbers such that $\sum_\alpha c_\alpha u^\alpha \geq 0$ for all $u \in \mathbb{R}^n$. In other words, $(\operatorname{cone} V(n, 2d))^* = P_{n,2d}$ naturally. By cone duality, therefore, $(P_{n,2d})^*$ is (linearly) isomorphic to the closure of the conic hull of $V(n, 2d)$. Since this conic hull is closed (Exercise 8.1.19), we have $(P_{n,2d})^* \cong \operatorname{cone} V(n, 2d)$, the convex cone generated by the real points on the affine Veronese variety.

There is a second remarkable interpretation of the dual cone $(P_{n,2d})^*$. Essentially we already saw it in the proof of Proposition 7.2.4. For $u = (u_1, \ldots, u_n) \in \mathbb{R}^n$ let $l_u = \sum_{i=1}^n u_i x_i$, and let $Q_{n,2d}$ denote the convex cone in $\mathbb{R}[x]_{2d}$ that is generated by all powers $(l_u)^{2d}$ with $u \in \mathbb{R}^n$. In other words, $Q_{n,2d}$ consists of all sums of $2d$-th powers of linear forms. It is easy to see (Exercise 8.1.19) that the convex cones $Q_{n,2d}$ and cone $V(n, 2d)$ are linearly isomorphic under the map $(l_u)^{2d} \mapsto (u^\alpha)_{|\alpha|=2d}$ $(u \in \mathbb{R}^n)$.

8.1.26 Proposition. *The convex cones $P_{n,2d}$ and $Q_{n,2d}$ are naturally duals of each other, via the apolarity pairing.*

Proof. The key property of the apolarity pairing (Proposition 7.2.2) says

$$\langle f, (l_u)^{2d} \rangle = f(u) \tag{8.4}$$

for every $f \in \mathbb{R}[x]_{2d}$ and every $u \in \mathbb{R}^n$. From (8.4) it follows immediately that f lies in $P_{n,2d}$ if and only if $\langle f, g \rangle \geq 0$ for every $g \in Q_{n,2d}$. In other words, $P_{n,2d}$ is linearly isomorphic to the dual cone of $Q_{n,2d}$. Since the convex cone $Q_{n,2d}$ is closed (Exercise 8.1.19), it follows from dualizing that $(P_{n,2d})^* \cong Q_{n,2d}$. □

The dual of the sums of squares cone $\Sigma_{n,2d}$ will be discussed in the next section (Example 8.2.6.8).

We briefly mention the concept of polar dual sets, which is the inhomogeneous version of cone duality:

8.1.27 Definition. The *polar dual* of a set $M \subseteq V$ is $M^o = \{f \in V^\vee : \forall x \in M\ f(x) \ge -1\}$. This is a closed convex subset of $V^\vee$, and $0 \in M^o$.

As usual, we identify $M^{oo} = (M^o)^o$ with a subset of $V^{\vee\vee} = V$. The bi-polar dual of a set is described as follows:

8.1.28 Proposition. *Let $M \subseteq V$ be a set, and let $K = \overline{\mathrm{conv}(M \cup \{0\})}$. Then $M^o = K^o$ and $M^{oo} = K$. In particular, $M^{oo} = M$ if and only if M is closed and convex and contains the origin.*

Proof. Equality $M^o = K^o$ is immediate, and $K \subseteq K^{oo}$ is trivial. Conversely let $v \in V$, $v \notin K$. By hyperplane separation 8.1.5 there exist $f \in V^\vee$ and $c \in \mathbb{R}$ with $f|_K \ge c$ and $f(v) < c$. If $c = 0$ we can assume $f(v) < -1$, by scaling f with a suitable positive factor. Otherwise $c < 0$ since $0 \in K$. Then we may scale f with $-\frac{1}{c} > 0$ and get $f|_K \ge -1$, $f(v) < -1$. In either case we have $f \in K^o$ and $v \notin K^{oo}$, which proves $K^{oo} \subseteq K$. □

8.1.29 Remarks.

1. The polar dual of a set $M \subseteq \mathbb{R}^n$ is $M^o = \{u \in \mathbb{R}^n : \langle u, x\rangle \ge -1 \text{ for all } x \in M\}$. Using the notation $P_M = \{p \in \mathbb{R}[x] : \deg(p) \le 1, p|_M \ge 0\}$ from 8.1.23, we see that M^o is naturally identified with $\{p \in P_M : p(0) = 1\}$.

2. In particular, when $M = K$ is convex and non-empty, the polar dual K^o is an affine section of the dual of the homogenization of K, namely $K^o = (K^h)^* \cap (1 \times \mathbb{R}^n)$, see 8.1.23. Together with the previous remark, this shows how the study of polar duals of sets can be reduced to the study of duals of convex cones.

3. See Exercises 8.1.17 and 8.1.18 for some more basic properties of polar dual sets.

Finally in this section, we mention the main theorem on polyhedra.

8.1.30 Let $P \subseteq V$ be a polyhedron, say $P = \bigcap_{i=1}^r \{x \in V : f_i(x) \ge 0\}$, where $f_i(x) = \lambda_i(x) - a_i$ with $0 \ne \lambda_i \in V^\vee$ and $a_i \in \mathbb{R}$ $(i = 1, \dots, r)$. For each i, the hypersurface $H_i = \{x \in V : f_i(x) = 0\}$ is either a supporting hyperplane of P, or else $H_i \cap P = \varnothing$. If $x \in P$, let $I = I(x) = \{i \in [r] : f_i(x) = 0\}$, the set of indices that are *active* at x. It is easy to see that the supporting face of x is $P_I := P \cap \bigcap_{i \in I} H_i$, and that $\mathrm{aff}(P_I) = \bigcap_{i \in I} H_i$ (Exercise 8.1.26). Hence every face of P has the form P_I for some subset $I \subseteq [r]$. In particular, a polyhedron has only finitely many faces, and they are all exposed.

Every compact polyhedron P is the convex hull of its set $\mathrm{Ex}(P)$ of extreme points (Theorem 8.1.13). Since $\mathrm{Ex}(P)$ is a finite set, compact polyhedra are polytopes. Conversely, every polytope is a polyhedron. This is part of the following general main theorem on polyhedra:

8.1.31 Theorem. (Minkowski, Weyl, Motzkin) *A subset $P \subseteq V$ is a polyhedron if, and only if, P is the Minkowski sum of a polytope K and a finitely generated convex cone C.*

In other words, a subset $K \subseteq V$ is a polyhedron if and only if

$$K = \operatorname{conv}(v_1, \ldots, v_r) + \operatorname{cone}(w_1, \ldots, w_r)$$

for suitable finitely many vectors v_i, $w_j \in V$. As a consequence of the theorem, note in particular that a convex cone is finitely generated if and only if it is the intersection of finitely many closed linear halfspaces. For this reason, finitely generated cones are often referred to as *polyhedral cones*.

An important feature of Theorem 8.1.31 is that its proof is constructive, by the Fourier–Motzkin elimination process. Theorem 8.1.31 won't be used in this book, and so we don't include its proof. It can be found in any textbook on polyhedra, such as [77] or [214].

8.1.32 Remark. Using Theorem 8.1.31, it is not hard to verify the following permanence properties of polyhedra (Exercise 8.1.27): The class of polyhedra is closed under forming finite intersections, finite Minkowski sums and affine-linear images or preimages. Dual cones and polar dual sets of polyhedra are again polyhedra. The closed convex hull of any finite union of polyhedra is again a polyhedron.

Exercises

Let V always be a finite-dimensional vector space over $\mathbb{R}$.

8.1.1 Show that the convex hull of any compact subset of $\mathbb{R}^n$ is compact. Give an example of a closed set in $\mathbb{R}^2$ whose convex hull fails to be closed.

8.1.2 Show that a closed and pointed convex cone contains no affine line.

8.1.3 Let $K \subseteq \mathbb{R}^n$ be a convex set.

(a) Given $x \in \operatorname{relint}(K)$ and $y \in \overline{K}$, show that $[x, y[\subseteq K$. Conclude that $[x, y[$ is contained in $\operatorname{relint}(K)$.

(b) Both $\operatorname{relint}(K)$ and $\overline{K}$ are convex.

8.1.4 Let $C \subseteq V$ be a convex cone, and let $C^* \subseteq V^\vee$ be the dual cone of C.

(a) Show that $\operatorname{supp}(C^*) = \operatorname{span}(C)^\perp$. In particular, $\dim(C) + \dim \operatorname{supp}(C^*) = \dim(V)$.

(b) Conclude that C^* is pointed if and only if C has non-empty interior, and that C^* has non-empty interior if and only if the closure of C is pointed.

8.1.5 Let $C \subseteq V$ be a closed convex cone, and let $f \neq 0$ be a boundary point of the dual cone $C^* \subseteq V^\vee$. Show that $f(u) = 0$ for some point $u \neq 0$ in C.

8.1.6 Let $S \subseteq \mathbb{R}^n$ be a non-empty set and let $P_S = \{f \in \mathbb{R}[x]_{\le 1} \colon \deg(f) \le 1, f|_S \ge 0\}$. Show that the closed convex hull of S is naturally identified with an affine-linear section of the dual cone P_S^*.

8.1.7 Let $K \subseteq \mathbb{R}^n$ be a convex set. Show that there exists a nested sequence $K_1 \subseteq K_2 \subseteq \cdots$ of compact convex sets whose union is $\operatorname{relint}(K)$.

8.1.8 If $K \subseteq V$ is compact and convex, show that every supporting hyperplane of K contains an extreme point of K.

8.1.9 A real matrix is *doubly stochastic* if all its entries are non-negative, and if all row and column sums are equal to one. Show that the doubly stochastic $n \times n$ matrices form a polytope B_n whose extreme points are the $n!$ permutation matrices (having exactly one entry 1 in every row and column and otherwise only zeros). B_n is called the *Birkhoff polytope*.

Hint: Let A be the affine hull of B_n. Show that $\dim(A) = (n-1)^2$, and that B_n consists of all matrices in A with non-negative entries. Argue by induction on n and use Exercise 8.1.26, to conclude that every extreme point of B_n is a permutation matrix.

8.1.10 In $\mathbb{R}^3$ let $S = \{(a, b, 0) : a^2 + b^2 = 1\}$ and $u = (1, 0, 1)$, $v = (1, 0, -1)$. Show that the convex hull of the set $S \cup \{u, v\}$ is compact, but that the set of its extreme points is not closed. On the other hand, show for every closed convex set K in $\mathbb{R}^2$ that $\text{Ex}(K)$ is closed.

8.1.11 Let $K \subseteq V$ be a closed convex set and let F be a face of K.

(a) If a convex combination $\sum_i a_i x_i$ of points $x_i \in K$ lies in F, and if $a_i > 0$ for all i, show that $x_i \in F$ for all i.
(b) $F = \text{aff}(F) \cap K$. In particular, every face of K is closed.

Hint for (b): To prove that $x \in \text{aff}(F) \cap K$ lies in F, write x as an affine combination of points of F and separate positive and negative coefficients.

8.1.12 Let $K \subseteq \mathbb{R}^n$ be a closed convex set.

(a) Show that any non-empty intersection of exposed faces of K is an exposed face of K.
(b) Show that every maximal proper face of K is exposed.
(c) Give an example of a closed convex set $K \subseteq \mathbb{R}^2$ and of faces $F' \subseteq F$ of K, such that F is an exposed face of K and F' is an exposed face of F, but F' is not exposed as a face of K.

8.1.13 Let K be a non-empty and closed convex subset of V. Show that K has an extreme point if and only if K contains no line. (To prove "$\Leftarrow$", consider a non-empty face of minimal dimension.)

8.1.14 Give an example of a closed convex cone $C \subseteq \mathbb{R}^3$ and a linear map $f : \mathbb{R}^3 \to \mathbb{R}^2$ such that the cone $f(C)$ in $\mathbb{R}^2$ is not closed.

8.1.15 Let $K \subseteq V$ be a closed convex set, $K \neq \emptyset$. For $u \in V$, show that any of the following conditions is equivalent to $u \in \text{rc}(K)$:

(i) $u + K \subseteq K$;
(ii) $x + \mathbb{R}_+ u \subseteq K$ for some $x \in K$;
(iii) there are a sequence (x_n) in K and a sequence (t_n) in $\mathbb{R}_+$ such that $t_n \to 0$ and $t_n x_n \to u$.

Conclude that the recession cone $\text{rc}(K)$ is closed, and that $\text{rc}(K) \neq \{0\}$ if and only if K is unbounded.

8.1.16 For any convex set K, show that its homogenization K^h satisfies $\dim(K^h) = 1 + \dim(K)$.

8.1.17 Let $n \geq 1$, and let $B_r(0) = \{x \in \mathbb{R}^n : |x| \leq r\}$ be the closed ball of radius r around the origin.

(a) Show that $B_r(0)^o = B_{1/r}(0)$.
(b) Prove that $M = B_1(0)$ is the only set in $\mathbb{R}^n$ with $M^o = -M$.

8.1.18 Let $M \subseteq \mathbb{R}^n$ be a set. Show that the polar dual M^o is bounded if and only if 0 is an interior point of M^o. Conclude that the polar dual operator $K \mapsto K^o$ maps the set $\mathcal{K} := \{K \subseteq \mathbb{R}^n : K$ is compact, convex and $0 \in \text{int}(K)\}$ to itself, and that $K^{oo} = K$ for every $K \in \mathcal{K}$.

8.1.19 For $n, m \geq 1$ let $Q_{n,m} \subseteq \mathbb{R}[x_1, \ldots, x_n]_m = \mathbb{R}[x]_m$ be the convex cone of all (finite) sums of m-th powers of linear forms in $\mathbb{R}[x]$. Let $V(n, m)$ denote the image set of the affine Veronese map $\mathbb{R}^n \to \mathbb{R}^N$, $u \mapsto (u^\alpha)_{|\alpha|=m}$.

(a) Show that $Q_{n,m}$ and $\text{cone}(V_{n,m})$ are linearly isomorphic.

(b) Show that $Q_{n,m} - Q_{n,m} = \mathbb{R}[x]_m$. Conclude that $Q_{n,m}$ has non-empty interior in $\mathbb{R}[x]_m$, and that $Q_{n,m} = \mathbb{R}[x]_m$ if m is odd. (*Hint*: Exercise 5.1.5)
(c) Show that the cone $Q_{n,m}$ is closed in $\mathbb{R}[x]_m$.
(d) If $m \geq 4$ is even and $n \geq 2$, show that the inclusion $Q_{n,m} \subseteq \Sigma_{n,m}$ is proper.

8.1.20 Let n, d be positive integers, let $x = (x_1, \dots, x_n)$, and let M be a closed semialgebraic subset of $\mathbb{P}^{n-1}(\mathbb{R})$. Let $P_{M,2d}$ denote the convex cone of all forms $p \in \mathbb{R}[x]_{2d}$ that are non-negative on M. Prove that the dual cone of $P_{M,2d}$ is the cone in $\mathbb{R}[x]_{2d}$ that is generated by all powers l_u^{2d} of linear forms for which $u \in \mathbb{R}^n$ satisfies $[u] \in M$.

8.1.21 Let $\mathbb{R}[x] = \mathbb{R}[x_1, \dots, x_n]$, let $d \geq 1$, and let $(f, g) \mapsto \langle f, g\rangle$ denote the apolarity pairing on $\mathbb{R}[x]_{2d}$. Given $f \in \mathbb{R}[x]_{2d}$, the symmetric bilinear form $\beta_f\colon (p, q) \mapsto \langle f, pq\rangle$ on $\mathbb{R}[x]_d$ is called the *catalecticant* of f. Let P_{2d}, Σ_{2d} and Q_{2d} denote the psd cone, the sos cone and the cone of sums of $2d$-th powers of linear forms in $\mathbb{R}[x]_{2d}$, respectively.

(a) Show that $f \in Q_{2d} \Rightarrow \beta_f \geq 0$, and that the converse is true if $\Sigma_{2d} = P_{2d}$. Similarly $f \in \mathrm{int}(Q_{2d}) \Rightarrow \beta_f > 0$ holds, with converse if $\Sigma_{2d} = P_{2d}$.
(b) Let $w(f)$ denote the smallest number $r \geq 0$ for which there exist $u_1, \dots, u_r \in \mathbb{R}^n$ with $f = \sum_{i=1}^r (l_{u_i})^{2d}$. Show for $f \in Q_{2d}$ that $w(f) \geq \mathrm{rk}(\beta_f)$, with equality if $\Sigma_{2d} = P_{2d}$.
(c) Let $f \in Q_{2d}$, and assume $\Sigma_{2d} = P_{2d}$. For $0 \neq u \in \mathbb{R}^n$, show that there is a real number $c > 0$ with $f - c(l_u)^{2d} \in Q_{2d}$ if, and only if, $p(u) = 0$ for every $p \in \ker(\beta_f)$. In this case, there exists $c > 0$ with $w(f - c(l_u)^{2d}) < w(f)$.
(d) Let $n = 2$, and let $f = \sum_{i=1}^r (l_{u_i})^{2d}$ with $0 \neq u_i \in \mathbb{R}^2$. If f lies in the boundary of Q_{2d}, conclude that the points $[u_1], \dots, [u_r]$ in $\mathbb{P}^1(\mathbb{R})$ are uniquely determined by f.

8.1.22 Decide for each of the following binary forms in $\mathbb{R}[x, y]$ if they are sums of powers of linear forms, and find such a representation in case it exists:

(a) $f = x^4 + 12x^2y^2 + 4y^4$,
(b) $f = 17x^4 - 88x^3y + 240x^2y^2 - 184xy^3 + 98y^4$,
(c) $f = 66x^6 - 180x^5y + 300x^4y^2 + 300x^2y^4 + 180xy^5 + 65y^6$.

8.1.23 Let n, $d \geq 1$, we are using the notation from the Exercise 8.1.21. For any psd form $g \in P_{2d}$, show that

$$F_g(Q_{2d}) = \Big\{\sum_{i=1}^{r} (l_{u_i})^{2d} \colon r \geq 1,\ u_1, \dots, u_r \in \mathcal{Z}(g)\Big\}$$

is an exposed face of the cone Q_{2d}, and that every exposed face of Q_{2d} is of this form. (Recall that $\mathcal{Z}(g)$ denotes the zero set of g in $\mathbb{R}^n$.) Conclude that a form $f \in \mathbb{R}[x]_{2d}$ lies in the boundary of Q_{2d} if, and only if, there exists a psd form $g \neq 0$ in P_{2d} such that $f = (l_{u_1})^{2d} + \cdots + (l_{u_r})^{2d}$ for suitable points $u_i \in \mathcal{Z}(g)$.

8.1.24 Let $P \subseteq \mathbb{R}^n$ be a non-empty polyhedron and let f be a linear polynomial that is bounded from below on P. Then f takes its minimum on P.

8.1.25 Prove *Farkas' Lemma* (cf. Example 8.1.22): Given a matrix $A \in \mathrm{M}_{n\times r}(\mathbb{R})$ and a vector $b \in \mathbb{R}^n$, exactly one of the following two statements holds:

(1) There is $u \in \mathbb{R}^r$ with $Au = b$ and $u \geq 0$;
(2) there is $v \in \mathbb{R}^n$ with $A^\top v \geq 0$ and $b^\top v < 0$.

(Here an inequality $x \geq y$ or $x > y$ between tuples is understood componentwise.)

8.1.26 Let $P = \bigcap_{i=1}^m \{\xi \in \mathbb{R}^n \colon f_i(\xi) \geq 0\}$ be a polyhedron, where $f_i \in \mathbb{R}[x]$ are non-constant linear polynomials. For $\xi \in P$ let $I(\xi) = \{i \in [m] \colon f_i(\xi) = 0\}$, corresponding to the set of inequalities that are active at ξ. Prove:

(a) The supporting face of ξ in P is $F := P \cap \bigcap_{i\in I(\xi)} \{u \in \mathbb{R}^n \colon f_i(u) = 0\}$;
(b) $\dim(F) \geq n - |I(\xi)|$.

8.1.27 Using the Minkowski–Weyl–Motzkin theorem 8.1.31, show that the class of polyhedra is closed under forming finite intersections or finite Minkowski sums, and under forming the polar dual set or the dual convex cone. The closed convex hull of a union of two polyhedra is again a polyhedron.

8.2 Spectrahedra

Since real symmetric matrices will be present almost everywhere in this chapter, we introduce a convenient short notation: The space of symmetric $n \times n$ matrices over $\mathbb{R}$ will be denoted $\mathsf{S}^n := \mathrm{Sym}_n(\mathbb{R})$.

8.2.1 We start with some basic facts on the psd symmetric matrix cone. As is well-known from linear algebra, every symmetric matrix $A \in \mathsf{S}^n$ can be diagonalized over $\mathbb{R}$ and has pairwise orthogonal eigenspaces. In other words, there exists an orthogonal matrix $U \in O(n)$, together with real numbers $a_1, \dots, a_n \in \mathbb{R}$, such that $UAU^\top = \mathrm{diag}(a_1, \dots, a_n)$. Recall that A is called positive semidefinite (psd for short), denoted $A \succeq 0$, if $a_i \geq 0$ for all i. We are going to study the convex cone

$$\mathsf{S}^n_+ := \{A \in \mathsf{S}^n \colon A \succeq 0\}$$

of all psd symmetric matrices By either Remark 1.3.25.2 or Exercise 1.3.1, S^n_+ is a basic closed semialgebraic set whose interior consists of the matrices A that are (strictly) positive definite, denoted $A \succ 0$. Note also that $\mathsf{S}^n_+ \cap (-\mathsf{S}^n_+) = \{0\}$, so the cone S^n_+ is pointed.

An important tool is the *trace inner product*, defined on $\mathrm{M}_{m\times n}(\mathbb{R})$ by $\langle A, B\rangle := \mathrm{tr}(AB^\top) = \mathrm{tr}(A^\top B)$. This is a Euclidean inner product (positive definite symmetric bilinear form) which is invariant under the left and right action of the orthogonal group. Indeed, $\langle UAV^\top, UBV^\top\rangle = \langle A, B\rangle$ holds for $A, B \in \mathrm{M}_{m\times n}(\mathbb{R})$ and $U \in O(m)$, $V \in O(n)$. We consider the restriction of this inner product to the space S^n of symmetric $n \times n$ matrices. If $a_1, \dots, a_n \in \mathbb{R}$ are the eigenvalues of $A \in \mathsf{S}^n$, note that $\langle A, A\rangle = \sum_{i=1}^n a_i^2$. An important property of the psd matrix cone is its self-duality:

8.2.2 Proposition. *The cone S^n_+ in S^n is self-dual with respect to the trace inner product. That is, for $A \in \mathsf{S}^n$ one has*

$$A \in \mathsf{S}^n_+ \quad \Leftrightarrow \quad \langle A, B\rangle \geq 0 \textit{ for all } B \in \mathsf{S}^n_+.$$

Moreover, for $A, B \in \mathsf{S}^n_+$ one has $\langle A, B\rangle = 0 \Leftrightarrow AB = 0 \Leftrightarrow BA = 0$.

Proof. Given $A, B \in \mathsf{S}^n_+$, diagonalize the matrices to see that there exist symmetric matrices $\sqrt{A}$, $\sqrt{B}$ with $(\sqrt{A})^2 = A$ and $(\sqrt{B})^2 = B$. So the cyclic invariance of the trace gives

$$\langle A, B\rangle = \mathrm{tr}((\sqrt{A})^2(\sqrt{B})^2) = \mathrm{tr}(\sqrt{A}\cdot(\sqrt{B})^2\cdot\sqrt{A}) = \langle \sqrt{A}\sqrt{B}, \sqrt{A}\sqrt{B}\rangle \geq 0. \quad (8.5)$$

This proves $\mathsf{S}^n_+ \subseteq (\mathsf{S}^n_+)^*$. The opposite inclusion is immediate from diagonalizing a given matrix $A \in (\mathsf{S}^n_+)^*$. Moreover, if $A, B \ge 0$ satisfy $\langle A, B\rangle = 0$, then (8.5) implies $\sqrt{A}\sqrt{B} = 0$, hence also $AB = 0$. □

8.2.3 The faces of the cone S^n_+ are in natural bijective correspondence with the linear subspaces U of $\mathbb{R}^n$, as follows. Given U, let $F_U = \{A \in \mathsf{S}^n_+ : \operatorname{im}(A) \subseteq U\}$, where $\operatorname{im}(A)$ denotes the linear subspace of $\mathbb{R}^n$ spanned by the columns of A. It is easy to see that F_U is a face of S^n_+, and conversely, that every non-empty face of S^n_+ has the form F_U for a unique linear subspace $U \subseteq \mathbb{R}^n$. Moreover all faces of S^n_+ are exposed, see Exercise 8.2.2 for the proofs. Note that F_U is linearly isomorphic to S^d_+ where $d = \dim(U)$. The supporting face of a matrix $A \in \mathsf{S}^n_+$ is $F_{\operatorname{im}(A)}$. The extreme rays of S^n_+, i.e. the one-dimensional faces, are the rays spanned by psd rank one matrices, which are the rays $\mathbb{R}_+ vv^\top$ with $0 \ne v \in \mathbb{R}^n$.

8.2.4 Another important device from linear algebra is the Schur complement. Suppose we are given a symmetric matrix of size $(m+n) \times (m+n)$ that is written in block form

$$M = \begin{pmatrix} A & C \\ C^\top & B \end{pmatrix}$$

with $A \in \mathsf{S}^m$, $B \in \mathsf{S}^n$ and $C \in \mathsf{M}_{m\times n}(\mathbb{R})$. Assume that $\det(A) \ne 0$. Then the matrix

$$S = B - C^\top \cdot A^{-1} \cdot C$$

in S^n is called the *Schur complement* of M (with respect to A). The matrix M is positive semidefinite if and only if both A and S are positive semidefinite, and likewise for positive definite. The easy proofs can be found in any textbook on linear algebra.

While polyhedra (that do not contain a line) are exactly the affine-linear slices of positive orthants $\mathbb{R}^n_+$, up to linear isomorphism, we will now consider affine-linear slices of psd matrix cones S^n_+. Let V always be an $\mathbb{R}$-vector space of finite dimension.

8.2.5 Definition. A set $S \subseteq V$ is a *spectrahedron* if there are a linear map $f : V \to \mathsf{S}^d$ and a matrix $A \in \mathsf{S}^d$ (for some $d \ge 1$) such that

$$S = \{x \in V : A + f(x) \ge 0\}.$$

8.2.6 Examples. Here are first examples and properties of spectrahedra.

1. Spectrahedra in $\mathbb{R}^n$ are the sets that can be written in the form

$$S = \{x \in \mathbb{R}^n : A_0 + x_1A_1 + \cdots + x_nA_n \ge 0\} \tag{8.6}$$

where $A_0, \dots, A_n$ are real symmetric matrices of some common size $d \times d$. The expression

$$A_0 + x_1A_1 + \cdots + x_nA_n \ge 0 \tag{8.7}$$

is called a *linear matrix inequality*, frequently abbreviated *LMI*, in the variables $x_1, \dots, x_n$, and (8.6) is called an *LMI representation* of the spectrahedron S. By definition, therefore, the spectrahedron S in (8.6) is the solution set, often called the *feasible set*, of the LMI (8.7). The terminology comes from semidefinite optimization, which is the task of optimizing linear functions over solution sets of LMIs. We'll say more on semidefinite optimization in Section 8.4. Every affine-linear slice $L \cap \mathsf{S}^d_+$ of the psd matrix cone (with $L \subseteq \mathsf{S}^d$ an affine-linear subspace) is a spectrahedron. Conversely, a spectrahedron is linearly isomorphic to such a slice if and only if it doesn't contain a line (Exercise 8.2.1).

2. Every spectrahedron S is a closed convex set in its surrounding vector space. In fact, the set S is basic closed since the psd matrix cone has this property (8.2.1). From the definition it is clear that the class of spectrahedra is stable under taking linear preimages, translations, finite intersections and direct products. A spectrahedron that is at the same time a convex cone is often called a *spectrahedral cone.*

3. Spectrahedra generalize polyhedra: Every polyhedron is the feasible set of an LMI that consists of diagonal matrices.

4. Every closed ball in Euclidean n-space is a spectrahedron. In a homogeneous setting, the Lorentz cone

$$L_n = \left\{(x_0, \dots, x_n) \in \mathbb{R}^{n+1} : \sqrt{x_1^2 + \cdots + x_n^2} \le x_0\right\}$$

is a spectrahedral cone, since it can be described by the LMI

$$\begin{pmatrix} x_0 & x_1 & x_2 & \cdots & x_n \\ x_1 & x_0 & 0 & \cdots & 0 \\ x_2 & 0 & x_0 & \cdots & 0 \\ \vdots & \vdots & & \ddots & \vdots \\ x_n & 0 & 0 & \cdots & x_0 \end{pmatrix} \succeq 0,$$

as one sees immediately using a Schur complement argument.

5. A correlation matrix is a psd matrix $A = (a_{ij}) \in \mathsf{S}^n_+$ with all diagonal entries equal to one. The set $E_n \subseteq \mathsf{S}^n_+$ of all $n \times n$ correlation matrices is a compact spectrahedron in $\binom{n}{2}$-dimensional space. For $n = 3$, E_3 is called the *elliptope* and looks like an "inflated tetrahedron":

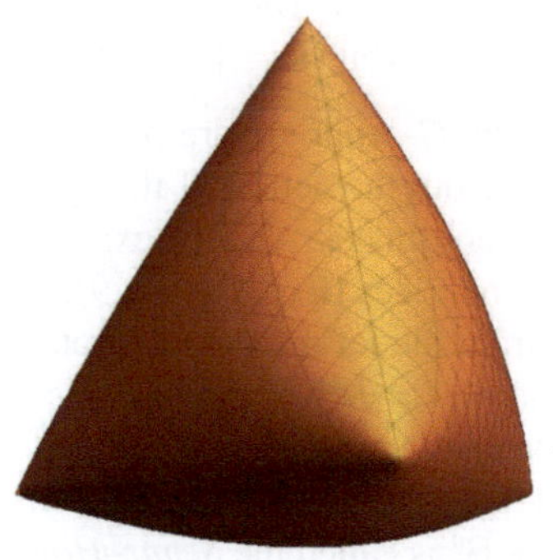

In fact, the vertices and the edges of the standard regular tetrahedron are faces of the elliptope E_3 (Exercise 8.2.7).

6. Let $x = (x_1, \dots, x_n)$ and let $d \geq 1$. Given a polynomial $f \in \mathbb{R}[x]$ of degree $2d$, recall (2.1.6) that a Gram matrix of f is a symmetric matrix of size $N = \binom{n+d-1}{d}$, with rows and columns indexed by the monomials x^α of degree $|\alpha| \leq d$, whose image under the linear Gram map $\gamma\colon \mathsf{S}^N \to \mathbb{R}[x]_{\leq 2d}$ is f. The set G_f^+ of all psd Gram matrices of f is therefore a spectrahedron, the *Gram spectrahedron* of f. Its elements correspond in a 1–1 fashion to the representations of f as a sum of squares, up to orthogonal equivalence (Corollary 2.1.13). By Exercise 2.1.9, the Gram spectrahedron of f is always compact.

7. For any choice of complex Hermitian $d \times d$ matrices $A_0, \dots, A_n$, the set $\{x \in \mathbb{R}^n \colon A_0 + \sum_{j=1}^n x_j A_j \geq 0\}$ is a spectrahedron that can be described by a ("real") linear matrix inequality (8.7) of size $2d \times 2d$. This follows from Exercise 8.2.5.

8. For positive integers $n, d \geq 1$ let $\Sigma_{n,2d}$ be the cone of all sums of squares forms of degree $2d$ in $\mathbb{R}[x_1, \dots, x_n]$. The dual cone $\Sigma_{n,2d}^*$ of $\Sigma_{n,2d}$ is a spectrahedral cone. Quite a bit more generally, let A be any $\mathbb{R}$-algebra and let $U \subseteq A$ be any linear subspace of finite dimension. Put $V = UU$ (the subspace spanned by all products of two elements from U), and let $\Sigma U^2 \subseteq V$ be the convex cone of all sums of squares of elements of U. Then the dual cone $(\Sigma U^2)^* \subseteq V^\vee$ is spectrahedral. Indeed, given a linear form $\lambda\colon V \to \mathbb{R}$, let $\beta_\lambda\colon U \times U \to \mathbb{R}$ denote the bilinear form defined by $\beta_\lambda(u, u') = \lambda(uu')$. Then, by definition, we have $\lambda \in (\Sigma U^2)^*$ if and only if $\beta_\lambda \geq 0$. Since the map $V^\vee \to S^2 U^\vee$, $\lambda \mapsto \beta_\lambda$ is linear, this shows the claim. For example, if $U = \operatorname{span}(1, t, \dots, t^d) \subseteq \mathbb{R}[t]$, the dual cone $(\Sigma U^2)^*$ consists of all psd real Hankel matrices

$$\begin{pmatrix} a_0 & a_1 & \cdots & a_d \\ a_1 & a_2 & \cdots & a_{d+1} \\ \vdots & \vdots & & \vdots \\ a_d & a_{d+1} & \cdots & a_{2d} \end{pmatrix} \geq 0.$$

For another prominent example of spectrahedra we need to elaborate a bit more. The following (up to and including the proof of 8.2.9) may be skipped, except for the first lemma (8.2.11) which will later be used again in another example.

8.2.7 Definition. Let V be a vector space over $\mathbb{R}$ with $\dim(V) < \infty$ and let $G \subseteq \mathrm{GL}(V)$ be a compact group of matrices. The *G-orbitope* of a vector $v \in V$ is the convex hull $\mathcal{O}(v) = \mathcal{O}_G(v)$ of the G-orbit $Gv = \{gv \colon g \in G\}$ in V.

8.2.8 When G is a finite group, G-orbitopes are polytopes. A prominent example are *permutahedra*. They arise from the action of the symmetric group $G = S_n$ on $V = \mathbb{R}^n$ by permutation of the coordinates. Another example is the Birkhoff polytope B_n of doubly stochastic $n \times n$ matrices (Exercise 8.1.9). Here the symmetric group $G = S_n$ acts on $V = \mathrm{M}_n(\mathbb{R})$ from the left by permutation of the rows, and B_n is the S_n-orbitope of the identity matrix $I = I_n$.

Let us consider the conjugation action $(S, A) \mapsto SAS^\top = SAS^{-1}$ of the orthogonal group $O(n) = \{S \in \mathrm{GL}_n(\mathbb{R}) \colon SS^\top = I\}$ on symmetric matrices $A \in \mathsf{S}^n$. The corresponding orbitope of $A \in \mathsf{S}^n$ is called the *Schur–Horn orbitope* of A.

8.2.9 Theorem. *Let $A \in \mathsf{S}^n$, and let $K_A = \text{conv}\{SAS^\top : S \in O(n)\}$ be the Schur–Horn orbitope of A. Then K_A is a spectrahedron.*

8.2.10 The proof requires several steps, some of which we'll only quote without proof. Given a symmetric matrix $A = (a_{ij}) \in \mathsf{S}^n$, let $\lambda_A = (\lambda_1, \dots, \lambda_n)$ with $\lambda_1 \geq \dots \geq \lambda_n$ be the tuple of eigenvalues of A in descending order. Moreover let $D(A) = (a_{11}, \dots, a_{nn}) \in \mathbb{R}^n$ denote the diagonal of A. For any vector $v \in \mathbb{R}^n$ let $\Pi(v)$ be the permutahedron of v, i.e. the convex hull in $\mathbb{R}^n$ of all the permutations of the coordinates of v.

8.2.11 Lemma. *For every $A \in \mathsf{S}^n$ and every $S \in O(n)$ we have $D(SAS^\top) \in \Pi(\lambda_A)$.*

Proof. We may assume $A = \text{diag}(\lambda)$ with $\lambda = (\lambda_1, \dots, \lambda_n)$. Let $S = (s_{ij})$ and $X = SAS^\top = (x_{ij})$, then

$$x_{ii} = \sum_{j,k=1}^{n} s_{ij}\delta_{jk}\lambda_j s_{ik} = \sum_{j=1}^{n} s_{ij}^2 \lambda_j.$$

So $D(X) = M\lambda$ where $M = (s_{ij}^2)$. The matrix M is doubly stochastic, so $M \in B_n$. Since B_n is a polytope whose extreme points are the permutation matrices (Exercise 8.1.9), it suffices to prove $M\lambda \in \Pi(\lambda)$ in the case where M is a permutation matrix. But then $M\lambda$ is a permutation of λ, and so the assertion is obvious. □

8.2.12 Theorem. (Schur, Horn) *Let $\lambda, w \in \mathbb{R}^n$ be given. There exists a symmetric matrix $X \in \mathsf{S}^n$ with eigenvalues $\lambda_1, \dots, \lambda_n$ and with diagonal $D(X) = w$ if, and only if, $w \in \Pi(\lambda)$.*

Proof. The easier direction "⇒" was just proved in Lemma 8.2.11, and is due to Schur (1923). The converse was proved by Horn in 1954, see [97]. We omit the proof. □

8.2.13 Another bit is needed for the proof of Theorem 8.2.9, namely the description of permutahedra by linear inequalities. Given $v, w \in \mathbb{R}^n$, let $\sigma, \tau \in S_n$ be permutations satisfying

$$v_{\sigma(1)} \geq \dots \geq v_{\sigma(n)} \quad \text{and} \quad w_{\tau(1)} \geq \dots \geq w_{\tau(n)}.$$

One says that w is *majorized* by v, denoted $w \trianglelefteq v$, if $\sum_{i=1}^n v_i = \sum_{i=1}^n w_i$ and

$$w_{\tau(1)} + \dots + w_{\tau(k)} \leq v_{\sigma(1)} + \dots + v_{\sigma(k)}$$

holds for $k = 1, \dots, n$. With this notation one has:

8.2.14 Proposition. $\Pi(v) = \{w \in \mathbb{R}^n : w \trianglelefteq v\}$ *for every* $v \in \mathbb{R}^n$.

Proposition 8.2.14 is due to Rado (1952). The proof is elementary but not obvious, and can be found in [11] (p. 257), for example.

Proof of Theorem 8.2.9. Let $A \in \mathsf{S}^n$ and let $\lambda_A = (\lambda_1, \dots, \lambda_n)$ be the ordered tuple of eigenvalues of A. By Theorem 8.2.12, a matrix $X \in \mathsf{S}^n$ lies in K_A if and only $D(X) \in \Pi(\lambda)$. By 8.2.14, this is equivalent to the condition that $\operatorname{tr}(X) = \operatorname{tr}(A)$, and that the sum of any k eigenvalues of X is at most $\lambda_1 + \cdots + \lambda_k$, for every $k = 1, \dots, n$. For k from 1 to n let $V_k := \Lambda^k(\mathbb{R}^n)$, the k-th exterior power of $\mathbb{R}^n$. If $v_1, \dots, v_n$ is any linear basis of $\mathbb{R}^n$, the wedge products $v_{i_1} \wedge \cdots \wedge v_{i_k}$ with $1 \le i_1 < \cdots < i_k \le n$ form a basis of V_k. In particular, $\dim(V_k) = \binom{n}{k}$. Any matrix $X \in \mathrm{M}_n(\mathbb{R})$ induces a linear endomorphism $L_k(X)$ of V_k via

$$L_k(X):\ v_1 \wedge \cdots \wedge v_k \ \mapsto\ \sum_{j=1}^{k} v_1 \wedge \cdots \wedge (Xv_j) \wedge \cdots \wedge v_k,$$

and the map $X \mapsto L_k(X)$ is clearly linear. Moreover, if X is symmetric then so is $L_k(X)$ (with respect to the standard basis of V_k consisting of the wedge products $e_{i_1} \wedge \cdots \wedge e_{i_k}$ of the canonical basis vectors). Therefore, if $\mu_1, \dots, \mu_n$ are the eigenvalues of X, the operator $L_k(X)$ has exactly the eigenvalues $\mu_{i_1} + \cdots + \mu_{i_k}$ where $1 \le i_1 < \cdots < i_k \le n$. In summary, the Schur–Horn orbitope of A can be described as

$$\begin{aligned} K_A \ &= \ \{X \in \mathsf{S}^n:\ \operatorname{tr}(X) = \operatorname{tr}(A)\} \\ &\cap \bigcap_{k=1}^{n-1} \{X \in \mathsf{S}^n:\ (\lambda_1 + \cdots + \lambda_k)\,\mathrm{id}_{V_k} - L_k(X) \succeq 0\}. \end{aligned}$$

This is clearly a spectrahedron, and it can be represented by a linear matrix inequality of size $2 + \sum_{k=1}^{n-1} \binom{n}{k} = 2^n$. □

8.2.15 Remarks. After this series of examples, we now start recording properties of spectrahedra in general.

1. Let $S \subseteq V$ be a non-empty spectrahedron, given as $S = \{x \in V\colon A + f(x) \succeq 0\}$ for some linear map $f\colon V \to \mathsf{S}^d$ and some matrix $A \in \mathsf{S}^d$. The recession cone of S (see 8.1.18) is $\operatorname{rc}(S) = f^{-1}(\mathsf{S}^d_+)$, since for $A \in \mathsf{S}^d_+$ and $B \in \mathsf{S}^d$ one has $A + tB \succeq 0$ for all $t \ge 0 \Leftrightarrow B \succeq 0$. Hence $\operatorname{rc}(S)$ is a spectrahedral cone. This also shows that every spectrahedral cone C in V has the form $C = \varphi^{-1}(\mathsf{S}^m_+)$ for some m and some linear map $\varphi\colon V \to \mathsf{S}^m$. In particular, spectrahedral cones in $\mathbb{R}^n$ can be represented by homogeneous LMIs.

2. Let $S \subseteq \mathbb{R}^n$ be a spectrahedron, given as the solution set of a linear matrix inequality $A_0 + \sum_{i=1}^n x_i A_i \succeq 0$ with $A_0, \dots, A_n \in \mathsf{S}^d$. Consider the extended symmetric (block) matrices

$$A_0' := \begin{pmatrix} 1 & 0 \\ 0 & A_0 \end{pmatrix}, \quad A_i' := \begin{pmatrix} 0 & 0 \\ 0 & A_i \end{pmatrix} \quad (i = 1, \dots, n)$$

in S^{d+1}, together with the spectrahedral cone $C = \{(x_0, x) \in \mathbb{R} \times \mathbb{R}^n\colon x_0 A_0' + \sum_{i=1}^n x_i A_i' \succeq 0\}$ in $\mathbb{R}^{n+1}$. If $S \neq \varnothing$, we claim that $C = S^h$, the homogenization of S. Indeed, the intersection of C with the hyperplane $x_0 = 0$ is the recession cone of

S, by the previous remark, while C intersected with $x_0 = 1$ is S. So $C = S^h$ follows from the description of S^h in 8.1.18. As a consequence, we see that a convex set $K \subseteq \mathbb{R}^n$ is a spectrahedron if and only if its homogenization $K^h \subseteq \mathbb{R}^{n+1}$ is a spectrahedral cone.

3. Neither linear images nor polar duals or dual cones of spectrahedra are usually spectrahedra any more. Examples will be given in the next section (see 8.3.3 and 8.3.14).

In 8.2.3 it was noted that all faces of the psd matrix cone are exposed. This fact carries over to arbitrary spectrahedra:

8.2.16 Proposition. *Let $g\colon V \to \mathsf{S}^d$ be an affine-linear map and let $S = g^{-1}(\mathsf{S}^d_+)$ be the corresponding spectrahedron in V. Every face of S has the form*

$$F_U(S) = \{x \in S : \operatorname{im} g(x) \subseteq U\}$$

where U is a linear subspace of $\mathbb{R}^d$. Every face of S is exposed.

For a given face F of S, there will usually be many different choices of a subspace $U \subseteq \mathbb{R}^d$ with $F_U(S) = F$. The assertions in Proposition 8.2.16 follow from the description of the faces of S^d_+ in 8.2.3, together with the following general observation:

8.2.17 Lemma. *Let $g\colon \mathbb{R}^n \to \mathbb{R}^m$ be an affine-linear map, let $K \subseteq \mathbb{R}^m$ be a closed convex set. The faces of $g^{-1}(K)$ are precisely the non-empty preimages $g^{-1}(F)$ where F is a face of K. If F is an exposed face of K, then $g^{-1}(F)$ is an exposed face of $g^{-1}(K)$.*

See Exercise 8.2.4 for the easy proof.

If $S = g^{-1}(\mathsf{S}^d_+)$ is a spectrahedron as in Proposition 8.2.16, and if $x \in V$ is a point for which the matrix $g(x)$ is positive definite, then x lies in the interior of S. This remark has the following converse (statement (b)):

8.2.18 Proposition. *Let $S \subseteq V$ be a spectrahedron, given as $S = g^{-1}(\mathsf{S}^d_+)$ where $g\colon V \to \mathsf{S}^d$ is an affine-linear map.*

(a) *If there exists $u \in V$ such that $g(u) > 0$, the interior and boundary of S are given as $\operatorname{int}(S) = \{x \in V\colon g(x) > 0\}$ and $\partial S = \{x \in S : \det g(x) = 0\}$, respectively.*
(b) *Conversely, if u is an interior point of S, there exists an affine-linear map $h\colon V \to \mathsf{S}^r$ with $r \le d$ such that $S = h^{-1}(\mathsf{S}^r_+)$ and $h(u)$ is the identity matrix I_r.*
(c) *If S is a spectrahedral cone, the map h in (b) may be chosen to be linear.*

Proof. We may assume $u = 0$ in (a) and (b).

(a) Since $u = 0$ we can write $g(x) = A + f(x)$ with $f\colon V \to \mathsf{S}^d$ linear and $A > 0$. Every $x \in V$ with $g(x) > 0$ lies in $\operatorname{int}(S)$, therefore $\partial S \subseteq \{x \in V\colon \det g(x) = 0\}$. On the other hand, let $x \in S$ with $\det g(x) = 0$, and let $0 \neq v \in \mathbb{R}^d$ be a vector with $g(x)v = 0$, which means $f(x)v = -Av$. For every $t > 0$ we have

$$\langle v, g((1+t)x)v\rangle = \langle v, g(x)v + tf(x)v\rangle = t\langle v, f(x)v\rangle = t\langle v, -Av\rangle < 0,$$

since $A > 0$ implies $\langle v, Av\rangle > 0$. So $g((1+t)x)$ has a negative eigenvalue for every $t > 0$, hence $cx \notin S$ for $c > 1$. This shows $x \in \partial S$.

(b) Let again $u = 0$, so we have $S = g^{-1}(\mathsf{S}^d_+)$ with $g(x) = A + f(x)$ and $A \ge 0$. After a linear base change in $\mathbb{R}^d$ we may assume $A = \left(\begin{smallmatrix} I_r & 0 \\ 0 & 0 \end{smallmatrix}\right)$ where $r = \operatorname{rk}(A)$. Write $f(x)$ as an $(r, d-r)$ block matrix

$$f(x) = \begin{pmatrix} f_1(x) & f_3(x) \\ f_3(x)^\top & f_2(x) \end{pmatrix}$$

where the matrices $f_1(x)$, $f_2(x)$, $f_3(x)$ depend linearly on x. So

$$g(x) = \begin{pmatrix} I + f_1(x) & f_3(x) \\ f_3(x)^\top & f_2(x) \end{pmatrix}.$$

By assumption, this matrix is psd for every $x \in V$ in a neighborhood of 0. In particular, $f_2(x) \ge 0$ for x near 0, which implies $f_2(x) \equiv 0$ since f_2 is linear. Therefore $f_3(x) \equiv 0$ as well, and we see that $S = h^{-1}(\mathsf{S}^r_+)$ for the map $h(x) = I + f_1(x)$.

(c) By Remark 8.2.15.1, we may assume that the map g is linear. Given an interior point u of S, we have $g(u) = \left(\begin{smallmatrix} I_r & 0 \\ 0 & 0 \end{smallmatrix}\right)$ after base change in $\mathbb{R}^d$, and the matrix $g(u)+g(x)$ is psd for $x \in V$ in a neighborhood of 0. As in (b), this implies $g(x) = \left(\begin{smallmatrix} g_1(x) & 0 \\ 0 & 0 \end{smallmatrix}\right)$, and so $S = g_1^{-1}(\mathsf{S}^r_+)$. □

A linear matrix inequality $A(x) = A_0 + \sum_{i=1}^n x_i A_i \ge 0$ is said to be *strictly feasible* if there exists $\xi \in \mathbb{R}^n$ with $A(\xi) > 0$.

8.2.19 Remark. Let $S \subseteq \mathbb{R}^n$ be a spectrahedron with non-empty interior, and let $X = \partial_a S \subseteq \mathbb{A}^n$ be its algebraic boundary. From Remark 4.6.14, recall that X is the Zariski closure of the topological boundary of S, and that X is a hypersurface with Zariski dense $\mathbb{R}$-points, except when $S = \mathbb{R}^n$. By 8.2.18(b), there exists a linear matrix polynomial $A(x) = A(x_1, \dots, x_n)$ whose feasible set is S, and such that $\det A(x)$ does not vanish identically. From 8.2.18(a) we see, for any such $A(x)$, that $X = \partial_a S$ is contained in the hypersurface $\det A(x) = 0$, and that $X(\mathbb{R}) \cap \operatorname{int}(S) = \varnothing$: The algebraic boundary of S does not meet the interior of S.

The most important property of spectrahedra is the following one, even though it is easy to prove:

8.2.20 Proposition. *Let $S \subseteq \mathbb{R}^n$ be a spectrahedron, let u be an interior point of S, and let $g \in \mathbb{R}[x]$ be a polynomial with $\partial_a S = \mathcal{V}(g)$. Then, for every $v \in \mathbb{R}^n$, the univariate polynomial $g(u - tv) \in \mathbb{R}[t]$ is real-rooted.*

Proof. By Proposition 8.2.18(b), S can be described by a linear matrix inequality $A(x) \ge 0$ satisfying $A(u) = I$. Since $\det A(x)$ vanishes identically on $\partial_a S$ (Remark 8.2.19), it suffices to show that every zero of the polynomial $\det A(u - tv)$ is real. Since $A(u + x) = I + L(x)$ with some homogeneous linear matrix polynomial $L(x)$, we have $\det A(u-tv) = \det(I - tL(v))$. This is the inverse characteristic polynomial of $L(v)$, its roots are therefore the inverses of the non-zero eigenvalues of the symmetric matrix $L(v)$. In particular, all roots of this polynomial are real. □

8.2.21 Definition. A set $K \subseteq \mathbb{R}^n$ is *rigidly convex* with respect to an interior point u of K if there exists a polynomial $p \in \mathbb{R}[x]$ with $p(u) \neq 0$ and with the following properties:

(1) K is the closure of the connected component of $\{\xi \in \mathbb{R}^n : p(\xi) \neq 0\}$ that contains u;
(2) for every $v \in \mathbb{R}^n$, the univariate polynomial $p(u - tv) \in \mathbb{R}[t]$ is real-rooted.

Observe that condition (2) requires that any real line through the point u meets the hypersurface $p = 0$ in real points only. It is not at all obvious from the definition, but every rigidly convex set is actually convex. See Remarks 8.2.26 below.

8.2.22 Corollary. *Every spectrahedron in $\mathbb{R}^n$ is rigidly convex with respect to any of its interior points.*

Proof. Let $S \subseteq \mathbb{R}^n$ be a spectrahedron and let u be an interior point of S. By Proposition 8.2.18 we can write $S = \{\xi \in \mathbb{R}^n : A(\xi) \succeq 0\}$ where $A(x)$ is a linear matrix polynomial that satisfies $A(u) = I$. Then the conditions in Definition 8.2.21 are satisfied for the polynomial $p(x) = \det A(x)$. Indeed, from the proof of Proposition 8.2.20 we see that $p(u - tv)$ is real-rooted for every $v \in \mathbb{R}^n$. And since the boundary of S satisfies $\partial S = \{\xi \in S : p(\xi) = 0\}$ by 8.2.18(a), the interior of S is a connected component of $\{\xi \in \mathbb{R}^n : p(\xi) \neq 0\}$. □

8.2.23 Remark. Let $K \subseteq \mathbb{R}^n$ be an arbitrary set that is rigidly convex with respect to a point $u \in \operatorname{int}(K)$, and let $p \in \mathbb{R}[x]$ be a polynomial as in 8.2.21. Clearly, K is connected and semialgebraic, and is the closure of its interior. Assuming $K \neq \mathbb{R}^n$, the algebraic boundary $\partial_a K$ of K is a hypersurface by Remark 4.6.14, say $\partial_a K = \mathcal{V}(g)$ with a polynomial $g \in \mathbb{R}[x]$. We may assume that g has no repeated factors. Clearly, p vanishes on the boundary of K, therefore p is a multiple of the polynomial g. In particular, property (2) in 8.2.21 implies that $g(u - tv)$ is real rooted for every $v \in \mathbb{R}^n$. Moreover, the connected component that contains u is the same for $\{\xi \in \mathbb{R}^n : p(\xi) \neq 0\}$ and for $\{\xi \in \mathbb{R}^n : g(\xi) \neq 0\}$. Together, this means that properties (1) and (2) remain true when the polynomial p is replaced by g. Therefore, condition (2) may be rephrased by saying that every (real) line through u intersects the algebraic boundary of K in $\mathbb{R}$-points only.

8.2.24 Examples.

1. The (real locus of the) plane affine curve $f(x, y) = 0$, with

$$f = 8x^4 + 2x^3y + 8y^4 + 12x^3 - 3xy^2 - 3x^2 - xy - 12y^2 - 5x + 1,$$

consists of two nested ovals. Since the curve $f = 0$ has degree four, the closed interior K of the inner oval is rigidly convex, with respect to any of its interior points (left picture):

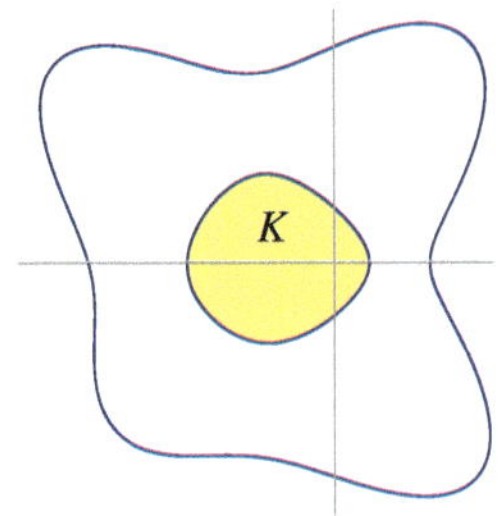
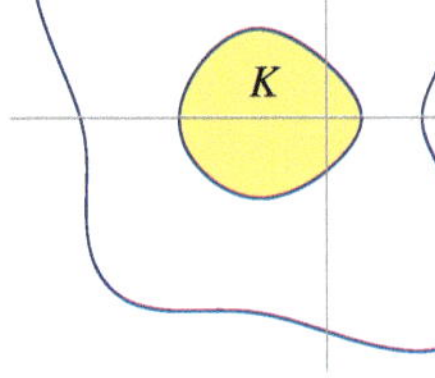

The curve $f(x, y) = 0$

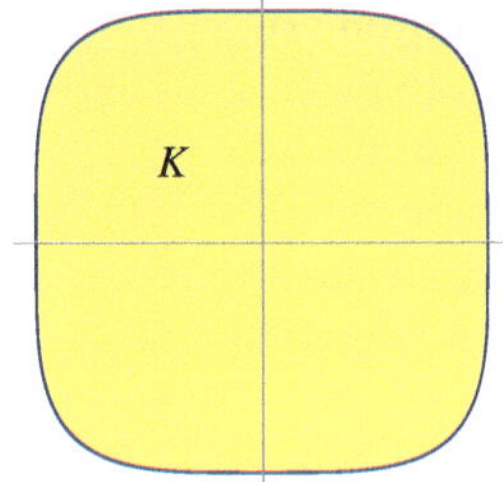

The tv-screen $x^4 + y^4 = 1$

2. Being rigidly convex is quite restrictive, as a property of convex sets. For example, consider the convex set $K = \{x \in \mathbb{R}^2 \colon x_1^{2d} + x_2^{2d} \le 1\}$ where $d \ge 1$. The algebraic boundary of K is the irreducible curve $x_1^{2d} + x_2^{2d} = 1$ of degree $2d$. Any line through the interior of K intersects this curve (transversally) in two $\mathbb{R}$-points only. Therefore, as soon as $2d \ge 4$, there exist non-real intersection points, and so K is not rigidly convex (and not a spectrahedron). For $2d = 4$, the set K has been dubbed the "tv-screen" in the literature (right picture).

8.2.25 Definition. A homogeneous polynomial $f \in \mathbb{R}[x_1, \dots, x_n]$ is *hyperbolic* with respect to $e \in \mathbb{R}^n$ if $f(e) \neq 0$, and if $f(te - v)$ is real-rooted for every $v \in \mathbb{R}^n$. The *(closed) hyperbolicity cone* of f (with respect to e) is

$$C_e(f) = \{v \in \mathbb{R}^n \colon f(te - v) \neq 0 \text{ for } t < 0\}.$$

8.2.26 Remarks.

1. Let $C \subseteq \mathbb{R}^n$ be a spectrahedral cone, represented by a homogeneous LMI $C = \{v \in \mathbb{R}^n \colon A(v) \succeq 0\}$, where $A(x) = \sum_{i=1}^n x_i A_i$ satisfies $A(e) \succ 0$ for some point $e \in \mathbb{R}^n$. Then the form $f(x) = \det A(x)$ is hyperbolic with respect to e, and C is the hyperbolicity cone $C_e(f)$ of f. Indeed, all roots t of $\det A(te - v)$ are non-negative if and only if $A(v) \succeq 0$, i.e. $v \in C$.

2. Hyperbolic forms are a fascinating object of current research. Unfortunately we do not have room in this course to go any deeper into the subject, but at least some remarks are in order. The notion of hyperbolic forms originates in the study of partial differential operators, and goes back to work of Petrovskii and Gårding, see [71]. Since the early 2000s, hyperbolic forms have moved into the focus of real algebraic geometry, of algebraic combinatorics (in particular, matroid theory) and of optimization. From the definition it is not obvious at all, but Gårding [70] proved that if the form f is hyperbolic with respect to $e \in \mathbb{R}^n$, the hyperbolicity cone $C_e(f)$ is a closed convex cone. Moreover, $C_e(f)$ is the closure of the connected component of $\{v \in \mathbb{R}^n \colon f(v) \neq 0\}$ that contains e.

3. Hyperbolic forms and their hyperbolicity cones are related to rigidly convex sets in the following way. Let the set $K \subseteq \mathbb{R}^n$ be rigidly convex with respect to $u \in \mathbb{R}^n$, and let $g \in \mathbb{R}[x]$ be the polynomial (unique up to scaling) without multiple factors whose zero set is the algebraic boundary $\partial_a K$ of K. Remark 8.2.23 shows

that the homogenization $f = g^h$ of g is a form in $\mathbb{R}[x_0, \ldots, x_n]$ that is hyperbolic with respect to $e = (1, u) = (1, u_1, \ldots, u_n) \in \mathbb{R}^{n+1}$. Moreover, K coincides with the "affine part" of the hyperbolicity cone $C_e(f)$, namely $K = \{v \in \mathbb{R}^n : (1, v) \in C_e(f)\}$. In other words, the rigidly convex sets in $\mathbb{R}^n$ are just the affine hyperplane sections of hyperbolicity cones in $\mathbb{R}^{n+1}$. In view of these facts it is usually more convenient to study rigidly convex sets in a homogeneous setting, via hyperbolic forms and their hyperbolicity cones.

4. Every hyperbolicity cone, and hence also every rigidly convex set, is a basic closed set (Exercise 8.2.15). This generalizes the corresponding property of spectrahedra (Example 8.2.6.2).

When $A(x) = A(x_1, \ldots, x_n)$ is a homogeneous linear (symmetric) matrix polynomial with $A(e) > 0$ for some $e \in \mathbb{R}^n$, we have seen that the form $f(x) = \det A(x)$ is hyperbolic with respect to e, and that $C_e(f)$ is the spectrahedral cone described by the LMI $A(x) \geq 0$. For forms in three variables there is a strong converse:

8.2.27 Theorem. (Helton–Vinnikov) *Let $f \in \mathbb{R}[x_1, x_2, x_3]$ be a form of degree $d \geq 1$ that is hyperbolic with respect to $e = (1, 0, 0)$ and satisfies $f(e) = 1$. Then there exist symmetric matrices A, $B \in \mathsf{S}^d$ such that $f(x_1, x_2, x_3) = \det(x_1 I + x_2 A + x_3 B)$.*

Theorem 8.2.27 had been conjectured by Peter Lax in 1958. The question was open for more than 40 years, until it was settled by Helton and Vinnikov in the early 2000s. A purely algebraic proof was later given by Hanselka. In view of the remarks in 8.2.26, Theorem 8.2.27 implies that every rigidly convex set in the plane $\mathbb{R}^2$ is a spectrahedron.

For more than three variables, an easy parameter count shows that a general hyperbolic form cannot be written as a symmetric linear determinant (this is already true for hyperbolic quadratic forms, Exercise 8.2.14). To what extent there might be a generalization of Theorem 8.2.27 to four or more variables is an intriguing question that is still open. Several candidate versions have been discussed, and dismissed again, over the years. Today, the following is widely considered to be a likely generalization:

8.2.28 Conjecture. (Generalized Lax Conjecture) *Every hyperbolicity cone is spectrahedral.*

If the conjecture is true, it implies that rigidly convex sets and spectrahedra are the same in all dimensions. See [211] for an overview of partial results in this direction, and see [144] for an excellent introduction with much more details.

Exercises

8.2.1 Let V be a finite-dimensional vector space over $\mathbb{R}$ and let $S \subseteq V$ be a spectrahedron. Show that S is linearly isomorphic to an affine-linear slice of the psd matrix cone S_+^d (for some $d \geq 0$) if, and only if, S contains no line.

8.2.2 If $U \subseteq \mathbb{R}^n$ is a linear subspace, let $F_U = \{A \in \mathsf{S}^n_+ : \operatorname{im}(A) \subseteq U\}$.

(a) Show for $A, B \in \mathsf{S}^n_+$ that $\operatorname{im}(A+B) = \operatorname{im}(A) + \operatorname{im}(B)$ and $\ker(A+B) = \ker(A) \cap \ker(B)$.
(b) Prove that $U \mapsto F_U$ is a lattice isomorphism from the lattice of linear subspaces of $\mathbb{R}^n$ onto the lattice of all non-empty faces of S^n_+.
(c) Show that every face of S^n_+ is exposed.

8.2.3 Let $S = L \cap \mathsf{S}^n_+$ be a spectrahedron, where L is an affine-linear subspace of S^n. For every face F of S, show that there is an integer $r = r(F) \geq 0$ such that $\operatorname{rk}(x) = r$ for every $x \in \operatorname{relint}(F)$. We call $r(F)$ the *rank* of the face F. Show also that $r(F') < r(F)$ whenever F' is a proper face of F.

8.2.4 Give the proof of Lemma 8.2.17.

8.2.5 Let $\mathsf{H}^n = \{B \in \mathrm{M}_n(\mathbb{C}) \colon B = \overline{B}^{\mathsf{T}}\}$, the $\mathbb{R}$-linear space of Hermitian $n \times n$ matrices.

(a) For $B \in \mathsf{H}^n$ put $\operatorname{Re}(B) = \frac{1}{2}(B + \overline{B})$ and $\operatorname{Im}(B) = \frac{1}{2i}(B - \overline{B})$. Then $\operatorname{Re}(B)$ is real symmetric and $\operatorname{Im}(B)$ is real skew-symmetric, so the real $2n \times 2n$ matrix

$$B' := \begin{pmatrix} \operatorname{Re}(B) & -\operatorname{Im}(B) \\ \operatorname{Im}(B) & \operatorname{Re}(B) \end{pmatrix}$$

is symmetric.
(b) If $a_1, \dots, a_n$ are the eigenvalues of B, show that B' has the eigenvalues $a_1, a_1, \dots, a_n, a_n$. Deduce that $\mathsf{H}^n_+ = \{B \in \mathsf{H}^n \colon B \succeq 0\}$ is linearly isomorphic to a linear section of the cone S^{2n}_+.
(c) Define the (Hermitian) trace inner product on $\mathrm{M}_{m \times n}(\mathbb{C})$, and prove similar properties as in Proposition 8.2.2.

8.2.6 Determine the faces of the Hermitian psd matrix cone H^n_+ (see Exercise 8.2.5).

8.2.7 Let $E = \{x \in \mathbb{R}^3 \colon A(x) \succeq 0\}$ be the elliptope, where

$$A(x) = \begin{pmatrix} 1 & x_1 & x_2 \\ x_1 & 1 & x_3 \\ x_2 & x_3 & 1 \end{pmatrix}.$$

(a) Show that $E = \{x \in \mathbb{R}^3 \colon |x|^2 \leq \min\{3,\ 1 + 2x_1x_2x_3\}\}$.
(b) Determine the faces of E of positive dimension.
(c) Show that the algebraic boundary V of E is a rational surface, and find a rational parametrization of V.

8.2.8 (A closed convex set that is not basic closed) Let $a > 0$, let $B, B' \subseteq \mathbb{R}^2$ be the circles with radius 1 and centre $\pm(a, 0)$, and let K be the convex hull of $B \cup B'$. Show that K is (closed but) not basic closed, as follows. Assume that $K = \mathcal{S}(g_1, \dots, g_r)$ for suitable polynomials $g_i \in \mathbb{R}[x, y]$, and let $f = (x-a)^2 + y^2 - 1$ be the equation of one of the two circles.

(a) Show that f divides one of the g_i with odd multiplicity.
(b) Why does this contradict the assumption $K = \mathcal{S}(g_1, \dots, g_r)$?

8.2.9 Show that the (compact and basic closed) set $K = \mathcal{S}(x^4 + y^4 + x^2y^2 - x^2 - y^2)$ in the plane is convex, but that the algebraic boundary of K intersects the interior of K. (Compare with Remark 8.2.19.)

8.2.10 Let $L \subseteq \mathsf{S}^n$ be a linear subspace and let $C = L \cap \mathsf{S}^n_+$. If a matrix $A \in C$ spans an extreme ray of C, show that the image $\operatorname{im}(A)$ of A is minimal in the following sense: Every matrix $B \in L$ with $\operatorname{im}(B) \subseteq \operatorname{im}(A)$ is a scalar multiple of A.

8.2.11 Let $n \geq 1$, let $S = L \cap \mathsf{S}^n_+$ where L is an affine subspace of S^n of codimension k, and assume that the spectrahedron S is non-empty. Prove that S contains a matrix whose rank r satisfies $r^2 + r \leq 2k$. (*Hint*: S contains an extreme point (why?).)

8.2.12 Use Exercise 8.2.11 to sharpen and generalize Corollary 2.1.16 as follows: Let A be an $\mathbb{R}$-algebra, let U be a finite-dimensional linear subspace of A and let UU denote the subspace generated by the products u_1u_2 (u_1, $u_2 \in U$). Then every sum of squares of elements of U is a sum of

$$\left\lfloor \frac{-1+\sqrt{8m+1}}{2} \right\rfloor$$

such squares, where $m = \dim(UU)$.

8.2.13 Let $f \in \mathbb{R}[x_1,\dots,x_n]$ be a quadratic form of rank r. Show that f is hyperbolic with respect to some point $e \in \mathbb{R}^n$ if, and only if, the Sylvester signature of f is $\pm(r-2)$.

8.2.14 Let $f \in \mathbb{R}[x_1,\dots,x_n]$ be a quadratic form of rank at least 4. Show that there do not exist symmetric 2×2 matrices $A_1,\dots,A_n$ with $f(x) = \det(x_1A_1 + \cdots + x_nA_n)$.

8.2.15 Let the form $f \in \mathbb{R}[x_0,\dots,x_n]$ be hyperbolic with respect to $e \in \mathbb{R}^{n+1}$, and assume $f(e) > 0$. Prove that the closed hyperbolicity cone of f is given by

$$C_e(f) = \{v \in \mathbb{R}^n : \partial_e^i f(v) \ge 0,\ i = 1,\dots,\deg(f)\}.$$

In particular, the set $C_e(f)$ is basic closed. (Here ∂_e denotes the (first) partial derivative in direction e, i.e. $\partial_e = \sum_{j=0}^n e_j \frac{\partial}{\partial x_j}$.)

8.2.16 Let $n = 2k \ge 0$ be even, and let K be the convex hull of $\{(t,t^2,\dots,t^n) : t \in \mathbb{R}\}$ in $\mathbb{R}^n$. Show that the closure of K is a spectrahedron, described by the linear matrix inequality

$$\begin{pmatrix} 1 & x_1 & \cdots & x_k \\ x_1 & x_2 & \cdots & x_{k+1} \\ \vdots & & & \vdots \\ x_k & x_{k+1} & \cdots & x_{2k} \end{pmatrix} \succeq 0.$$

Moreover show either that K is closed, or exhibit a point in $\overline{K} \setminus K$. What changes when $n = 2k+1$ is odd? (*Hint*: Use Examples 8.1.23 and 8.2.6.8.)

8.3 Spectrahedral Shadows

8.3.1 Definition. Let V be an $\mathbb{R}$-vector space of finite dimension, and let $K \subseteq V$ be a set. A *semidefinite representation* of K is a pair (f, g) of affine-linear maps $f : W \to V$, $g : W \to \mathsf{S}^d$ (where $d \ge 1$ and W is some vector space of finite dimension) such that $K = f(g^{-1}(\mathsf{S}^d_+))$. If K has a semidefinite representation, K is said to be a *projected spectrahedron*, or a *spectrahedral shadow*. By definition, therefore, the spectrahedral shadows are precisely the linear images of spectrahedra.

8.3.2 Remarks.

1. A set $K \subseteq \mathbb{R}^n$ is a spectrahedral shadow if, and only if, it can be written in the form

$$K = \left\{x \in \mathbb{R}^n : \exists y \in \mathbb{R}^m\ A + M(x) + N(y) \succeq 0\right\} \tag{8.8}$$

where $m \ge 0$ and

$$A + M(x) + N(y) = A + \sum_{i=1}^n x_iM_i + \sum_{j=1}^m y_jN_j \tag{8.9}$$

(with $A, M_i, N_j \in \mathsf{S}^d$) is a linear (symmetric) matrix polynomial of some size $d \times d$ in the variables $x = (x_1, \dots, x_n)$ and $y = (y_1, \dots, y_m)$. A representation of the form (8.8) is also called a *lifted LMI representation* of K.

2. By the Tarski–Seidenberg projection theorem, and since linear images of convex sets are convex, every spectrahedral shadow is a convex semialgebraic set. Beyond this, there are no other general properties of spectrahedral shadows that are obvious. But we will see in Section 8.7 that the class of spectrahedral shadows is much more restricted.

8.3.3 Examples. Here are first examples of spectrahedral shadows:

1. For $d \geq 2$, the convex set $K = \{x \in \mathbb{R}^2 \colon x_1^{2d} + x_2^{2d} \leq 1\}$ fails to be a spectrahedron (Example 8.2.24.2). But K is a spectrahedral shadow. Indeed, the following is an explicit semidefinite representation in case $2d = 4$:

$$K = \left\{x \in \mathbb{R}^2 \colon \exists y \in \mathbb{R}^2 \text{ s.t. } \begin{pmatrix} y_1 & x_1 \\ x_1 & 1 \end{pmatrix} \succeq 0, \begin{pmatrix} y_2 & x_2 \\ x_2 & 1 \end{pmatrix} \succeq 0, \begin{pmatrix} 1 + y_1 & y_2 \\ y_2 & 1 - y_1 \end{pmatrix} \succeq 0\right\}.$$

Of course, the three LMIs of size 2×2 can be combined into a single LMI of size 6×6. In Example 8.5.23, an explicit semidefinite representation of K will be given for all $d \geq 1$.

2. Other than spectrahedra, their shadows need not be closed. For example, consider the set $K \subseteq \mathbb{R}^2$ that consists of all $x \in \mathbb{R}^2$ for which there is $y \in \mathbb{R}$ with

$$\begin{pmatrix} 1 + x_1 & x_2 & 1 \\ x_2 & 1 - x_1 & 0 \\ 1 & 0 & y \end{pmatrix} \succeq 0. \tag{8.10}$$

Then $K = \{x \colon x_1^2 + x_2^2 < 1\} \cup \{(1, 0)\}$. Indeed, K is the set of points x in the closed unit disk for which there exists $y > 0$ making the determinant of (8.10) non-negative. This says $y(1 - |x|^2) \geq 1 - x_1$. If $|x| < 1$, there clearly exists such y, but when $|x| = 1$ this only holds if $x = (1, 0)$.

3. It is easy to see that the class of spectrahedral shadows is closed under forming finite intersections or Minkowski sums, finite direct products, and under taking linear images or preimages (Exercise 8.3.1). Other permanence properties will be proved below.

8.3.4 Examples.

1. More examples of spectrahedral shadows arise from sums of squares cones. For any n and d, the cone $\Sigma_{n,\leq 2d}$ in $\mathbb{R}[x]_{\leq 2d} = \mathbb{R}[x_1, \dots, x_n]_{\leq 2d}$ is a spectrahedral shadow. An explicit semidefinite representation is given by the Gram matrix construction (Proposition 2.1.7): $f \in \mathbb{R}[x]_{\leq 2d}$ is a sum of squares if, and only if, there is a psd symmetric matrix S, whose rows and columns are indexed by the monomials of degree $\leq d$, such that $f = X^\top S X$ where X is the vector of these monomials. In particular, a univariate polynomial $f \in \mathbb{R}[t]_{\leq 2d}$ is sos if and only if there exists a symmetric $(d + 1) \times (d + 1)$ matrix $S \succeq 0$ with $f = (1, \dots, t^d) \cdot S \cdot (1, \dots, t^d)^\top$.

2. The generalization to cones of weighted sums of squares is immediate, if we impose bounds on the degrees of the summands: Given polynomials $g_1, \dots, g_r$ in $\mathbb{R}[x] = \mathbb{R}[x_1, \dots, x_n]$ and an integer $k \geq 1$, the set of all $f \in \mathbb{R}[x]_{\leq k}$ which admit a representation $f = \sum_{i=1}^r s_i g_i$ with sums of squares $s_i \in \mathbb{R}[x]$ and $\deg(s_i g_i) \leq k$ for all i, is a spectrahedral shadow.

3. The previous remarks also generalize in a different direction: If A is an arbitrary $\mathbb{R}$-algebra and $U \subseteq A$ is a finite-dimensional linear subspace, the sos cone ΣU^2 in UU (cf. Example 8.2.6.8) is a spectrahedral shadow. Indeed, if $u_1, \dots, u_n$ is a linear basis of U, the linear map

$$\phi \colon \mathsf{S}^n \to UU, \quad (a_{ij}) \mapsto \sum_{i,j=1}^n a_{ij} u_i u_j$$

satisfies $\phi(\mathsf{S}^n_+) = \Sigma U^2$, since $\phi(ww^\top) = \left(\sum_i w_i u_i\right)^2$ for $w \in \mathbb{R}^n$.

8.3.5 Lemma. *A convex cone that is a spectrahedral shadow is a linear image of a spectrahedral cone.*

Proof. We may assume $C = f(S)$ where $f \colon W \to V$ is a linear map and $S \subseteq W$ is a spectrahedron with $0 \in S$. The homogenization S^h of S is a spectrahedral cone in $\mathbb{R} \times W$, by Remark 8.2.15.2. The linear map $f' \colon \mathbb{R} \times W \to V$, $f'(t, w) = f(w)$ satisfies $f'(S^h) = f(S)$. Indeed, "$\supseteq$" is obvious since $w \in S$ implies $(1, w) \in S^h$ and $f'(1, w) = f(w)$. To prove "$\subseteq$", recall that elements in S^h either have the form (t, tw) where $t > 0$ and $w \in S$, or else $(0, w)$ where $w \in \mathrm{rc}(S)$. If $w \in S$ and $t \geq 0$ then $f'(t, tw) = tf(w) \in C = f(S)$ since C is a cone. Since $0 \in S$, we have $\mathrm{rc}(S) \subseteq S$, and so $f'(0, w) = f(w) \in f(S)$ for every $w \in \mathrm{rc}(S)$. □

8.3.6 Proposition. *If $K \subseteq V$ is a spectrahedral shadow, the same is true for the conic hull* $\mathrm{cone}(K)$ *of K.*

Proof. We may assume $V = \mathbb{R}^n$ and

$$K = \left\{x \in \mathbb{R}^n \colon \exists y \in \mathbb{R}^m \; A + \sum_{i=1}^n x_i M_i + \sum_{j=1}^m y_j N_j \succeq 0\right\}$$

where A, M_i, $N_j \in \mathsf{S}^d$ for some d. Let C be the set of $x \in \mathbb{R}^n$ for which there are $y \in \mathbb{R}^m$ and $s, t \in \mathbb{R}$ such that $sA + \sum_{i=1}^n x_i B_i + \sum_{j=1}^m y_j C_j \succeq 0$ and

$$\begin{pmatrix} s & x_i \\ x_i & t \end{pmatrix} \succeq 0 \quad (i = 1, \dots, n). \tag{8.11}$$

We claim that $C = \mathrm{cone}(K)$. Indeed, C is a convex cone that contains K, so $\mathrm{cone}(K) \subseteq C$ is clear. Conversely let $(x, y, s, t) \in \mathbb{R}^n \times \mathbb{R}^m \times \mathbb{R} \times \mathbb{R}$ be such that the above LMIs are satisfied. If $s > 0$ then $\frac{x}{s} \in K$ and hence $x \in \mathrm{cone}(K)$. If $s = 0$ we have $x = 0$, since otherwise there would not be any $t \in \mathbb{R}$ satisfying the LMIs (8.11). Hence $x \in \mathrm{cone}(K)$ is true in this case as well. □

8.3.7 Proposition. *If K_1, $K_2 \subseteq V$ are spectrahedral shadows, the convex hull of $K_1 \cup K_2$ is a spectrahedral shadow as well.*

Proof. Write $K_i^c = \text{cone}(\{1\} \times K_i) \subseteq \mathbb{R}^{n+1}$ $(i = 1, 2)$ as in 8.1.18. Then K_1^c, K_2^c are shadows by Proposition 8.3.6. Hence the same is true for their Minkowski sum $K_1^c + K_2^c$ (Exercise 8.3.1). Since $\text{conv}(K_1 \cup K_2) = \{x \in \mathbb{R}^n : (1, x) \in K_1^c + K_2^c\}$, this implies the assertion, see again Exercise 8.3.1. □

8.3.8 Proposition. *Let $C \subseteq V$ be a spectrahedral cone. Then the dual cone C^* in $V^\vee$ is a linear image of a spectrahedral cone.*

Let $U = C \cap (-C)$, and let $W = C - C$ be the linear hull of C in V. Replacing V by W/U and C by C/U is harmless, so we may assume that C is pointed and has non-empty interior in V. Using Proposition 8.2.18, it then suffices to prove the following more explicit version:

8.3.9 Proposition. *Let $V \subseteq \mathsf{S}^d$ be a linear subspace and let $C = V \cap \mathsf{S}_+^d$. Assume that V contains a matrix A_0 that is positive definite. Then the natural linear map $\phi \colon \mathsf{S}^d \to V^\vee$ satisfies $\phi(\mathsf{S}_+^d) = C^*$.*

Proof. ϕ is the map $B \mapsto \phi_B \in V^\vee$ defined by $\phi_B(A) = \langle A, B\rangle$ $(B \in \mathsf{S}^d, A \in V)$. For $B \in \mathsf{S}_+^d$ and $A \in C$ we have $\phi_B(A) = \langle A, B\rangle \geq 0$, so the inclusion $\phi(\mathsf{S}_+^d) \subseteq C^*$ is clear. Moreover, if $B \in \mathsf{S}_+^d$ satisfies $\phi_B = 0$, then in particular $\langle A_0, B\rangle = 0$, which implies $B = 0$ since $A_0 > 0$ (Proposition 8.2.2). This shows $\ker(\phi) \cap \mathsf{S}_+^d = \{0\}$, and so the image cone $\phi(\mathsf{S}_+^d)$ is closed in $V^\vee$ by Proposition 8.1.15. To prove the remaining inclusion $C^* \subseteq \phi(\mathsf{S}_+^d)$, it therefore suffices to show the dual inclusion $\phi(\mathsf{S}_+^d)^* \subseteq C$. For this let $A \in \phi(\mathsf{S}_+^d)^*$, so $A \in V$ is a matrix that satisfies $\langle A, B\rangle \geq 0$ for every $B \in \mathsf{S}_+^d$. But then $A \geq 0$ since S_+^d is self-dual, and therefore $A \in C$. □

8.3.10 Corollary. *If $C \subseteq V$ is a convex cone which is a spectrahedral shadow, the dual cone C^* is a spectrahedral shadow as well.*

Proof. By Lemma 8.3.5 there are a spectrahedral cone $D \subseteq W$ and a linear map $f \colon W \to V$ with $C = f(D)$. The dual cone $D^* \subseteq W^\vee$ is a spectrahedral shadow by Proposition 8.3.8. On the other hand, $C^* = (f^\vee)^{-1}(D^*)$ where $f^\vee \colon V^\vee \to W^\vee$ denotes the linear map dual to f. So the claim follows using Exercise 8.3.1. □

8.3.11 Corollary. *If $K \subseteq V$ is a spectrahedral shadow, the same is true for the cone $(K^h)^*$ (dual of the homogenization of K), and for the polar dual K^o of K. It is also true for K^h if K is closed.*

Proof. The cone K^c is dense in K^h (Proposition 8.1.19), so $(K^h)^* = (K^c)^* = \text{cone}(1 \times K)^*$, which is a shadow by Propositions 8.3.6 and 8.3.10. If K is closed then K^h is closed as well (8.1.19), and so $K^h = (K^h)^{**}$ is a shadow. The polar dual K^o is an affine section of $(K^h)^*$ (Remarks 8.1.29), which gives the assertion for K^o. □

8.3.12 Corollary. *The closure of any spectrahedral shadow is again a spectrahedral shadow.*

Proof. We may assume that $K \subseteq \mathbb{R}^n$ is a spectrahedral shadow that contains the origin. Then $\overline{K} = K^{oo}$ (8.1.28), which is a shadow by 8.3.11. □

Netzer [143] has shown that removing certain families of faces from a spectrahedral shadow gives again spectrahedral shadows (cf. also Exercise 8.3.4). In particular, he proved that the relative interior of every spectrahedral shadow is again a spectrahedral shadow.

8.3.13 Example. Let $\Sigma_{n,2d}$ be the cone of sos forms of degree $2d$ in n variables, a spectrahedral shadow by Example 8.3.4. For $2d = 2$, the cone $\Sigma_{n,2}$ is self-dual (isomorphic to the psd matrix cone), and hence is spectrahedral. However as soon as $2d \geq 4$ (and $n \geq 2$), $\Sigma_{n,2d}$ has non-exposed faces, so it is not a spectrahedron (Proposition 8.2.16). To see this, note that the form x_1^{2d} generates an extreme ray in $\Sigma_{n,2d}$ by Exercise 3.5.1(a). Assume that this ray is exposed, which means assume that there is a linear form $\lambda \in \Sigma_{n,2d}^*$ satisfying

$$\lambda(p^2) = 0 \;\Leftrightarrow\; p \in \mathbb{R}x_1^d$$

for all forms p of degree d. Consider the form $p_t = x_1^{d-2}(tx_1^2 - x_2^2)$ where t is a real parameter. From $\lambda(x_1^{2d-2}x_2^2) > 0$ and from

$$\lambda(p_t^2) = \lambda(x_1^{2d-4}x_2^4) - 2t\,\lambda(x_1^{2d-2}x_2^2)$$

we see $\lambda(p_t^2) < 0$ for sufficiently large $t > 0$, a contradiction. To summarize:

8.3.14 Proposition. *For all n and d, the sos cone $\Sigma_{n,2d}$ is a spectrahedral shadow, and its dual $(\Sigma_{n,2d})^*$ is a spectrahedron. If $n \geq 2$ and $2d \geq 4$, the cone $\Sigma_{n,2d}$ has non-exposed faces and is not a spectrahedron.* □

Exercises

8.3.1 Let V, W be real vector spaces of finite dimension and let $f\colon W \to V$ be a linear map. For any spectrahedral shadow K in V, show that the preimage $f^{-1}(K)$ is a spectrahedral shadow in W. The class of spectrahedral shadows is stable under finite intersections, Minkowski sums and direct products.

8.3.2 Consider the set $K = \{x \in \mathbb{R}_+^3 : x_1x_2x_3 \geq 1\}$. Show that K is not a spectrahedron, but that K is a linear image of a spectrahedron that is described by three LMIs of size 2×2.

8.3.3 Generalize Exercise 8.3.2 to show: For every $n \geq 1$ and any integers $k_1, \dots, k_n \geq 1$, the set $\{x \in \mathbb{R}_+^n : x_1^{k_1} \cdots x_n^{k_n} \geq 1\}$ is a spectrahedral shadow.

8.3.4 Let $K \subseteq \mathbb{R}^n$ be a closed set that is a spectrahedral shadow, and let F be a face of K. Show that the difference set $K \setminus F$ is a spectrahedral shadow as well. (*Hint*: First do the case where K is a spectrahedron.)

8.4 A (Very) Brief Introduction to Semidefinite Programming

A semidefinite program, often abbreviated SDP, is the task of optimizing a linear function over a spectrahedron. Under mild assumptions, semidefinite programs can be solved in polynomial time, up to any prescribed accuracy. They have numerous applications from a wide range of areas, including optimization, discrete and combinatorial mathematics, geometry, signal processing, electrical engineering and others.

To introduce semidefinite programs, we start by discussing the setting of conic programming, which is more general. Every conic program has a dual conic program, whose dual in turn is the original program. In the usual formulation, the complete symmetry between primal and dual problem gets somewhat hidden. Therefore we prefer to start with a formulation that is entirely symmetric. After that we'll pass to the standard setup.

Many excellent textbooks and survey articles are available that provide much more background on theoretical or practical aspects of conic programming in general, or of semidefinite optimization in particular. See for example (in historical order) [213], [17], [108], [31], [142], [12], [4], [123], to mention just a few standard references.

8.4.1 Let V be a finite-dimensional $\mathbb{R}$-vector space and $C \subseteq V$ a closed convex cone. As usual let $V^\vee$ denote the dual vector space and $C^* \subseteq V^\vee$ the dual convex cone, and write $\langle x, y\rangle$ for the natural pairing between $x \in V$ and $y \in V^\vee$. Further let $v \in V$ and $w \in V^\vee$ be given, let $U \subseteq V$ be a linear subspace, and let $U^\perp = \{y \in V^\vee : \forall\, u \in U\ \langle u, y\rangle = 0\}$ be the subspace of $V^\vee$ that is orthogonal to U. With this data we associate the *primal program*

$$p^* \;=\; \inf \langle x, w\rangle \quad \text{s.t. } x \in C \cap (v + U), \tag{P}$$

which means the problem of minimizing the linear function w on V (the *(primal) objective function*) over the set $C \cap (v + U) \subseteq V$. Likewise, the *dual program*

$$p'^* \;=\; \inf \langle v, y\rangle \quad \text{s.t. } y \in C^* \cap (w + U^\perp) \tag{P'}$$

is the problem of minimizing v, considered as a linear function on $V^\vee$, over the set $C^* \cap (w + U^\perp) \subseteq V^\vee$. We point out that both primal and dual program are of the same form, and that the dual of the dual problem is again the original primal problem.

8.4.2 Usually, one of (P) or (P') is written as the task of maximizing (instead of minimizing) a suitable linear function. If we let $y = w - z$ in (P'), then p'^* becomes

$$\inf\{\langle v, w - z\rangle : z \in U^\perp,\ w - z \in C^*\} \;=\; \langle v, w\rangle - \sup\{\langle v, z\rangle : z \in U^\perp \cap (w - C^*)\}.$$

Thus we may consider the dual problem in the form

$$d^* \;=\; \sup \langle v, y\rangle \quad \text{s.t. } \; y \in U^\perp \; \text{ and } \; w - y \in C^*, \tag{D}$$

giving

$$d^* = \langle v, w \rangle - p'^*.$$

It is common to consider (P) and (D) above as the primal and dual problem pair, and we will adopt this convention in what follows.

The subset $K := C \cap (v + U)$ of V is closed and convex and is called the *set of feasible points* for the primal program (P). The program (P) is *feasible* if $K \neq \emptyset$, and is *strictly feasible* if K contains an interior point of C. A point $x \in K$ is *optimal* for (P) if $\langle x, w \rangle = p^*$. Dually, the set $S := U^\perp \cap (w - C^*)$ of feasible points for (D) is closed and convex in $U^\perp$. The dual program (D) is feasible (or strictly feasible) if $S \neq \emptyset$ (or if $w - S$ contains an interior point of C^*, respectively). An optimal point for (D) is a point $y \in S$ with $\langle v, y \rangle = d^*$.

The general duality theory of conic programs is summarized in the following theorem:

8.4.3 Theorem. *Let K and S be the sets of feasible points for (P) and (D), respectively.*

(a) (Weak duality) *For every $x \in K$ and $y \in S$ one has $\langle v, y \rangle \leq \langle x, w \rangle$. Therefore $d^* \leq p^*$, and the* duality gap *$\Delta := p^* - d^*$ is non-negative and satisfies $\Delta = \inf\{\langle x, w \rangle - \langle v, y \rangle : x \in K,\ y \in S\}$.*
(b) *Given $x \in K$ and $y \in S$, equality $\langle x, w \rangle = \langle v, y \rangle$ holds if and only if x is optimal for (P) and y is optimal for (D).*
(c) *Assume that (P) is strictly feasible. Then (D) is feasible if and only if (P) is bounded, i.e. $p^* \neq -\infty$. If this holds, (D) has an optimal point and $\Delta = 0$.*
(d) *Dually, if (D) is strictly feasible, then (P) is feasible if and only if $d^* \neq \infty$. If this holds, (P) has an optimal point and $\Delta = 0$.*

Assertions (c) and (d) are dual to each other and are commonly referred to as *strong duality*.

Proof. To prove (a), write $y = w - z$ with $z \in C^*$. Then $\langle x, z \rangle \geq 0$, hence $\langle x, w \rangle \geq \langle x, w \rangle - \langle x, z \rangle = \langle x, y \rangle = \langle v, y \rangle$, where the last equality holds since $x - v \in U$ and $y \in U^\perp$. Clearly this implies the remaining assertions in (a), and it also implies (b). We now prove (d). This will also imply (c), since (c) is dual to (d).

Thus assume that (D) is strictly feasible. From (a) it is clear that (P) can be feasible only if (D) is bounded. Hence assume $d^* < \infty$. If $v \in U$, then $x = 0$ lies in K and $\langle v, y \rangle = 0$ for every $y \in S$. So $d^* = p^* = 0$, and $x = 0$ is an optimal point for (P).

We are left with the case $v \notin U$. Consider the set $H := \{y \in U^\perp : \langle v, y \rangle \geq d^*\}$, which is a closed halfspace in $U^\perp$ since $v \notin U$. Any point y in $H \cap (w - C^*) = H \cap S$ satisfies $\langle v, y \rangle = d^*$. If there were $y \in H$ such that $w - y \in \operatorname{int}(C^*)$ (and hence $\langle v, y \rangle = d^*$), we could find $y' \in U^\perp$ close to y such that $\langle v, y' \rangle > \langle v, y \rangle$ (using $v \notin U$) and still $w - y' \in C^*$ holds. This would mean $y' \in H$ and $\langle v, y' \rangle > d^*$, a contradiction. Therefore the intersection $(w - H) \cap \operatorname{int}(C^*)$ must be empty. By Theorem 8.1.5(a) there exists a hyperplane in $V^\vee$ that separates the two convex sets, namely an element $x \neq 0$ in V such that

$$\sup\{\langle x,z\rangle : z \in w - H\} \le \inf\{\langle x,\mu\rangle : \mu \in C^*\}. \tag{8.12}$$

The right-hand infimum is 0 since otherwise it would be $-\infty$, contradicting $H \neq \emptyset$. This shows that $x \in (C^*)^* = C$, and it means that every $y \in H$ satisfies $\langle x, w - y\rangle \le 0$. In other words, the closed halfspace $H = \{y \in U^\perp : \langle v,y\rangle \ge d^*\}$ of $U^\perp$ is contained in $\{y \in U^\perp : \langle x,y\rangle \ge \langle x,w\rangle\}$.

This means that there is some constant $c \ge 0$ for which $x \in cv + U$, and also that

$$\langle x,w\rangle = cd^* \tag{8.13}$$

in case $c > 0$. We claim that $c = 0$ leads to a contradiction. Indeed, this would mean $x \in U$, thus implying $\langle x,y\rangle = 0$ for every $y \in H$, and hence $\langle x,w\rangle \le 0$ by (8.12). On the other hand, the assumption that (D) is strictly feasible means that there is $y \in U^\perp$ with $w - y \in \mathrm{int}(C^*)$. Thus $\langle x, w-y\rangle > 0$ since $x \in C$, $x \neq 0$. But $\langle x,y\rangle = 0$, so $\langle x,w\rangle > 0$, contradicting $\langle x,w\rangle \le 0$.

We conclude that $c > 0$, and so $x^* = \frac{x}{c} \in C \cap (v + U)$ is a feasible point for (P). Moreover (8.13) gives $\langle x^*, w\rangle = d^*$. In particular, x^* is an optimal point for (P), and the duality gap is zero. The theorem is proved. □

8.4.4 Remark. If (P) is strictly feasible in Theorem 8.4.3, strong duality (c) gives an a priori criterion for optimality of a feasible point x of (P). Namely, x is optimal for (P) (if and) only if there exists a feasible point y for (D) with $\langle x,w\rangle = \langle v,y\rangle$.

8.4.5 Remark. If we use coordinates, the conic program takes the following more "concrete" form. If $C, C^* \subseteq \mathbb{R}^n$ are closed convex cones dual to each other, the primal program (P) has the form

$$p^* = \inf\langle x,w\rangle \quad \text{s.t.} \quad x \in C \ \text{ and } \ \langle x,u_j\rangle = b_j \ \ (j = 1,\dots,m) \tag{8.14}$$

where $w, u_1,\dots,u_m \in \mathbb{R}^n$ and $b = (b_1,\dots,b_m) \in \mathbb{R}^m$, and where we may assume that the u_i are linearly independent. The corresponding dual program (D) is

$$d^* = \sup\langle b,y\rangle \quad \text{s.t.} \quad y \in \mathbb{R}^m \ \text{ and } \ w - \sum_{j=1}^m y_j u_j \in C^*. \tag{8.15}$$

Indeed, choose $v \in \mathbb{R}^n$ with $\langle v,u_j\rangle = b_j$ for all j (such v exists by linear independence of the u_i), and let $U = \mathrm{span}(u_1,\dots,u_m)^\perp \subseteq \mathbb{R}^n$. Then any element $\mu \in U^\perp = \mathrm{span}(u_1,\dots,u_m)$ has the form $\mu = \sum_{j=1}^m y_j u_j$ with $y = (y_1,\dots,y_m) \in \mathbb{R}^m$. Such μ satisfies $\langle v,\mu\rangle = \sum_j b_j y_j = \langle b,y\rangle$, which in view of 8.4.2 gives the dual program (8.15).

8.4.6 Remark. The best known instance of conic programming is *linear programming (LP)*. Here the cone in question is the positive orthant $C = \mathbb{R}^n_+$, which is a self-dual cone in $\mathbb{R}^n$. A linear program may be given as $p^* = \inf\{\langle x,w\rangle : x \ge 0, Ax = b\}$ where $w \in \mathbb{R}^n$, $b \in \mathbb{R}^m$ are (column) vectors and A is a matrix of size $m \times n$, and where $\ge$ denotes componentwise inequality between two vectors. The corresponding dual linear program is $d^* = \sup\{\langle b,y\rangle : y \in \mathbb{R}^m, A^\top y \le w\}$ (Remark

8.4.2). As a consequence of the duality theory of polyhedra, linear programs have stronger duality properties than general conic programs:

8.4.7 Proposition. (Strong duality for LP) *If a linear program (P) is feasible and bounded (meaning that $p^* \neq \pm\infty$), then the dual linear program (D) is feasible (and bounded) as well. Moreover then, both (P) and (D) have optimal points, and the duality gap vanishes, i.e. $d^* = p^*$.*

Proof. We consider the linear program $p^* = \inf\{\langle x, w\rangle : Ax = b,\ x \geq 0\}$ with dual program $d^* = \sup\{\langle b, y\rangle : A^\top y \leq w\}$, as in 8.4.6. Assume $p^* \in \mathbb{R}$, so (P) is feasible and bounded. By Exercise 8.1.24, the infimum p^* is a minimum, so (P) has an optimal point x^*. Let $c \leq p^*$ be a real number and let $\tilde{A} = \binom{A}{w^\top} \in \mathrm{M}_{(m+1)\times n}$, $\tilde{b} = \binom{b}{c} \in \mathbb{R}^{m+1}$. We use Farkas' lemma (Exercise 8.1.25) twice, as follows. If $c = p^*$ then, since $x^* \geq 0$ satisfies $\tilde{A}x^* = \tilde{b}$, this lemma implies that there is no $\tilde{y} \in \mathbb{R}^{m+1}$ with $\tilde{A}^\top \tilde{y} \geq 0$ and $\tilde{b}^\top \tilde{y} < 0$. Writing $\tilde{y} = \binom{y}{a}$ with $y \in \mathbb{R}^m$ and $a \in \mathbb{R}$, this means for $(y, a) \in \mathbb{R}^m \times \mathbb{R}$ that

$$A^\top y + aw \geq 0 \text{ implies } \langle b, y\rangle + ap^* \geq 0. \tag{8.16}$$

Now let $c < p^*$. Since there is no $x \in \mathbb{R}^n$ with $x \geq 0$ and $\tilde{A}x = \tilde{b}$ Farkas' lemma implies that there is $\tilde{y} \in \mathbb{R}^{m+1}$ with $\tilde{A}^\top \tilde{y} \geq 0$ and $\tilde{b}^\top \tilde{y} < 0$. Rewriting this as before, it means that there is $(y, a) \in \mathbb{R}^m \times \mathbb{R}$ with $A^\top y + aw \geq 0$ and $\langle b, y\rangle + ac < 0$. By (8.16) we have $\langle b, y\rangle + ap^* \geq 0$, which implies $a > 0$ since $c < p^*$. Therefore $A^\top(-\frac{y}{a}) \leq w$ and $\langle b, -\frac{y}{a}\rangle > c$ hold, showing that $-\frac{y}{a}$ is a feasible point for (D) and that $d^* > c$. Since this argument works for every $c < p^*$ we conclude $d^* = p^*$. Finally (D) has an optimal point, again by Exercise 8.1.24. □

8.4.8 Remark. *Semidefinite programming (SDP)* is conic programming with respect to the psd symmetric matrix cone S^d_+. Recall that this cone is self-dual with respect to the trace inner product on S^d (8.2.2). A semidefinite program is usually written in the form

$$p^* = \inf_{X\in K} \langle X, A\rangle, \quad K = \Big\{X \in \mathsf{S}^d : X \succeq 0,\ \langle X, A_i\rangle = b_i\ (i = 1, \ldots, n)\Big\} \tag{P}$$

where $A, A_1, \ldots, A_n \in \mathsf{S}^d$ and $b = (b_1, \ldots, b_n) \in \mathbb{R}^n$, and where $A_1, \ldots, A_n$ are linearly independent. The dual semidefinite program is

$$d^* = \sup_{y\in S} \langle b, y\rangle, \quad S = \Big\{y \in \mathbb{R}^n : \sum_{i=1}^n y_i A_i \preceq A\Big\}. \tag{D}$$

Both the primal and the dual feasible set are spectrahedra. Clearly, semidefinite programs generalize linear programs, the latter corresponding to semidefinite programs described by diagonal matrices. Note that (P) is strictly feasible if and only if K contains a matrix $X \succ 0$, and (D) is strictly feasible if and only if $\sum_i y_i A_i \prec A$ for some vector $y \in \mathbb{R}^n$. Apart from the general features of conic duality 8.4.3, we point out that if $X \in \mathsf{S}^d$ resp. $y \in \mathbb{R}^n$ are primal resp. dual feasible points, then both are optimal if and only if $\langle X, A\rangle = \sum_i y_i\langle X, A_i\rangle$, or equivalently, $XA = \sum_i y_i XA_i$. This is statement 8.4.3(b), combined with the last statement in Proposition 8.2.2.

Strong duality, in the form 8.4.7 that holds for linear programs, is usually not satisfied for semidefinite programs. A semidefinite program may be feasible and bounded while its dual is still infeasible. Also, when both the program and its dual are feasible, there may still be a non-zero duality gap, and/or the programs need not have optimal points. For examples we refer to Exercise 8.4.1.

The use of semidefinite programming for polynomial optimization will be discussed in Section 8.5 in greater detail. Here we only sketch two sample applications of SDP:

8.4.9 (*The max-cut problem*) Let $G = (V, E, w)$ be a finite weighted graph. So V is a finite set (the vertices), $E \subseteq \binom{V}{2}$ is a set of two-element subsets of V (the edges), and $w_{ij} = w_{ji} \ge 0$ is a non-negative real number for every edge $\{i, j\} \in E$. The *max-cut problem* is the task of partitioning the vertices V into two disjoint subsets V_1, V_2, in such a way that

$$\gamma = \sum_{i\in V_1,\, j\in V_2} w_{ij}$$

is maximized. For example, if G is given by

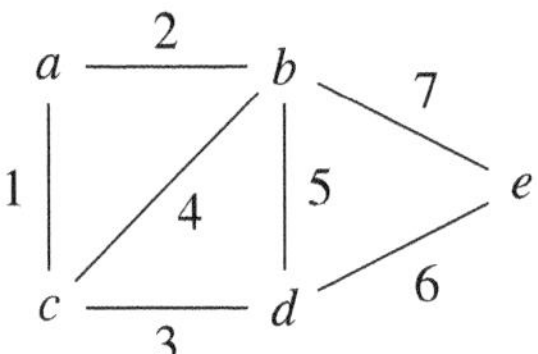

then $(V_1, V_2) = (\{a, b, e\}, \{c, d\})$ gives $\gamma = 16$, while $(V_1, V_2) = (\{a, c, d\}, \{b, e\})$ gives $\gamma = 17$. The maximum $\gamma = 18$ is achieved for $(V_1, V_2) = (\{a, c, d, e\}, \{b\})$. The general problem of finding a maximal cut is known to be computationally hard (technically speaking it is NP-hard).

To model the problem mathematically we assume $V = \{1, \dots, n\}$. So a partition (V_1, V_2) of V can be identified with a vector $x = (x_1, \dots, x_n) \in \mathbb{R}^n$ satisfying $x_i^2 = 1$ for $i = 1, \dots, n$. Setting $w_{ij} = 0$ if $\{i, j\} \notin E$, the partition x gives the value

$$\gamma(x) = \frac{1}{4}\sum_{i,j\in V}(1 - x_i x_j)w_{ij},$$

and this expression has to be maximized over all vectors $x \in \{1, -1\}^n$.

A celebrated result of Goemans and Williamson [73] uses semidefinite programming to give a polynomial-time approximation for this problem. Indeed, we may write

$$\mathrm{maxcut}(G) = \frac{1}{4}\max\Big\{\langle E_n - X, W\rangle\colon X = (x_{ij}) \succeq 0,\ \mathrm{rk}(X) = 1,\ x_{ii} = 1\ \forall\, i\Big\} \quad (8.17)$$

where $W = (w_{ij}) \in \mathsf{S}^n$ and E_n is the all-one matrix. Removing the rank condition turns this into an SDP:

$$\sigma(G) = \frac{1}{4}\max\Big\{\langle E_n - X, W\rangle : X = (x_{ij}) \succeq 0,\ x_{ii} = 1\ \forall\, i\Big\}. \tag{8.18}$$

Thereby the maximum gets increased, $\operatorname{maxcut}(G) \le \sigma(G)$. Goemans and Williamson proved that the increase is universally bounded by a constant factor not much bigger than one: $\sigma(G) \le c \cdot \operatorname{maxcut}(G)$ holds with $c = \frac{\pi}{2}\max_{\theta\in\mathbb{R}} \frac{1-\cos\theta}{\theta} \approx 1.1382$. It is easy to see that both the semidefinite program 8.18 and its dual are strictly feasible (check this).

8.4.10 For another application of semidefinite programming, let $A \in \mathsf{S}^n$ be a symmetric matrix, and let $\lambda_A = (\lambda_1, \ldots, \lambda_n)$ be the vector of eigenvalues of A, listed in decreasing order $\lambda_1 \ge \cdots \ge \lambda_n$. We show how to find the largest eigenvalue of A by a semidefinite program. More generally, we can determine the sum

$$s_k(A) := \lambda_1 + \cdots + \lambda_k$$

of the k largest eigenvalues in this way:

8.4.11 Theorem. *For every real symmetric matrix $A \in \mathsf{S}^n$ and for $k = 1, \ldots, n$, one has*

$$s_k(A) = \max\{\operatorname{tr}(AX) : X \in \mathsf{S}^n,\ \operatorname{tr}(X) = k,\ 0 \preceq X \preceq I\}. \tag{8.19}$$

Proof. Let $\lambda_1 \ge \cdots \ge \lambda_n$ be the eigenvalues of A. Neither side in (8.19) changes if A is replaced with $SAS^\top$ where $S \in O(n)$. So we may assume that $A = \operatorname{diag}(\lambda)$ is diagonal, with $\lambda = (\lambda_1, \ldots, \lambda_n)$. Let $K = \{X \in \mathsf{S}^n : \operatorname{tr}(X) = k,\ 0 \preceq X \preceq I\}$ be the feasible set of the semidefinite program. Since $P := E_{11} + \cdots + E_{kk}$ lies in K and satisfies $\operatorname{tr}(AP) = s_k(A)$, the inequality "$\le$" is clear in (8.19). For the converse note that K is compact, so the maximum in (8.19) will be taken at an extreme point of K (Exercise 8.1.8). Let $D_n \subseteq \mathsf{S}^n$ be the subspace of diagonal matrices. Every extreme point X of K is $O(n)$-conjugate to an extreme point Y of $D_n \cap K$. Now $D_n \cap K$ is a polyhedron, and hence a polytope since it is compact. It is easy to see that the extreme points of $D_n \cap K$ are the diagonal matrices with exactly k entries one and all other entries zero. Indeed, if $X = \operatorname{diag}(x_1, \ldots, x_n)$ is an extreme point of $D_n \cap K$ then $x_i \in \{0, 1\}$ for at least $n - 1$ indices i (Exercise 8.1.26), and therefore for all i. We conclude that the extreme points of K are the $O(n)$-conjugates of the matrix P, and hence K is the convex hull of the $O(n)$-orbit of P under this action. In other words, K is the $O(n)$-orbitope of P.

In particular, the right-hand maximum in (8.19) is equal to $\operatorname{tr}(AX)$ for $X = SPS^\top$ with some matrix $S \in O(n)$. Since $\operatorname{tr}(AX) = \operatorname{tr}(S^\top AS \cdot P)$, and since the diagonal $D(S^\top AS)$ is contained in the permutahedron $\Pi(\lambda)$ by Lemma 8.2.11, we have

$$\operatorname{tr}(AX) \le \max_{\substack{I\subseteq[n]\\ |I|=k}} \sum_{i\in I} \lambda_i = \sum_{i=1}^{k} \lambda_i = s_k(A). \qquad \square$$

8.4.12 Corollary. *$s_k(A+B) \le s_k(A) + s_k(B)$ for all $A,\ B \in \mathrm{S}^n$ and $1 \le k \le n$.*

Proof. Let $K = \{X \in \mathrm{S}^n\colon \operatorname{tr}(X) = k,\ 0 \preceq X \preceq I\}$, and let $A,\ B \in \mathrm{S}^n$. By the previous theorem there is $X \in K$ with $s_k(A+B) = \langle A+B, X\rangle$. On the other hand, $\langle A, X\rangle \le s_k(A)$ and $\langle B, X\rangle \le s_k(B)$ hold by the same theorem, which implies the assertion. □

8.4.13 Remark. If the tuples λ_A and λ_B of eigenvalues of A and B are given, Corollary 8.4.12 gives some restrictions for the possible eigenvalues of $A+B$. For a long time, the question of describing, for given λ_A and λ_B, all possible eigenvalue tuples of $A+B$ was a famous open problem (the *Horn Conjecture*, 1962). The conjecture was finally settled in the affirmative by Knutson and Tao in 1999 [110].

Exercises

8.4.1 In the following let $V = \mathrm{S}^d$ and $C = \mathrm{S}^d_+$ with $d \ge 1$. For each of the following data, consider the semidefinite program (P): $p^* = \inf\{\langle x, w\rangle\colon x \in C \cap (v+U)\}$. Formulate the dual program (D) (see Remark 8.4.8), decide for both (P) and (D) whether they are feasible or even strictly feasible, and whether they have an optimal point, and determine the duality gap:

(a) $d = 2$, $U = \operatorname{span}(E_{11}, E_{22})$, $v = E_{12} + E_{21}$, $w = E_{11}$;
(b) $d = 2$, $U = \operatorname{span}(E_{22})$, $v = E_{12} + E_{21}$, $w = cE_{22}$ with $c \in \mathbb{R}$;
(c) $d = 3$, $U = \operatorname{span}(E_{11}, E_{13} + E_{22} + E_{31})$, $v = E_{22}$, $w = E_{13} + E_{31}$.

8.4.2 For $1 \le k \le n$, write the right-hand side of (8.19) (Theorem 8.4.11) as a semidefinite program in the standard form (P) of 8.4.1, and formulate the dual program (D) as in 8.4.2. Decide whether (P) and/or (D) are strictly feasible, and find primal and dual optimal points as far as they exist.

8.4.3 What optimum do you get in (8.19) (Theorem 8.4.11) if the condition $\operatorname{tr}(X) = k$ is replaced by $\operatorname{tr}(X) = \alpha$ for some real number $0 < \alpha < n$?

8.4.4 Let A be a real matrix of size $m{\times}n$ where $m \le n$, and let $\lambda_1 \ge \cdots \ge \lambda_m \ge 0$ be the eigenvalues of $AA^\top$. Then $\sigma_i(A) := \sqrt{\lambda_i} \ge 0$ is called the *i-th singular value* of A $(i = 1, \ldots, m)$. For $k = 1, \ldots, m$ let $\|A\|_k := \sigma_1(A) + \cdots + \sigma_k(A)$, and prove:

(a) The symmetric matrix $\tilde{A} = \left(\begin{smallmatrix} 0 & A \\ A^\top & 0 \end{smallmatrix}\right)$ (of size $(m+n) \times (m+n)$) has eigenvalues $\pm\sqrt{\lambda_i}$ $(i = 1, \ldots, m)$, together with the $(n-m)$-fold eigenvalue 0.
(b) $\|A + B\|_k \le \|A\|_k + \|B\|_k$ holds for arbitrary $A,\ B \in \mathrm{M}_{m\times n}(\mathbb{R})$.

(b) implies that $\|\cdot\|_k$ is a norm on the vector space $\mathrm{M}_{m\times n}(\mathbb{R})$, called the *k-th Ky Fan norm.*

8.5 Polynomial Optimization via Moment Relaxation

We now turn to an important application of semidefinite programming, namely optimization of a polynomial over a given (semialgebraic) set. A key technique for this task is the moment relaxation approach, and we are going to describe it in detail. Under Archimedean conditions, we will prove strong convergence results.

8.5.1 Let $M \subseteq \mathbb{R}^n$ be a given semialgebraic set. Optimizing a polynomial $f \in \mathbb{R}[x] = \mathbb{R}[x_1, \dots, x_n]$ over M is the task of finding the infimum $f^* := \inf_{\xi \in M} f(\xi)$ of f over M. When the infimum is known to be a minimum (e.g. when M is compact), one may in addition ask for an explicit optimizer, namely a point $u \in M$ with $f^* = f(u)$.

In principle one could hope to proceed as follows, using semidefinite optimization. Let $f = \sum_{|\alpha| \le d} c_\alpha x^\alpha$ with $d = \deg(f)$, and let $v \colon \mathbb{R}^n \to \mathbb{R}^N$ be the Veronese map $v(\xi) = (\xi^\alpha)_{|\alpha| \le d}$ with $N = \binom{n+d}{d}$. Replacing the set M with its image $S := v(M) \subseteq \mathbb{R}^N$, we have linearized the objective function, since f^* is the infimum of the linear function

$$z = (z_\alpha)_{|\alpha| \le d} \mapsto \sum_\alpha c_\alpha z_\alpha$$

over $S = v(M)$, or equivalently, over the convex hull $K = \text{conv}(S)$ of S. Assuming that we know how to represent K as a spectrahedral shadow, the original problem has been transformed into a semidefinite program, and as such can be solved.

All practical problems notwithstanding that come with such an approach, there remains a serious theoretical issue: In general, the convex hull $K = \text{conv}(S)$ will not allow any semidefinite representation. This question will be taken up in detail in Section 8.7.

8.5.2 Instead there exists a much more manageable and explicit approach. Generally known as the *moment relaxation* method, it was developed around the turn of the millennium by Lasserre, and also independently by Parrilo. The basic idea is as follows. Using a Veronese-type argument as before, we may assume that the problem is to optimize a *linear* polynomial $f \in \mathbb{R}[x]$ over a semialgebraic set $S \subseteq \mathbb{R}^n$, or equivalently, over its convex hull $K = \text{conv}(S)$. Assuming that the set S is basic closed (which is essentially harmless), one constructs a decreasing sequence $K_1 \supseteq K_2 \supseteq \cdots \supseteq K$ of convex outer approximations of K, together with an explicit semidefinite representation for each K_d. Under suitable assumptions (see below), every neighborhood of K contains some K_d, and under still stronger assumptions, the convex hull $K = \text{conv}(S)$ even coincides with some K_d. Since each set K_d comes with an explicit semidefinite representation, we may find $f^*_{(d)} := \inf_{\xi \in K_d} f(\xi)$ for each d by an explicit semidefinite program. We thus get a hierarchy of semidefinite programs, often referred to as the *Lasserre hierarchy*. Under favorable conditions as above, the sequence $f^*_{(1)} \le f^*_{(2)} \le \cdots$ of their optima will converge to $f^* = \inf_{\xi \in S} f(\xi)$. Responsible for this are the fundamental results discussed in Chapter 5, and in particular, the positivstellensätze of Schmüdgen and Putinar.

Usually, with d increasing, the complexity of the semidefinite programs in the Lasserre hierarchy grows rapidly. Therefore, when d gets too large, performing a semidefinite program over K_d tends to become too expensive, or even practically infeasible altogether. Still, moment relaxation has become an indispensable tool, both in practice (with many refinements and ramifications in detail), and also for theoretical questions, as we will see.

8.5.3 Now the details. Let $S \subseteq \mathbb{R}^n$ be a semialgebraic set and let $I \subseteq \mathbb{R}[x] = \mathbb{R}[x_1, \dots, x_n]$ be the full vanishing ideal of S (or of the Zariski closure of S). By

$L = \mathbb{R}[x]_{\le 1}$ we denote the space of linear polynomials. Let $g_0 = 1$, fix a tuple $\boldsymbol{g} = (g_1, \dots, g_r)$ of polynomials in $\mathbb{R}[x]$ with $g_i \ge 0$ on S, and let $U, W_0, \dots, W_r \subseteq \mathbb{R}[x]$ be linear subspaces of finite dimension such that $L \subseteq U$ and $g_i W_i W_i \subseteq U$ holds for $i = 0, \dots, r$. For a standard choice of such U and W_i see Remark 8.5.7 below. Consider the convex cone

$$P_U = g_0 \cdot \Sigma W_0^2 + \cdots + g_r \cdot \Sigma W_r^2 + (I \cap U) \tag{8.20}$$

in U, where ΣW_i^2 denotes the cone of sums of squares of elements of W_i. Note that P_U consists of polynomials that are non-negative on S. As usual let $P_U^* = \{\mu \in U^\vee \colon \mu(f) \ge 0 \text{ for every } f \in P_U\}$ denote the dual cone of P_U, contained in the dual linear space $U^\vee$ of U.

Comparing P_U against the quadratic module $M = QM(g_1, \dots, g_r) + I$ in $\mathbb{R}[x]$, we observe that $P_U \subseteq M \cap U$ holds by construction. Usually this inclusion will be *strict*. We are going to use the cone P_U for constructing an explicit outer spectrahedral shadow approximation K_U of $K = \mathrm{conv}(S)$.

8.5.4 Lemma. *P_U^* is a spectrahedral cone in $U^\vee$.*

Proof. For every (finite-dimensional) vector space V over $\mathbb{R}$, let $S^2V^\vee$ denote the space of symmetric bilinear forms $V \times V \to \mathbb{R}$. After fixing a basis of V, we may identify $S^2V^\vee$ with the space of symmetric $N \times N$ matrices where $N = \dim(V)$. Any linear form μ on U gives symmetric bilinear forms

$$\beta_i(\mu)\colon\ W_i \times W_i \to \mathbb{R}, \quad (p, q) \mapsto \mu(pqg_i)$$

for $i = 0, \dots, r$. Moreover the maps $\beta_i \colon U^\vee \to S^2W_i^\vee$, $\mu \mapsto \beta_i(\mu)$ are linear. By definition of the dual cone we have $P_U^* = (I \cap U)^\perp \cap \bigcap_{i=0}^r \{\mu \in U^\vee \colon \forall\, p \in W_i\ \mu(g_i p^2) \ge 0\} = (I \cap U)^\perp \cap \bigcap_{i=0}^r \{\mu \in U^\vee \colon \beta_i(\mu) \ge 0\}$. This shows that P_U^* is a spectrahedral cone in $U^\vee$. □

8.5.5 As before let $L = \mathbb{R}[x]_{\le 1}$, let $\rho \colon U^\vee \to L^\vee$ be the restriction map between the dual linear spaces (recall $L \subseteq U$), and consider the affine-linear subspaces $U_1^\vee := \{\mu \in U^\vee \colon \mu(1) = 1\}$ of $U^\vee$ and $L_1^\vee := \{\lambda \in L^\vee \colon \lambda(1) = 1\}$ of $L^\vee$. Clearly $\rho(U_1^\vee) \subseteq L_1^\vee$ holds. For $\xi \in \mathbb{R}^n$ let $\varphi_\xi \in \mathbb{R}[x]^\vee$ be point evaluation in ξ, i.e. $\varphi_\xi(f) = f(\xi)$ for $f \in \mathbb{R}[x]$. Identify $L_1^\vee$ with $\mathbb{R}^n$ via

$$\mathbb{R}^n \xrightarrow{\approx} L_1^\vee, \quad \xi \mapsto (\varphi_\xi)|_L, \tag{8.21}$$

and consider $S \subseteq \mathbb{R}^n$ as a subset of $L_1^\vee$ via (8.21). Note that the map inverse to (8.21) is $\lambda \mapsto (\lambda(x_1), \dots, \lambda(x_n))$.

Since P_U^* is a spectrahedral cone in $U^\vee$ (Lemma 8.5.4), the intersection $P_U^* \cap U_1^\vee$ is a spectrahedron in $U_1^\vee \subseteq U^\vee$. Therefore the set

$$K_U := \rho(P_U^* \cap U_1^\vee) = L_1^\vee \cap \rho(P_U^*) \tag{8.22}$$

is a spectrahedral shadow in $L_1^\vee = \mathbb{R}^n$. By definition, K_U consists of all $\xi \in \mathbb{R}^n$ for which there exists $\mu \in U^\vee$ with $\mu|_{P_U} \ge 0$ and $\mu(f) = f(\xi)$ for all $f \in L$. Or else,

$$K_U = \{(\mu(x_1), \dots, \mu(x_n)) : \mu \in P_U^* \subseteq U^\vee,\ \mu(1) = 1\}.$$

Tautologically, every linear polynomial in P_U is non-negative on K_U.

The spectrahedral cone $P_U \subseteq U^\vee$, together with its shadow $K_U \subseteq \mathbb{R}^n$, depends on the choice of U, and also of the tuples $\boldsymbol{g} = (g_1, \dots, g_r)$ and $W = (W_0, \dots, W_r)$. To indicate this dependence explicitly, we temporarily write $P(\boldsymbol{g}, U, W)$ for P_U and $K(\boldsymbol{g}, U, W)$ for K_U, respectively. We record:

8.5.6 Lemma. *The spectrahedral shadow $K_U = K(\boldsymbol{g}, U, W)$ contains the convex hull K of S. If U' and $W' = (W_0', \dots, W_r')$ are other choices of finite-dimensional spaces satisfying the conditions in 8.5.3, and if $U \subseteq U'$ and $W_i \subseteq W_i'$ hold for all i, then $K(\boldsymbol{g}, U', W') \subseteq K(\boldsymbol{g}, U, W)$.*

Proof. Point evaluation φ_ξ in $\xi \in S$ is a linear form on U that is non-negative on P_U. This shows $S \subseteq K_U$, and therefore also $\mathrm{conv}(S) = K \subseteq K_U$ since K_U is convex. The second assertion is immediate from $P(\boldsymbol{g}, U, W) \subseteq P(\boldsymbol{g}, U', W')$. □

8.5.7 Remark. To approximate the convex hull $K = \mathrm{conv}(S)$ of a basic closed set $S = \mathcal{S}(g_1, \dots, g_r)$ in $\mathbb{R}^n$, the standard choice is to put $g_0 = 1$, to fix a degree $d \geq 1$ and to let $U = \mathbb{R}[x]_{\leq d}$ and $W_i = \{p \in \mathbb{R}[x] : \deg(p^2 g_i) \leq d\}$ $(i = 0, \dots, r)$ in 8.5.3. The resulting convex cone P_U, that we denote by M_d here, is

$$M_d = \Sigma W_0^2 + g_1 \cdot \Sigma W_1^2 + \cdots + g_r \cdot \Sigma W_r^2 + I_{\leq d},$$

where again I is the vanishing ideal of S. The cone M_d is a truncated version of the quadratic module $M = QM(g_1, \dots, g_r) + I$, often called the truncation of M at degree d.[1] As pointed out before, the inclusion $M_d \subseteq M \cap \mathbb{R}[x]_{\leq d}$ will usually be strict. Let $K_d := L_1^\vee \cap \rho(M_d^*)$ as in (8.22). Increasing the degree $d = 1, 2, \dots$ results in a nested sequence $K_1 \supseteq K_2 \supseteq \cdots$ of outer approximations K_d of K, each of them coming with an explicit semidefinite representation (Lemma 8.5.6).

Although this is the standard choice, it may be preferable to work with different $\boldsymbol{g}$, U and W_i, depending on the situation. This is why we allow for the greater flexibility in our discussion. For a concrete example illustrating this point, see 8.5.23 below.

8.5.8 Remark. The spectrahedral shadow $K_U = K(\boldsymbol{g}, U, W)$ in $\mathbb{R}^n$ is described by a lifted LMI that is entirely explicit. To amplify this remark assume $I = \{0\}$, and take the standard truncated quadratic module M_d for some $d \geq 0$, as in Remark 8.5.7, for the sake of concreteness. Let $g \in \mathbb{R}[x]$, say $g = \sum_{|\alpha| \leq \deg(g)} g_\alpha x^\alpha$, put $e = \lfloor \frac{1}{2}(d - \deg(g)) \rfloor$, let $U = \mathbb{R}[x]_{\leq d}$ and $W = \mathbb{R}[x]_{\leq e}$. A linear basis for the dual space $U^\vee$ is given by the linear forms μ_α defined by $\mu_\alpha(x^\beta) = \delta_{\alpha,\beta}$ for $|\alpha|, |\beta| \leq d$. Following the proof of Lemma 8.5.4 we consider, for every $\mu \in U^\vee$, the symmetric bilinear form $\beta_g(\mu)$ on W defined by

[1] Note that, in general, M_d will depend not only on M and d, but also on the choice of the generators $g_1, \dots, g_r$ of M.

$$\beta_g(\mu)\colon\ W \times W \to \mathbb{R}, \quad (p,q) \mapsto \mu(pqg).$$

Let $S_\alpha(g)$ denote the matrix of $\beta_g(\mu_\alpha)$ with respect to the monomial basis of W, arranged in some fixed order. This is the matrix

$$S_\alpha(g) = \left(g_{\alpha-\chi-\eta}\right)_{|\chi|,\,|\eta|\le e}$$

since

$$\beta_g(\mu_\alpha)(x^\chi, x^\eta) = \mu_\alpha(x^{\chi+\eta}g) = \sum_\sigma g_\sigma\,\mu_\alpha(x^{\chi+\eta+\sigma}) = g_{\alpha-\chi-\eta}$$

for $|\chi|$, $|\eta| \le e$. Using the abbreviations $S_0(g) := S_{(0,\dots,0)}(g)$ and $S_i(g) := S_{e_i}(g)$ for $i = 1,\dots,n$, the spectrahedral cone in $U^\vee$ defined by the linear map $\beta_g\colon U^\vee \to S^2W^\vee$ is

$$\left\{\mu \in U^\vee\colon \beta_g(\mu) \ge 0\right\} = \left\{\sum_{|\alpha|\le d} u_\alpha\mu_\alpha\colon u_\alpha \in \mathbb{R},\ \sum_{|\alpha|\le d} u_\alpha S_\alpha(g) \ge 0\right\}.$$

And its projection to $\mathbb{R}^n = L_1^\vee$ is the set of $\xi \in \mathbb{R}^n$ for which there exist real numbers $u_\alpha \in \mathbb{R}$ $(1 < |\alpha| \le d)$ with

$$S_0(g) + \sum_{i=1}^n \xi_i S_i(g) + \sum_{1<|\alpha|\le d} u_\alpha S_\alpha(g) \ \ge\ 0.$$

Denoting the left-hand matrix by $S_g(\xi, u)$, the spectrahedral shadow K_d is therefore given as

$$K_d = \left\{\xi \in \mathbb{R}^n\colon \exists\, u_\alpha \in \mathbb{R}\ (1 < |\alpha| \le d) \text{ with } S_{g_i}(\xi, u) \ge 0\ (i = 0,\dots,r)\right\}.$$

This is an explicit semidefinite representation of K_d.

8.5.9 Example. For an illustration let $g_0 = 1$, $g_1 = x$, $g_2 = x^2 - y^2 - x^3$. The basic closed set $S = \mathcal{S}(g_1, g_2)$ in $\mathbb{R}^2$ (drawn in yellow below) is compact and convex, so $S = K$ in the previous setup:

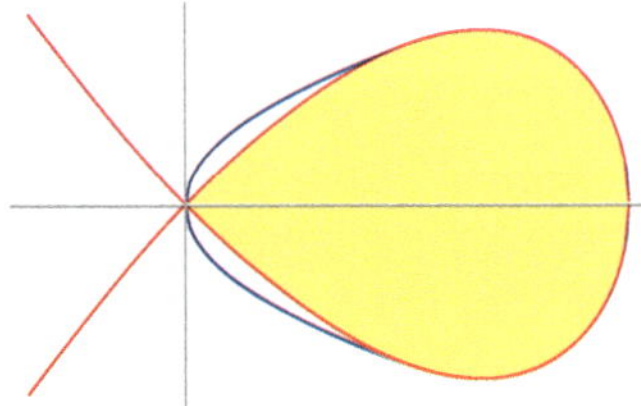

The blue curve shows the (boundary of the) moment relaxation K_3, obtained for degree $d = 3$. So $U = \mathbb{R}[x,y]_{\le 3}$ here, and $W_0 = W_1 = \mathbb{R}[x,y]_{\le 1}$ and $W_2 = \mathbb{R}$. Writing a general element of $U^\vee$ as

$$\mu = w\mu_1 + \xi\mu_x + \eta\mu_y + a\mu_{x^2} + b\mu_{xy} + c\mu_{y^2} + d\mu_{x^3} + e\mu_{x^2y} + f\mu_{xy^2} + g\mu_{y^3} \quad (8.23)$$

with real scalars $w, \xi, \eta, \dots, g \in \mathbb{R}$, the relaxation K_3 consists of all points $(\xi, \eta) \in \mathbb{R}^2$ for which there exist real numbers $a, b, \dots, f$ with

$$\begin{pmatrix} 1 & \xi & \eta \\ \xi & a & b \\ \eta & b & c \end{pmatrix} \succeq 0, \quad \begin{pmatrix} \xi & a & b \\ a & d & e \\ b & e & f \end{pmatrix} \succeq 0, \quad a - c - d \geq 0. \tag{8.24}$$

The three inequalities in (8.24) represent the conditions $\beta_{g_i}(\mu) \succeq 0$ for $i = 0, 1, 2$, for μ a linear form as in (8.23). From the drawing we see $K_3 \neq K$. In fact, it is not hard to show (Exercise 8.5.3) that $K_d \neq K$ for every $d \geq 1$. That is, moment relaxation for K does not become exact at any finite level. On the other hand, we'll see a little later (Example 8.5.20) that a slight variation of the present approach leads easily to an exact semidefinite representation of K.

8.5.10 Remark. Construction 8.5.5 arises from relaxing truncated moment problems. We explain this using the standard degree relaxation, as in 8.5.7. Let $S \subseteq \mathbb{R}^n$ be a semialgebraic set, let $g_1, \dots, g_r \in \mathbb{R}[x]$ with $g_i|_S \geq 0$, and let $M = QM(g_1, \dots, g_r) \subseteq \mathbb{R}[x]$. A linear form $\lambda \colon \mathbb{R}[x] \to \mathbb{R}$ is an *S-moment functional* if there exists a positive Borel measure μ on S with $\lambda(f) = \int_S f \, d\mu =: \lambda_\mu(f)$ for all $f \in \mathbb{R}[x]$. In particular, it is required that all moments of μ exist, i.e. that $\int_S \left| x_1^{\alpha_1} \cdots x_n^{\alpha_n} \right| \mu(dx) < \infty$ for all $\alpha \in \mathbb{Z}_+^n$. Let $\mathcal{M}(S) \subseteq \mathbb{R}[x]^\vee$ be the convex cone consisting of all S-moment functionals (the *moment cone* of S), and let $\mathcal{M}_1(S) = \{\lambda \in \mathcal{M}(S) \colon \lambda(1) = 1\}$, corresponding to probability measures on S. It is not hard to see (Exercise 8.5.1) that the points in the convex hull of S are precisely the expectations of measures in $\mathcal{M}_1(S)$, i.e.,

$$K = \operatorname{conv}(S) = \left\{ (\mu(x_1), \dots, \mu(x_n)) \colon \mu \in \mathcal{M}_1(S) \right\} \subseteq \mathbb{R}^n. \tag{8.25}$$

For any $d \geq 1$, let M_d be the truncated quadratic module, as defined in 8.5.7. We have the obvious inclusions

$$\mathcal{M}(S) \subseteq M^* \subseteq M_d^*$$

of convex cones in $\mathbb{R}[x]^\vee$, where the first holds since $\int_S f \, d\mu \geq 0$ for any $f \in M$ and any measure μ on S. The intersections of these cones with the monic functionals give rise to outer approximations of K, via (8.25). The last cone corresponds to K_d as constructed before, since $K_d = \{(\mu(x_1), \dots, \mu(x_n)) \colon \mu \in M_d^*, \, \mu(1) = 1\}$ by definition.

We can only mention the important classical theorem of Haviland [85], [86]: If $S \subseteq \mathbb{R}^n$ is any closed subset, then $\mathcal{M}(S) = \mathcal{P}(S)^*$. In other words, a linear form $\lambda \colon \mathbb{R}[x] \to \mathbb{R}$ is integration with respect to some measure supported on S if (and only if) $\lambda(f) \geq 0$ for every polynomial f that is non-negative on S.

For $d \to \infty$, the construction in 8.5.5 gives a nested sequence of smaller and smaller sets K_d that all contain $K = \operatorname{conv}(S)$, and that all are projected spectrahedra. We ask, when is it true that $K = \bigcap_{d \geq 1} K_d$? Or even $K = K_d$ for some $d \geq 1$? To answer these questions we take the dual point of view and study the cones $M_d \cap L$

of linear polynomials in the truncated modules M_d. For the following discussion we return to the general setup introduced in 8.5.3. So we fix a semialgebraic set $S \subseteq \mathbb{R}^n$, together with its convex hull $K = \mathrm{conv}(S)$ and its vanishing ideal $I = \mathcal{I}(S) \subseteq \mathbb{R}[x]$. Moreover, put $g_0 = 1$ and fix a sequence $\boldsymbol{g} = (g_1, \dots, g_r)$ of polynomials that are non-negative on S, together with linear subspaces $U, W_0, \dots, W_r \subseteq \mathbb{R}[x]$ satisfying $g_i W_i W_i \subseteq U$ for all i. Let the convex cone $P_U \subseteq U$ be defined as in (8.20), and let $K_U \subseteq \mathbb{R}^n$ be the resulting projected spectrahedron, see (8.22).

8.5.11 Proposition. *The cone P_U is closed in U.*

Proof. This was proved in Lemma 6.6.18. Indeed, writing $M = QM(g_1, \dots, g_r)$, the ideal $I = \mathcal{I}(S)$ is M-convex, since $p + q \in I$ and $p, q \in M$ imply $p, q \geq 0$ on S, and hence $p, q \in I$. So the assertion follows from Lemma 6.6.18. □

8.5.12 Lemma. *The cone P_U contains every linear polynomial that is non-negative on K_U, and so $P_U \cap L = \mathcal{P}(K_U) \cap L$.*

Proof. When $S = \varnothing$ we have $I = \langle 1 \rangle$, and therefore $P_U = U$. So we may assume that S is non-empty. Every linear polynomial in P_U is non-negative on K_U, see 8.5.5. Conversely, let $f \in L$ be linear with $f \geq 0$ on K_U. We have to show $f \in P_U$. Since P_U is closed, it suffices to prove that $\mu(f) \geq 0$ for every linear form μ in $P_U^* \subseteq U^\vee$. Since $1 \in P_U$, note that $\mu(1) \geq 0$ for any such μ. We consider two cases.

1st case: If $\mu(1) > 0$, we may scale μ to have $\mu(1) = 1$. The point $\xi := (\mu(x_1), \dots, \mu(x_n))$ in $\mathbb{R}^n$ satisfies $f(\xi) = \mu(f)$. Since $\xi \in K_U$ by the definition of K_U, this implies $\mu(f) = f(\xi) \geq 0$.

2nd case: $\mu(1) = 0$. Choose a point $\xi \in S$, let φ_ξ be point evaluation in ξ, and put $\mu_t := \frac{1}{t}(\mu + t\varphi_\xi) \in U^\vee$ for $t > 0$. Since clearly $\mu_t \in P_U^*$ and $\mu_t(1) = 1$, the point $\xi_t := (\mu_t(x_1), \dots, \mu_t(x_n)) \in \mathbb{R}^n$ lies in K_U and satisfies $\mu_t(f) = f(\xi_t)$. So again $\mu_t(f) \geq 0$, and hence $\mu(f) + tf(\xi) \geq 0$. Now pass to the limit $t \to 0$ to conclude $\mu(f) \geq 0$. □

8.5.13 Corollary. *Let K be the convex hull of S, as before. Then*

(a) $\overline{K_U} = \{\xi \in \mathbb{R}^n \colon f(\xi) \geq 0$ *for every* $f \in P_U \cap L\}$.
(b) $\overline{K_U} = \overline{K}$ *holds if and only if P_U contains every linear polynomial that is non-negative on S.*

Proof. (a) Since K_U is convex, its closure is $\overline{K_U} = \bigcap_f \{\xi \in \mathbb{R}^n \colon f(\xi) \geq 0\}$, intersection over all linear f with $f \geq 0$ on K_U (hyperplane separation, Corollary 8.1.6). By Lemma 8.5.12, the intersection is running over all $f \in P_U \cap L$.

(b) Both sets $K \subseteq K_U$ are convex. So their closures agree if, and only if, every linear polynomial non-negative on S is also non-negative on K_U. This means precisely $L \cap \mathcal{P}(S) \subseteq P_U$, again by Lemma 8.5.12. □

In particular we see:

8.5.14 Corollary. *Assume that S is compact. Then $K_U = K$ holds if, and only if, P_U contains every linear polynomial that is non-negative on S.*

Proof. Since S compact implies K compact, this is clear from 8.5.13(b). □

Let us exploit the previous conclusions in the situation of 8.5.7 (standard degree relaxation). We adopt the notation introduced there.

8.5.15 Proposition. *Let $S = \mathcal{S}(g_1, \dots, g_r) \subseteq \mathbb{R}^n$ with $g_i \in \mathbb{R}[x]$, let $K = \mathrm{conv}(S)$ as before, and put $M = QM(g_1, \dots, g_r) + I$ where I is the vanishing ideal of S. If every linear polynomial f with $f|_S \geq 0$ can be approximated by linear polynomials in M, then*

$$\overline{K} = \bigcap_{d \geq 1} \overline{K_d} \tag{8.26}$$

holds. Under the same assumption, if in addition S is compact, then even $K = \bigcap_{d \geq 1} K_d$ holds.

Proof. By hypothesis, $L \cap M$ is dense in $L \cap \mathcal{P}(S)$. The inclusion "$\subseteq$" in (8.26) holds anyway. To prove equality, we have to show $f(\xi) \geq 0$ for every $f \in L \cap \mathcal{P}(S)$ and every $\xi \in \bigcap_d \overline{K_d}$. Since $f \in \overline{L \cap M}$ by hypothesis, and since $L \cap M$ is convex, there exists $g \in L$ with $f + tg \in M$ for all $t > 0$, see 8.1.4. Thus, for every $t > 0$, there is an integer $d(t) \geq 1$ with $f + tg \in L \cap M_{d(t)}$. Since $\xi \in \overline{K_{d(t)}}$ we have $(f + tg)(\xi) = f(\xi) + tg(\xi) \geq 0$. Letting $t \to 0$ we see $f(\xi) \geq 0$, as desired. If in addition S is compact, the last claim follows from (8.26) since $\overline{K} = K \subseteq \bigcap_d K_d$. □

If S is compact in 8.5.15, the convex hull K of S is in fact approximated arbitrarily closely by the projected spectrahedra K_d, according to the following general observation:

8.5.16 Proposition. *If $K_1 \supseteq K_2 \supseteq \cdots \supseteq \bigcap_{m \geq 1} K_m = K$ is a sequence of convex sets in $\mathbb{R}^n$, and if K is compact and has non-empty interior, then for every $\varepsilon > 0$ there is an index $m \geq 1$ such that $d_K(u) < \varepsilon$ for every $u \in K_m$.*

Here $d_K(u) = \mathrm{dist}(u, K) = \inf\{|u - v| \colon v \in K\}$, the distance from u to K (see 4.3.6.2). The proposition asserts that K_m is contained in an ε-tube around K, for m large enough.

Proof. Choose a point u in the interior of K, and let $v \in \bigcap_m \overline{K_m}$. Then $[u, v[\subseteq K_m$ for all m since K_m is convex (8.1.4). So $[u, v[\subseteq K$, and hence $v \in K$ since K is closed. This shows that $K = \bigcap_m \overline{K_m}$ holds as well. Let $\varepsilon > 0$, then $U := \{v \in \mathbb{R}^n \colon d_K(v) < \varepsilon\}$ is an open neighborhood of K since the map d_K is continuous. Assume that $K_m \not\subseteq U$ for all m. Then for every $m \geq 1$ there is $v_m \in K_m$ with $d_K(v_m) \geq \varepsilon$. Since K_m is convex, there also exists $v_m \in K_m$ with $d_K(y_m) = \varepsilon$. The set $X := \{v \in \mathbb{R}^n \colon d_K(v) = \varepsilon\}$ is compact and satisfies $X \cap K_m \neq \varnothing$ for all m. Hence $X \cap \bigcap_m \overline{K_m}$ is non-empty, being a filtering intersection of non-empty compact sets, and so $X \cap K \neq \varnothing$ by the argument at the beginning of the proof. This is a contradiction. □

The most important case where the previous results apply is when the quadratic module M is Archimedean:

8.5.17 Corollary. *With assumptions as in 8.5.15, assume that the quadratic module M is Archimedean. Then $K = \bigcap_{d\ge 1} K_d$, and every ε-tube around K contains K_d for some $d \ge 1$.*

Proof. M Archimedean implies that S is compact, and hence $K = \operatorname{conv}(S)$ is compact as well. We may assume that K is full-dimensional in $\mathbb{R}^n$. For every $f \in \mathcal{P}(S)$, and for every real number $\varepsilon > 0$, we have $f + \varepsilon \in M$ by the Archimedean positivstellensatz. So the hypothesis $L \cap \mathcal{P}(S) \subseteq \overline{L \cap M}$ of Proposition 8.5.15 holds. The assertion follows from this proposition and from 8.5.16. □

8.5.18 Remark. For any compact and basic closed set $S \subseteq \mathbb{R}^n$, we can therefore construct a sequence $K_1 \supseteq K_2 \supseteq \cdots$ of projected spectrahedra with explicit lifted LMI representations, whose intersection is the convex hull K of S. For this take any finite system $\boldsymbol{g}$ of polynomial inequalities that describes S, and add the inequality $r^2 - \sum_i x_i^2 \ge 0$ if $S \subseteq B_r(0)$, to make sure that the quadratic module M generated by the chosen inequalities is Archimedean (Putinar's positivstellensatz 5.5.10). If there exists $d \ge 1$ with

$$L \cap \mathcal{P}(S) \subseteq M_d, \tag{8.27}$$

we even have equality $\operatorname{conv}(S) = K_d$ (Corollary 8.5.14). The condition $K = K_d$ is usually expressed by saying that the *relaxation gets exact at level d*. Note that condition (8.27) means both a "partial saturatedness" and a "partial stability" condition on M, each for polynomials of degree one.

8.5.19 Remark. Given a semialgebraic set $S \subseteq \mathbb{R}^n$, we constructed a family of outer approximations to the convex hull of S, see 8.5.3–8.5.5. The essential point was that these approximations come with explicit semidefinite representations, which in turn can be used to approximate the problem of minimizing linear functions over S by explicit semidefinite programs. At least under a compactness assumption on S, we proved that these approximations can be made arbitrarily close.

Instead of working in the polynomial ring $\mathbb{R}[x]$ and with the space L of linear polynomials, one can take advantage of a more general approach that we briefly want to sketch. Let V be an arbitrary affine $\mathbb{R}$-variety, let $S_0 \subseteq V(\mathbb{R})$ be a semialgebraic set that we assume to be Zariski dense in V, and let $L \subseteq \mathbb{R}[V]$ be a linear subspace of finite dimension with $1 \in L$. Given a linear basis $1, y_1, \dots, y_n$ of L, we consider the Veronese-type map $\phi\colon V \to \mathbb{A}^n$ whose components are $y_1, \dots, y_n$. Working in the coordinate ring $\mathbb{R}[V]$ and mimicking the procedure explained in 8.5.3–8.5.5, it is then straightforward to construct outer approximations for the convex hull of $S := \phi(S_0)$ in $\mathbb{R}^n$, which again come with explicit semidefinite representations:

Let $1 = g_0, g_1, \dots, g_r \in \mathbb{R}[V]$ with $g_i \ge 0$ on S_0, and let U, $W_i \subseteq \mathbb{R}[V]$ be linear subspaces of finite dimension with $L \subseteq U$ and $g_i W_i W_i \subseteq U$ $(0 \le i \le r)$. Similar to Lemma 8.5.4 and its proof, we use the linear maps

$$\beta_i\colon U^\vee \to S^2 W_i^\vee, \quad \beta_i(\mu)(p, q) = \mu(pqg_i) \quad (i = 0, \dots, r).$$

Extend the given basis of L to a linear basis $1, y_1, \dots, y_n, z_1, \dots, z_m$ of U, and denote the corresponding dual basis of $U^\vee$ by $\mu_1, \mu_{y_1}, \dots, \mu_{z_m}$. The projected spectrahedron associated with this data is

$$K_U = \Big\{\xi \in \mathbb{R}^n \colon \ \exists b \in \mathbb{R}^m \text{ with } \beta_i\Big(\mu_1 + \sum_{j=1}^n \xi_j \mu_{y_j} + \sum_{k=1}^m b_k \mu_{z_k}\Big) \geq 0 \ (i = 0, \dots, r)\Big\},$$

and we have $\phi(S_0) \subseteq K_U$. If S_0 is compact and $L \cap \mathcal{P}(S_0) \subseteq \sum_i g_i \Sigma W_i^2$, then relaxation is exact as in Corollary 8.5.14, i.e. $K_U = \operatorname{conv}(\phi(S_0))$.

8.5.20 Examples.

1. To illustrate the usefulness of this generalized approach, consider again the compact convex set $S = K = \mathcal{S}(x,\ x^2 - x^3 - y^2) \subseteq \mathbb{R}^2$ from Example 8.5.9. As noted in Exercise 8.5.3, standard moment relaxation for K does not become exact at any finite level. Responsible for this is the singularity of the boundary curve at the origin, as already suggested by the drawing in 8.5.9. Still K is a spectrahedral shadow, and the approach from 8.5.19 offers a cheap way of producing an exact semidefinite representation:

Let $C = \mathcal{V}(x^2 - x^3 - y^2) \subseteq \mathbb{A}^2$, the algebraic boundary curve of K. We consider the normalization of C, which is the affine line $\mathbb{A}^1$ together with the morphism $\phi\colon \mathbb{A}^1 \to C \subseteq \mathbb{A}^2$ given by $\phi(t) = (1 - t^2, t - t^3)$. Let $\varphi = \phi^*\colon \mathbb{R}[x,y] \to \mathbb{R}[t]$ be the corresponding ring homomorphism, so $\varphi(x) = 1 - t^2$ and $\varphi(y) = t - t^3$. Accordingly, let $L = \operatorname{span}(1, \varphi(x), \varphi(y)) \subseteq \mathbb{R}[t]$. Since $\partial K = \phi(I)$ with $I = [-1, 1]$, we want to express the polynomials $p \in L$ with $p|_I \geq 0$ by weighted sums of squares in $\mathbb{R}[t]$. For this we use Exercise 6.6.1, which implies that every such p can be written $p(t) = \sigma_0(t) + (1 - t^2)\sigma_1(t)$, where $\sigma_0(t)$, $\sigma_1(t)$ are sums of squares in $\mathbb{R}[t]$ with $\deg(\sigma_0) \leq 4$ and $\deg(\sigma_1) \leq 2$. Accordingly put $g_0 = 1$, $g_1 = 1 - t^2$ and $U = \mathbb{R}[t]_{\leq 4}$, $W_0 = \mathbb{R}[t]_{\leq 2}$, $W_1 = \mathbb{R}[t]_{\leq 1}$, in the notation of 8.5.19. This gives the following semidefinite representation: K consists of all points $(\xi, \eta) \in \mathbb{R}^2$ for which there exist a, $b \in \mathbb{R}$ such that

$$\begin{pmatrix} 1 & a+\eta & 1-\xi \\ a+\eta & 1-\xi & a \\ 1-\xi & a & b \end{pmatrix} \succeq 0 \quad \text{and} \quad \begin{pmatrix} \xi & \eta \\ \eta & 1-\xi-b \end{pmatrix} \succeq 0.$$

(These are the matrices of $\beta_0(\mu)$ and $\beta_1(\mu)$ for the linear form $\mu = \mu_1 + \xi\mu_{\varphi(x)} + \eta\mu_{\varphi(y)} + a\mu_{t^3} + b\mu_{t^4}$ in $U^\vee$, with notation as in 8.5.19.)

2. Independently from the previous approach, one can do even better for this particular example. The convex set K is rigidly convex (with respect to any of its interior points). Therefore, according to the former Lax conjecture, as proved by Helton and Vinnikov (Theorem 8.2.27), K must be a spectrahedron that can be represented by a linear matrix inequality of size 3×3. In this concrete case, it is not hard to find such an LMI directly, see Exercise 8.5.4.

We close with a few more examples in which we obtain explicit semidefinite representations for given convex sets.

8.5.21 Example. Let $X \subseteq \mathbb{A}^2$ be the plane affine curve with equation $x^2 - y^2 - x^4 = 0$. The curve X is rational and can be parametrized by the circle curve $C = \mathcal{V}(u^2 + v^2 - 1)$, via the (normalization) map $\varphi\colon C \to X$, $(u, v) \mapsto (u, uv)$. Note that $\varphi(C(\mathbb{R})) = X(\mathbb{R})$.

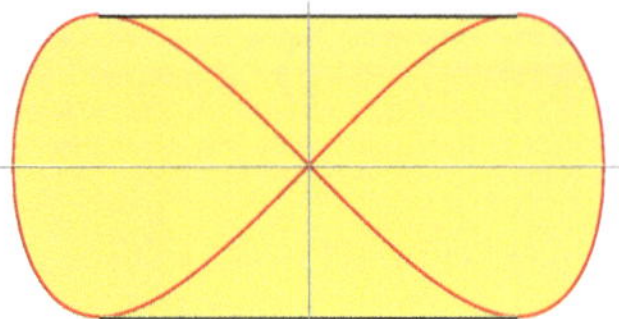

The curve $y^2 = x^2 - x^4$ (red)
and its convex hull

To describe the convex hull of $X(\mathbb{R})$, we construct relaxations with respect to the subspace $L = \text{span}(1, u, uv)$ of $\mathbb{R}[C]$. Choose $W = \text{span}(1, u, v)$ and $U = WW = \text{span}(1, u, v, u^2, uv)$ in $\mathbb{R}[C]$, then $L \subseteq U$, and K_W is the set of all $(\xi, \eta) \in \mathbb{R}^2$ with

$$\exists\, a, b \in \mathbb{R} \quad \begin{pmatrix} 1 & \xi & a \\ \xi & b & \eta \\ a & \eta & 1-b \end{pmatrix} \succeq 0.$$

(The matrix corresponds to $\beta(\mu_1 + \xi\mu_u + a\mu_v + b\mu_{u^2} + \eta\mu_{uv})$ with respect to the basis $1, u, v$ of W, since $v^2 = 1 - u^2$ implies $\mu(v^2) = 1 - b$. Compare the discussion in Remark 8.5.19.) In fact, the resulting spectrahedral shadow K coincides with the convex hull of $X(\mathbb{R})$, so this relaxation is already exact. Indeed, this follows from Corollary 8.5.14 since $L \cap \mathbb{R}[C]_+ \subseteq \Sigma W^2$, either by the S-Lemma 7.2.10 or by Fejér-Riesz 2.3.1. Note that this convex hull K is not a spectrahedron, by Remark 8.2.19. Indeed, its algebraic boundary intersects the interior of K.

8.5.22 Example. For another example consider the cuspidal curve $X \subseteq \mathbb{A}^2$ with equation $x^4 + y^2 = x^3$. Again the curve is rational, with normalization $\varphi\colon C \to X$,

$$\varphi(u, v) = \Big(\frac{1-v}{2}, \frac{u(1-v)}{4}\Big)$$

where $C = \mathcal{V}(u^2 + v^2 - 1)$ as before. To describe or approximate the convex hull of $X(\mathbb{R})$, we need to do relaxation with respect to the subspace $L = \text{span}(1,\ v,\ u(1-v))$ of $\mathbb{R}[C]$:

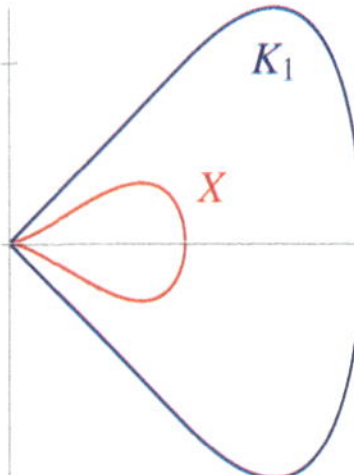

The curve $x^4 + y^2 = x^3$ (red)
and its first convex approximation

Using the substitution $w = 1 - v$ we have $\mathbb{R}[C] = \mathbb{R}[u, w]/\langle u^2 + w^2 - 2w\rangle$ and $L = \text{span}(1, w, uw)$, and the normalization map is $\varphi(u, w) = (\frac{w}{2}, \frac{uw}{4})$. For a first approximation take $W_1 = \text{span}(1, u, w)$ and $U_1 = W_1W_1 = \text{span}(1, u, w, u^2, uw)$. Then $L \subseteq U_1$ is satisfied, and K_{W_1} is the set of $(\xi, \eta) \in \mathbb{R}^2$ such that

$$\exists\, a, b \in \mathbb{R} \quad \begin{pmatrix} 1 & a & \xi \\ a & b & \eta \\ \xi & \eta & 2\xi - b \end{pmatrix} \succeq 0.$$

The set $K_1 := K_{W_1}$ is much larger than the convex hull of $X(\mathbb{R})$, as shown by the illustration. To improve the approximation we need to enlarge W_1. Taking $W_2 = \mathrm{span}(1, u, w, u^2, uw)$ and $U_2 = W_2W_2 = \mathrm{span}(1, u, w, u^2, uw, u^3, u^2w, u^4, u^3w)$, $K_2 = K_{W_2}$ is the set of $(\xi, \eta) \in \mathbb{R}^2$ for which real numbers $a, \dots, f$ exist with

$$\begin{pmatrix} 1 & a & \xi & b & \eta \\ a & b & \eta & c & d \\ \xi & \eta & 2\xi - b & d & 2\eta - c \\ b & c & d & e & f \\ \eta & d & 2\eta - c & f & 2d - e \end{pmatrix} \succeq 0.$$

It can be shown that this relaxation is exact, so K_2 agrees with the convex hull of $C(\mathbb{R})$.

8.5.23 Example. Let $m_1, \dots, m_n$ be even positive integers, and consider the polynomial $f(x) = f(x_1, \dots, x_n) = 1 - x_1^{m_1} - \cdots - x_n^{m_n}$. The set $S = \{\xi \in \mathbb{R}^n \colon f(\xi) \geq 0\}$ is compact and convex, and is a multi-dimensional generalization of the "tv-screen" in 8.2.24. We construct an explicit semidefinite representation of S. For $u \in \mathbb{R}^n$ let

$$t_u(x) = \sum_{i=1}^{n} \frac{\partial f}{\partial x_i}(u) \cdot (x_i - u_i) = \sum_{i=1}^{n} m_i u_i^{m_i - 1}(u_i - x_i),$$

the tangent to the level set $f = f(u)$ at u. In particular, when $f(u) = 0$ then t_u is the tangent to ∂S at u, and $t_u \geq 0$ on S. Therefore, the cone $\mathcal{P}(S) \cap L$ of linear polynomials which are non-negative on S is generated by 1 and the tangents $t_u(x)$, for $u \in \mathbb{R}^n$ with $f(u) = 0$. For $m \geq 1$ let p_m denote the binary form

$$p_m(x, y) = x^m - mxy^{m-1} + (m-1)y^m.$$

Since $my^{m-1}(y - x) = y^m - x^m + p_m(x, y)$, we have

$$t_u(x) = f(x) - f(u) + \sum_{i=1}^{n} p_{m_i}(x_i, u_i) \tag{8.28}$$

for all $x, u \in \mathbb{R}^n$.

For every even number $m > 1$, the univariate polynomial $p_m(t, 1) = t^m - mt + (m-1)$ (and hence the binary form $p_m(x, y)$ as well) is psd. Indeed, $p_m(t, 1)$ has a double root at $t = 1$ and is positive for $t > 0$, $t \neq 1$ by Descartes' rule of signs 1.3.12. For $t < 0$ it is obvious that $p_m(t, 1) > 0$. Therefore, the binary form $p_m(x, y)$ is a sum of (two) squares of binary forms of degree $\frac{m}{2}$.

With the notation as in 8.5.3, consider the finite-dimensional linear subspaces of $\mathbb{R}[x] = \mathbb{R}[x_1, \dots, x_n]$ defined by $U = \mathrm{span}(x_i^k \colon i = 1, \dots, n,\ k = 0, \dots, m_i)$ and $W_i = \mathrm{span}(x_i^k \colon k = 0, \dots, \frac{m_i}{2})$. Moreover, let $P \subseteq U$ be the truncated quadratic module $P = \Sigma W_1^2 + \cdots + \Sigma W_n^2 + \mathbb{R}f$. Then we have $L \cap \mathcal{P}(S) \subseteq \Sigma W_1^2 + \cdots + \Sigma W_n^2$, in view of the previous discussion. Therefore Corollary 8.5.14 gives (see Exercise 8.5.5 for full details):

8.5.24 Proposition. *Let $m_1, \ldots, m_n > 1$ be even integers. A point $\xi \in \mathbb{R}^n$ satisfies $\xi_1^{m_1} + \cdots + \xi_n^{m_n} \le 1$ if, and only if, there exist real numbers c_{ij} for $i = 1, \ldots, n$ and $2 \le j \le m_i$, such that*

$$c_{1m_1} + \cdots + c_{nm_n} = 1$$

and, writing $c_{i0} := 1$ and $c_{i1} := \xi_i$, the n LMIs

$$\left(c_{i,\lambda+\nu}\right)_{0 \le \lambda,\nu \le \frac{m_i}{2}} \succeq 0 \quad (i = 1, \ldots, n)$$

(of sizes $1 + \frac{m_1}{2}, \ldots, 1 + \frac{m_n}{2}$) are satisfied. □

For example, the convex set $S \subseteq \mathbb{R}^2$ described by the inequality $x^6 + y^4 \le 1$ consists of all $(\xi, \eta) \in \mathbb{R}^2$ such that

$$\begin{pmatrix} 1 & \xi & a_2 & a_3 \\ \xi & a_2 & a_3 & a_4 \\ a_2 & a_3 & a_4 & a_5 \\ a_3 & a_4 & a_5 & c \end{pmatrix} \succeq 0 \quad \text{and} \quad \begin{pmatrix} 1 & \eta & b_2 \\ \eta & b_2 & b_3 \\ b_2 & b_3 & 1-c \end{pmatrix} \succeq 0$$

hold for suitable real numbers $a_2, a_3, a_4, a_5, b_2, b_3$ and c.

It has been proved [186] that the closed convex hull of every one-dimensional semialgebraic set S in $\mathbb{R}^n$ has a semidefinite representation. Using a parametrization of S by a non-singular curve, such a representation can be found by some generalized moment relaxation, similar to Example 8.5.21. That this is possible is a consequence of a stability result, proved in [186] for compact subsets on non-singular affine curves C. Basically, when $C(\mathbb{R})$ is compact, it comes down to the existence of uniform degree bounds for sum of squares representations of non-negative polynomials on C.

Exercises

8.5.1 Let $S \subseteq \mathbb{R}^n$ be a semialgebraic set. With the notation of Remark 8.5.10, prove that

$$\operatorname{conv}(S) = \left\{(\mu(x_1), \ldots, \mu(x_n)) \colon \mu \in \mathcal{M}_1(K)\right\}.$$

In other words, the convex hull of S consists precisely of the expectations of probability measures with support in S, all of whose moments exist. (*Hint*: Let $K = \operatorname{conv}(S)$. To show that the right-hand set is contained in K, and not just in $\overline{K}$, use hyperplane separation and induction on $\dim(K)$.)

8.5.2 If $A = (a_{i+j})_{0 \le i,j \le d}$ is a real Hankel matrix of size $(d+1) \times (d+1)$, let A' denote the upper left $d \times d$ submatrix of A, so $A' = (a_{i+j})_{0 \le i,j \le d-1}$. The matrix A is said to be flat if $\operatorname{rk}(A) = \operatorname{rk}(A')$. A Hankel matrix B of size $(d+2) \times (d+2)$ is said to be a Hankel extension of A if $B' = A$.

(a) Assume that A is positive definite. Show that A has a positive definite Hankel extension.
(b) Give an example of a positive semidefinite Hankel matrix A that does not have a positive semidefinite Hankel extension.
(c) Show that A is flat if and only if (1) $A'u' = 0$ implies $A\binom{u'}{0} = 0$ for all $u' \in \mathbb{R}^d$, and (2) there is $u = (u_0, \dots, u_d)^\top \in \mathbb{R}^{d+1}$ with $Au = 0$ and $u_d \neq 0$.
(d) Assume that A is positive semidefinite and flat. Show that A has a unique Hankel extension that is again positive semidefinite and flat.

Hint: It helps to identify A with a symmetric bilinear form on $\mathbb{R}[t]_{\le d}$.

8.5.3 For the quadratic module $M = QM(x, x^2 - y^2 - x^3)$ and the compact convex set $K = \mathcal{S}(M) \subseteq \mathbb{R}^2$ discussed in Example 8.5.9, show that moment relaxation does not become exact in any finite degree, i.e. show that $K_d \neq K$ for all $d \ge 1$. (*Hint*: Consider supporting hyperplanes (lines) of K through the origin.)

8.5.4 Find a linear matrix inequality of size 3×3 that represents the compact convex set K from Exercise 8.5.3. In other words, find symmetric matrices $A_0, A_1, A_2 \in \mathsf{S}^3$ such that $K = \{(x, y) \in \mathbb{R}^2 : A_0 + xA_1 + yA_2 \succeq 0\}$. (*Hint*: Writing $A_0 + xA_1 + yA_2 = (a_{ij}(x, y))$, the A_i may be chosen in such a way that $a_{11}(x, y) = 1$ and $a_{12}(x, y) = 0$.)

8.5.5 Fill in the missing details for the proof of Proposition 8.5.24.

8.6 The Helton–Nie Theorems

8.6.1 Projected spectrahedra are the feasible sets in semidefinite programming (SDP). To use SDP for optimizing a linear function over a set $S \subseteq \mathbb{R}^n$, one needs a semidefinite representation of the convex hull K of S, or at least of the closure of K. To understand the expressive power of SDP, it is therefore essential to understand the nature of spectrahedral shadows. Specifically, what properties does a set need to have to make it a spectrahedral shadow?

Apart from a number of ad hoc constructions, not very much was known in this direction before roughly 2010. Clearly, spectrahedral shadows are convex semialgebraic sets, but no other general restrictions are visible. In 2009 and 2010, Helton and Nie published two long and technical papers in which they established the existence of semidefinite representations under quite general conditions. Roughly, their results say that if a compact convex semialgebraic set has sufficiently regular boundary of strictly positive curvature, it has a semidefinite representation. We are going to prove these results in the most important cases.

8.6.2 This needs a series of preparations. The first ingredient is an Archimedean positivstellensatz with degree bounds, and we even need it in a matrix version. First recall the usual Archimedean positivstellensatz for polynomial rings. Let $\mathbb{R}[x] = \mathbb{R}[x_1, \dots, x_n]$, let $\boldsymbol{g} = (g_1, \dots, g_r)$ be a tuple of polynomials in $\mathbb{R}[x]$, let $S = \mathcal{S}(\boldsymbol{g}) = \{\xi \in \mathbb{R}^n : g_1(\xi) \ge 0, \dots, g_r(\xi) \ge 0\}$ be the associated basic closed set and $M = QM(\boldsymbol{g}) = \Sigma + g_1\Sigma + \cdots + g_r\Sigma$ the associated quadratic module, with $\Sigma := \Sigma\mathbb{R}[x]^2$. If M is Archimedean then M contains every $f \in \mathbb{R}[x]$ with $f > 0$ on S, by the results of Section 5.3. Recall also (Proposition 5.2.7) that M Archimedean means $c - \sum_{i=1}^n x_i^2 \in M$ for some constant $c \in \mathbb{R}$.

We need a version with degree bounds, whereby we only require the existence of such bounds. Put $g_0 := 1$, and let

$$M_k = \Big\{\sum_{i=0}^{r} s_i g_i \colon s_i \in \Sigma,\ \deg(s_i g_i) \le k\ (0 \le i \le r)\Big\}$$

be the truncated quadratic module for $k \ge 1$, as in 8.5.7. We have $M_k \subseteq \mathbb{R}[x]_{\le k}$ and $M_1 \subseteq M_2 \subseteq \cdots \subseteq \bigcup_k M_k = M$. In general there will exist degrees d for which $M \cap \mathbb{R}[x]_{\le d}$ is not contained in M_k for any k. Given a polynomial $f = \sum_\alpha c_\alpha x^\alpha$ in $\mathbb{R}[x]$, let $\|f\| = \max_\alpha |c_\alpha|$, the L^∞-norm of the coefficient vector of f. We are going to prove:

8.6.3 Theorem. *Let $\mathbf{g}$, S, M be as above, and assume that M is Archimedean. For every $d \ge 1$ and every real number $c > 0$, there exists a positive integer $k = k(\mathbf{g}, d, c)$ such that M_k contains every $f \in \mathbb{R}[x]_{\le d}$ with $f \ge \frac{1}{c}$ on S and with $\|f\| \le c$.*

8.6.4 Let us see how to convert such a statement into coordinate-free form. Let A be any finitely generated $\mathbb{R}$-algebra, let $M = QM(\mathbf{g})$ be an Archimedean quadratic module in A where $\mathbf{g} = (g_1, \dots, g_r)$ is a finite sequence in A, and let $S = X_M = \{\alpha \in \mathrm{Hom}(A, \mathbb{R}) \colon \alpha(g_i) \ge 0\ (i = 1, \dots, r)\}$, the associated basic closed semialgebraic set, see 5.2.14. We want to prove the following, thereby generalizing Theorem 8.6.3: Given any $\mathbb{R}$-linear subspace $U \subseteq A$ with $\dim(U) < \infty$, and any real number $c > 0$, there exists an $\mathbb{R}$-linear subspace $V \subseteq A$ with $\dim(V) < \infty$ such that the following is true: Every $f \in U$ with $f \ge \frac{1}{c}$ on S and $\|f\| \le c$ has a representation

$$f = s_0 + s_1 g_1 + \cdots + s_r g_r$$

with $s_0, \dots, s_r \in \Sigma V^2$ (i.e., the s_i are sums of squares of elements of V). Here the L^∞-norm $\|f\|$ may be taken with respect to any fixed $\mathbb{R}$-linear basis of A.

Clearly, this statement implies Theorem 8.6.3. But it also follows from the latter: Write $A = \mathbb{R}[x]/I$ with some ideal I, and observe that every finite-dimensional subspace of $\mathbb{R}[x]$ is contained in $\mathbb{R}[x]_{\le d}$ for some d, and that conversely $\dim \mathbb{R}[x]_{\le d} < \infty$ for every d.

8.6.5 As remarked before, we need a matrix version of Theorem 8.6.3, generalizing Theorem 5.5.21. A *matrix polynomial* (of size $r \times s$) is an $r \times s$ matrix with entries in $\mathbb{R}[x]$. Given such a matrix polynomial $T = (t_{ij})$, define degree and L^∞-norm of T by $\deg(T) := \max_{i,j} \deg(t_{ij})$ and $\|T\| := \max_{i,j} \|t_{ij}\|$, respectively. Recall (2.1.3) that a symmetric matrix polynomial T of size $m \times m$ is called a (matrix) sum of squares if $T = \sum_\nu T_\nu T_\nu^\top$ for suitable $m \times m$ matrix polynomials T_ν, or equivalently, if $T = UU^\top$ for some (rectangular) matrix polynomial U with m rows. We claim that Theorem 8.6.3 implies:

8.6.6 Theorem. *Let $\mathbf{g} = (g_1, \dots, g_r)$, $M = QM(\mathbf{g}) \subseteq \mathbb{R}[x]$ and $S = \mathcal{S}(\mathbf{g}) \subseteq \mathbb{R}^n$ be as above, and assume that M is Archimedean. For any d, $m \ge 1$ and any real number $c > 0$, there exists a positive integer $k = k(\mathbf{g}, d, m, c)$ with the following property: Whenever $T \in \mathrm{Sym}_m(\mathbb{R}[x])$ is such that* $\deg(T) \le d$, *$\|T\| \le c$ and $T \ge \frac{1}{c} I_m$ on S,* *there exist matrix sums of squares* $T_0, \dots, T_r \in \mathrm{Sym}_m(\mathbb{R}[x])$ *with* $\deg(T_i) \le k$ *and with*

$$T = T_0 + g_1 T_1 + \cdots + g_r T_r.$$

8.6.7 Again, Theorem 8.6.6 can be stated in coordinate-free form (that we won't bother to make explicit). To prove 8.6.6 we go back to the proof of Theorem 5.5.21. What we did there was to consider the commutative subring $B = \mathbb{R}[x, T]$ of the ring of matrix polynomials. The ring B is a finite extension of $\mathbb{R}[x]$, and the quadratic module $M^B = QM_B(\boldsymbol{g})$, generated by M in B, is Archimedean (see the proof of 5.5.21). The hypothesis $T \ge \frac{1}{c} I$ on S implies that $T \ge \frac{1}{c}$ on X_{M^B}, the basic closed set in $\mathrm{Hom}(B, \mathbb{R})$ associated with the quadratic module M^B of B. Assume that Theorem 8.6.3 has been proved. Then apply the coordinate-free version 8.6.4 of this result to the quadratic module M^B in B. This gives Theorem 8.6.6, after converting back to coordinates. As in 5.5.21, we see that the matrices T_i can in fact be chosen to be sums of squares in the ring B, and in particular, to be sums of squares of matrices that are polynomials in T. We leave it to the assiduous reader to write out full details.

8.6.8 To prove Theorem 8.6.6, it therefore suffices to prove Theorem 8.6.3, see the previous discussion. Let the quadratic module $M = QM(\boldsymbol{g}) \subseteq \mathbb{R}[x]$ be Archimedean, fix a degree $d \ge 1$ and a real number $c > 0$, and let $S = \mathcal{S}(\boldsymbol{g}) \subseteq \mathbb{R}^n$. We put $P_{d,c} := \{f \in \mathbb{R}[x]_{\le d} \colon f|_S \ge \frac{1}{c}, \|f\| \le c\}$ and note that this is a semialgebraic subset of $\mathbb{R}[x]_{\le d}$. By the Archimedean positivstellensatz we have $P_{d,c} \subseteq M$, and therefore

$$P_{d,c} \subseteq \bigcup_{k \ge 1} \big(M_k \cap \mathbb{R}[x]_{\le d}\big). \tag{8.29}$$

The right-hand side is an ascending union of semialgebraic sets in $\mathbb{R}[x]_{\le d}$, and we want to prove that $P_{d,c} \subseteq M_k$ holds for some $k \ge 1$. It is equivalent to prove that inclusion (8.29) remains true after extending to an arbitrary real closed base field $R \supseteq \mathbb{R}$, see Exercise 4.1.6. (A similar reasoning was used in Exercise 3.3.5, to prove the existence of degree bounds for Hilbert 17.) This means that we have to show: Given any real closed field R containing $\mathbb{R}$, the inclusion

$$(P_{d,c})_R \subseteq \bigcup_{k \ge 1} \big((M^{R[x]})_k \cap R[x]_{\le d}\big)$$

holds, where $M^{R[x]}$ is the quadratic module in $R[x]$ generated by $\boldsymbol{g}$. Putting it simpler, we have to show: If $f \in R[x]_{\le d}$ satisfies $f \ge \frac{1}{c}$ on S_R and $\|f\| \le c$, then $f \in M^{R[x]}$.

Let $\mathcal{O} \subseteq R$ denote the convex hull of $\mathbb{R}$ in R, a valuation subring of R, and let $M^{\mathcal{O}} \subseteq \mathcal{O}[x]$ be the quadratic module generated by $\boldsymbol{g}$ in $\mathcal{O}[x]$. Then $M^{\mathcal{O}}$ is Archimedean, as a quadratic module in $\mathcal{O}[x]$. Indeed, the subring

$$O(M^{\mathcal{O}}) = \{f \in \mathcal{O}[x] \colon \exists n \in \mathbb{N}\; n \pm f \in M^{\mathcal{O}}\}$$

of $\mathcal{O}[x]$ (consisting of the $M^{\mathcal{O}}$-bounded elements in $\mathcal{O}[x]$, see 5.2.3) contains both $\mathbb{R}[x]$ and $\mathcal{O}$. Therefore $O(M^{\mathcal{O}}) = \mathcal{O}[x]$, which means that $M^{\mathcal{O}}$ is Archimedean.

So we can apply the Archimedean positivstellensatz to $M^{\mathcal{O}}$. Any polynomial $f \in R[x]$ with $\|f\| \le c$ lies in $\mathcal{O}[x]$, since $c \in \mathbb{R}$. Moreover, if $f \ge \frac{1}{c}$ on $S_R = \{\xi \in R^n : g_i(\xi) \ge 0\ (i = 1, \dots, r)\}$, then $f \ge \frac{1}{c}$ holds on $X_{M^{\mathcal{O}}} \subseteq \mathrm{Hom}(\mathcal{O}[x], \mathbb{R})$. Indeed, any ring homomorphism $\alpha\colon \mathcal{O}[x] \to \mathbb{R}$ in $X_{M^{\mathcal{O}}}$ factors as $\mathcal{O}[x] \xrightarrow{can.} \mathbb{R}[x] \xrightarrow{u} \mathbb{R}$ for some point u in $X_M = S$. In particular we have $\alpha(f) = \overline{f(u)} > 0$ for f as before. So $f > 0$ on $X_{M^{\mathcal{O}}}$, and so $f \in M^{\mathcal{O}}$ by the Archimedean positivstellensatz 5.3.3 applied to $M^{\mathcal{O}}$. In particular, $f \in M^{R[x]}$. □

8.6.9 Next we recall a few basic facts on convex functions that should be familiar from calculus. Let $K \subseteq \mathbb{R}^n$ be a convex set. A function $f\colon K \to \mathbb{R}$ is *convex* if

$$f((1-t)u + tv) \le (1-t)f(u) + tf(v) \tag{8.30}$$

for all $u, v \in K$ and $0 \le t \le 1$. If strict inequality holds whenever $u \ne v$ and $0 < t < 1$, then f is *strictly convex*. The function f is *(strictly) concave* if $-f$ is (strictly) convex. If f is convex on K, the sublevel sets $\{u \in K : f(u) \le c\}$ $(c \in \mathbb{R})$ of f are all convex. Similarly, if f is concave on K, the superlevel sets $\{u \in K : f(u) \ge c\}$ are all convex.

To relate convexity of f to (partial) derivatives of f, assume that K is open in $\mathbb{R}^n$. A C^1-function $f\colon K \to \mathbb{R}$ is convex if and only if

$$\langle \nabla f(u), v - u\rangle \le f(v) - f(u) \tag{8.31}$$

holds for all $u, v \in K$. If f is C^2, then f is convex on K if and only if the Hesse matrix of f is positive semidefinite at any point of K, i.e. $D^2 f(u) \succeq 0$ for every $u \in K$. If $D^2 f(u) \succ 0$ for $u \in K$ then f is strictly convex on K, the converse being false in general. For concave, the same hold with opposite inequalities.

8.6.10 Definition. A polynomial $f \in \mathbb{R}[x] = \mathbb{R}[x_1, \dots, x_n]$ is said to be *sos-convex* if the Hessian $D^2 f(x)$ of f, considered as a matrix polynomial, is a matrix sum of squares. The polynomial f is *sos-concave* if $-f$ is sos-convex.

8.6.11 Remarks.

1. The concept of sos-convexity for polynomials was introduced in Helton–Nie [88]. For brevity, let us say that a polynomial $f \in \mathbb{R}[x] = \mathbb{R}[x_1, \dots, x_n]$ is *convex* if the function represented by f is convex on all of $\mathbb{R}^n$. Every sos-convex polynomial f is convex since $D^2 f \succeq 0$ on $\mathbb{R}^n$. The converse is not true. A first example of a homogeneous convex polynomial that is not sos-convex appears in [1]. More explicit examples, and a systematic study of the difference set, can be found in [2].

2. It is easy to see that every non-linear homogeneous convex polynomial is nonnegative on $\mathbb{R}^n$. So it is natural to ask whether convex forms are sums of squares of forms. Blekherman [22] proved that there exist convex forms that are not sos, in fact of any even degree $2d \ge 4$, as long as there are sufficiently many variables. But no single explicit example of such a form seems to be known, see also [66].

3. On the other hand, it is not hard to see that every sos-convex form of degree at least two is a sum of squares of forms (Exercise 8.6.2). Together with the previously mentioned result, this gives another proof for the existence of convex forms

that are not sos-convex. Altogether, the following chain of implications holds for homogeneous polynomials of degree ≥ 2:

$$\text{sos-convex} \Rightarrow \text{sos and convex} \Rightarrow \text{convex} \Rightarrow \text{psd},$$

and none of them can be reverted in general.

8.6.12 Lemma. *If f is a concave function on the non-empty compact convex set $K \subseteq \mathbb{R}^n$, then f has a minimizer $u \in K$ which is an extreme point of K.*

Proof. Let $u \in K$ be a minimizer of f. By Theorem 8.1.13, u is a convex combination $u = \sum_{i=0}^r a_i u_i$ of extreme points $u_0, \dots, u_r$ of K. Since f is concave we have $f(u) \geq \sum_{i=0}^r a_i f(u_i)$, and since u is a minimizer we have $f(u_i) = f(u)$ for any index i with $a_i \neq 0$. □

8.6.13 Lemma. *Let $g_1, \dots, g_r \in \mathbb{R}[x]$ be such that $K = \mathcal{S}(g_1, \dots, g_r)$ is compact and convex with nonempty interior, and assume that the g_i are concave on K. If $f \in \mathbb{R}[x]$ and $u \in K$ is a minimizer of f on K, there exist real numbers $b_i \geq 0$ with $\nabla f(u) = \sum_{i=1}^r b_i \nabla g_i(u)$ and satisfying $b_i g_i(u) = 0$ for all i.*

Such b_i are called *Lagrange multipliers*. The condition $b_i g_i(u) = 0$, usually referred to as *complementary slackness*, means that $\nabla f(u)$ lies in the convex cone generated by the gradients at u of those g_i which are *active* at u, meaning that $g_i(u) = 0$.

Proof. Since K has non-empty interior, there is $v \in K$ with $g_i(v) > 0$ for all i. Let $u \in K$ be the given minimizer of f and put $w = v - u$. If i is an index with $g_i(u) = 0$ then $0 < g_i(v) \leq \langle \nabla g_i(u), w \rangle$ since g_i is concave on K (see (8.31)). We may assume that $g_i(u) = 0$ for $i = 1, \dots, p$ and $g_i(u) > 0$ for $i = p+1, \dots, r$. The convex cone C generated by $\nabla g_1(u), \dots, \nabla g_p(u)$ in $\mathbb{R}^n$ is closed (Proposition 8.1.17), and we need to show $\nabla f(u) \in C$. Assuming that this fails, there is $z \in C^*$ with $\langle z, \nabla f(u) \rangle < 0$. Choose $s > 0$ so small that $\langle z + sw, \nabla f(u) \rangle < 0$. Going a little bit into direction $z + sw$ from u, we stay in K since $\langle \nabla g_i(u), z + sw \rangle \geq 0$ for $i = 1, \dots, p$, and since

$$g_i(u + t(z + sw)) = t \cdot \langle \nabla g_i(u), z + sw \rangle + (\text{higher order terms in } t)$$

by expanding the left-hand function into a Taylor series with respect to the variable t. On the other hand,

$$f(u + t(z + sw)) - f(u) = t \langle \nabla f(u), z + sw \rangle + (\text{higher order terms in } t)$$

has negative leading coefficient, and so f decreases strictly for small $t > 0$ along this path. This contradicts the assumption that u was a minimizer of f on K. □

8.6.14 Recall that the moment relaxation scheme is a systematic way to find semidefinite representations for outer approximations of convex sets. Let $\boldsymbol{g} = (g_1, \dots, g_r)$ be a tuple in $\mathbb{R}[x]$, consider the quadratic module $M = QM(\boldsymbol{g})$ and its truncations M_d ($d \geq 1$) as before (8.6.2). Put $S = S(\boldsymbol{g})$ and $K = \text{conv}(S)$. In the previous section (8.5.4, 8.5.6) it was shown that the dual cone M_d^* is a spectrahedral cone, and that a natural linear projection K_d of $\{\mu \in M_d^* \colon \mu(1) = 1\}$ is an outer approximation of K, with an explicit semidefinite representation. Moreover, assuming that S is compact and Zariski dense in $\mathbb{R}^n$ (to force $\mathrm{supp}(M) = \{0\}$), equality $K = K_d$ holds if and only if M_d contains every linear polynomial that is non-negative on S (Corollary 8.5.13(b)). In this case we say that K has an *exact moment relaxation* (of order d) with respect to $\boldsymbol{g}$. We are going to employ this method.

8.6.15 Lemma. *Let $F \in \mathrm{Sym}_m\mathbb{R}[x]$ be a matrix polynomial that is a matrix sum of squares, and let $u \in \mathbb{R}^n$. Then the matrix polynomial*

$$G_u(x) = \int_0^1 \int_0^t F(u + s(x-u))\, ds\, dt$$

is again a matrix sum of squares. Moreover $\deg(G_u) = \deg(F)$.

Proof. Integration of matrix-valued functions is carried out entry-wise. For $\alpha \in \mathbb{Z}_+^n$, $d = |\alpha|$ and $u \in \mathbb{R}^n$ we have $(u + s(x-u))^\alpha = s^d x^\alpha + h_u(s, x)$ with h_u a polynomial of x-degree at most $d-1$. So $\int_0^1 \int_0^t (u+s(x-u))^\alpha \, ds\, dt = \frac{x^\alpha}{(d+1)(d+2)} + \tilde{h}_u(x)$ with $\deg(h_u) \le d-1$, which already proves the last statement. For the first we may assume that F is a single square, i.e. $F(x) = v(x)v(x)^\top$ with $v(x) = (v_1(x), \dots, v_m(x))^\top$ a column vector over $\mathbb{R}[x]$. Let $d = \deg v(x)$, and let $y = (y_1, \dots, y_m)^\top$ be a tuple of new variables, considered as a column. We have to show that the polynomial $g(x, y) = y^\top G_u(x) y$ is a sum of squares of polynomials in $\mathbb{R}[x, y]$ (each of them necessarily homogeneous and linear in y, compare Exercise 2.1.1). Clearly $y^\top F(x) y = \left(\sum_{i=1}^m y_i v_i(x)\right)^2$, and so

$$g(x, y) = \int_0^1 \int_0^t \Big(\sum_{i=1}^m y_i v_i(u + s(x-u))\Big)^2 ds\, dt. \tag{8.32}$$

For $i = 1, \dots, m$ write $v_i(u+s(x-u)) = \sum_{k=0}^d p_{ik}(x, u)\, s^k$ with polynomial coefficients $p_{ik} \in \mathbb{R}[x, u]$, and note that

$$g(x, y) = \sum_{i,j=1}^m \sum_{k,l=0}^d y_i y_j\, p_{ik}(x, u) p_{jl}(x, u) \int_0^1 \int_0^t s^{k+l}\, ds\, dt. \tag{8.33}$$

The real symmetric $(d+1) \times (d+1)$-matrix

$$A_d = \left(\int_0^1 \int_0^t s^{k+l}\, ds\, dt\right)_{0 \le k, l \le d} = \left(\frac{1}{(k+l+1)(k+l+2)}\right)_{0 \le k, l \le d}$$

is positive definite since $z^\top A_d z = \int_0^1 \int_0^t (z_0 + z_1 s + \dots + z_d s^d)^2 \, ds\, dt > 0$ for every $z = (z_0, \dots, z_d)^\top \in \mathbb{R}^{d+1}$, $z \ne 0$. So we can factor it as $A_d = BB^\top$, i.e. $\int_0^1 \int_0^t s^{k+l}\, ds\, dt = \sum_r b_{kr} b_{lr}$ with $B = (b_{rs})$. It follows that

$$g(x, y) = \sum_{i,j=1}^m \sum_{k,l=0}^d y_i y_j\, p_{ik} p_{jl} \sum_r b_{kr} b_{lr} = \sum_r \Big(\sum_{i=1}^m \sum_{k=0}^d b_{kr} y_i p_{ik}(x, u)\Big)^2$$

is a sum of squares as desired. □

Let $K = \mathcal{S}(g_1, \dots, g_r) \subseteq \mathbb{R}^n$ be a basic closed set, where $g_1, \dots, g_r \in \mathbb{R}[x]$ are polynomials. For a convenient way of speaking, let us say that $K = \mathcal{S}(g_1, \dots, g_r)$ is an *Archimedean description* of the set K if the quadratic module $QM(g_1, \dots, g_r)$ is Archimedean. Of course, K has an Archimedean description if and only if K is compact.

8.6.16 Theorem. *Let $K \subseteq \mathbb{R}^n$ be a compact and convex basic closed set with non-empty interior, let $K = \mathcal{S}(\mathbf{g})$ with $\mathbf{g} = (g_1, \dots, g_r)$ be an Archimedean description of K. For each $i = 1, \dots, r$, assume that at least one of the following two conditions holds:*

(1) *g_i is sos-concave, i.e. the negative Hessian $-D^2 g_i$ is a matrix sum of squares;*
(2) *g_i is concave on K, and $D^2 g_i(u) < 0$ for every u in the closure of $\mathcal{Z}(g_i) \cap \mathrm{Ex}(K)$.*

Then K has an exact moment relaxation (8.6.14) with respect to $\mathbf{g}$. In particular, K is a spectrahedral shadow.

Recall that $\mathcal{Z}(g_i)$ denotes the zero set of g_i in $\mathbb{R}^n$, and $\mathrm{Ex}(K)$ is the set of extreme points of K. For convenience write $g_0 := 1$ in the following. As a first step we prove the following lemma:

8.6.17 Lemma. *Under the assumptions of Theorem 8.6.16 there exists, for every $i = 1, \dots, r$, a positive integer N_i such that, for every $u \in \mathrm{Ex}(K)$, the $n \times n$ matrix polynomial*

$$G_{i,u}(x) := -\int_0^1 \int_0^t D^2 g_i(u + s(x-u))\, ds\, dt$$

can be written

$$G_{i,u}(x) = \sum_{j=0}^{r} g_j(x)\, S_{i,u,j}(x)$$

where each matrix polynomial $S_{i,u,j}(x)$ is a matrix sum of squares of degree $\le N_i$.

8.6.18 Suppose for a moment that Lemma 8.6.17 has been proved. Then the proof of Theorem 8.6.16 is completed as follows. According to Corollary 8.5.14, see also Remark 8.5.18, we have to find an integer $d \ge 0$ such that the truncated quadratic module M_d contains every linear polynomial f with $f|_K \ge 0$. Such f has a minimizer $u \in K$ on K that is an extreme point of K (Lemma 8.6.12). By Lemma 8.6.13, and since the g_i are concave on K, we can write $\nabla f(u) = \sum_{i=1}^r b_i \cdot \nabla g_i(u)$ with real numbers $b_i \ge 0$ satisfying $b_i g_i(u) = 0$ for all i. The polynomial $h_u(x) := f(x) - f(u) - \sum_{i=1}^r b_i g_i(x)$ satisfies $h_u(u) = 0$, $\nabla h_u(u) = 0$ and $D^2 h_u(x) = -\sum_{i=1}^r b_i D^2 g_i(x)$. By Exercise 8.6.1, the matrix polynomial

$$H_u(x) := \int_0^1 \int_0^t D^2 h_u(u + s(x-u))\, ds\, dt$$

(where $x = (x_1, \dots, x_n)$) satisfies $h_u(x) = (x-u)^\top \cdot H_u(x) \cdot (x-u)$. By linearity of the integral, Lemma 8.6.17 gives

$$H_u(x) = \sum_{i=1}^{r} b_i G_{i,u}(x) = \sum_{i=1}^{r}\sum_{j=0}^{r} b_i g_j(x) S_{i,u,j}(x)$$

where the $S_{i,u,j}(x)$ are sos matrix polynomials of degrees $\le N := \max\{N_1, \dots, N_r\}$. Since $f(u) \ge 0$, we see from

$$f(x) = f(u) + \sum_{i=1}^{r} b_i g_i(x) + (x-u)^\top \cdot H_u(x) \cdot (x-u)$$

that f lies in the truncated quadratic module $M_{\delta+N+2}$ where $\delta := \max_i \deg(g_i)$ (use Lemma 8.6.15). Indeed, we have written f as a weighted sum of squares with weights $1 = g_0, g_1, \dots, g_r$, and with each summand of degree $\le \delta + N + 2$.

8.6.19 It remains to prove Lemma 8.6.17, so fix $i \in \{1, \dots, r\}$. First assume that g_i is sos-concave (condition (1) in 8.6.16). Then the negative Hessian $-D^2 g_i$ is an sos matrix polynomial, of degree $\deg(D^2 g_i) = \deg(g_i) - 2$. By Lemma 8.6.15, the same is true for $G_{i,u}(x)$ and for any $u \in \mathbb{R}^n$, and we are already done with this case.

Now assume that g_i satisfies condition (2). Let Z_i be the closure of $\mathcal{Z}(g_i) \cap \mathrm{Ex}(K)$. The matrix polynomial $G_{i,u}(x) \in \mathrm{Sym}_n(\mathbb{R}[x])$ has $\deg_x(G_{i,u}) = \deg(g_i) - 2$ (Exercise 8.6.1) and depends polynomially on $u \in \mathbb{R}^n$. If $u \in K$ then $D^2 g_i(u) \preceq 0$ since g_i is concave on K, see 8.6.9. For $u \in Z_i$ we even have $D^2 g_i(u) \prec 0$, by assumption (2). For $u \in Z_i$ and $v \in K$ it follows that $G_{i,u}(v) \succ 0$. Indeed, for $w \ne 0$ in $\mathbb{R}^n$ we have

$$w^\top G_{i,u}(v) w = \int_0^1 \int_0^t w^\top \cdot (-D^2 g_i)(u + s(v-u)) \cdot w \, ds\, dt > 0$$

since the polynomial under the integral is strictly positive for $s = 0$ and otherwise non-negative. By compactness of K and Z_i, there exists $\varepsilon > 0$ such that $G_{i,u}(x) \succeq \varepsilon I$ for all $(u, x) \in Z_i \times K$. By Theorem 8.6.6 we find, for every $u \in Z_i$, a weighted matrix sos representation

$$G_{i,u}(x) = \sum_{j=0}^{r} g_j(x) S_{i,u,j}(x)$$

where each $S_{i,u,j} \in \mathrm{Sym}_n(\mathbb{R}[x])$ is a matrix sum of squares of degree

$$\deg(S_{i,u,j}) \le k\Big(\mathbf{g}, \deg(g_i) - 2, n, \max\{\varepsilon^{-1}, \|G_{i,u}\|\}\Big),$$

and the right-hand number is the integer k whose existence is guaranteed by Theorem 8.6.6. By compactness of Z_i, the number $\|G_i\| := \max\{\|G_{i,u}\| : u \in Z_i\}$ is finite. So altogether we have the uniform bound

$$\deg(S_{i,u,j}) \le k\Big(\mathbf{g}, \deg(g_i) - 2, n, \max\{\varepsilon^{-1}, \|G_i\|\}\Big) =: N_i$$

that is independent of $u \in Z_i$. This completes the proof of Lemma 8.6.17, and therefore of Theorem 8.6.16 as well. □

8.6.20 Remark. A function $f\colon K \to \mathbb{R}$ defined on a convex set $K \subseteq \mathbb{R}^n$ is called *quasi-concave* if, for every $c \in \mathbb{R}$, the superlevel set $\{u \in K\colon f(u) \geq c\}$ is convex (see Section 3.4 in [31]). If $u \in K$ and f is twice continuously differentiable in a neighborhood of u, such f satisfies $v^\top \cdot D^2 f(u) \cdot v \leq 0$ for every $v \in \mathbb{R}^n$ with $\langle v, \nabla f(u)\rangle = 0$ ([31] p. 101). Following [88] we consider the following stronger property:

8.6.21 Definition. A C^2-function $f\colon U \to \mathbb{R}$ defined on an open set $U \subseteq \mathbb{R}^n$ is *strictly quasi-concave* at a point $u \in U$ if $v^\top \cdot D^2 f(u) \cdot v < 0$ for every $0 \neq v \in \mathbb{R}^n$ with $\langle v, \nabla f(u)\rangle = 0$. We say that f is strictly quasi-concave (throughout) if f is strictly quasi-concave at every point $u \in U$.

8.6.22 Examples.

1. A polynomial $f \in \mathbb{R}[x]$ is strictly quasi-concave at a point $u \in \mathbb{R}^n$ if, and only if, the Hessian $D^2 f(u)$ is negative definite when restricted to the linear tangent space of the level hypersurface $f(x) = f(u)$ at u. It is not hard to see that the set of points at which a given C^2-function is strictly quasi-concave, is open (Exercise 8.6.3).

2. If a function f, defined on an open convex set $U \subseteq \mathbb{R}^n$, is strictly quasi-concave on U, then f is quasi-concave on U, i.e. all superlevel sets $\{u \in U\colon f(u) \geq c\}$ are convex. For the proof see Exercise 8.6.4.

3. A C^2-function f defined on an open interval $K \subseteq \mathbb{R}$ is strictly quasi-concave if, and only if, the derivative f' has at most one zero u in K, and $f''(u) < 0$ if u exists. In particular, a strictly quasi-concave function need not be concave. For an example in dimension two, the polynomial $f(x_1, x_2) = x_1 x_2$ is quasi-concave on the closed positive orthant $\mathbb{R}^2_+$, and is strictly quasi-concave on its interior.

8.6.23 Lemma. *Given a real number $c > 0$, there exists a (univariate) sum of squares $h \in \mathbb{R}[t]$ that satisfies*

$$(1)\quad h(t) > 0, \qquad (2)\quad h(t) + th'(t) > 0, \qquad (3)\quad \frac{2h'(t) + th''(t)}{h(t) + th'(t)} \leq -c$$

for every $t \in \mathbb{R}$ with $|t| \leq 1$.

For the proof see Exercise 8.6.6.

8.6.24 Proposition. *Let $K = \mathcal{S}(g_1, \ldots, g_r)$ be an Archimedean description of a compact convex set in $\mathbb{R}^n$, and assume that the g_i are strictly quasi-concave at every point of K. Then there exists a second Archimedean description $K = \mathcal{S}(h_0, \ldots, h_r)$ of K with $h_0, h_1, \ldots, h_r \in QM(\mathbf{g})$, and such that $D^2 h_i < 0$ on K for $i = 0, \ldots, r$.*

Proof. There is a real number $b > 0$ such that $h_0 := b^2 - \sum_{i=1}^n x_i^2 \in QM(\mathbf{g})$. By scaling the g_i we may assume that $|g_i(u)| \leq 1$ for $|u| \leq b$. Let $c > 0$ be a (large) constant, to be determined later. By Lemma 8.6.23 there exists a sum of squares $h \in \mathbb{R}[t]$ satisfying

$$(1)\ h(t) > 0, \quad (2)\ h(t) + th'(t) > 0, \quad (3)\ \frac{2h'(t) + th''(t)}{h(t) + th'(t)} \leq -c \tag{8.34}$$

for every $|t| \le 1$. Using this polynomial we consider the tuple $\boldsymbol{h} = (h_0, \dots, h_r)$, with h_0 as before and with

$$h_i(x) = g_i(x) \cdot h(g_i(x)) \quad (i = 1, \dots, r).$$

We have $h_0, \dots, h_r \in QM(\boldsymbol{g})$ since $h(t)$ is sos, and so $K \subseteq S(\boldsymbol{h})$. Conversely let $u \in \mathbb{R}^n$ with $u \notin K$. If $|u| > b$ then $h_0(u) < 0$. If $|u| \le b$ then $-1 \le g_i(u) < 0$ for some index $i \in \{1, \dots, r\}$, and so $h_i(u) < 0$. Altogether $K = \mathcal{S}(\boldsymbol{h})$.

Clearly $D^2(h_0) = -2I \prec 0$ everywhere. For $i = 1, \dots, r$ we show that $D^2(h_i)$ is negative definite on K, if c was chosen sufficiently large. Calculating the Hessian of $h_i = g_i \cdot (h \circ g_i)$, the product rule gives

$$D^2(h_i) = (h \circ g_i) \cdot D^2(g_i) + \left(\nabla(g_i) \cdot \nabla(h \circ g_i)^\top\right)^{sym} + g_i \cdot D^2(h \circ g_i).$$

Here gradients ∇ are considered as column vectors, and we write $M^{sym} := M + M^\top$ for the symmetrization of a square matrix M. Further $\nabla(h \circ g_i) = (h' \circ g_i) \cdot \nabla g_i$ and $D^2(h \circ g_i) = (h'' \circ g_i) \cdot \nabla(g_i)\nabla(g_i)^\top + (h' \circ g_i) \cdot D^2(g_i)$, hence

$$D^2(h_i) = p_i \cdot D^2(g_i) + q_i \cdot (\nabla g_i)(\nabla g_i)^\top$$

with $p_i := (h \circ g_i) + g_i \cdot (h' \circ g_i)$ and $q_i := 2(h' \circ g_i) + g_i \cdot (h'' \circ g_i)$. Note that $p_i > 0$ on K, by property (2) in (8.34). Since g_i is strictly quasi-concave on K, Exercise 8.6.5 implies that there exists a constant $\kappa_i > 0$ such that

$$D^2 g_i \prec \kappa_i \cdot (\nabla g_i)(\nabla g_i)^\top, \tag{8.35}$$

uniformly on K. Take $c > 0$ in (8.34) so large that $c > \max\{\kappa_1, \dots, \kappa_r\}$, and choose h accordingly. Then $\frac{q_i}{p_i} \le -c \le -\kappa_i$ on K, by condition (3) in (8.34). So $\kappa_i p_i + q_i \le 0$ holds on K for all i, and hence $D^2(h_i) = p_i \cdot D^2(g_i) + q_i \cdot (\nabla g_i)(\nabla g_i)^\top \prec (\kappa_i p_i + q_i) \cdot (\nabla g_i)(\nabla g_i)^\top \preceq 0$ holds on K, where $\prec$ follows from (8.35). Hence each of the h_i is strictly concave on K. □

As a consequence we get the following extension of Theorem 8.6.16:

8.6.25 Corollary. *Let $K \subseteq \mathbb{R}^n$ be a compact convex set, and let $K = \mathcal{S}(\boldsymbol{g})$ with $\boldsymbol{g} = (g_1, \dots, g_r)$ be an Archimedean description of K. For each $i = 1, \dots, r$, assume that g_i either satisfies condition (1) or (2) from 8.6.16, or else*

(3) *g_i is strictly quasi-concave on K.*

Then it remains true that K has an exact moment relaxation with respect to $\boldsymbol{g}$.

Proof. Suppose that $g_1, \dots, g_s$ satisfy (3) and that each of $g_{s+1}, \dots, g_r$ satisfies (1) or (2). Choose $b \in \mathbb{R}$ with $b^2 - \sum_i x_i^2 \in QM(\boldsymbol{g})$. By Proposition 8.6.24 we may replace the sequence $b^2 - \sum_i x_i^2, g_1, \dots, g_s$ by a sequence of polynomials in $QM(\boldsymbol{g})$ whose Hessian in every point of K is negative definite, without changing the basic closed set they define. Together with $g_{s+1}, \dots, g_r$, this gives a new Archimedean description $K = \mathcal{S}(h_1, \dots, h_p)$, with $h_i \in QM(\boldsymbol{g})$ for each i, such that each h_i satisfies condition (1) or (2). So the hypotheses of Theorem 8.6.16 are fulfilled, and this theorem gives the desired conclusion. □

8.6.26 Examples.

1. Let $m_1, \dots, m_n$ be even positive integers. The "tv hyperscreen" $K = \{\xi \in \mathbb{R}^n : \xi_1^{m_1} + \cdots + \xi_n^{m_n} \le 1\}$ is a spectrahedral shadow, as we saw in 8.5.24 by an explicit argument. We also get this conclusion from Theorem 8.6.16, since $g = 1 - \sum_{i=1}^n x_i^{m_i}$ is sos-concave. On the other hand, neither condition (2) in 8.6.16 nor condition (3) in 8.6.25 is satisfied, since they do not hold in boundary points u of K with $u_i = 0$ for some i.

2. Let $g = x^a y^b - 1 \in \mathbb{R}[x, y]$ with $a, b \ge 1$. Calculating we find

$$\nabla g = x^{a-1}y^{b-1}\begin{pmatrix} ay \\ bx \end{pmatrix}, \quad D^2 g = x^{a-2}y^{b-2}\begin{pmatrix} a(a-1)y^2 & abxy \\ abxy & b(b-1)x^2 \end{pmatrix}.$$

At $(x, y) \ne (0, 0)$, the linear tangent space to the level set of g is spanned by $u = (-bx, ay)^\top$. Since $u^\top \cdot D^2 g \cdot u = -ab(a+b)x^a y^b$, we see that g is strictly quasi-concave (but not concave) on the open positive quadrant. Let $h = 1 - (x-1)^2 - (y-1)^2$ and consider the compact convex set $K = \mathcal{S}(g, h) \subseteq \mathbb{R}^2$. From condition (3) of Corollary 8.6.25, we see that K has an exact moment relaxation. But this example is not covered by Corollary 8.6.16.

3. Exercise 8.6.7 shows that the conditions in Theorem 8.6.16 or Corollary 8.6.25 (sufficient for existence of an exact relaxation with respect to $\mathbf{g}$) are not necessary: K may have an exact relaxation with respect to $\mathbf{g}$, even if none of these conditions is satisfied.

4. In Exercise 8.6.8, a compact, convex and basic closed set K is considered for which no moment relaxation gets exact with respect to any finite sequence $\mathbf{g}$ with $K = \mathcal{S}(\mathbf{g})$. Still K is a spectrahedral shadow. The example is taken from [145], and is an instance of the following result proved there: Let $\mathbf{g}$ be a finite tuple in $\mathbb{R}[x]$ such that the basic closed set $K = \mathcal{S}(\mathbf{g}) \subseteq \mathbb{R}^n$ is convex. If K has a face that is not exposed, then no moment relaxation of K with respect to $\mathbf{g}$ is exact.

We conclude this section with another general existence result for semidefinite representations. It is a consequence of Corollary 8.6.25.

8.6.27 Definition. Let $K \subseteq \mathbb{R}^n$ be a closed convex semialgebraic set with nonempty interior, and let $u \in \partial K$.

(a) We say that u is a *smooth boundary point* of K if there are a polynomial $g \in \mathbb{R}[x]$ with $\nabla g(u) \ne 0$ and a neighborhood U of u in $\mathbb{R}^n$, such that $\partial K \cap U = \{v \in U : g(v) = 0\}$. If every boundary point of K is smooth, we say that K has *smooth boundary*.

In this situation we have $K \cap U = \{v \in U : g(v) \ge 0\}$, after possibly replacing g with $-g$ (Exercise 8.6.10). Such a polynomial g will be called a *positive inequality* for K at u.

(b) If u is a smooth boundary point and g is a positive inequality for K at u, we say that the boundary ∂K is *strictly positively curved* at u if g is strictly quasi-concave at u.

8.6.28 Examples.

1. Let K be as in the previous definition, let $V = \partial_a K$ be the algebraic boundary of K (the Zariski closure of ∂K, 4.6.14). Then $u \in \partial K$ is a smooth boundary point of K if and only if u is a non-singular point of the hypersurface V (Exercise 8.6.10).

2. The boundary points P, $P' = (0, \pm 1)$ of the convex set $K_1 = ([0,1]\times[-1,1]) \cup \{(u,v)\colon u^2 + v^2 \le 1\} \subseteq \mathbb{R}^2$ are not smooth. Neither is the origin $Q = (0,0)$ as a boundary point of $K_2 = \{(u,v)\colon u^4 + u^2v^2 + v^4 \le u(u^2+v^2)\}$, despite appearances to the contrary:

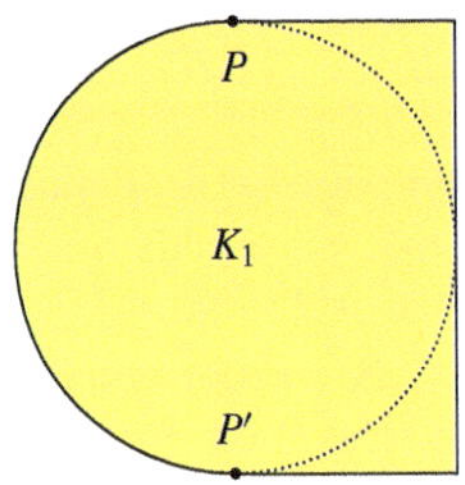

Two non-smooth boundary points

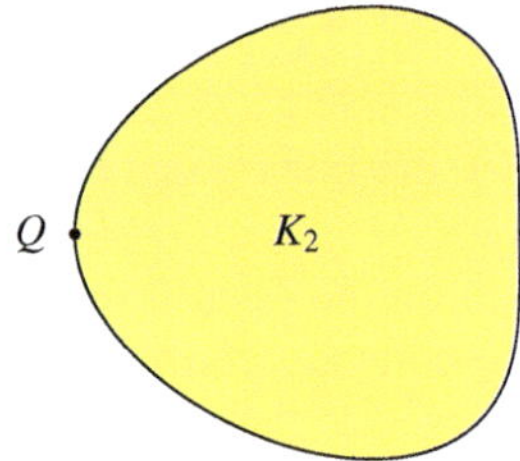

The curve $x^4 + x^2y^2 + y^4 = x(x^2+y^2)$ and a non-smooth boundary point

3. Let g be a positive inequality for K in the smooth boundary point u. One can show that convexity of K implies $w^\top \cdot D^2g(u) \cdot w \le 0$ for all $w \in \mathbb{R}^n$ with $\langle w, \nabla g(u)\rangle = 0$ (non-strict positive curvature of the ∂K at u), see Exercise 8.6.10. The restriction of the symmetric bilinear form $D^2g(u)$ to $\nabla g(u)^\perp$ is called the second fundamental form of the hypersurface $g = 0$ at its smooth point u.

8.6.29 Theorem. *Let $K \subseteq \mathbb{R}^n$ be a semialgebraic set that is compact and convex, and assume that the boundary ∂K is smooth and strictly positively curved everywhere. Then K is a spectrahedral shadow.*

Proof. For every $u \in \partial K$ there are a polynomial $g \in \mathbb{R}[x]$ and a real number $r > 0$ with $B_r(u) \cap K = \{\xi \in B_r(u)\colon g(\xi) \ge 0\}$, and such that g is strictly quasi-concave on $B_r(u)$ (see Exercise 8.6.3 for the latter). So $K \cap B_r(u) = \mathcal{S}(g(x), r^2 - |x-u|^2)$ is a spectrahedral shadow for any such u and r, by Corollary 8.6.25. There are finitely many such pairs (u_i, r_i) such that ∂K is covered by the balls $B_{r_i}(u_i)$. The convex hull of $\bigcup_i (K \cap B_{r_i}(u_i))$ contains ∂K, so it is all of K. And it is a spectrahedral shadow by Proposition 8.3.7. □

By their results, as discussed in this section, Helton and Nie [87] were led to conjecture that every convex semialgebraic set in $\mathbb{R}^n$ is a spectrahedral shadow (this was subsequently called the *Helton–Nie conjecture*). Eventually it turned out that the conjecture is false. In fact there exist plenty of prominent counter-examples, as we shall see in the next section.

Exercises

8.6.1 Let $f \in \mathbb{R}[x] = \mathbb{R}[x_1, \dots, x_n]$ and $u \in \mathbb{R}^n$. The matrix polynomial

$$F(x) := \int_0^1 \int_0^t D^2 f(u + s(x-u))\, ds\, dt$$

satisfies

$$f(x) = f(u) + (x-u)^\top \cdot \nabla f(u) + (x-u)^\top \cdot F(x) \cdot (x-u),$$

and also $\deg(F) = \deg(f) - 2$ unless f is constant.

8.6.2 Show that every sos-convex homogeneous polynomial of degree ≥ 2 in $\mathbb{R}[x] = \mathbb{R}[x_1, \dots, x_n]$ is a sum of squares of polynomials. (*Hint*: Use Exercise 8.6.1 and Lemma 8.6.15)

8.6.3 If $f\colon U \to \mathbb{R}$ is a C^2-function on an open set $U \subseteq \mathbb{R}^n$, and if f is strictly quasi-concave in $u \in U$, prove that f is strictly quasi-concave in a neighborhood of u.

8.6.4 Let $K \subseteq \mathbb{R}^n$ be a convex set, and let $f \in \mathbb{R}[x]$ be strictly quasi-concave on a neighborhood of K in $\mathbb{R}^n$. Prove that all superlevel sets $\{x \in K\colon f(x) \geq c\}$ $(c \in \mathbb{R})$ are convex.

8.6.5 Show that a C^2-function $f\colon \mathbb{R}^n \to \mathbb{R}$ is strictly quasi-concave at $u \in \mathbb{R}^n$ if and only if $D^2 f(u) < c \cdot \nabla f(u) \nabla f(u)^\top$ for some (positive) constant $c \in \mathbb{R}$.

8.6.6 Give the proof of Lemma 8.6.23. *Hint*: The analytic function $f(t) = \frac{1}{at}(1 - e^{-at})$ on $\mathbb{R}$ has properties (1)–(3) from Lemma 8.6.23, for some $a > 0$. Show that a suitable Taylor approximation of f satisfies the lemma.

8.6.7 Let $\boldsymbol{g} = (g_1, \dots, g_5)$ with $g_1 = y - x^3$, $g_2 = x$, $g_3 = 1 - x$, $g_4 = y$, $g_5 = 1 - y$ in $\mathbb{R}[x, y]$, and let $K = S(\boldsymbol{g})$. Show that $QM(\boldsymbol{g})$ is Archimedean and that the third moment relaxation of K with respect to $\boldsymbol{g}$ is exact. But g_1 is neither sos-concave, nor strictly quasi-concave in all points of $\mathcal{Z}(g_1) \cap \mathrm{Ex}(K)$.

8.6.8 Consider the basic closed set $S = \mathcal{S}(y - x^3,\ x + 1,\ y - y^2)$ in $\mathbb{R}^2$, and show that S is compact and convex. Then do the following:

(a) Let $\boldsymbol{g} = (g_1, \dots, g_r)$ be a finite sequence in $\mathbb{R}[x, y]$ with $S = \mathcal{S}(\boldsymbol{g})$, put $M = QM(\boldsymbol{g})$, and assume that moment relaxation with respect to $\boldsymbol{g}$ gets exact at level $d < \infty$. For any real number $c > 0$, show that there is $b \in \mathbb{R}$ such that $c - x + by$ is contained in the truncated quadratic module M_d.

(b) Let $N \subseteq \mathbb{R}[x]$ be the image of M under the ring homomorphism $\mathbb{R}[x, y] \to \mathbb{R}[x]$, $p(x, y) \mapsto p(x, 0)$. Show that the assumption in (a) implies $-x \in N$, and deduce a contradiction. Conclude that no moment relaxation for S gets exact at any finite level.

(c) On the other hand, show that S is a spectrahedral shadow (use Exercise 8.6.7).

8.6.9 Find all boundary points of the elliptope (Exercise 8.2.7) that are not smooth.

8.6.10 Let $K \subseteq \mathbb{R}^n$ be a closed convex semialgebraic set with non-empty interior, and let $f \in \mathbb{R}[x]$ be a polynomial without multiple factors and with $\partial_a K = \mathcal{V}(f)$.

(a) A point $u \in \partial K$ is a smooth boundary point of K (Definition 8.6.27) if, and only if, u is a non-singular point of the hypersurface $\mathcal{V}(f) = \partial_a K$. In this case, one of $\pm f$ is a positive inequality for K at u.

(b) If u is a smooth boundary point of K, and if g is a positive inequality for K at u, show that $w^\top \cdot D^2 g(u) \cdot w \leq 0$ holds for all $w \in \mathbb{R}^n$ with $\langle \nabla f(v), w \rangle = 0$.

Hint: Use the implicit function theorem, plus the fact that K is the closure of a union of connected components of $\{\xi \in \mathbb{R}^n\colon f(\xi) \neq 0\}$.

8.6.11 Let $K \subseteq \mathbb{R}^n$ be a closed convex semialgebraic set with non-empty interior, let $u \in \partial K$ be a smooth boundary point of K, and let $g \in \mathbb{R}[x]$ be a positive inequality for K at u. Show that there exists a unique supporting hyperplane of K that contains u, and that it has the equation $\sum_{i=1}^n \frac{\partial g}{\partial x_i}(u) \cdot (x_i - u_i) = 0$.

8.6.12 Let $C \subseteq \mathbb{R}^n$ be a semialgebraic convex cone which is closed and pointed and has non-empty interior. Let $\partial_a C = \mathcal{V}(f) \subseteq \mathbb{A}^n$ be the algebraic boundary of C, where $f \in \mathbb{R}[x] = \mathbb{R}[x_1, \dots, x_n]$ has no multiple factors. The polynomial f is homogeneous (verify this).

(a) Let the morphism $g \colon \mathcal{V}(f) \to \mathbb{A}^n$ be defined by $g(x) = (\partial_1 f(x), \dots, \partial_n f(x))$ (with $\partial_i = \frac{\partial}{\partial x_i}$, $i =, \dots, n$), and let W be the Zariski closure of the image of g. Show that W is contained in the algebraic boundary of the dual cone C^*.
(b) If $\nabla f(u) \neq 0$ for every boundary point $u \neq 0$ of C, show that $\partial_a(C^*) = W$.
(c) Give an example of a cone $C \subseteq \mathbb{R}^3$ as above for which $\partial_a(C^*)$ is strictly larger than W.

Considering $\mathcal{V}(f)$ as a hypersurface in $\mathbb{P}^{n-1}$, the map g corresponds to a rational map $\gamma \colon \mathcal{V}(f) \dashrightarrow \mathbb{P}^{n-1}$ (the *Gauss map*). The Zariski closure of the image of γ in $\mathbb{P}^{n-1}$ is called the *dual variety* of the projective hypersurface $\mathcal{V}(f)$. The affine cone over this projective variety is the variety W above. Loosely speaking, this exercise compares the dual variety of the algebraic boundary of C with the algebraic boundary of the dual cone C^*.

8.7 Convex Sets That are not Spectrahedral Shadows

We show that there exist convex semialgebraic sets that fail to be spectrahedral shadows, thereby disproving the Helton–Nie conjecture. Once more, sums of squares play a key role. This section is based on [187].

8.7.1 If V is an affine $\mathbb{R}$-variety and $S \subseteq V(\mathbb{R})$ is a subset, we keep writing $\mathcal{P}(S) = \{f \in \mathbb{R}[V] \colon f|_S \geq 0\}$, as in Chapter 6. If S is a subset of $\mathbb{R}^n$, let $P_S = \mathbb{R}[x]_{\leq 1} \cap \mathcal{P}(S)$ (as in 8.1.23), and put $P_{S,0} = \{f \in P_S : f(0) = 0\}$. So P_S is a closed convex cone in $\mathbb{R}[x]_{\leq 1} \cong \mathbb{R}^{n+1}$, and $P_{S,0}$ consists of the homogeneous polynomials in P_S. The closed convex hull and the closed conic hull of S are respectively given by $\overline{\mathrm{conv}(S)} = \{u \in \mathbb{R}^n \colon \forall f \in P_S \ f(u) \geq 0\}$ and $\overline{\mathrm{cone}(S)} = \{u \in \mathbb{R}^n \colon \forall f \in P_{S,0} \ f(u) \geq 0\}$, see 8.1.6. From this we see:

8.7.2 Lemma. *Let $S \subseteq \mathbb{R}^n$ be a set.*

(a) *The closed convex hull K of S is an affine-linear section of the dual cone $(P_S)^*$. If P_S is a spectrahedral shadow then so is K, and vice versa.*
(b) *The closed conic hull C of S is equal to the dual cone $(P_{S,0})^*$. If $P_{S,0}$ is a spectrahedral shadow then so is C, and vice versa.*

Proof. (a) For a linear polynomial $f = a_0 + \sum_{i=1}^n a_i x_i$ let $\tilde{f} = \sum_{i=0}^n a_i x_i \in \mathbb{R}[x_0, x]$ be its degree one homogenization. By definition, $(P_S)^*$ consists of the tuples $b = (b_0, \dots, b_n)$ in $\mathbb{R}^{n+1}$ that satisfy $\tilde{f}(b) \geq 0$ for every $f \in P_S$. Therefore $K = \{u \in \mathbb{R}^n \colon (1, u) \in (P_S)^*\}$. The second part of (a) holds by Corollaries 8.3.10 and 8.3.11, since $P_S = P_K = (K^h)^*$ by Remark 8.1.23. In (b), the first assertion is clear, and so 8.3.10 implies the second. □

Our first step is a reformulation of the spectrahedral shadow property. To this end we define:

8.7.3 Definition. Let V be an affine $\mathbb{R}$-variety, let $S \subseteq V(\mathbb{R})$ be a subset and $L \subseteq \mathbb{R}[V]$ a linear subspace of finite dimension. We say that S admits *uniform* *sos representations for* L if there exists a morphism $\phi\colon X \to V$ of affine $\mathbb{R}$-varieties, together with a linear subspace U of $\mathbb{R}[X]$, such that

(1) $\dim(U) < \infty$,
(2) $S \subseteq \phi(X(\mathbb{R}))$,
(3) $\phi^*(L \cap \mathcal{P}(S)) \subseteq \Sigma U^2$.

Here $\phi^*\colon \mathbb{R}[V] \to \mathbb{R}[X]$ is the ring homomorphism dual to ϕ. So (3) requires that the ϕ-pullback of any $f \in L$ with $f|_S \geq 0$ can be written as a sum of squares of elements from U. In view of (2), note that every $f \in \mathbb{R}[V]$, for which $\phi^*(f)$ is a sum of squares in $\mathbb{R}[X]$, will be non-negative on S.

8.7.4 Remark. We may reformulate Definition 8.7.3 in different terms. Given a semialgebraic set $S \subseteq V(\mathbb{R})$, let $\mathcal{A}_0(S)$ denote the ring of all definable functions $S \to \mathbb{R}$ (i.e. functions with semialgebraic graph, but not necessarily continuous, see 4.3.1). Then S admits uniform sos representations for a given finite-dimensional linear space $L \subseteq \mathbb{R}[V]$ (8.7.3) if, and only if, the following holds:

(∗) *There exist finitely many definable functions* $h_1, \dots, h_N$ *on* S *such that every* $f \in L \cap \mathcal{P}(S)$ *can be written as a sum of squares of linear combinations of* $h_1, \dots, h_N$.

Indeed, assume that the conditions of 8.7.3 hold, so we are given a morphism of varieties $\phi\colon X \to V$ together with a linear subspace U of $\mathbb{R}[X]$ as in 8.7.3. Since $S \subseteq \phi(X(\mathbb{R}))$, there exists a definable section $\sigma\colon S \to X(\mathbb{R})$ of ϕ over S, by Proposition 4.5.9. If $g_1, \dots, g_N$ is a basis of U then the definable functions $h_i = g_i \circ \sigma$ ($i = 1, \dots, N$) on S satisfy condition (∗). Conversely, if $h_1, \dots, h_N \in \mathcal{A}_0(S)$ are given as in (∗), let X be the Zariski closure of

$$\operatorname{graph}(h_1, \dots, h_N) = \{(s, h_1(s), \dots, h_N(s))\colon s \in S\} \subseteq S \times \mathbb{R}^N$$

in $V \times \mathbb{A}^N$, and let $\phi\colon X \to V$ be the natural morphism. Then $h_i = g_i \circ \sigma$, where $\sigma\colon S \to X(\mathbb{R})$ is the obvious section and $g_i \in \mathbb{R}[X]$ is projection to the i-th component of $\mathbb{A}^N$. So $\phi\colon X \to V$, together with the linear subspace $U \subseteq \mathbb{R}[X]$ spanned by $g_1, \dots, g_N$, satisfies the conditions in Definition 8.7.3.

8.7.5 Theorem. (Scheiderer) *Let* $S \subseteq \mathbb{R}^n$ *be a semialgebraic set and let* $K = \overline{\operatorname{conv}(S)}$ *be its closed convex hull. The following are equivalent:*

(i) K *is a spectrahedral shadow;*
(ii) S *admits uniform sos representations for* $L_1 = \operatorname{span}(1, x_1, \dots, x_n) \subseteq \mathbb{R}[x]$.

Similarly, the closed conic hull $C = \overline{\operatorname{cone}(S)}$ *of* S *is a spectrahedral shadow if, and only if,* S *admits uniform sos representations for* $L = \operatorname{span}(x_1, \dots, x_n) \subseteq \mathbb{R}[x]$.

We start by proving (ii) ⇒ (i).

8.7.6 Proposition. *If* $S \subseteq \mathbb{R}^n$ *is a set that admits uniform sos representations for* L_1 *(resp. for* L*), then* $K = \overline{\operatorname{conv}(S)}$ *(resp.* $C = \overline{\operatorname{cone}(S)}$*) is a spectrahedral shadow.*

Proof. We first give the proof for L_1 and K. By assumption we have a morphism $\phi\colon X \to \mathbb{A}^n$ with $S \subseteq \phi(X(\mathbb{R}))$ and a finite-dimensional linear subspace $U \subseteq \mathbb{R}[X]$, such that $\phi^*(f) \in \Sigma U^2$ for every $f \in L_1 = \mathbb{R}[x]_{\le 1}$ with $f|_S \ge 0$. Let $W := \phi^*(L_1) + UU$, a finite-dimensional subspace of $\mathbb{R}[X]$, and consider the restriction $\varphi := \phi^*|_{L_1}\colon L_1 \to W$ of $\phi^*\colon \mathbb{R}[x] \to \mathbb{R}[X]$. By assumption we have $\varphi(P_S) \subseteq \Sigma U^2$, or equivalently, $P_S \subseteq \varphi^{-1}(\Sigma U^2)$. The opposite inclusion holds anyway since $S \subseteq X(\mathbb{R})$. So P_S is the preimage of ΣU^2 under the linear map φ. Since ΣU^2 is a spectrahedral shadow (in $UU \subseteq W$) by Example 8.3.4.3, and since the class of spectrahedral shadows is closed under taking linear preimages (Exercise 8.3.1), P_S is a spectrahedral shadow as well. Therefore K is a spectrahedral shadow by Lemma 8.7.2. The proof for the closed conic hull C is identical, up to replacing L_1, P_S and K by L, $P_{S,0}$ and C, respectively. □

8.7.7 Remarks.

1. Assume that the set $S \subseteq \mathbb{R}^n$ satisfies the conditions of Proposition 8.7.6. If corresponding $\phi\colon X \to \mathbb{A}^n$ and $U \subseteq \mathbb{R}[X]$ are given explicitly, we get an explicit semidefinite representation of the cone P_S, and therefore of the closed convex hull of S as well. The matrices in the corresponding lifted LMI have size $\dim(U)$, as we see from the proof of 8.7.6 and from 8.3.4.3. The construction can be seen as generalizing the moment relaxation construction (Section 8.5), performed however in $\mathbb{R}[X]$, which is a ring extension of the polynomial ring $\mathbb{R}[x]$.

2. It is natural to ask whether Proposition 8.7.6 remains true if the condition $\dim(U) < \infty$ in 8.7.3 is dropped. That is, assume that there is a morphism $\phi\colon X \to \mathbb{A}^n$ of affine varieties with $S \subseteq \phi(X(\mathbb{R}))$ such that, for every $f \in P_S$, the pullback $\phi^*(f)$ of f is a sum of squares in $\mathbb{R}[X]$. We'll see a little later (Remark 8.7.21.1) that this weaker condition does not suffice to conclude that K is a spectrahedral shadow.

Now we prove the converse (i) ⇒ (ii) in Theorem 8.7.5, starting with the cone version.

8.7.8 Proposition. *Assume that $S \subseteq \mathbb{R}^n$ is a semialgebraic set for which the closed conic hull $C = \overline{\mathrm{cone}(S)} \subseteq \mathbb{R}^n$ is a spectrahedral shadow. Then S (or C) admits uniform sos representations for $L = \mathrm{span}(x_1, \dots, x_n)$.*

Proof. We may assume that $\mathbb{R}^n$ is affinely spanned by S. By assumption, C is the image of a spectrahedral cone $T \subseteq \mathbb{R}^p$ under a linear map $\pi\colon \mathbb{R}^p \to \mathbb{R}^n$, for some p (use Lemma 8.3.5). We may assume that $\mathbb{R}^p$ is the linear hull of T, which implies that T has non-empty interior. So T can be represented by a homogeneous LMI that is strictly feasible (Proposition 8.2.18). This means, there are linear matrix pencils $M(x) = \sum_{i=1}^n x_i M_i$ and $N(y) = \sum_{j=1}^m y_j N_j$ in S^d (for some $m \ge 0$ and $d \ge 1$) such that

$$T = \{(\xi, \eta) \in \mathbb{R}^n \times \mathbb{R}^m \colon M(\xi) + N(\eta) \succeq 0\},$$

such that $C = \pi(T)$ where $\pi(\xi, \eta) = \xi$, and such that there exists $(\xi, \eta) \in T$ with $M(\xi) + N(\eta) \succ 0$.

Consider the closed subvariety X of $\mathbb{A}^n \times \mathbb{A}^m \times \mathrm{Sym}_d$, defined over $\mathbb{R}$, whose $\mathbb{C}$-points are the triples (ξ, η, A) where A is a symmetric $d \times d$-matrix satisfying

$$A^2 = \sum_{i=1}^{n} \xi_i M_i + \sum_{j=1}^{m} \eta_j N_j.$$

We shall denote the coordinate functions on X by

$$(x_1, \ldots, x_n, y_1, \ldots, y_m, (z_{\mu\nu})_{1 \le \mu,\nu \le d}) = (x, y, Z)$$

with $z_{\mu\nu} = z_{\nu\mu}$ for $1 \le \mu, \nu \le d$. Let $\phi \colon X \to \mathbb{A}^n$ be the projection $\phi(\xi, \eta, A) = \xi$. Then $\phi(X(\mathbb{R})) = \pi(T) = C$, since a real symmetric matrix is psd if and only if it is the square of a real symmetric matrix. Let $U \subseteq \mathbb{R}[X]$ be the linear subspace spanned by the coefficient functions $z_{\mu\nu} = z_{\nu\mu}$ $(1 \le \mu, \nu \le d)$ of Z. We claim that property (3) of Definition 8.7.3 holds with these choices of ϕ and U.

To see this, let $f = \sum_{i=1}^{n} a_i x_i \in L$ be a linear homogeneous polynomial with $f \ge 0$ on S, and hence $f \ge 0$ on C. So the tuple $(a, 0) = (a_1, \ldots, a_n, 0, \ldots, 0) \in \mathbb{R}^n \times \mathbb{R}^m$ lies in the dual cone T^* of T. (If $(\xi, \eta) \in T$, then $\xi \in C$, hence $0 \le f(\xi) = \langle a, \xi \rangle = \langle (a, 0), (\xi, \eta) \rangle$.) Since the linear matrix inequality is strictly feasible, there exists a psd matrix $B \in \mathsf{S}_+^d$ with $a_i = \langle B, M_i \rangle$ $(1 \le i \le n)$ and $0 = \langle B, N_j \rangle$ $(1 \le j \le m)$. Indeed, it was proved in Proposition 8.3.9 that T^* consists of all tuples

$$\big(\langle B, M_1 \rangle, \ldots, \langle B, M_n \rangle;\ \langle B, N_1 \rangle, \ldots, \langle B, N_m \rangle\big)$$

with $B \in \mathsf{S}_+^d$. Let $W = (w_{kl}) \in \mathsf{S}^d$ be a symmetric matrix with $B = W^2$. Then, as an element of $\mathbb{R}[X]$, $\phi^*(f)$ is equal to

$$\sum_{i=1}^{n} \langle B, M_i \rangle x_i + \sum_{j=1}^{n} \langle B, N_j \rangle y_j = \langle B,\ M(x) + N(y) \rangle = \langle W^2, Z^2 \rangle = \langle ZW, ZW \rangle$$

since $\langle W^2, Z^2 \rangle = \operatorname{tr}(W^2 Z^2) = \operatorname{tr}(Z W^2 Z) = \operatorname{tr}((ZW)(ZW)^\top) = \langle ZW, ZW \rangle$. This means that

$$\phi^*(f) = \sum_{\mu,\nu=1}^{d} ((ZW)_{\mu\nu})^2 = \sum_{\mu,\nu=1}^{d} \Big(\sum_{k} z_{\mu k} w_{k\nu}\Big)^2$$

is a sum of squares in $\mathbb{R}[X]$ of elements from the linear subspace $U \subseteq \mathbb{R}[X]$. □

Here is the inhomogeneous version of Proposition 8.7.8:

8.7.9 Corollary. *Let $S \subseteq \mathbb{R}^n$ be a semialgebraic set and let $K = \overline{\operatorname{conv}(S)}$ be its closed convex hull. If K is a spectrahedral shadow then S (or K) admits uniform sos representations for $L_1 = \operatorname{span}(1, x_1, \ldots, x_n)$.*

Proof. Since K is a spectrahedral shadow, the same is true for the homogenization $K^h \subseteq \mathbb{R} \times \mathbb{R}^n$ of K (Corollary 8.3.11). The latter is the closure of $\operatorname{cone}(\{1\} \times K)$ in $\mathbb{R} \times \mathbb{R}^n$ (Proposition 8.1.19). By Proposition 8.7.8, $\{1\} \times K$ admits uniform sos representations for $\operatorname{span}(x_0, x_1, \ldots, x_n)$. Dehomogenizing, we directly get the assertion of the corollary. □

With this, Theorem 8.7.5 has been proved.

8.7.10 Examples.

1. Let $S = \mathcal{S}(g_1,\dots,g_r) \subseteq \mathbb{R}^n$ be a basic closed set, and assume that standard moment relaxation for the convex hull $\mathrm{conv}(S)$ of S (Remark 8.5.7) is exact in high degrees. Let $\mathbb{R}[x] \subseteq B$ be the ring extension arising from adjoining square roots of $g_1,\dots,g_r$ to $\mathbb{R}[x]$, and let $\phi\colon X \to \mathbb{A}^n$ be the morphism of affine varieties that is dual to $\mathbb{R}[x] \subseteq B$. Then the conditions of Definition 8.7.3 are satisfied for $\phi\colon X \to \mathbb{A}^n$, and for some subspace $U \subseteq \mathbb{R}[X]$ of finite dimension. Indeed, this follows from Corollary 8.5.13(b). The semidefinite representation constructed from ϕ and U agrees with the representation that was constructed with the moment relaxation method in Section 8.5.

2. The "tv hyperscreen" $K = \{u \in \mathbb{R}^2\colon u_1^{2d_1} + u_2^{2d_2} \le 1\}$ in the plane is a spectrahedral shadow, as we saw in Example 8.5.23. To verify this using Theorem 8.7.5, note that K is the convex hull of its boundary $S = \partial K$. Let $X \subseteq \mathbb{A}^2$ be the Zariski closure of S, i.e. $X = \mathcal{V}(f)$ with $f = 1 - x_1^{2d_1} - x_2^{2d_2}$, and let $\phi\colon X \to \mathbb{A}^2$ be the inclusion. As remarked in 8.5.23, the convex cone $P_K = P_S$ in $L_1 = \mathbb{R}[x_1,x_2]_{\le 1}$ is generated by 1 together with the positive tangents $t_u = 2d_1u_1^{2d_1-1}(u_1 - x_1) + 2d_2u_2^{2d_2-1}(u_2 - x_2)$ for $u = (u_1,u_2) \in S$. By identity (8.28) and the remark made after it, $\phi^*(t_u)$ is a sum squares of elements from the subspace $U = \mathrm{span}(1, x_1,\dots,x_1^{d_1}, x_2,\dots,x_2^{d_2})$ of $\mathbb{R}[X]$. Hence S admits uniform sos representations for $L_1 = \mathrm{span}(1,x_1,x_2)$, with these choices of X, ϕ and U. Of course, this reasoning generalizes to the higher-dimensional versions of K as in 8.5.23.

Before proceeding further, we state an easy generalization that is more convenient and flexible to apply. Given a finite-dimensional linear subspace $L \subseteq \mathbb{R}[x]$ with linear basis $p_1,\dots,p_m$, let $\varphi_L\colon \mathbb{R}^n \to \mathbb{R}^m$ be the Veronese-type map defined by $u \mapsto (p_1(u),\dots,p_m(u))$. (A basis-free definition of φ_L would be the map $\mathbb{R}^n \to L^\vee = \mathrm{Hom}(L,\mathbb{R})$, $u \mapsto (\varphi_u\colon L \to \mathbb{R},\ p \mapsto p(u))$, but for the sake of concreteness we stick to the version in coordinates.)

8.7.11 Corollary. *Let $S \subseteq \mathbb{R}^n$ be a semialgebraic set and let $L \subseteq \mathbb{R}[x]$ with $\dim(L) = m < \infty$. The closed convex hull of $\varphi_L(S)$ in $\mathbb{R}^m$ is a spectrahedral shadow if, and only if, S admits uniform sos representations for $L_1 = L + \mathbb{R}1$. The analogous statement for the closed conic hull holds as well, replacing L_1 with L.*

Proof. We do the inhomogeneous case (sketch). If S admits uniform sos representations for L_1, there are $\phi\colon X \to \mathbb{A}^n$ and $U \subseteq \mathbb{R}[X]$ with $S \subseteq \phi(X(\mathbb{R}))$, $\dim(U) < \infty$ and $\phi^*(L_1 \cap \mathcal{P}(S)) \subseteq \Sigma U^2$. Then the composition $\varphi_L \circ \phi\colon X \to \mathbb{A}^m$, together with U, satisfies the conditions of 8.7.3 for the set $\varphi_L(S)$ in $\mathbb{R}^m$ and the space of linear polynomials in $\mathbb{R}^m$. So $K = \overline{\mathrm{conv}\,\varphi_L(S)}$ is a spectrahedral shadow, by the backward implication of Theorem 8.7.5. Conversely, if K is a spectrahedral shadow in $\mathbb{R}^m$, then by the forward implication of 8.7.5 there exists $\phi\colon X \to \mathbb{A}^m$ and $U \subseteq \mathbb{R}[X]$ with $\dim(U) < \infty$, $\varphi_L(S) \subseteq \phi(X(\mathbb{R}))$ and $\phi^*(f) \in \Sigma U^2$ for every $f \in \mathbb{R}[y_1,\dots,y_m]_{\le 1}$ with $f \ge 0$ on $\varphi_L(S)$. Consider the fibre product

$$\begin{array}{ccc} Y & \xrightarrow{\varphi'} & X \\ {\scriptstyle\psi}\downarrow & & \downarrow{\scriptstyle\phi} \\ \mathbb{A}^n & \xrightarrow{\varphi_L} & \mathbb{A}^m \end{array}$$

so $\mathbb{R}[Y] = \mathbb{R}[x] \otimes_{\mathbb{R}[y]} \mathbb{R}[X]$. Then $S \subseteq \psi(Y(\mathbb{R}))$, so ψ together with the subspace $(\varphi')^*(U)$ of $\mathbb{R}[Y]$ satisfies the conditions of 8.7.3 for S and L_1.

For both implications, the conic case is completely analogous. □

8.7.12 It may not yet be obvious, but the necessary conditions for spectrahedral shadows that result from Proposition 8.7.8 (resp. Corollary 8.7.9) are quite restrictive. To see this, we use the reformulation exhibited in Remark 8.7.4.

From Section 4.5, recall the concept of Nash functions (Remark 4.5.3.3). If $U \subseteq \mathbb{R}^n$ is an open semialgebraic set, let $\mathcal{N}(U) \subseteq \mathcal{A}_0(U)$ denote the ring of all Nash functions $U \to \mathbb{R}$. For $u \in \mathbb{R}^n$ let $\mathcal{O}_u = \mathbb{R}[x]_{\mathfrak{m}_u}$, the local ring at u, and let $\widehat{\mathcal{O}}_u$ be its completion. Then $\widehat{\mathcal{O}}_u = \mathbb{R}[\![x-u]\!] = \mathbb{R}[\![x_1-u_1,\dots,x_n-u_n]\!]$, the ring of formal power series in $x_1-u_1,\dots,x_n-u_n$. Given any Nash function $f \in \mathcal{N}(U)$ and any point $u \in U$, we may consider the formal Taylor series expansion of f around u, viz.

$$\tau_u(f) = \sum_\alpha \frac{1}{\alpha!}\frac{\partial^\alpha f}{\partial x^\alpha}(u)\cdot(x-u)^\alpha \in \widehat{\mathcal{O}}_u.$$

(This series actually converges in a neighborhood of u, since every Nash function is analytic. We didn't prove this fact however, and we won't need it.) The map

$$\tau_u\colon \mathcal{N}(U) \to \widehat{\mathcal{O}}_u, \quad f \mapsto \tau_u(f)$$

is a ring homomorphism (in fact injective when U is connected, Exercise 4.5.14).

8.7.13 Recall that the (vanishing) order $\omega(f)$ of a formal power series $f = \sum_\alpha c_\alpha x^\alpha$ in $\mathbb{R}[\![x]\!]$ is $\omega(f) = \inf\{|\alpha|\colon c_\alpha \ne 0\}$. If $\omega(f) = d$, the leading form of f is $L(f) = \sum_{|\alpha|=d} c_\alpha x^\alpha$ (compare A.4.7). To give counter-examples to the Helton–Nie conjecture, we are going to use the easy fact that the leading form of any sum of squares f in $\mathbb{R}[\![x]\!]$ is a sum of squares of homogeneous polynomials (this was remarked in Example 6.1.11).

For $n, d \in \mathbb{N}$, recall that $\dim \mathbb{R}[x_1,\dots,x_n]_{\le d} = \binom{n+d}{n}$. Consider the Veronese polynomial map

$$\varphi_{n,d}\colon \mathbb{R}^n \to \mathbb{R}^{\binom{n+d}{n}-1}, \quad u \mapsto (u^\alpha)_{1\le|\alpha|\le d}$$

8.7.14 Theorem. *Let $S \subseteq \mathbb{R}^n$ be any semialgebraic set with non-empty interior. If $n \ge 3$ and $d \ge 6$, or if $n \ge 4$ and $d \ge 4$, the closed convex hull K of $\varphi_{n,d}(S)$ in $\mathbb{R}^{\binom{n+d}{n}-1}$ fails to be a spectrahedral shadow.*

Proof. Assume that K is a spectrahedral shadow. By Corollary 8.7.11 and Remark 8.7.4, there exists a finite-dimensional linear subspace U of $\mathcal{A}_0(S)$ such that, for every polynomial $f \in \mathbb{R}[x]$ with $f|_S \ge 0$ and $\deg(f) \le d$, the function $f|_S$ is a sum of squares of elements of U.

By Theorem 4.5.7, every definable function $h\colon S \to \mathbb{R}$ is Nash on some open dense semialgebraic subset of the interior of S. Hence there is an open non-empty subset W of S such that all members of U are Nash on W. Choose a point $u \in W$, and let $p \in \mathbb{R}[x_1, \dots, x_n]$ be a psd form of degree 6 (or 4, if $n \geq 4$) that is not sos, for example the Motzkin form or the Choi–Lam form (Examples 2.2.10). Since the polynomial $f := p(x_1 - u_1, \dots, x_n - u_n)$ is psd on $\mathbb{R}^n$ and has degree $\deg(f) \leq d$, its restriction to S is a sum of squares of functions in U. Since every element of U is Nash on a neighborhood of u, this implies that the Taylor series $\tau_u(f) \in \widehat{\mathcal{O}}_u$ is a sum of squares in $\widehat{\mathcal{O}}_u$. In particular, the leading form of $\tau_u(f)$ is a sum of squares of polynomials (cf. Example 6.1.11). This is a contradiction, since this leading form is f itself. □

Using $(n, d) = (3, 6)$, we get examples of convex non-shadows of dimension $83 = \binom{9}{3} - 1$. Using $(n, d) = (4, 4)$ gives examples of dimension $69 = \binom{8}{4} - 1$. These examples are closed convex hulls of sets of dimension at least three. We'll now refine our approach and construct examples of smaller dimensions, and also examples that are convex hulls of two-dimensional sets.

8.7.15 Let $R \supseteq \mathbb{R}$ be a fixed extension of real closed fields, and let $B \subseteq R$ be the canonical valuation ring of R, i.e. the convex hull of $\mathbb{R}$ in R. Let $b \mapsto \overline{b}$ denote the residue map $B \to B/\mathfrak{m}_B = \mathbb{R}$. If A is an $\mathbb{R}$-algebra, write $A_R := A \otimes R$ and $A_B := A \otimes B$ (tensor product over $\mathbb{R}$), and note that A_B is a subring of A_R. The homomorphism $A_B \to A$, $f \otimes b \mapsto \overline{b} \cdot f$ will be called reduction modulo $\mathfrak{m}_B$.

Recall that a ring A is *real reduced* (Definition 3.2.17) if $\sum_{i=1}^r a_i^2 = 0$ and $a_1, \dots, a_r \in A$ implies $a_1 = \dots = a_r = 0$.

8.7.16 Lemma. *Let A be an $\mathbb{R}$-algebra that is real reduced. If $f_1, \dots, f_m \in A_R$ are such that $f := \sum_i f_i^2$ lies in A_B, it follows that $f_i \in A_B$ for all i.*

Proof. There exist finitely many linearly independent elements $g_1, \dots, g_r \in A$ such that each f_i can be written (uniquely) as $f_i = \sum_{j=1}^r c_{ij} g_j$ with $c_{ij} \in R$. We have to show $c_{ij} \in B$ for all i, j. Assume this is false, then $c := \max_{i,j} |c_{ij}|$ doesn't lie in B. The element $h_i = \frac{1}{c} f_i$ lies in A_B for each i, and $\frac{1}{c^2} f = \sum_{i=1}^m h_i^2$ has coefficients in $\mathfrak{m}_B$. Reducing both sides modulo $\mathfrak{m}_B$ gives the identity $0 = \sum_{i=1}^m \overline{h}_i^{\,2}$ in A. But $\overline{h}_i \neq 0$ for at least one index i, contradicting the assumption that A is real reduced. □

Here is the key observation for the refinement:

8.7.17 Proposition. *Let $f \in \mathbb{R}[t, x] = \mathbb{R}[t, x_1, \dots, x_n]$ be homogeneous of even degree d in (t, x), and assume that f is not a sum of squares in $\mathbb{R}[t, x]$. If $\varepsilon > 0$ is infinitesimal in B, the polynomial $f(\varepsilon, x) \in B[x]$ is not sos in $B[x]/\langle x \rangle^{d+1} B[x]$.*

Proof. Here $\langle x \rangle^{d+1} := \langle x_1, \dots, x_n \rangle^{d+1}$, the $(d+1)$-st ideal power of the ideal generated by $x_1, \dots, x_n$. Recall that ε infinitesimal means $\varepsilon \in \mathfrak{m}_B$. Assume we have an identity

$$f(\varepsilon, x) + g(x) = \sum_j p_j(x)^2, \tag{8.36}$$

where $g(x) \in \langle x\rangle^{d+1}B[x]$ and $p_j(x) \in B[x]$ for all j. Let the polynomial $g_1(x) \in B[x]$ be defined by $g_1(x) = \varepsilon^{-(d+1)}g(\varepsilon x)$. Replacing x by εx in (8.36) and dividing by ε^d, we get

$$f(1,x) + \varepsilon g_1(x) = \sum_j \varepsilon^{-d} p_j(\varepsilon x)^2. \tag{8.37}$$

By Lemma 8.7.16, the polynomials $q_j(x) = \varepsilon^{-d/2}p_j(\varepsilon x) \in R[x]$ have coefficients in B. So we may reduce (8.37) modulo $\mathfrak{m}_B$, thereby concluding that $f(1,x)$ is a sum of squares in $\mathbb{R}[x]$. Since $\deg(f) = d$ is even, this implies that $f(t,x)$ is sos in $\mathbb{R}[t,x]$ (Lemma 2.4.4), contradicting the hypothesis. □

Using Proposition 8.7.17, we get the following refined version of 8.7.14:

8.7.18 Theorem. *Let $p(t,x) \in \mathbb{R}[t,x] = \mathbb{R}[t,x_1,\dots,x_n]$ be a psd form that is not sos, and let $L \subseteq \mathbb{R}[x]$ be a subspace with $\dim(L) = m < \infty$ such that $p(c, x+u) \in \mathbb{R} + L$ for all $c \in \mathbb{R}$ and $u \in \mathbb{R}^n$. Whenever $S \subseteq \mathbb{R}^n$ is a semialgebraic set with non-empty interior, the closed convex hull of $\varphi_L(S)$ in $\mathbb{R}^m$ fails to be a spectrahedral shadow.*

Proof. Assume this is false. As in the proof of Theorem 8.7.14, we use Corollary 8.7.11 to find an open non-empty semialgebraic subset W of S and a subspace $U \subseteq \mathcal{N}(W)$ with $\dim(U) < \infty$, such that $f|_W \in \Sigma U^2$ for every $f \in \mathbb{R}+L$ with $f|_W \ge 0$. Let $R \supsetneq \mathbb{R}$ be a proper real closed field extension, let $B \subseteq R$ be the convex hull of $\mathbb{R}$ in R as in 8.7.15, and choose an infinitesimal $\varepsilon > 0$ in R. For every $f \in R + L_R \subseteq R[x]$ which is non-negative on $W_R \subseteq R^n$, the restriction of f to W_R is a sum of squares of elements from $U_R = U \otimes R$, by Tarski's transfer principle.

Let $d = \deg(p)$, and note that d is even since p is psd. Fix $u \in W \subseteq \mathbb{R}^n$. The polynomial

$$f = p(\varepsilon,\, x_1 - u_1,\, \dots,\, x_n - u_n) \in R[x]$$

lies in $R + L_R$, in fact in $B + L_B$, and is psd on all of R^n. So $f|_W \in \Sigma(U_R)^2$. Taking formal Taylor expansions at u, we see that f is a sum of squares in the ring $\widehat{\mathcal{O}}_u \otimes R$, where $\widehat{\mathcal{O}}_u = \mathbb{R}[\![x-u]\!]$ as before. (Caution: $\mathbb{R}[\![y]\!] \otimes R$ is a *proper* subring of $R[\![y]\!]$!) The ring $\widehat{\mathcal{O}}_u$ is clearly real reduced, so applying Lemma 8.7.16 we conclude that f is a sum of squares even in the subring $\widehat{\mathcal{O}}_u \otimes B$ of $\widehat{\mathcal{O}}_u \otimes R$. In particular, f is sos in $\widehat{\mathcal{O}}_u \otimes B$ modulo the ideal $\langle x-u\rangle^{d+1}$, which means that f is sos in $B[x]/\langle x-u\rangle^{d+1}B[x]$. Now we have a contradiction to Proposition 8.7.17. □

8.7.19 Remark. In Theorem 8.7.18, let L be spanned by all non-constant monomials of degree at most d, where $n = 2$ and $d \ge 6$, or $n \ge 3$ and $d \ge 4$ (corresponding to the cases when $P_{n,\le d} \ne \Sigma_{n,\le d}$). For $p(t,x)$ we can take the Motzkin form (if $n = 2$) or the Choi–Lam form (if $n \ge 3$). Then by 8.7.18, if $S \subseteq \mathbb{R}^n$ is semialgebraic with $\mathrm{int}(S) \ne \emptyset$, the closed convex hull of $\varphi_{n,d}(S)$ in $\mathbb{R}^{\binom{n+d}{n}-1}$ is not a shadow. For $(n,d) = (2,6)$ and $(3,4)$, this is a convex set of dimension $27 = \binom{8}{2} - 1$ and $34 = \binom{7}{3} - 1$, respectively. These dimensions are already much smaller than the dimensions obtained from Theorem 8.7.14. Moreover, we have the following interesting consequence:

8.7.20 Theorem. *Let $P_{n,2d} \subseteq \mathbb{R}[x_1, \ldots, x_n]_{2d}$ be the cone of psd n-ary forms of degree d. Then $P_{n,2d}$ is a spectrahedral shadow if and only if $P_{n,2d} = \Sigma_{n,2d}$, i.e., if and only if $n \le 2$ or $2d = 2$ or $(n, 2d) = (3, 4)$.*

Proof. Since $\Sigma_{n,2d}$ is always a spectrahedral shadow (Proposition 8.3.14), the "if" part is clear. Conversely let $(n, 2d)$ be such that $\Sigma_{n,2d} \neq P_{n,2d}$. Under the linear (dehomogenization) isomorphism

$$\mathbb{R}[x_1, \ldots, x_n]_{2d} \xrightarrow{\sim} \mathbb{R}[x_1, \ldots, x_{n-1}]_{\le 2d}, \quad f(x_1, \ldots, x_n) \mapsto f(x_1, \ldots, x_{n-1}, 1),$$

$P_{n,2d}$ is mapped onto $P_{n-1,\le 2d}$ (Lemma 2.4.4). On the other hand, $P_{n-1,\le 2d}$ is identified with the cone of linear (inhomogeneous) polynomials that are psd on $\varphi_{n-1,2d}(\mathbb{R}^{n-1}) \subseteq \mathbb{R}^{\binom{n+2d-1}{n-1}-1}$. By 8.7.19 (and Lemma 8.7.2), this cone is not a spectrahedral shadow. Therefore $P_{n,2d}$ isn't one either. □

8.7.21 Remarks.

1. Let $g = 1 - x^2 - y^2 \in \mathbb{R}[x, y]$ and $S = \mathcal{S}(g) \subseteq \mathbb{R}^2$, the unit disk. By Theorem 6.5.22, the preordering $PO(g)$ in $\mathbb{R}[x, y]$ contains every polynomial that is non-negative on the disk. On the other hand, for $L \subseteq \mathbb{R}[x, y]$ sufficiently large (e.g. $L = \mathbb{R}[x, y]_{\le 6}$), the convex hull of $\varphi_L(S)$ in $\mathbb{R}^{\dim(L)}$ fails to be a spectrahedral shadow (Remark 8.7.19). What goes wrong in the conditions of 8.7.3 is *uniformity*: Although it is true for suitable $\phi\colon X \to \mathbb{A}^2$ with $S \subseteq \phi(X(\mathbb{R}))$ that $\phi^*(L \cap \mathcal{P}(S))$ consists of sums of squares in $\mathbb{R}[X]$ (namely, when $\sqrt{g}$ gets adjoined), there never exists a subspace $U \subseteq \mathbb{R}[X]$ of *finite dimension* with $\phi^*(L \cap \mathcal{P}(S)) \subseteq \Sigma U^2$.

2. Refining the arguments, one can further improve on the dimension of convex semialgebraic sets that fail to be spectrahedral shadows. For a construction that leads to examples of dimension 12, see Exercises 8.7.2 and 8.7.3. Using a more sophisticated argument, an example of dimension 11 can be constructed (Exercise 8.7.4). That's it at the time of writing these lines (2024), no examples of smaller dimension are currently known.

3. For every $d \ge 2$, we have seen counter-examples to the Helton–Nie conjecture that are convex hulls of d-dimensional sets. For $d = 1$, no such counter-examples exist. Indeed, the closed convex hull of any semialgebraic set $S \subseteq \mathbb{R}^n$ of dimension one is a spectrahedral shadow. In the plane $\mathbb{R}^2$, even the full Helton–Nie conjecture is true: Every convex semialgebraic subset of $\mathbb{R}^2$ (closed or not) is a spectrahedral shadow. These results are proven in [186]. For $3 \le d \le 10$ there remains an embarrassing gap: It is unknown whether there exists any convex semialgebraic set in $\mathbb{R}^d$ that is not a spectrahedral shadow.

It also seems not to be known whether a counterexample to the Helton–Nie conjecture exists that has smooth boundary.

4. A different approach was taken in [29]. If $E \subseteq \mathbb{Z}_+^n$ is a finite set of multi-indices, let $P_+(E)$ denote the cone of all psd polynomials $f \in \mathbb{R}[x] = \mathbb{R}[x_1, \ldots, x_n]$ with $\mathrm{supp}(f) \subseteq E$. The authors prove that $P_+(E)$ is a spectrahedral shadow if, and only if, there exists an integer $d \ge 1$ such that, for every $f \in P_+(E)$, the polynomial $f(x_1^d, \ldots, x_n^d)$ is a sum of squares of polynomials. Using this criterion, they are able

to show that the cone

$$C_n = \{A \in \mathsf{S}^n : x^\top A x \geq 0 \text{ for every } \mathbb{R}^n_+\}$$

of *copositive matrices* is not a spectrahedral shadow for $n \geq 5$. Whether or not C_n is a spectrahedral shadow had been a well-known open problem before. (For $n \leq 4$, it has long been known that C_n is a spectrahedral shadow.)

5. Without proof we remark that there exists yet another, and quite different, characterization of spectrahedral shadows, as follows. If $\xi = (\xi_1, \dots, \xi_n) \in R^n$ where R is a real closed overfield of $\mathbb{R}$, and if $f = a_0 + \sum_{i=1}^n a_i x_i \in R[x]$ is a linear polynomial, let the "tensor evaluation" of f at ξ be the element

$$f^{\otimes}(\xi) = a_0 \otimes 1 + \sum_{i=1}^{n} a_i \otimes \xi_i$$

in the ring $R \otimes R := R \otimes_{\mathbb{R}} R$. Let $S \subseteq \mathbb{R}^n$ be a semialgebraic set, let $K \subseteq \mathbb{R}^n$ be its closed convex hull. Then, for every $\xi \in S_R$ and every linear polynomial $f \in R[x]$ that is non-negative on S_R, the element $f^{\otimes}(\xi)$ is psd in $R \otimes R$ (i.e., maps to a non-negative element for every ring homomorphism $R \otimes R \to R'$ into a real closed field R'). With these notations, the following theorem holds [188]: The set K is a spectrahedral shadow if, and only if, $f^{\otimes}(\xi)$ is a sum of squares in the ring $R \otimes R$, for all choices of R, ξ and f.

Exercises

8.7.1 Show that the closed convex hull of $\{(s^2 t, st^2, st) : s, t \geq 0\}$ in $\mathbb{R}^3$ is a spectrahedral shadow. (*Hint*: Use Exercise 2.4.3)

8.7.2 Let $f \in \mathbb{R}[t, x] = \mathbb{R}[t, x_1, \dots, x_n]$ be a polynomial in $n + 1$ variables, and write f in the form

$$f(t, x) = \sum_{i \geq 0} t^i f_i(x)$$

with polynomials $f_i \in \mathbb{R}[x]$ $(i \geq 0)$. Let $L \subseteq \mathbb{R}[x]$ be the linear subspace spanned by all iterated partial derivatives of $f_0, f_1, \dots$ with respect to the x-variables:

$$L = \operatorname{span}\left\{\frac{\partial^\alpha f_i(x)}{\partial x^\alpha} : i \geq 0,\ \alpha \in \mathbb{Z}^n_+\right\}.$$

Show that L contains $f(c, x + u)$ for every $c \in \mathbb{R}$ and $u \in \mathbb{R}^n$, and that L is the smallest subspace of $\mathbb{R}[x]$ with this property.

8.7.3 Use Theorem 8.7.18 in combination with Exercise 8.7.2 to find a polynomial map $\varphi : \mathbb{R}^2 \to \mathbb{R}^{12}$ such that the closed convex hull of $\varphi(S)$ in $\mathbb{R}^{12}$ fails to be a spectrahedral shadow, whenever $S \subseteq \mathbb{R}^2$ is a semialgebraic set with non-empty interior. (*Hint*: Use the Motzkin form.)

8.7.4 In this exercise we construct convex semialgebraic sets of dimension 11 that are not spectrahedral shadows. For this consider the polynomial

$$p = p(t, x, y) = t^6 + t^5 x + (y^2 - x^3)^2$$

in $\mathbb{R}[t, x, y]$.

(a) Show that $p(t, x, y) > 0$ for every $(t, x, y) \in \mathbb{R}^3$ with $t > 0$.
(b) Calculate a basis for the linear span L_1 of $\{\partial_x^i \partial_y^j p \colon i, j \geq 0\}$, and conclude that $\dim(L_1) = 12$. (Here $\partial_x = \partial/\partial x$, $\partial_y = \partial/\partial y$.)

Let $L \subseteq L_1$ be of dimension 11 with $L_1 = \mathbb{R} + L$, and let $S \subseteq \mathbb{R}^2$ be a semialgebraic set with non-empty interior. Prove that the closed convex hull of $\varphi_L(S)$ in $\mathbb{R}^{11}$ is not a spectrahedral shadow, following the arguments in the proof of Theorem 8.7.18 and observing Exercise 8.7.2. For the key step, do the following:

(c) Let $R \supseteq \mathbb{R}$ be a proper extension of real closed fields, let B denote the convex hull of $\mathbb{R}$ in R, and let $\varepsilon \in \mathfrak{m}_B$ with $\varepsilon > 0$. Show that $p(\varepsilon, x, y)$ is not a sum of squares in $\mathbb{R}[[x, y]] \otimes B$, by reducing modulo the ideal generated by $y^2 - x^3$.

8.8 Notes

The first use of the term *spectrahedron* seems to be in [163], where the faces of spectrahedra are determined. Proposition 8.3.7 is taken from [146]. The Lax conjecture was raised in 1958 by Peter Lax [126]. Helton and Vinnikov's proof of the conjecture is in [89], and Hanselka's algebraic proof appeared in [81]. An excellent and accessible source for more information on hyperbolic forms is the recent book [144].

The moment relaxation method appears in Lasserre [121] and Parrilo [148] for the first time. Meanwhile hundreds of papers (and several books) have been published in which various aspects are studied. For comprehensive introductions the reader may consult [122] or [90]. Laurent [124] gives an extensive overview which also surveys some of the foundations from real algebraic geometry.

The results in Section 8.6 are taken from Helton and Nie [88], [87]. Lemma 8.6.23 is from [112]. The Helton–Nie conjecture is stated in [87]. The question for a characterization of the class of spectrahedral shadows goes back to Nemirovski [142]. The disproof of the conjecture, together with the counter-examples presented here, is due to Scheiderer [187].

Appendix A: Commutative Algebra and Algebraic Geometry

The purpose of this appendix is to provide quick access to definitions, notations and basic facts from general topology, commutative algebra and algebraic geometry, as far as they are used in the main text. As a rule, motivational remarks are rarely made here and proofs are almost never given, since there exists ample references in the literature (see below for some suggestions).

A.1 Topological Spaces

A.1.1 For ease of reference we fix some notation from general point set topology, and recall a few basic concepts. Let X be a topological space. If Y is a subset of X, the closure (or interior, or boundary) of Y in X is denoted $\overline{Y}$ (or $\operatorname{int}(Y)$ or ∂Y, respectively). The subset Y is *locally closed* in X if it is relatively open in its closure, or equivalently, if $Y = U \cap Z$ with $U \subseteq X$ open and $Z \subseteq X$ closed. We also recall that a map $f\colon X \to Y$ between topological spaces is *open* if $f(U)$ is an open set in Y for every open subset U of X. Similarly, f is *closed* if $f(A)$ is closed in Y for every closed subset A of X.

A.1.2 The space X has property T_0 (resp. T_1) if, for any $x \neq y$ in X, there is an open set $U \subseteq X$ with $|U \cap \{x, y\}| = 1$ (resp., with $U \cap \{x, y\} = \{x\}$). The T_2 (alias *Hausdorff*) property requires that for $x \neq y$ there always exist open neighborhoods U of x and V of y with $U \cap V = \varnothing$. The space X is *quasi-compact* if every covering of X by open sets has a finite subcovering. If, in addition, X is Hausdorff, then X is said to be *compact*. Tikhonov's theorem states that an arbitrary direct product of compact topological spaces is compact.

A.1.3 The space X is *connected* if X cannot be written as a disjoint union of two non-empty open subsets. X is *irreducible* if X is non-empty and cannot be written as a union of two proper closed subsets. Connected or irreducible components of X are defined to be the maximal connected or maximal irreducible subsets of X, respectively. Every space is the disjoint union of its connected components, and is

C. Scheiderer, *A Course in Real Algebraic Geometry*, Graduate Texts in Mathematics 303,
https://doi.org/10.1007/978-3-031-69213-0

the union of its irreducible components. The space X is *totally disconnected* if every singleton $\{x\}$ ($x \in X$) is a connected component of X.

A *Boolean space* is a Hausdorff space that is compact and totally disconnected. In such a space, every open subset is a union of compact open subsets, or equivalently, every closed subset is an intersection of compact open sets.

A.1.4 The *Krull* (or *combinatorial*) *dimension* of a topological space X, denoted $\dim(X)$, is the supremum of all lengths d of chains $Y_0 \subsetneq Y_1 \subsetneq \cdots \subsetneq Y_d$ of irreducible closed subsets Y_i of X (proper inclusions). The empty space has $\dim(\emptyset) = -1$. Note that this definition is not suitable for all topological spaces (for example, every non-empty Hausdorff space has Krull dimension 0).

A.2 General Rings

As a general reference for basic commutative algebra, the old book [7] by Atiyah–Macdonald is still an excellent choice. Other very good options with a more advanced scope are the books by Matsumura [140], Eisenbud [63] and Bruns–Herzog [36], for example.

All rings in this text are assumed to be commutative and unital, i.e. to have a multiplicative unit 1 (with $1 = 0$ allowed), except when explicitly mentioned otherwise. Subrings of a ring A are required to contain the unit of A. All ring homomorphisms send 1 to 1. The group of units (elements which have a multiplicative inverse) is denoted A^*. In the following let A always be a ring.

A.2.1 The ideal of A generated by a set $M \subseteq A$ is denoted $\langle M\rangle$, or $\langle f_1, \dots, f_r\rangle$ if $M = \{f_1, \dots, f_r\}$. The *radical* of the ideal $I \subseteq A$ is $\sqrt{I} = \{a \in A\colon \exists n \geq 1\ a^n \in I\}$, and the ideal I is said to be a *radical ideal* if $I = \sqrt{I}$. The *nilradical* of A is $\mathrm{Nil}(A) = \sqrt{\{0\}}$. This is also the intersection of all prime ideals of A. The ring A is *reduced* if $\mathrm{Nil}(A) = \{0\}$. The residue field of a prime ideal $\mathfrak{p}$ of A is written $\kappa(\mathfrak{p})$.

If $S \subseteq A$ is a multiplicative set (always assumed to contain 1) then $A_S = \{\frac{a}{s}\colon a \in A,\ s \in S\}$ denotes the ring of fractions with denominators in S. If $\mathfrak{p}$ is a prime ideal of A, the localization of A at $\mathfrak{p}$ is $A_{\mathfrak{p}} := A_S$ where $S = A \setminus \mathfrak{p}$.

A.2.2 A is an *integral domain*, or briefly a *domain*, if $A \neq \{0\}$, and if $ab = 0$ implies $a = 0$ or $b = 0$, for $a, b \in A$. The field of fractions of an integral domain A is denoted $\mathrm{qf}(A)$. A *prime element* in a domain A is an element $p \neq 0$ in A such that the principal ideal $\langle p\rangle$ is a prime ideal in A. The domain A is a *unique factorization domain*, abbreviated *ufd*, if every non-unit $a \neq 0$ in A is a product of finitely many prime elements. If A is a ufd, this decomposition is unique up to permutation of the factors and up to multiplying them with units. For every ufd A, the polynomial ring $A[x]$ is a ufd as well (Gauss's lemma).

A.2.3 An A-algebra is a ring homomorphism $\varphi\colon A \to B$. Mention of φ will often be suppressed if φ is either clear from the context or not explicitly important. Accordingly, it is customary for $a \in A$ to write just a instead of $\varphi(a)$. Homomorphisms of

A-algebras are defined in the obvious way. The A-algebra B is *finitely generated* (or of finite type) if B is generated as a ring by A together with finitely many elements. The A-algebra B is *finite* if B is finitely generated as an A-module.

The ring A is *Noetherian* if every ideal of A can be generated by finitely many elements. Hilbert's basis theorem states that every finitely generated algebra over a Noetherian ring is again Noetherian. If A is a Noetherian ring then so is A_S for every multiplicative set S in A.

The Zariski spectrum $\mathrm{Spec}(A)$ of A is the set of all prime ideals of A, equipped with the Zariski topology. Every ring homomorphism $\varphi\colon A \to B$ induces a continuous map $\varphi^*\colon \mathrm{Spec}(B) \to \mathrm{Spec}(A)$ via $\mathfrak{q} \mapsto \varphi^{-1}(\mathfrak{q})$. The *dimension* $\dim(A)$ of the ring A is defined to be the Krull dimension (A.1.4) of the topological space $\mathrm{Spec}(A)$. So $\dim(A)$ is the supremum of all lengths d of chains of prime ideals $\mathfrak{p}_0 \subsetneq \mathfrak{p}_1 \subsetneq \cdots \subsetneq \mathfrak{p}_d$ (with proper inclusions) in A. The nullring $A = \{0\}$ has $\dim(A) = -1$.

A.2.4 Let $\varphi\colon A \to B$ be a ring homomorphism. An element $b \in B$ is *integral* over A if there exists a monic polynomial $p \in A[t]$ with $p(b) = 0$. The A-algebra B, or the homomorphism φ, is said to be integral if every $b \in B$ is integral over A. When B is an integral A-algebra, *going-up* holds: Given $\mathfrak{p} \in \mathrm{Spec}(A)$ and $\mathfrak{q}' \in \mathrm{Spec}(B)$ with $\varphi^{-1}(\mathfrak{q}') \subseteq \mathfrak{p}$, there exists $\mathfrak{q} \in \mathrm{Spec}(B)$ with $\mathfrak{q}' \subseteq \mathfrak{q}$ and $\varphi^{-1}(\mathfrak{q}) = \mathfrak{p}$. In particular, the map $\varphi^*\colon \mathrm{Spec}(B) \to \mathrm{Spec}(A)$ sends closed sets to closed sets. If φ is in addition injective then $\dim(A) = \dim(B)$.

If $A \to B$ is any A-algebra, the (relative) *integral closure* of A in B is $\{b \in B\colon b$ is integral over $A\}$, and is a subring of B. A domain A is *integrally closed* if it coincides with its integral closure in $\mathrm{qf}(A)$.

A.2.5 Let A be a Noetherian ring. The *height* of a prime ideal $\mathfrak{p} \in \mathrm{Spec}(A)$ is $\mathrm{ht}(\mathfrak{p}) = \dim(A_\mathfrak{p})$. The height of an arbitrary ideal $I \subsetneq A$ is $\mathrm{ht}(I) = \min_\mathfrak{p} \mathrm{ht}(\mathfrak{p})$, minimum over all prime ideals $\mathfrak{p} \supseteq I$ of A. Every ideal I satisfies $\mathrm{ht}(I) + \dim(A/I) \le \dim(A)$, with equality failing in general. Krull's principal ideal theorem states, for any ideal $I = \langle a_1, \dots, a_n\rangle$ generated by n elements, that $\mathrm{ht}(\mathfrak{p}) \le n$ holds for every minimal prime ideal $\mathfrak{p} \supseteq I$.

A.3 Affine Algebras

A.3.1 Let k be a field. Finitely generated k-algebras are also called *affine k-algebras*. The single most important general result for affine algebras is Hilbert's nullstellensatz. In its algebraic version, it asserts that if a field extension K/k is finitely generated as a k-algebra, then K/k is finite algebraic. In A.6.3 below we recall geometric formulations.

A.3.2 Let A be a k-algebra. Elements $a_1, \dots, a_n$ of A are *k-algebraically independent* if the only polynomial $p \in k[x_1, \dots, x_n]$ with $p(a_1, \dots, a_n) = 0$ is the zero polynomial. An arbitrary family of elements of A is k-algebraically independent if every finite subfamily is. The *transcendence degree* of A over k, denoted $\mathrm{trdeg}_k(A)$, is the maximal cardinality of a k-algebraically independent family in A. We do not distinguish between different infinite cardinalities and regard $\mathrm{trdeg}_k(A)$ as a non-negative integer or ∞. If $A = K$ is a field extension of k, the maximal k-algebraically independent families in K are called *transcendence bases* for K/k. Any two transcendence bases of K/k have the same cardinality $\mathrm{trdeg}_k(K)$.

A.3.3 Let A be an affine k-algebra. Every chain of prime ideals of A that cannot be extended to a longer sequence has the same length, which is $\dim(A)$. When A is a domain, every ideal I satisfies $\mathrm{ht}(I) + \dim(A/I) = \dim(A)$. Moreover, $\dim(A) = \mathrm{trdeg}_k(K)$ holds in this case, with $K = \mathrm{qf}(A)$.

Noether normalization, in its most basic version, says that every affine k-algebra A is a finite extension of a polynomial ring: There exist algebraically independent elements $x_1, \dots, x_n \in A$ such that the ring extension $k[x_1, \dots, x_n] \subseteq A$ is finite. Here necessarily $n = \dim(A)$.

A.3.4 If A, B are affine k-algebras then so is $A \otimes_k B$, and $\dim(A \otimes_k B) = \dim(A) + \dim(B)$. If K/k is a field extension then $A \otimes_k K$ is an affine K-algebra, of dimension $\dim(A \otimes_k K) = \dim(A)$.

A.4 Local rings

A.4.1 A ring A is *local* if it has a unique maximal ideal $\mathfrak{m}$. The field $k = A/\mathfrak{m}$ is called the *residue field* of the local ring A. We often use a phrase like "let $(A, \mathfrak{m})$ [or $(A, \mathfrak{m}, k)$] be a local ring". It is meant to indicate that A is a local ring with maximal ideal $\mathfrak{m}$ [and residue field k]. The ring A is *semilocal* if $A \neq \{0\}$ and A has only finitely many maximal ideals.

A.4.2 An important basic tool for working with local rings is the *Nakayama Lemma*. In its general form, the lemma asserts that if A is a ring, $I \subseteq A$ is an ideal and M is a finitely generated A-module with $M = IM$, there exists $a \in I$ with $(1 - a)M = 0$. A frequently used consequence is this: Let $(A, \mathfrak{m}, k)$ be a local ring and M a finitely generated A-module, and let $\overline{M} = M \otimes_A k = M/\mathfrak{m}M$, a k-vector space of finite dimension. If $x_1, \dots, x_n \in M$ are such that their residue classes $\overline{x}_1, \dots, \overline{x}_n$ generate the k-vector space $\overline{M}$, then $x_1, \dots, x_n$ generate the A-module M.

A.4.3 Let A be any (base) ring. The ring of formal power series over A in the variable x is denoted $A[\![x]\!]$. It consists of all formal infinite sums $f = \sum_{i=0}^{\infty} a_i x^i$ with $a_i \in A$ $(i \geq 0)$, with natural definition of addition and multiplication. One writes $f(0) = a_0$. Given any $g \in A[\![x]\!]$ with $g(0) = 0$, the series $1 - g$ has an inverse, given by the geometric series $(1 - g)^{-1} = \sum_{i=0}^{\infty} g^i$. Therefore, the power series f is a unit in $A[\![x]\!]$ if, and only if, $f(0)$ is a unit in A. In particular, if A is a local ring with maximal ideal $\mathfrak{m}$, the power series ring $A[\![x]\!]$ is again local, with maximal ideal generated by $\mathfrak{m}$ and x. Iterating the construction, one writes $A[\![x_1, \dots, x_n]\!] := A[\![x_1, \dots, x_{n-1}]\!]\,[\![x_n]\!]$ for any $n \geq 1$.

When $A = k$ is a field, $k[\![x]\!]$ is a discrete valuation ring. Its field of fractions is the field $k(\!(x)\!)$ of formal Laurent series. The elements of $k(\!(x)\!)$ are the formal sums $f = \sum_{i \ge m}^{\infty} a_i x^i$ where $m \in \mathbb{Z}$ and $a_i \in k$.

A.4.4 Let $(A, \mathfrak{m}, k)$ be a Noetherian local ring. The sequence of natural ring homomorphisms $A/\mathfrak{m} \xleftarrow{\pi_1} A/\mathfrak{m}^2 \xleftarrow{\pi_2} \cdots \leftarrow A/\mathfrak{m}^\nu \xleftarrow{\pi_\nu} \cdots$ forms an inverse system whose inverse (projective) limit is $\widehat{A}$, the *completion* of A. The ring $\widehat{A}$ therefore consists of all sequences $(b_\nu)_{\nu \ge 1}$ in $\prod_{\nu \ge 1} A/\mathfrak{m}^\nu$ that satisfy $\pi_\nu(b_{\nu+1}) = b_\nu$ for all $\nu \ge 1$. The ring $\widehat{A}$ is again local, its maximal ideal $\widehat{\mathfrak{m}}$ being the kernel of the natural homomorphism $\widehat{A} \to A/\mathfrak{m} = k$. The natural homomorphisms $\widehat{A}/\widehat{\mathfrak{m}}^\nu \to A/\mathfrak{m}^\nu$ ($\nu \ge 1$) are isomorphisms, and in particular, the residue fields of A and $\widehat{A}$ coincide. The natural homomorphism $i \colon A \to \widehat{A}$ is injective, and A is *complete* if i is an isomorphism. The completion $\widehat{A}$ is known to be Noetherian as well, and both rings A and $\widehat{A}$ have the same dimension: $\dim(A) = \dim(\widehat{A})$.

A.4.5 Let $(A, \mathfrak{m}, k)$ be a Noetherian local ring of dimension d. Then $\dim_k(\mathfrak{m}/\mathfrak{m}^2) \ge d$ holds, and both are finite. The local ring A is *regular* if equality holds, which means that the maximal ideal $\mathfrak{m}$ can be generated by d elements. In this case, a *regular system of parameters* of A is a sequence $x_1, \dots, x_d$ (of length d) that generates the ideal $\mathfrak{m}$. By A.4.2, it is equivalent that the residue classes $x_i + \mathfrak{m}^2$ ($i = 1, \dots, d$) form a basis of the k-vector space $\mathfrak{m}/\mathfrak{m}^2$. If $x_1, \dots, x_d$ is a regular system of parameters, then for any $m \le d$ the ideal $\mathfrak{p} = \langle x_1, \dots, x_m \rangle$ of A is prime, and the quotient ring $A/\mathfrak{p}$ is regular local of dimension $d - m$.

Every regular local ring is a unique factorization domain (Auslander–Buchsbaum theorem). In particular, regular local rings are integrally closed domains. The localization of a regular local ring at any prime ideal is again a regular local ring. This is a consequence of the homological characterization of regular local rings (Auslander–Buchsbaum–Serre theorem). In the case of local rings of algebraic varieties, an easier proof is available via the Jacobian criterion for regularity (A.6.17).

A Noetherian ring A, not necessarily local, is said to be regular if the localization $A_\mathfrak{p}$ at every prime ideal $\mathfrak{p}$ is a regular local ring. By the Auslander–Buchsbaum–Serre theorem, it suffices that $A_\mathfrak{m}$ is regular for every maximal ideal $\mathfrak{m}$ of A.

A.4.6 Let A be a Noetherian local ring. If A is regular then the same is true for the completion $\widehat{A}$ of A, and vice versa. If $(A, \mathfrak{m}, k)$ is a complete regular local ring, and if A has equal characteristic (meaning that $\mathrm{char}(K) = \mathrm{char}(k)$ for $K = \mathrm{qf}(A)$), it is known that $A \cong k[\![x_1, \dots, x_d]\!]$ with $d = \dim(A)$. This is a particular case of the Cohen structure theorem.

A.4.7 Let $(A, \mathfrak{m}, k)$ be a local Noetherian ring. The *graded ring associated with* A is $\mathrm{gr}(A) = \bigoplus_{\nu \ge 0} \mathfrak{m}^\nu/\mathfrak{m}^{\nu+1}$ where $\mathfrak{m}^0 := A$. This is a finitely generated graded k-algebra of dimension $\dim \mathrm{gr}(A) = \dim(A)$. The natural homomorphism $A \to \widehat{A}$ induces an isomorphism $\mathrm{gr}(A) \xrightarrow{\sim} \mathrm{gr}(\widehat{A})$ of the associated graded rings. The local ring A is regular of dimension d if, and only if, $\mathrm{gr}(A)$ is a polynomial ring over k in d variables. In fact, if $a_1, \dots, a_d$ is a regular parameter sequence in A, the k-homomorphism $k[x_1, \dots, x_d] \to \mathrm{gr}(A)$ defined by $x_i \mapsto a_i + \mathfrak{m}^2$ ($i = 1, \dots, d$)

is an isomorphism of graded k-algebras. For regular A, the *(vanishing) order* of $f \in A$ is $\omega(f) = \sup\{\nu \geq 0\colon f \in \mathfrak{m}^\nu\}$. The map ω extends to a discrete valuation of the quotient field $\mathrm{qf}(A)$ of A. In particular, $\omega(f) < \infty$ if $f \neq 0$. The *leading form* of $f \neq 0$ is the coset $L(f) := f + \mathfrak{m}^{n+1}$ in $\mathrm{gr}_n(A)$, where $n = \omega(f)$.

A.4.8 A local ring $(A, \mathfrak{m}, k)$ is *Henselian* if, for every monic polynomial $f \in A[t]$ and every $a \in A$ whose residue class $\overline{a}$ in $A/\mathfrak{m} = k$ is a simple root of $\overline{f} \in k[t]$, there exists $b \in A$ with $f(b) = 0$ and $b \equiv a \pmod{\mathfrak{m}}$. For example, if A is Henselian and $f \in \mathfrak{m}$, and if $n \geq 1$ is an integer that is relatively prime to $\mathrm{char}(k)$, there is a unique element $g \in \mathfrak{m}$ with $(1+g)^n = 1 + f$. Every complete Noetherian local ring is Henselian.

A.4.9 In Section 6.4 we are using some standard theorems for rings of formal power series. Let k be a field and $n \geq 1$, let $x = (x_1, \dots, x_n)$ and $x' = (x_1, \dots, x_{n-1})$. A formal power series $g = g(x) \in k[\![x]\!]$ is a *Weierstrass polynomial* (of order $m \geq 0$) with respect to x_n if

$$g(x) = x_n^m + \sum_{i=0}^{m-1} a_i(x')x_n^i$$

where $a_i(x') \in k[\![x']\!]$ are power series with $a_i(0, \dots, 0) = 0$ for $0 \leq i \leq m-1$. For such g, the natural ring homomorphism $k[\![x']\!][x_n]/\langle g\rangle \to k[\![x]\!]/\langle g\rangle$ is an isomorphism. This is a consequence of the Weierstrass division theorem (that we omit). In particular, $k[\![x]\!]/\langle g\rangle$ is a finite $k[\![x']\!]$-algebra then. The *Weierstrass preparation theorem* states that every $f \in k[\![x]\!]$, after a linear change of coordinates, has the form $f = ug$ with $u, g \in k[\![x]\!]$ where u is a unit and g is a Weierstrass polynomial with respect to x_n. (These are just basic versions. More precise statements hold, and they hold in rings of convergent power series as well (over $\mathbb{R}$ or $\mathbb{C}$).)

A.4.10 Let k be an algebraically closed field of characteristic zero. Then the field

$$k(\!(x^{1/\infty})\!) := \bigcup_{n \geq 1} k(\!(x^{1/n})\!)$$

is again algebraically closed. It is called the field of formal *Puiseux series*.

A.5 Valuation Rings

A.5.1 Let K be a field. A subring B of K is a *valuation ring of* K if, for every $a \in K^*$, (at least) one of $a \in B$ or $a^{-1} \in B$ holds. Clearly this implies $K = \mathrm{qf}(B)$. An integral domain B is a *valuation ring* if it is a valuation ring of its quotient field. If B is a valuation ring of K then the same is true for every overring of B in K. Every valuation ring B is a local ring, with maximal ideal $\mathfrak{m}_B = \{0\} \cup \{a \in B\colon a \neq 0, a^{-1} \notin B\}$. The residue field $B/\mathfrak{m}_B$ of B will usually be denoted k_B. Every valuation ring is integrally closed in its quotient field.

A.5.2 An *ordered abelian group* is an abelian group $(G, +)$ together with a total (linear) ordering $\le$ that is compatible with the group structure, i.e. that satisfies $a \le b \Rightarrow a+c \le b+c$ for all $a, b, c \in G$. A subgroup H of G is *convex*[2] if $0 < b < a$ and $a \in H$, $b \in G$ implies $b \in H$. If H is a convex subgroup of G then G/H becomes an ordered abelian group by the ordering induced from G.

A.5.3 Let B be a valuation ring of K. The (multiplicative) abelian group $\Gamma = K^*/B^*$ is called the *value group* of B. Usually it is written additively. The *valuation* of K associated with B is the map $v\colon K \to \Gamma \cup \{\infty\}$ defined by $v(a) = aB^*$ $(a \ne 0)$ and $v(0) = \infty$. Here ∞ is an extra symbol not in Γ that satisfies $\alpha+\infty = \infty+\alpha = \infty+\infty = \infty$ for every $\alpha \in \Gamma$. The abelian group Γ is ordered by $v(a) \le v(b) \Leftrightarrow ba^{-1} \in B$ $(a, b \in K^*)$. Extend this ordering from the group Γ to the set $\Gamma \cup \{\infty\}$ by defining $\alpha \le \infty$ for every $\alpha \in \Gamma$, then $v(ab) = v(a) + v(b)$ and $v(a+b) \ge \min\{v(a), v(b)\}$ hold for all $a, b \in K$. The last inequality is an equality whenever $v(a) \ne v(b)$. Note that $B = \{x \in K\colon v(x) \ge 0\}$ and $\mathfrak{m} = \{x \in K\colon v(x) > 0\}$.

Conversely, a (Krull) *valuation* of a field K is a map $v\colon K \to \Gamma \cup \{\infty\}$, where Γ is an ordered abelian group and $\infty \notin \Gamma$ is an extra symbol as above, such that $v(1) = 0$, $v(ab) = v(a) + v(b)$ and $v(a+b) \ge \min\{v(a), v(b)\}$ hold for all $a, b \in K$. For such v, the set $O_v := \{a \in K\colon v(a) \ge 0\}$ is a valuation ring of K, called the valuation ring associated with v. If v was surjective then v coincides with the valuation associated with the ring O_v, up to an order isomorphism of the value group. The residue field of O_v is also called the residue field of the valuation v.

A *discrete valuation ring* is a valuation ring whose value group is infinite cyclic. At the same time, discrete valuation rings are precisely the regular local rings of dimension one.

A.5.4 Let A, B be local subrings of a field K. Then A *dominates* B if $B \subseteq A$ and $\mathfrak{m}_B \subseteq \mathfrak{m}_A$ (hence $\mathfrak{m}_B = B \cap \mathfrak{m}_A$) hold. Note that this implies a natural embedding $k_B \to k_A$ of the residue fields. Domination is a (partial) order relation on the set of all local subrings of K. The domination-maximal local subrings of K are exactly the valuation rings of K.

A.5.5 Proposition. *Given a regular local ring $(A, \mathfrak{m})$ with field of fractions K, there exists a valuation ring B of K that dominates A, and such that the induced embedding $k_A \to k_B$ of the residue fields is an isomorphism.*

Proof. We include the proof since this important result is not standardly included in textbooks on commutative algebra. Induction on $d = \dim(A)$. If $d = 0$ then $A = K$, if $d = 1$ then A is a discrete valuation ring. In both cases we may take $B = A$. Let $d \ge 2$ and choose an element $a \in \mathfrak{m} \setminus \mathfrak{m}^2$. Then $\mathfrak{p} := Aa$ is a prime ideal of A, and the local ring $A/\mathfrak{p}$ is regular of dimension $d-1$. The localized ring $A_\mathfrak{p}$ is a domain of dimension one whose maximal ideal $\mathfrak{p}A_\mathfrak{p} = aA_\mathfrak{p}$ is principal. Therefore $A_\mathfrak{p}$ is a discrete valuation ring, with residue field $F := A_\mathfrak{p}/\mathfrak{p}A_\mathfrak{p} = \mathrm{qf}(A/\mathfrak{p})$.

[2] Some authors (e.g. Bourbaki) use the term *isolated subgroup* for convex subgroup.

Let $\pi\colon A_\mathfrak{p} \to F$ denote the residue map. Since $A/\mathfrak{p}$ is a regular local subring of F of dimension $d-1$, the inductive hypothesis gives a valuation subring C of F that dominates $A/\mathfrak{p}$ and has the same residue field. Then $B := \{y \in A_\mathfrak{p} : \pi(y) \in C\}$ is a valuation ring of K. Indeed, if $x \in K^*$ with $x \notin B$, two cases are possible. Either $x \notin A_\mathfrak{p}$, in which case x^{-1} lies in the maximal ideal of $A_\mathfrak{p}$, and so $\pi(x^{-1}) = 0$ and $x^{-1} \in B$. Or else $x \in A_\mathfrak{p}$ but $\pi(x) \notin C$, in which case x is a unit of $A_\mathfrak{p}$ and $\pi(x^{-1}) \in C$, and so again $x^{-1} \in B$.

It is clear that $A \subseteq B$ and $A \cap \mathfrak{m}_B \subseteq \mathfrak{m}_A$, and that the induced map $A/\mathfrak{m}_A \to B/\mathfrak{m}_B$ is an isomorphism. Therefore the valuation ring B has the desired properties. □

A.6 Algebraic Geometry

In order to keep the requirements for this course as basic as reasonably possible, we do not assume familiarity with the language of schemes. The view point of schemes is nowhere truly necessary in this book, although its use would have simplified some of the exposition. Instead, a "naive" understanding of varieties is sufficient, and all varieties that occur may be assumed to be quasi-projective. For the reader who has not (yet) learned about schemes so far, this means that he or she can read the entire book on the basis of the definitions outlined below, and may assume throughout that "k-variety" means "quasi-projective k-variety". On the other hand, a reader who is familiar with the basics of schemes may everywhere understand the term "k-variety" as *reduced and separated k-scheme of finite type*, in the sense of Grothendieck. Then everything will remain true in this more general sense. We point out that, in our use, the term "variety" does not imply irreducibility.

When we adopt naive language, there is one subtle point that is important: We need a systematic way for speaking of algebraic varieties over *non-closed* ground fields (think of the field $k = \mathbb{R}$ of real numbers!). This point of view is missing in many, if not most introductory books. A notable exception is Kunz's textbook [117]. If one is willing to accept the restriction to algebraically closed ground fields, there exist numerous great choices for a first introduction to algebraic geometry, ranging from very broad expositions like Cox, Little, O'Shea ([50] and [49]) to texts that proceed at a much more demanding pace, like Harris [83]. The two volumes by Shafarevich [198], [197] offer a thorough treatment of algebraic varieties over algebraically closed fields in the first volume, before they turn to the view point of schemes in the second.

For more than the first half of this course, familiarity with just the most basic notions of algebraic geometry is sufficient: Quasi-projective algebraic varieties and their morphisms, Zariski topology, regular and rational functions, dictionary between affine varieties and their coordinate rings. Towards the end of Chapter 5, and more so in the remaining chapters, concepts are required that are slightly more advanced. This includes tangent spaces, regular and singular points, and others. Below we review these most important notions and facts in a naive language setup, and fix the terminology.

Ultimately however, we strongly advise the reader to become friends with the modern language of schemes. It has become the standard in algebraic geometry for a long time. The classic choice for an introduction is still Hartshorne's textbook [84]. But meanwhile several excellent alternatives are available as well, for example [127], [210] or [75].

A.6.1 Let k be an arbitrary (base) field. For the entire discussion we fix an algebraically closed field extension E of k. Definitions and facts below are essentially independent of the choice of E, and for most purposes it suffices to take $E = \overline{k}$, an algebraic closure of k.

For $n \geq 0$, affine n-space is defined to be $\mathbb{A}^n := E^n$. Similarly, projective n-space is $\mathbb{P}^n := (E^{n+1} \setminus \{0\})/{\sim}$, where two points $u, v \neq 0$ in E^{n+1} are considered equivalent ($u \sim v$) iff there exists $0 \neq a \in E$ with $au = v$. The equivalence class of $u = (u_0, \dots, u_n)$ in $\mathbb{P}^n$ is denoted $[u]$ or $(u_0 : \cdots : u_n)$ (homogeneous coordinates). Note that $\mathbb{A}^0 = \mathbb{P}^0$ is just a point.

Let $x = (x_1, \dots, x_n)$ be an n-tuple of indeterminates. For $p \in k[x]$ let $D(p) = \{u \in \mathbb{A}^n : p(u) \neq 0\}$. The *k-Zariski topology* on $\mathbb{A}^n$ has the sets $D(p)$ (with $p \in k[x]$) as a basis of open sets. The closed subsets of $\mathbb{A}^n$ are therefore the common zero sets of families of k-polynomials. Similarly, if $p \in k[x_0, x]$ is a homogeneous polynomial in $n + 1$ variables, put $D_+(p) = \{[u] \in \mathbb{P}^n : p(u) \neq 0\}$. The k-Zariski topology on $\mathbb{P}^n$ has the sets $D_+(p)$ (with p homogeneous) as a basis of open sets. The closed subsets of $\mathbb{P}^n$ are the common zero sets of families of homogeneous k-polynomials. We will always consider $\mathbb{A}^n$ and $\mathbb{P}^n$ with the k-Zariski topology.

A.6.2 Recall that a subset of a topological space X is locally closed if it has the form $U \cap Z$ with $U \subseteq X$ open and $Z \subseteq X$ closed. A quasi-projective *k-variety* is a locally closed subset V of $\mathbb{A}^n$ or $\mathbb{P}^n$, equipped with the relative k-Zariski topology (A.6.1) and with the sheaf $\mathcal{O}_V$ of regular functions (see A.6.4 below). The qualifier "quasi-projective" will be dropped in the sequel since no other varieties are considered here. If V is a variety, a closed (or open, or locally closed) *subvariety* of V is a variety that is a closed (or open, or locally closed, respectively) subset W of V. Note that, according to our conventions, a variety need not be irreducible.

A.6.3 If $P \subseteq k[x]$ is a set of polynomials, $\mathcal{V}(P) = \{u \in \mathbb{A}^n : \forall\, p \in P\ p(u) = 0\}$ denotes the zero set of P. This is a closed subset of $\mathbb{A}^n$. If $X \subseteq \mathbb{A}^n$ is a set of points in affine n-space, $\mathcal{I}(X) = \{p \in k[x] : \forall\, u \in X\ p(u) = 0\}$ denotes the ideal of all polynomials in $k[x]$ that vanish on X. Hilbert's nullstellensatz A.3.1 is equivalent to saying that $\mathcal{V}(P) = \varnothing$ implies $1 \in \langle P \rangle$. More generally, it implies $\mathcal{I}(\mathcal{V}(P)) = \sqrt{\langle P \rangle}$ for every subset $P \subseteq k[x]$. Altogether, this means that the operators $\mathcal{V}$ and $\mathcal{I}$ define a bijective and inclusion-reversing correspondence between the radical ideals of $k[x] = k[x_1, \dots, x_n]$ and the closed k-subvarieties of $\mathbb{A}^n$.

For projective varieties, the picture is similar. If $P \subseteq k[x_0, x]$ is a set of *homogeneous* polynomials, $\mathcal{V}(P) = \{u \in \mathbb{P}^n : \forall\, p \in P\ p(u) = 0\}$ is[3] a closed subset of $\mathbb{P}^n$. If $I \subseteq k[x_0, x]$ is a homogeneous ideal one puts $\mathcal{V}(I) := \mathcal{V}(P) \subseteq \mathbb{P}^n$ where P is the set

[3] We use the same symbol $\mathcal{V}$ for zero varieties in affine or projective space. Which one is actually meant should always be clear from the context.

of homogeneous members of I. Then $I \mapsto \mathcal{V}(I)$ is a bijective and inclusion-reversing correspondence between the homogeneous radical ideals $I \subseteq k[x_0, x]$ with $I \neq \langle 1 \rangle$, and the closed k-subvarieties of $\mathbb{P}^n$. The inverse operator sends a subset $X \subseteq \mathbb{P}^n$ to $\mathcal{I}(X)$, the ideal of $k[x_0, x]$ that is generated by all homogeneous polynomials p that vanish identically on X.

A.6.4 Next we recall the notion of regular functions on a k-variety. Let V be a locally closed subset of $\mathbb{A}^n$. A map $f\colon V \to E$ is a *regular function* on V if the following holds: For every $u \in V$ there exist $p, q \in k[x_1, \dots, x_n]$ and an open neighborhood W of u in V, such that $q(w) \neq 0$ and $f(w) = \frac{p(w)}{q(w)}$ for all $w \in W$. Similarly, if V is a locally closed subset of $\mathbb{P}^n$, a regular function on V is a map $f\colon V \to E$ that, locally around every point $u \in V$, has the form $f(w) = \frac{p(w)}{q(w)}$ where $p, q \in k[x_0, \dots, x_n]$ are *homogeneous* polynomials of the *same degree*, and with $q(u) \neq 0$.

In either case, the set of all regular functions on V is a k-algebra, with ring operations defined pointwise, and is denoted $\mathcal{O}(V)$. The *structural sheaf* $\mathcal{O}_V$ of a k-variety V is defined by $\mathcal{O}_V(U) := \mathcal{O}(U)$ for every open subset U of V. Saying that $\mathcal{O}_V$ is a sheaf means two things: (1) For any open subsets $U' \subseteq U$ of V and any $f \in \mathcal{O}_V(U)$, the restriction $f|_{U'}$ of f lies in $\mathcal{O}_V(U')$; (2) if U_i ($i \in I$) are open subsets of V, if $U = \bigcup_i U_i$, and if $f\colon U \to E$ is a map with $f|_{U_i} \in \mathcal{O}_V(U_i)$ for every i, then $f \in \mathcal{O}_V(U)$.

A.6.5 Let V, W be k-varieties. A *morphism* (of k-varieties) from V to W is a continuous map $f\colon V \to W$ with the property that, for every open subset W' of W and every $g \in \mathcal{O}_W(W')$, the pull-back $f^*(g) = g \circ f$ (which is a map $f^{-1}(W') \to E$) lies in $\mathcal{O}_V(f^{-1}(W'))$. This defines the category (Var_k) of (quasi-projective) k-varieties, and in particular, the according notion of isomorphism of k-varieties.

A morphism $f\colon V \to W$ of k-varieties is an open (or closed) *immersion* if f induces an isomorphism from V onto an open (or closed, respectively) subvariety of W.

Since $\mathbb{A}^n$ is isomorphic to an open subset of $\mathbb{P}^n$ as a k-variety (for example, to $D_+(x_0)$), every (quasi-projective) k-variety is isomorphic to a locally closed subset of $\mathbb{P}^n$, for some n.

A.6.6 A k-variety V is *irreducible* if $V \neq \varnothing$ and V cannot be written as the union of two proper closed subsets. Otherwise V is *reducible*. The *irreducible components* of V are the maximal closed irreducible subset of V; there are only finitely many, and V is their union.

Beware that the notion of irreducibility depends strongly on the base field k, since the k-Zariski topology does. It may very well happen that the k-variety V is irreducible, but that it becomes reducible when considered as a variety over a larger field, e.g. over $\overline{k}$. For a simple example, the equation $x^2 + y^2 = 0$ defines a closed subvariety V of the affine plane $\mathbb{A}^2$ that is irreducible as an $\mathbb{R}$-variety. But seen as a $\mathbb{C}$-variety, V becomes a union of two proper closed subsets (lines), and therefore V is reducible. Of course, what is behind this example is the fact that the polynomial $x^2 + y^2$ is irreducible over $\mathbb{R}$ but splits over $\mathbb{C}$: $x^2 + y^2 = (x + iy)(x - iy)$ with $i = \sqrt{-1}$.

A.6.7 The *dimension* of V is the maximum length d of a chain $V_0 \subsetneq V_1 \subsetneq \cdots \subsetneq V_d$ of irreducible closed subsets of V (with proper inclusions). The empty variety has $\dim(\varnothing) = -1$. The *local dimension* of V at a point $u \in V$ is $\dim_u(V) = \min\{\dim(U) \colon U \subseteq V$ is an open neighborhood of u in $V\}$. While irreducibility of V may depend on the base field k, the dimension of V is independent of the base field.

A.6.8 Let $k \subseteq K \subseteq E$ be an intermediate field. A K-rational point of $\mathbb{A}^n$ is a point in K^n. A K-rational point of $\mathbb{P}^n$ is a point $[u] \in \mathbb{P}^n$ that can be represented by a tuple $u \in K^{n+1}$, $u \neq 0$. In general, when V is a locally closed subvariety of $\mathbb{A}^n$ or $\mathbb{P}^n$, a *K-rational point* of V is a point in V that is K-rational as a point of $\mathbb{A}^n$ or of $\mathbb{P}^n$, respectively. The set of K-rational points of V is denoted $V(K)$. Every morphism $f \colon V \to W$ of k-varieties sends K-rational points to K-rational points.

In the main text, we often consider the condition that a subset M of $V(k)$ be *Zariski dense* in the k-variety V. What is meant by this is that M, considered as a subset of $V = V(E)$, should be dense in V with respect to the k-Zariski topology.

It is essential not to confuse the set of k-rational points of V with the k-variety V itself. For example, when $k = \mathbb{R}$ and V is a closed $\mathbb{R}$-subvariety of $\mathbb{A}^n$, the set $V(\mathbb{R}) \subseteq \mathbb{R}^n$ will be called an ($\mathbb{R}$-) *algebraic set* in this book, but will never be called an algebraic $\mathbb{R}$-variety.

A.6.9 A k-variety V is *affine* if it is isomorphic to a closed subvariety of $\mathbb{A}^n$, for some $n \geq 0$. For such V one usually writes $k[V] := \mathcal{O}_V(V)$. The ring $k[V]$, called the *affine coordinate ring* of V, is a reduced and finitely generated k-algebra, and depends functorially on V. Given a morphism $f \colon V \to W$ of affine k-varieties, the associated (pull-back) homomorphism between the coordinate rings is denoted $f^* \colon k[W] \to k[V]$. Thus, $V \mapsto k[V]$ is a contravariant functor from the category of affine k-varieties to the category of reduced affine k-algebras.

The coordinate ring $k[V]$ determines the affine k-variety V up to isomorphism. To recover V from $k[V]$, choose any finite system $p_1, \ldots, p_n$ of generators of the k-algebra $k[V]$ and let I be the kernel of the homomorphism of k-algebras $k[x_1, \ldots, x_n] \to k[V]$, $x_i \mapsto p_i$. The zero set $\mathcal{V}(I) \subseteq \mathbb{A}^n$ of I is an affine k-variety that is canonically isomorphic to V. In a similar way one sees, for affine k-varieties V and W, that every k-homomorphism $k[W] \to k[V]$ between their coordinate rings has the form f^*, for a unique morphism $f \colon V \to W$ of the k-varieties. In other words, the functor $V \mapsto k[V]$ is an anti-equivalence from the category of affine k-varieties to the category of reduced affine k-algebras.

Given an affine variety V, the ideal–subvariety correspondence A.6.3 restricts to a bijective correspondence between radical ideals of $k[V]$ and closed subsets of V. We use the notations $\mathcal{V}_V(P)$ for $P \subseteq k[V]$ (the vanishing set of P in V) and $\mathcal{I}_V(X)$ for $X \subseteq V$ (the vanishing ideal of X in $k[V]$), respectively. Note that, under this correspondence, the prime ideals of $k[V]$ correspond exactly to the irreducible closed subsets of V. In particular, $\dim(V) = \dim k[V]$.

We remark that the K-rational points of an affine k-variety V are in natural bijection with the set $\mathrm{Hom}_k(k[V], K)$ of k-algebra homomorphisms.

On every k-variety V, the open affine subvarieties U of V form a basis of open sets on V. Together with quasi-compactness of V, this fact makes it often possible to reduce proofs from general varieties to the case of affine varieties.

A.6.10 The k-variety V is *projective* if it is a closed subvariety of $\mathbb{P}^n$, for some $n \geq 0$. The projective coordinate ring of V is the graded k-algebra $k[V] = k[x_0, x]/\mathcal{I}(V)$. We'll write $k[V]_d$ for the graded piece of degree d of $k[V]$, so $k[V]_d$ consists of the elements in $k[V]$ that can be represented by a form of degree d in $k[x_0, x]$. The correspondence between subvarieties and ideals from A.6.3 restricts to a bijection between homogeneous radical ideals $I \neq \langle 1 \rangle$ of $k[V]$ and closed subvarieties Z of V, denoted $I \mapsto \mathcal{V}_V(I)$ and $Z \mapsto \mathcal{I}_V(Z)$. Homogeneous prime ideals properly contained in $k[V]_+ = \bigoplus_{d \geq 1} k[V]_d$ correspond to irreducible subvarieties of V under this bijection, and vice versa.

If $f \in k[x_0, x]$ is a homogeneous polynomial, the complement $D_+(f) = \mathbb{P}^n \setminus \mathcal{V}(f)$ of the projective hypersurface $\mathcal{V}(f)$ is an affine k-variety. Its affine coordinate ring is naturally isomorphic to the ring $k[x_0, x]_{(f)}$ of all homogeneous fractions $\frac{g}{f^m}$, meaning that $m \geq 0$ and $g \in k[x_0, x]$ is homogeneous of degree $\deg(g) = \deg(f^m)$.

Given a closed (projective) subvariety V of $\mathbb{P}^n$, the set $\widehat{V} = \{v \in \mathbb{A}^{n+1} \setminus \{0\} : [v] \in V\} \cup \{0\}$ is a closed subvariety of $\mathbb{A}^{n+1}$, called the *affine cone* over V. The affine coordinate ring of $\widehat{V}$ is the homogeneous coordinate ring of V, stripped of its grading.

A.6.11 For every n, $d \geq 1$, the d-th *Veronese embedding* $v_d : \mathbb{P}^n \to \mathbb{P}^N$ is the morphism defined by $v_d(u) = (u^\alpha)_{|\alpha|=d}$ for $u \in \mathbb{P}^n$ (multinomial notation $u^\alpha = u_0^{\alpha_0} \cdots u_n^{\alpha_n}$). Here $N = \binom{n+d}{d} - 1$, and coordinates in $\mathbb{P}^N$ correspond to monomials $\alpha \in \mathbb{Z}_+^{n+1}$ of degree d, in some fixed order. The morphism v_d is an isomorphism onto the image variety $V_{n,d} := v_d(\mathbb{P}^n)$, which is called a *Veronese variety*. Moreover, v_d induces an isomorphism $k[\mathbb{P}^n] \to k[V_{n,d}]$ of the homogeneous coordinate rings that is compatible with the gradings and multiplies degrees by d. In particular, hypersurfaces of degree d in $\mathbb{P}^n$ correspond naturally to hyperplane sections of $V_{n,d}$.

A.6.12 *Direct products* of varieties can be constructed as follows. The direct product $\mathbb{P}^m \times \mathbb{P}^n$ of two projective spaces is the Segre variety $S_{m,n}$. By definition, this is the projective variety of all rank one matrices of size $(m+1) \times (n+1)$, considered as a closed subset of projective space $\mathbb{P}^{mn+m+n}$. Sending a rank one matrix to its column span or row span defines a morphism $\pi_1 : S_{m,n} \to \mathbb{P}^m$ or $\pi_2 : S_{m,n} \to \mathbb{P}^n$, respectively. With these morphisms, the Segre variety satisfies the usual universal property for the direct product $\mathbb{P}^m \times \mathbb{P}^n$, in the category (Var_k) of k-varieties.

To construct the direct product of a general pair of varieties we may assume that $V \subseteq \mathbb{P}^m$ and $W \subseteq \mathbb{P}^n$ are locally closed. Then $V \times W = \{u \in \mathbb{P}^m \times \mathbb{P}^n : \pi_1(u) \in V, \pi_2(u) \in W\}$, together with the restrictions of π_1 and π_2, is a direct product of V and W in (Var_k). For affine varieties, the construction of the direct product simplifies: If $V \subseteq \mathbb{A}^m$ and $W \subseteq \mathbb{A}^n$ are closed subvarieties, the product variety $V \times W$ is simply the cartesian product of V and W in $\mathbb{A}^m \times \mathbb{A}^n = \mathbb{A}^{m+n}$. Assuming that the base field k is perfect, the affine coordinate ring of $V \times W$ is the tensor product $k[V] \otimes_k k[W]$.[4] Of

[4] If k fails to be perfect, this tensor product need not be reduced.

course, the assumption k perfect is harmless in the context of this book since every field with an ordering has characteristic zero.

For any k-variety V, the diagonal $\Delta_V = \{(u, u): u \in V\}$ of V is a closed subvariety of $V \times V$. (If one considers varieties more general than quasi-projective ones, this property is not automatic and means that the variety V is separated.) As a consequence, the intersection of any two open affine subsets of V is again (open and) affine.

A.6.13 *Rational functions and maps.* Let V, W be k-varieties. A *rational map* $f: V \dashrightarrow W$ is an equivalence class of morphisms $\varphi: U \to W$ with $U \subseteq V$ open and dense. Here φ and $\varphi': U' \to W$ are said to be equivalent if both agree on an open dense subset U'' of $U \cap U'$. The equivalence class of φ is denoted $[\varphi]$. A rational map $V \dashrightarrow \mathbb{A}^1$ is also called a *rational function* on V.

Given a rational map $f: V \dashrightarrow W$, the *domain* $\mathrm{dom}(f)$ of f is the union of all open dense sets $U \subseteq V$ for which a representative $\varphi: U \to W$ of f exists. There exists a morphism $\varphi_0: \mathrm{dom}(f) \to W$ that represents f, and every other representative of f is a restriction of φ_0. Given $u \in V$, one says that f is *defined at* u if $u \in \mathrm{dom}(f)$.

We usually consider rational maps on irreducible varieties only. If V is irreducible, the set of all rational functions on V forms the *function field* $k(V)$ of V (with naturally defined sum and product). Let all varieties in the following be irreducible. A rational map $f: V \dashrightarrow W$ is *dominant* if the image set of some (equivalently, any) representative of f is dense in W. Dominant rational maps between irreducible varieties can be composed, and so there is a category (Rat_k) of irreducible k-varieties, with the dominant rational maps as morphisms. The rational dominant map $f: V \dashrightarrow W$ induces an embedding $f^*: k(W) \to k(V)$ of the function fields, and the functor $V \mapsto k(V)$ defined in this way is an anti-equivalence from (Rat_k) to the category of finitely generated field extensions of k, with k-embeddings as morphisms. The rational map $V \dashrightarrow W$ is *birational*, or a birational equivalence, if it has a rational inverse, or equivalently, if the induced map $k(W) \to k(V)$ between the function fields is an isomorphism.

An important example of rational maps is given by linear projections in projective space. If $L \subseteq \mathbb{P}^n$ is an m-dimensional linear k-subvariety, *linear projection* $\pi_L: \mathbb{P}^n \dashrightarrow \mathbb{P}^{n-m-1}$ *with centre* L is a rational map defined outside of L. Identifying $\mathbb{P}^{n-m-1}$ with a fixed linear k-subvariety $L' \subseteq \mathbb{P}^n$, disjoint to L, the image point $\pi_L(\xi)$ of $\xi \notin L$ is the unique point of intersection between L' and the linear space spanned by L and ξ.

A.6.14 Let V be an irreducible k-variety. If V is affine, the function field of V is the quotient field of the coordinate ring of V. In other words, a rational function on V is a quotient of two regular functions on V (with non-zero denominator). If V is projective, then $k(V)$ is the subfield of $\mathrm{qf}(k[V])$ that consists of all fractions $\frac{p}{q}$ with p, q homogeneous of the same degree (and $q \neq 0$).

A.6.15 Let k'/k be a field extension and let V be a k-variety. So $V \subseteq \mathbb{P}^n$ is a set that is locally closed in the k-Zariski topology, which means that V can be expressed in the form

$$V = V(f_1, \dots, f_r) \cap (D_+(g_1) \cup \dots \cup D_+(g_s)) \tag{A.1}$$

with homogeneous polynomials f_i, $g_j \in k[x]$. The same expression defines a k'-variety that is denoted $V_{k'}$. It is easy to see that the k'-variety $V_{k'}$ does not depend on the particular choice of (A.1). The operator $V \mapsto V_{k'}$ extends to a functor $(Var_k) \to (Var_{k'})$, called *base field extension* (from k to k'). If the k-variety V is affine, with coordinate ring $k[V]$, the coordinate of the affine k'-variety $V_{k'}$ is $k[V] \otimes_k k'$. The analogous statement is true for the projective coordinate ring when the k-variety V is projective. For both statements we are again assuming that k is perfect.

A.6.16 Let V be a k-variety and let $u \in V$ be a point. The local ring of V at u is $\mathcal{O}_{V,u} := \varinjlim_{u \in U} \mathcal{O}_V(U)$, the inductive limit over the directed set of all open neighborhoods of u in V, with restriction maps as transition homomorphisms. Clearly, this ring doesn't change when V is replaced by an open neighborhood of u in V. For the study of $\mathcal{O}_{V,u}$ one may therefore assume that V is affine. If $V \subseteq \mathbb{A}^n$ is a closed subset then $\mathfrak{p} := \mathcal{I}_V(\{u\}) = \{f \in k[V] \colon f(u) = 0\}$ is a prime ideal of $k[V]$ (which is maximal if $u \in \overline{k}^n$), and $\mathcal{O}_{V,u}$ is naturally isomorphic to the localization $k[V]_\mathfrak{p}$ of $k[V]$. When V is irreducible, the local ring $\mathcal{O}_{V,u}$ can also be characterized as the subring of $k(V)$ that consists of all rational functions f with $u \in \mathrm{dom}(f)$.

A.6.17 Let V be a k-variety. A point $u \in V$ is a *non-singular point* (or *regular point*) of V if the local ring $\mathcal{O}_{V,u}$ is regular (A.4.5). Otherwise u is a *singular point* of V. Assume that the field k is perfect, for example $\mathrm{char}(k) = 0$. Then the set V_{reg} of all non-singular points of V is open and dense in V, and therefore the set V_{sing} of singular points is a proper closed subvariety of V. The variety V is *non-singular* if it has no singular points. Moreover, when k is perfect, there is a convenient way to determine the singular locus of a variety. Assume that $V \subseteq \mathbb{A}^n$ is a closed subvariety, and let $f_1, \dots, f_r \in k[x] = k[x_1, \dots, x_n]$ be a generating system for the full vanishing ideal $\mathcal{I}(V)$ of V in $k[x]$. Let $u \in V$, and let $d = \dim_u(V)$ be the local dimension of V at u. The Jacobian matrix

$$\Bigl(\frac{\partial f_i}{\partial x_j}(u)\Bigr)_{1 \le i \le r,\, 1 \le j \le n},$$

evaluated at u, has rank $\le n - d$. The point u is a non-singular point of V if and only if this rank is equal to $n - d$. Observe that from this criterion, it is clear that V_{sing} is a closed subset of V.

A.6.18 For a systematic discussion of (co-) tangent spaces on algebraic varieties, one should use the sheaf-theoretic approach via the sheaf of k-differentials and its dual. Or even better, their relative versions for morphisms of varieties (or schemes). In this course, only (co-) tangent spaces at k-rational points are used, and only in very few places. For this it suffices to introduce them in a more elementary way, as follows.

Let V be an affine k-variety, let $A := k[V]$ denote its affine coordinate ring, and let ξ be a k-rational point of V. The *cotangent space* of V at ξ is the k-vector space $T_\xi^\vee(V) := \mathfrak{m}/\mathfrak{m}^2$, where $\mathfrak{m}$ is the maximal ideal of A corresponding to ξ. Accordingly, the *tangent space* of V at ξ is the k-linear dual of $T_\xi^\vee(V)$, viz. $T_\xi(V) = (\mathfrak{m}/\mathfrak{m}^2)^\vee$. Note that this may also be written as $T_\xi(V) = \mathrm{Hom}_A(\mathfrak{m}, A/\mathfrak{m})$. An element of $T_\xi(V)$ is therefore a map $\tau \colon \mathfrak{m} \to k$ satisfying $\tau(fg) = f(\xi)\tau(g)$ for $f \in A$, $g \in \mathfrak{m}$. Such a map τ should be thought of as the directional derivative at ξ corresponding to the specified tangent direction.

Since the definition of (co-) tangent space at $\xi \in V(k)$ is not affected when V is replaced by an open neighborhood of ξ, the definition carries over immediately to k-varieties that are not necessarily affine. Note that always $\dim T_\xi(V) \geq \dim_\xi(V)$, and that equality holds if and only if ξ is a non-singular point of V. The tangent space is functorial (covariantly) for morphisms of k-varieties in an obvious way.

Let W be a closed subvariety of V, and let $\xi \in W(k)$. The inclusion $i\colon W \to V$ induces an embedding $i_*\colon T_\xi(W) \to T_\xi(V)$ of the tangent spaces. The *normal space* $N_\xi(W, V)$ (of W in V at ξ) is defined to be the cokernel, so by definition the sequence

$$0 \to T_\xi(W) \xrightarrow{i_*} T_\xi(V) \to N_\xi(W, V) \to 0 \tag{A.2}$$

of k-vector spaces is exact. For an algebraic description, assume that V is affine and I is the vanishing ideal of W in $A = k[V]$. If $\mathfrak{m} \subseteq A$ is the maximal ideal corresponding to ξ, we have the *conormal exact sequence* at ξ, which is the dual of (A.2):

$$0 \to \frac{I}{I \cap \mathfrak{m}^2} \to \frac{\mathfrak{m}}{\mathfrak{m}^2} \to \frac{\mathfrak{m}}{I + \mathfrak{m}^2} \to 0. \tag{A.3}$$

When V and W are both non-singular at ξ, the inclusion $I\mathfrak{m} \subseteq I \cap \mathfrak{m}^2$ is an equality, and so (A.3) reads

$$0 \to \frac{I}{I\mathfrak{m}} \to \frac{\mathfrak{m}}{\mathfrak{m}^2} \to \frac{\mathfrak{m}}{I + \mathfrak{m}^2} \to 0. \tag{A.4}$$

In particular, $N_\xi(W, V) = \mathrm{Hom}_A(I, A/\mathfrak{m})$ in this case.

A.6.19 For use in Chapter 7, we recall the notion of degree of projective varieties. Let $V \subseteq \mathbb{P}^n$ be a projective k-variety, with vanishing ideal $I = \mathfrak{I}(V) \subseteq k[x] = k[x_0, \ldots, x_n]$ and homogeneous coordinate ring $k[V] = k[x]/I$. The *Hilbert series* of V is the formal power series $H_V(t) = \sum_{i=0}^{\infty} \dim(k[V]_i)\, t^i$ in the variable t. There exists a unique polynomial $P_V(t) \in \mathbb{Q}[t]$ with $P_V(i) = \dim(k[V]_i)$ for all sufficiently large integers i, the *Hilbert polynomial* of V. By the Hilbert–Serre theorem, $H_V(t)$ is a rational function of the form $H_V(t) = p(t)(1-t)^{-r}$ with a (unique) polynomial $p \in \mathbb{Z}[t]$ satisfying $p(1) \neq 0$, and with $r \geq 0$. The Hilbert polynomial has degree $r-1$ and has leading (highest) coefficient $\frac{p(1)}{(r-1)!}$. It is known that $\deg(P_V) = \dim(V)$, so the pole order r of $H_V(t)$ at $t = 1$ is $r = \dim(V) + 1$. The *degree* of the projective variety V is defined to be $\deg(V) = p(1)$. So if the Hilbert polynomial is $P_V(t) = ct^m +$(lower degree summands) with $c \in \mathbb{Q}^*$, the degree of V is $\deg(V) = m! \cdot c$. Note that $\deg(V)$ is always a positive integer.

Appendix B: Convex Sets in Real Infinite-Dimensional Vector Spaces

We briefly recall the notions of topological $\mathbb{R}$-vector spaces in general, and of locally convex vector spaces in particular. Then we state the two single most important theorems for the latter, which are the Hahn–Banach separation theorem and the Krein–Milman theorem. For more background one may consult any textbook on functional analysis, like [173]. We also state the Eidelheit–Kakutani separation theorem, which applies to arbitrary $\mathbb{R}$-vector spaces without topology. Here we refer to Köthe's monograph [111]. Finally we explain how to extend this last result to vector spaces over the field $\mathbb{Q}$ of rational numbers. Unless otherwise said, all $\mathbb{R}$-vector spaces may have arbitrary dimension.

B.1 Let V be an $\mathbb{R}$-vector space. The full dual space of all linear maps $V \to \mathbb{R}$ is denoted $V^\vee$. Recall a few basic notions from convexity. A set $K \subseteq V$ is *convex* if $(1-t)x + ty \in K$ holds whenever $x, y \in K$ and $0 \le t \le 1$ is a real number. If in addition $K \ne \varnothing$ and $tK \subseteq K$ holds for all $t \ge 0$, then K is a *convex cone*. The *convex hull* $\mathrm{conv}(M)$ of a set $M \subseteq V$ is the smallest convex set in V that contains M. It consists of all convex combinations of points in M, so $\mathrm{conv}(M) = \{\sum_{i=0}^n a_i x_i \colon n \ge 0, x_i \in M, 0 \le a_i \in \mathbb{R}, \sum_{i=1}^n a_i = 1\}$.

An affine hyperplane in V is a set of the form $H = \{x \in V \colon f(x) = c\}$ where $0 \ne f \in V^\vee$ and $c \in \mathbb{R}$. If $K \subseteq V$ is a convex set then H is a *supporting hyperplane* of K if $K \cap H \ne \varnothing$, and if either $f(y) \ge c$ or $f(y) \le c$ holds for every $y \in K$. A point $x \in K$ is an *extreme point* of K if $x = (1-t)y + tz$ with $0 < t < 1$ and $y, z \in K$ implies $y = z = x$. The set of extreme points of K is denoted $\mathrm{Ex}(K)$. If K is a convex cone and $0 \ne x \in K$, the half-line $\mathbb{R}_+ x$ is an *extreme ray* of K if $x = y + z$ with $y, z \in K$ implies $y, z \in \mathbb{R}_+ x$.

B.2 A *topological vector space* (over $\mathbb{R}$) is an $\mathbb{R}$-vector space V together with a Hausdorff topology on V, such that addition $V \times V \to V$ and scalar multiplication $\mathbb{R} \times V \to V$ are continuous maps. Given such V, the vector space of all *continuous* linear forms $V \to \mathbb{R}$ will be denoted V'. We don't consider a topology on V'.

C. Scheiderer, *A Course in Real Algebraic Geometry*, Graduate Texts in Mathematics 303,
https://doi.org/10.1007/978-3-031-69213-0

B.3 Remarks.

1. A finite-dimensional $\mathbb{R}$-vector space V has a unique vector space topology. If we use a linear basis of V to identify V with $\mathbb{R}^n$, this is the Euclidean topology on $\mathbb{R}^n$.

2. A *normed vector space* is an $\mathbb{R}$-vector space V together with a map $V \to \mathbb{R}$, $x \mapsto ||x||$ (called the *norm*) that satisfies $||x|| > 0$ for $x \neq 0$, $||x+y|| \le ||x|| + ||y||$ and $||ax|| = |a| \cdot ||x||$ for $a \in \mathbb{R}$ and $x, y \in V$. The norm defines a metric d on V via $d(x,y) = ||x-y||$ $(x, y \in V)$, and the associated topology makes V a topological vector space. Well-known examples of normed vector spaces are $\mathbb{R}^n$ (with the Euclidean norm), the space $\mathcal{C}(X,\mathbb{R})$ of continuous real-valued functions on a compact space X (with the sup-norm), or the L^p-space of a measure μ (for $1 \le p \le \infty$) with the L^p-norm. A *Banach space* is a normed vector space that is complete.

3. If V is a topological vector space then closure $\overline{K}$ and interior $\operatorname{int}(K)$ of any convex set $K \subseteq V$ are again convex (easy exercise). For $M \subseteq V$ an arbitrary subset, the set $M^* := \{f \in V' : f|_M \ge 0\}$ is a convex cone in V'.

B.4 Definition. A topological $\mathbb{R}$-vector space V is *locally convex* if every neighborhood of 0 contains a convex such neighborhood.

B.5 Examples.

1. Every normed vector space is locally convex, since the open balls $\{x \in V: ||x|| < r\}$ are convex. In particular, finite-dimensional vector spaces are locally convex. Linear subspaces or arbitrary direct products of locally convex vector spaces are again locally convex (in the subspace topology and the product topology, respectively). In particular, $\prod_I \mathbb{R}$ is a locally convex vector space for any index set I.

2. In the complex case, a locally convex $\mathbb{C}$-vector space is required to have a neighborhood basis of 0 that consists of convex sets U which are *balanced*, meaning that $cU \subseteq U$ for every $c \in \mathbb{C}$ with $|c| = 1$. For $\mathbb{R}$-vector spaces, Definition B.4 automatically implies the existence of arbitrarily small (real) balanced convex neighborhoods, since $U \cap (-U)$ is (real) balanced for every convex set U. For both the real and the complex case, there is an equivalent characterization of locally convex vector spaces in terms of seminorms, that is technically more convenient to work with. For our modest purposes, the above definition suffices.

3. Let V be an $\mathbb{R}$-vector space with a linear basis that is (at most) countable. Define a topology τ on V as follows: A subset $M \subseteq V$ is τ-closed if, for every linear subspace $U \subseteq V$ of finite dimension, $M \cap U$ is closed in U (with respect to the unique vector space topology of U). Then (V, τ) is a topological vector space which is locally convex, and for which every linear form $V \to \mathbb{R}$ is continuous. This topology τ is the finest among all vector space topologies on V. When $\dim(V)$ is infinite (and countable), the topology τ cannot be defined by a metric on V (cf. Exercise 6.6.5). When the vector space dimension of V is uncountable, τ is not a vector space topology any more [21]. Regardless of the dimension, there always exists a unique finest locally convex vector space topology on V (which agrees with τ when the dimension is at most countable).

A fundamental fact that holds in every locally convex vector space is the separation theorem:

B.6 Theorem. (Hahn–Banach) *Let V be a locally convex $\mathbb{R}$-vector space, and let $A, B \subseteq V$ be two disjoint convex sets, where A is compact and B is closed in V. Then there exists a continuous linear form $f \in V'$ and a real number $c \in \mathbb{R}$ such that $f(b) < c < f(a)$ holds for all $a \in A$, $b \in B$.*

In other words, there exists a closed affine hyperplane H in V which separates A and B, in the sense that A is contained in one of the two open halfspaces defined by H, and B in the other. For the proof see any book on functional analysis, e.g. [173]. The Hahn–Banach theorem exists in many formulations. For this course, the one stated above is sufficient.

Let V be a locally convex vector space. We record a few direct consequences.

B.7 Corollary. *Given any convex set $K \subseteq V$, the closure $\overline{K}$ is the intersection of a family of closed halfspaces in V.* □

Here, of course, a closed halfspace is a set of the form $\{x \in V : f(x) \geq c\}$ with $0 \neq f \in V'$ and $c \in \mathbb{R}$.

B.8 Corollary. *Let $C \subseteq V$ be a convex cone, and let $C^{**} := \{x \in V : \forall f \in C^* \; f(x) \geq 0\}$. Then C^{**} is the closure of C.*

Proof. C^{**} is a closed convex set in V that contains C. If $f \in V'$ and $c \in \mathbb{R}$ are such that $f \geq c$ on C, then $c \leq 0$, and $f|_C \geq 0$ since C is a cone. So $C^{**} = \overline{C}$ follows from Corollary B.7. □

B.9 Theorem. (Krein–Milman) *Let V be a locally convex vector space, and let $K \subseteq V$ be a compact convex set. Then K is the closed convex hull of its extreme points, i.e. $K = \overline{\mathrm{conv}(\mathrm{Ex}(K))}$.*

B.10 Remarks.

1. Clearly, the Krein–Milman theorem implies that every non-empty compact convex set has an extreme point. Conversely, this weak version already implies the full theorem. Indeed, put $K' = \overline{\mathrm{conv}(\mathrm{Ex}(K))}$ and assume that there is a point $x \in K$ with $x \notin K'$. Apply Hahn–Banach to K' and $\{x\}$, to get a linear form $f \in V'$ with $f(x) < \min f(K')$. Put $c = \min f(K)$, then $H := f^{-1}(c)$ is a supporting hyperplane of K, and $K \cap H$ is a non-empty compact convex set in H. By the weak version of Krein–Milman, $K \cap H$ has an extreme point y. Clearly y is an extreme point of K as well, and so $y \in K'$. But $f(y) = c < \min f(K')$, a contradiction.

2. When $\dim(V) < \infty$ then Theorem B.9 is true even without taking the closure. This is proved in Theorem 8.1.13.

B.11 Corollary. *Let V be locally convex and let $K \subseteq V$ be a non-empty compact convex set.*

(a) *Every closed supporting hyperplane of K contains an extreme point of K.*
(b) *If $f \in V'$ then* $\min f(K)$ *and* $\max f(K)$ *are taken in extreme points of K.* □

Lastly, we introduce the Eidelheit–Kakutani separation theorem, which applies to arbitrary real vector spaces without topology. To prepare for this, consider the following definitions. We use the general notation 8.1.1 for line segments in vector spaces.

B.12 Definition. Let V be an $\mathbb{R}$-vector space and let $K \subseteq V$ be a subset.

(a) $x \in K$ is an *algebraic interior point* of K if for every $v \in V$ there is a real number $c > 0$ with $]x - cv, x + cv[\subseteq K$. The set K^i of all algebraic interior points of K is the *algebraic interior* of K.
(b) The *algebraic closure* K^a of K consists of all $x \in V$ for which there exists $y \in K$ with $[y, x[\subseteq K$.

B.13 Remarks.

1. For any set $K \subseteq V$ one has $K^i \subseteq K \subseteq K^a$. If K is convex then so are K^i and K^a. When K is convex and $\dim(V)$ is finite, K^i and K^a are the usual topological interior and closure of K.

2. If $\dim(V) = \infty$, there may exist convex sets $K \subseteq V$ for which $(K^a)^a \neq K^a$ ([111] pp. 177–178). So there need not exist any topology on V that satisfies $K^a = \overline{K}$ for every convex set $K \subseteq V$. If $\dim(V) = \infty$, there always exists a proper convex set $K \neq V$ with $K^a = V$.

B.14 Theorem. (Eidelheit–Kakutani) *Let V be an arbitrary $\mathbb{R}$-vector space, and let K_1, K_2 be non-empty convex subsets of V satisfying $K_1^i \neq \varnothing$ and $K_1^i \cap K_2 = \varnothing$. Then there exists a linear form $0 \neq f \in V^\vee$ and a real number c such that $f(x_2) \leq c \leq f(x_1)$ for all $x_1 \in K_1$ and $x_2 \in K_2$.*

In other words, there exists an affine hyperplane $H \subseteq V$ such that K_1 is contained in one of the two "closed" halfspaces defined by H, and K_2 in the other. The original papers are [62] and [105], a more accessible reference is [111] pp. 186–187.

B.15 Corollary. *Let V be an $\mathbb{R}$-vector space, let $C \subseteq V$ be a convex cone with an algebraic interior point u, and let $x \in V$ with $x \notin C^i$. Then there exists a linear form $f \in V^\vee$ with $f|_C \geq 0$, $f(u) = 1$ and $f(x) \leq 0$.*

Proof. Applying Theorem B.14 to $K_1 = C$ and $K_2 = \{x\}$ we get $0 \neq f \in V^\vee$ and $c \in \mathbb{R}$ with $f|_C \geq c$ and $f(x) \leq c$. Since $0 \in C$ we have $c \leq 0$. There cannot exist $y \in C$ with $f(y) < 0$ since then f would not be bounded below on C. So we can take $c = 0$. Moreover $f(u) > 0$ since $u \in C^i$, and replacing f with $\frac{1}{f(u)} f$ does the job. □

B.16 Theorem B.14 is also useful in vector spaces over $\mathbb{Q}$, as we will now explain. Let V be a $\mathbb{Q}$-vector space. For $x, y \in V$ write $[x,y]_{\mathbb{Q}} = \{(1-t)x + ty : t \in \mathbb{Q}, 0 \le t \le 1\}$, and similarly for open or half-open intervals. A subset $K \subseteq V$ is $\mathbb{Q}$-*convex* if $x, y \in K$ implies $[x,y]_{\mathbb{Q}} \subseteq K$. If $M \subseteq V$ is a set, an element $u \in M$ is a $\mathbb{Q}$-*algebraic interior point* of M, if for every $v \in V$ there exists $t > 0$ in $\mathbb{Q}$ with $[u - tv, u + tv]_{\mathbb{Q}} \subseteq M$.

Let $V_{\mathbb{R}} = V \otimes_{\mathbb{Q}} \mathbb{R}$, and consider V as a subset of $V_{\mathbb{R}}$ via $V \to V_{\mathbb{R}}$, $v \mapsto v \otimes 1$. Given a $\mathbb{Q}$-convex set K in V let $K_{\mathbb{R}}$ be the ($\mathbb{R}$-) convex hull of K in $V_{\mathbb{R}}$. If K is a $\mathbb{Q}$-convex cone (meaning that $0 \in K$, $K + K \subseteq K$ and $\mathbb{Q}_+ K \subseteq K$), $K_{\mathbb{R}}$ is a ($\mathbb{R}$-) convex cone in $V_{\mathbb{R}}$. The following lemma is not hard not prove:

B.17 Lemma. *Let $K \subseteq V$ be a $\mathbb{Q}$-convex set. Then $K_{\mathbb{R}} \cap V = K$.*

B.18 Lemma. *Let K be a $\mathbb{Q}$-convex set in V. A point $x \in K$ is a $\mathbb{Q}$-algebraic interior point of K (in V) if, and only if, x is an $\mathbb{R}$-algebraic interior point of the $\mathbb{R}$-convex hull $K_{\mathbb{R}}$ (in $V_{\mathbb{R}}$).*

Proof. If x is an $\mathbb{R}$-algebraic interior point of $K_{\mathbb{R}}$ then x is a $\mathbb{Q}$-algebraic interior point of K, by Lemma B.17. Conversely let x be a $\mathbb{Q}$-algebraic interior point of K, and let $v \in V_{\mathbb{R}}$. Since $K_{\mathbb{R}}$ is $\mathbb{R}$-convex, it suffices to find a real number $t > 0$ with $x + tv \in K_{\mathbb{R}}$. Write $v = \sum_{i=1}^{n} a_i v_i$ with real numbers $a_i > 0$ and with $v_i \in V$. Scaling v by a positive real number we may assume $\sum_{i=1}^{n} a_i = 1$. By hypothesis there is $t > 0$ in $\mathbb{Q}$ with $x + tv_i \in K$ for $i = 1, \dots, n$. So $x + tv = \sum_i a_i(x + tv_i)$ is in $K_{\mathbb{R}}$, being an $\mathbb{R}$-convex combination of the points $x + tv_i \in K$. □

From the previous lemma we get a version of Eidelheit–Kakutani for $\mathbb{Q}$-vector spaces. We content ourselves with stating the analogue of Corollary B.15:

B.19 Corollary. *Let V be a $\mathbb{Q}$-vector space and let $C \subseteq V$ be a $\mathbb{Q}$-convex cone with $\mathbb{Q}$-algebraic interior point u. If $x \in V$ is not a $\mathbb{Q}$-algebraic interior point of C, there is a $\mathbb{Q}$-linear map $f : V \to \mathbb{R}$ such that $f \ge 0$ on C, $f(u) = 1$ and $f(x) \le 0$.*

Proof. Let $C_{\mathbb{R}}$ be the $\mathbb{R}$-convex hull of C in $V_{\mathbb{R}} = V \otimes_{\mathbb{Q}} \mathbb{R}$. This is a convex cone in $V_{\mathbb{R}}$. Using Lemma B.18 we see that u is an algebraic interior point of $C_{\mathbb{R}}$, and that x is not. So the assertion follows from Corollary B.15. □

Bibliography

1. Ahmadi, A.A., Parrilo, P.A.: A convex polynomial that is not sos-convex. Math. Program. **135**(1-2, Ser. A), 275–292 (2012)
2. Ahmadi, A.A., Parrilo, P.A.: A complete characterization of the gap between convexity and sos-convexity. SIAM J. Optim. **23**(2), 811–833 (2013)
3. Andradas, C., Bröcker, L., Ruiz, J.M.: Constructible sets in real geometry, *Ergebnisse der Mathematik und ihrer Grenzgebiete (3)*, vol. 33. Springer-Verlag, Berlin (1996)
4. Anjos, M.F., Lasserre, J.B. (eds.): Handbook on semidefinite, conic and polynomial optimization, *International Series in Operations Research & Management Science*, vol. 166. Springer, New York (2012)
5. Artin, E.: Über die Zerlegung definiter Funktionen in Quadrate. Abh. Math. Sem. Univ. Hamburg **5**(1), 100–115 (1927)
6. Artin, E., Schreier, O.: Algebraische Konstruktion reeller Körper. Abh. Math. Sem. Univ. Hamburg **5**(1), 85–99 (1927)
7. Atiyah, M.F., Macdonald, I.G.: Introduction to commutative algebra. Addison-Wesley Publishing Co., Reading, Mass.-London-Don Mills, Ont. (1969)
8. Averkov, G., Bröcker, L.: Minimal polynomial descriptions of polyhedra and special semialgebraic sets. Adv. Geom. **12**(3), 447–459 (2012)
9. Baer, R.: Über nicht-archimedisch geordnete Körper. (Beiträge zur Algebra 5.). Sitzungsberichte Heidelberg 1927, 8. Abh., 3-13 (1927).
10. Baldi, L., Slot, L.: Degree bounds for Putinar's Positivstellensatz on the hypercube. SIAM J. Appl. Algebra Geom. **8**(1), 1–25 (2024)
11. Barvinok, A.: A course in convexity, *Graduate Studies in Mathematics*, vol. 54. American Mathematical Society, Providence, RI (2002)
12. Basu, S.: Algorithms in real algebraic geometry: a survey. In: Real algebraic geometry, *Panor. Synthèses*, vol. 51, pp. 107–153. Soc. Math. France, Paris (2017)
13. Basu, S., Pollack, R., Roy, M.F.: Algorithms in real algebraic geometry, *Algorithms and Computation in Mathematics*, vol. 10, second edn. Springer-Verlag, Berlin (2006)
14. Becker, E.: Valuations and real places in the theory of formally real fields. In: Real algebraic geometry and quadratic forms (Rennes, 1981), *Lecture Notes in Math.*, vol. 959, pp. 1–40. Springer, Berlin-New York (1982)
15. Becker, E.: On the real spectrum of a ring and its application to semialgebraic geometry. Bull. Amer. Math. Soc. (N.S.) **15**(1), 19–60 (1986)
16. Becker, E., Schwartz, N.: Zum Darstellungssatz von Kadison-Dubois. Arch. Math. (Basel) **40**(5), 421–428 (1983)
17. Ben-Tal, A., Nemirovski, A.: Lectures on modern convex optimization. MPS-SIAM Series on Optimization. Society for Industrial and Applied Mathematics (SIAM), Philadelphia, PA; Mathematical Programming Society (MPS), Philadelphia, PA (2001)

C. Scheiderer, *A Course in Real Algebraic Geometry*, Graduate Texts in Mathematics 303,
https://doi.org/10.1007/978-3-031-69213-0

18. Benoist, O.: On the bad points of positive semidefinite polynomials. Math. Z. **300**(4), 3383–3403 (2022)
19. Berr, R., Wörmann, T.: Positive polynomials and tame preorderings. Math. Z. **236**(4), 813–840 (2001)
20. Berr, R., Wörmann, T.: Positive polynomials on compact sets. Manuscripta Math. **104**(2), 135–143 (2001)
21. Bisgaard, T.M.: The topology of finitely open sets is not a vector space topology. Arch. Math. (Basel) **60**(6), 546–552 (1993)
22. Blekherman, G.: Convex forms that are not sums of squares (2009). URL `https://arxiv.org/pdf/0910.0656`
23. Blekherman, G., Plaumann, D., Sinn, R., Vinzant, C.: Low-rank sum-of-squares representations on varieties of minimal degree. Int. Math. Res. Not. **2019**(1), 33–54
24. Blekherman, G., Smith, G.G., Velasco, M.: Sums of squares and varieties of minimal degree. J. Amer. Math. Soc. **29**(3), 893–913 (2016)
25. Bochnak, J., Coste, M., Roy, M.F.: Géométrie algébrique réelle, *Ergebnisse der Mathematik und ihrer Grenzgebiete (3)*, vol. 12. Springer-Verlag, Berlin (1987)
26. Bochnak, J., Coste, M., Roy, M.F.: Real algebraic geometry, *Ergebnisse der Mathematik und ihrer Grenzgebiete (3)*, vol. 36. Springer-Verlag, Berlin (1998). Translated from the 1987 French original. Revised by the authors.
27. Bochnak, J., Efroymson, G.: Real algebraic geometry and the 17th Hilbert problem. Math. Ann. **251**(3), 213–241 (1980)
28. Bochnak, J., Risler, J.J.: Le théorème des zéros pour les variétés analytiques réelles de dimension 2. Ann. Sci. École Norm. Sup. (4) **8**(3), 353–363 (1975)
29. Bodirsky, M., Kummer, M., Thom, A.: Spectrahedral shadows and completely positive maps on real closed fields. J. Eur. Math. Soc. (to appear)
30. Bonsall, F.F., Lindenstrauss, J., Phelps, R.R.: Extreme positive operators on algebras of functions. Math. Scand. **18**, 161–182 (1966)
31. Boyd, S., Vandenberghe, L.: Convex optimization. Cambridge University Press, Cambridge (2004)
32. Bröcker, L.: Positivbereiche in kommutativen Ringen. Abh. Math. Sem. Univ. Hamburg **52**, 170–178 (1982)
33. Bröcker, L.: Real spectra and distributions of signatures. In: Real algebraic geometry and quadratic forms (Rennes, 1981), *Lecture Notes in Math.*, vol. 959, pp. 249–272. Springer, Berlin-New York (1982)
34. Bröcker, L.: Spaces of orderings and semialgebraic sets. In: Quadratic and Hermitian forms (Hamilton, Ont., 1983), *CMS Conf. Proc.*, vol. 4, pp. 231–248. Amer. Math. Soc., Providence, RI (1984)
35. Brumfiel, G.W.: Partially ordered rings and semi-algebraic geometry, *London Mathematical Society Lecture Note Series*, vol. 37. Cambridge University Press, Cambridge-New York (1979)
36. Bruns, W., Herzog, J.: Cohen-Macaulay rings, *Cambridge Studies in Advanced Mathematics*, vol. 39. Cambridge University Press, Cambridge (1993)
37. Burgdorf, S., Scheiderer, C., Schweighofer, M.: Pure states, nonnegative polynomials and sums of squares. Comment. Math. Helv. **87**(1), 113–140 (2012)
38. Cassels, J.W.S.: On the representation of rational functions as sums of squares. Acta Arith. **9**, 79–82 (1964)
39. Cassels, J.W.S., Ellison, W.J., Pfister, A.: On sums of squares and on elliptic curves over function fields. J. Number Theory **3**, 125–149 (1971)
40. Choi, M.D.: Positive semidefinite biquadratic forms. Linear Algebra Appl. **12**(2), 95–100 (1975)
41. Choi, M.D., Dai, Z.D., Lam, T.Y., Reznick, B.: The Pythagoras number of some affine algebras and local algebras. J. Reine Angew. Math. **336**, 45–82 (1982)
42. Choi, M.D., Lam, T.Y.: An old question of Hilbert. In: Conference on Quadratic Forms—1976 (Proc. Conf., Queen's Univ., Kingston, Ont., 1976), pp. 385–405. Queen's Papers in Pure and Appl. Math., No. 46 (1977)

43. Choi, M.D., Lam, T.Y., Reznick, B.: Even symmetric sextics. Math. Z. **195**(4), 559–580 (1987)
44. Choi, M.D., Lam, T.Y., Reznick, B.: Sums of squares of real polynomials. In: K-theory and algebraic geometry: Connections with quadratic forms and division algebras (Santa Barbara, CA, 1992), *Proc. Sympos. Pure Math.*, vol. 58, pp. 103–126. Amer. Math. Soc., Providence, RI (1995)
45. Chua, L., Plaumann, D., Sinn, R., Vinzant, C.: Gram spectrahedra. In: Ordered algebraic structures and related topics, *Contemp. Math.*, vol. 697, pp. 81–105. Amer. Math. Soc., Providence, RI (2017)
46. Coste, M., Roy, M.F.: La topologie du spectre réel. In: Ordered fields and real algebraic geometry (San Francisco, Calif., 1981), *Contemp. Math.*, vol. 8, pp. 27–59. Amer. Math. Soc., Providence, R.I. (1982)
47. Coste, M.F., Coste, M.: Topologies for real algebraic geometry. In: Topos theoretic methods in geometry, *Various Publications Series*, vol. 30, pp. 37–100. Aarhus Univ., Aarhus (1979)
48. Coste-Roy, M.F., Coste, M.: Le spectre étale réel d'un anneau est spatial. C. R. Acad. Sci. Paris Sér. A-B **290**(2), A91–A94 (1980)
49. Cox, D.A., Little, J., O'Shea, D.: Using algebraic geometry, *Graduate Texts in Mathematics*, vol. 185, second edn. Springer, New York (2005)
50. Cox, D.A., Little, J., O'Shea, D.: Ideals, varieties, and algorithms. An introduction to computational algebraic geometry and commutative algebra, fourth edn. Undergraduate Texts in Mathematics. Springer, Cham (2015)
51. Delzell, C.N.: A constructive, continuous solution to Hilbert's 17th problem, and other results in semi-algebraic geometry. Ph.D. thesis, Stanford University (1980)
52. Delzell, C.N., Madden, J.J.: A completely normal spectral space that is not a real spectrum. J. Algebra **169**(1), 71–77 (1994)
53. Dickmann, M., Schwartz, N., Tressl, M.: Spectral spaces, *New Mathematical Monographs*, vol. 35. Cambridge University Press, Cambridge (2019)
54. Dines, L.L.: On the mapping of quadratic forms. Bull. Amer. Math. Soc. **47**, 494–498 (1941)
55. Dressler, M., Iliman, S., de Wolff, T.: A Positivstellensatz for sums of nonnegative circuit polynomials. SIAM J. Appl. Algebra Geom. **1**(1), 536–555 (2017)
56. van den Dries, L.: Tame topology and o-minimal structures, *London Mathematical Society Lecture Note Series*, vol. 248. Cambridge University Press, Cambridge (1998)
57. Dritschel, M.A., Rovnyak, J.: The operator Fejér-Riesz theorem. In: A glimpse at Hilbert space operators, *Oper. Theory Adv. Appl.*, vol. 207, pp. 223–254. Birkhäuser Verlag, Basel (2010)
58. Dubois, D.W.: A note on David Harrison's theory of preprimes. Pacific J. Math. **21**, 15–19 (1967)
59. Dubois, D.W.: A nullstellensatz for ordered fields. Ark. Mat. **8**, 111–114 (1969)
60. Dubois, D.W., Efroymson, G.: Algebraic theory of real varieties. I. In: Studies and essays (presented to Yu-Why Chen on his 60th birthday, April 1, 1970), pp. 107–135. Math. Res. Center, Nat. Taiwan Univ., Taipei (1970)
61. Effros, E.G., Handelman, D.E., Shen, C.L.: Dimension groups and their affine representations. Amer. J. Math. **102**(2), 385–407 (1980)
62. Eidelheit, M.: Zur Theorie der konvexen Mengen in linearen normierten Räumen. Stud. Math. **6**, 104–111 (1936)
63. Eisenbud, D.: Commutative algebra. With a view toward algebraic geometry, *Graduate Texts in Mathematics*, vol. 150. Springer-Verlag, New York (1995)
64. Eisenbud, D., Goto, S.: Linear free resolutions and minimal multiplicity. J. Algebra **88**(1), 89–133 (1984)
65. Eisenbud, D., Harris, J.: On varieties of minimal degree. (A centennial account). In: Algebraic geometry, Bowdoin, 1985 (Brunswick, Maine, 1985), *Proc. Sympos. Pure Math.*, vol. 46, pp. 3–13. Amer. Math. Soc., Providence, RI (1987)
66. El Khadir, B.: On sum of squares representation of convex forms and generalized Cauchy-Schwarz inequalities. SIAM J. Appl. Algebra Geom. **4**(2), 377–400 (2020)

67. Fejér, L.: Über trigonometrische Polynome. J. Reine Angew. Math. **146**, 53–82 (1916)
68. Finsler, P.: Über das Vorkommen definiter und semidefiniter Formen in Scharen quadratischer Formen. Comment. Math. Helv. **9**(1), 188–192 (1936)
69. Gantmacher, F.R.: The theory of matrices. Vol. 1. AMS Chelsea Publishing, Providence, RI (1998). Translated from the Russian by K. A. Hirsch
70. Gårding, L.: An inequality for hyperbolic polynomials. J. Math. Mech. **8**, 957–965 (1959)
71. Gårding, L.: Hyperbolic equations in the twentieth century. In: Matériaux pour l'histoire des mathématiques au XXe siècle (Nice, 1996), *Sémin. Congr.*, vol. 3, pp. 37–68. Soc. Math. France, Paris (1998)
72. Gel'fand, I.M., Kapranov, M.M., Zelevinsky, A.V.: Discriminants, resultants, and multidimensional determinants. Mathematics: Theory & Applications. Birkhäuser Boston, Inc., Boston, MA (1994)
73. Goemans, M.X., Williamson, D.P.: Improved approximation algorithms for maximum cut and satisfiability problems using semidefinite programming. J. Assoc. Comput. Mach. **42**(6), 1115–1145 (1995)
74. Goodearl, K.R.: Partially ordered abelian groups with interpolation, *Mathematical Surveys and Monographs*, vol. 20. American Mathematical Society, Providence, RI (1986)
75. Görtz, U., Wedhorn, T.: Algebraic geometry I. Schemes. With examples and exercises. Springer Studium Mathematik—Master. Springer Spektrum, Wiesbaden (2020)
76. Grenier-Boley, N., Hoffmann, D.W.: Isomorphism criteria for Witt rings of real fields. Forum Math. **25**(1), 1–18 (2013). With an appendix by C. Scheiderer.
77. Grünbaum, B.: Convex polytopes, *Graduate Texts in Mathematics*, vol. 221, second edn. Springer-Verlag, New York (2003). Prepared and with a preface by V. Kaibel, V. Klee and G.M. Ziegler.
78. Habicht, W.: Über die Zerlegung strikte definiter Formen in Quadrate. Comment. Math. Helv. **12**, 317–322 (1940)
79. Handelman, D.E.: Positive polynomials, convex integral polytopes, and a random walk problem, *Lecture Notes in Mathematics*, vol. 1282. Springer-Verlag, Berlin (1987)
80. Handelman, D.E.: Representing polynomials by positive linear functions on compact convex polyhedra. Pacific J. Math. **132**(1), 35–62 (1988)
81. Hanselka, C.: Characteristic polynomials of symmetric matrices over the univariate polynomial ring. J. Algebra **487**, 340–356 (2017)
82. Hanselka, C., Sinn, R.: Positive semidefinite univariate matrix polynomials. Math. Z. **292**(1-2), 83–101 (2019)
83. Harris, J.: Algebraic geometry. A first course, *Graduate Texts in Mathematics*, vol. 133. Springer-Verlag, New York (1995)
84. Hartshorne, R.: Algebraic geometry. Graduate Texts in Mathematics, No. 52. Springer-Verlag, New York-Heidelberg (1977)
85. Haviland, E.K.: On the momentum problem for distribution functions in more than one dimension. Amer. J. Math. **57**(3), 562–568 (1935)
86. Haviland, E.K.: On the momentum problem for distribution functions in more than one dimension. II. Amer. J. Math. **58**(1), 164–168 (1936)
87. Helton, J.W., Nie, J.: Sufficient and necessary conditions for semidefinite representability of convex hulls and sets. SIAM J. Optim. **20**(2), 759–791 (2009)
88. Helton, J.W., Nie, J.: Semidefinite representation of convex sets. Math. Program. **122**(1, Ser. A), 21–64 (2010)
89. Helton, J.W., Vinnikov, V.: Linear matrix inequality representation of sets. Comm. Pure Appl. Math. **60**(5), 654–674 (2007)
90. Henrion, D., Korda, M., Lasserre, J.B.: The moment-SOS hierarchy. Lectures in probability, statistics, computational geometry, control and nonlinear PDEs, *Series on Optimization and its Applications*, vol. 4. World Scientific Publishing Co. Pte. Ltd., Hackensack, NJ (2021)
91. Hilbert, D.: über die Darstellung definiter Formen als Summe von Formenquadraten. Math. Ann. **32**(3), 342–350 (1888)
92. Hilbert, D.: Über ternäre definite Formen. Acta Math. **17**(1), 169–197 (1893)

93. Hilbert, D.: Mathematische Probleme. Vortrag, gehalten auf dem internationalen Mathematiker-Congress zu Paris 1900. Nachr. Ges. Wiss. Göttingen, Math.-Phys. Kl. **1900**, 253–297 (1900)
94. Hilbert, D.: Mathematical problems. Bull. Amer. Math. Soc. **8**(10), 437–479 (1902)
95. Hilbert, D.: Gesammelte Abhandlungen. Band III: Analysis, Grundlagen der Mathematik, Physik, Verschiedenes, Lebensgeschichte. Springer-Verlag, Berlin-New York (1970)
96. Hochster, M.: Prime ideal structure in commutative rings. Trans. Amer. Math. Soc. **142**, 43–60 (1969)
97. Horn, A.: Doubly stochastic matrices and the diagonal of a rotation matrix. Amer. J. Math. **76**, 620–630 (1954)
98. Iliman, S., de Wolff, T.: Amoebas, nonnegative polynomials and sums of squares supported on circuits. Res. Math. Sci. **2016**, 3:9 (35 pp)
99. Iliman, S., de Wolff, T.: Lower bounds for polynomials with simplex Newton polytopes based on geometric programming. SIAM J. Optim. **26**(2), 1128–1146 (2016)
100. Jacobi, T.: A representation theorem for certain partially ordered commutative rings. Math. Z. **237**(2), 259–273 (2001)
101. Jacobi, T., Prestel, A.: Distinguished representations of strictly positive polynomials. J. Reine Angew. Math. **532**, 223–235 (2001)
102. Jakubović, V.A.: Factorization of symmetric matrix polynomials. Dokl. Akad. Nauk SSSR **194**, 532–535 (1970). Russian
103. Jakubović, V.A.: The S-procedure in nonlinear control theory. Vestnik Leningrad. Univ. (1), 62–77 (1971). Russian
104. Kadison, R.V.: A representation theory for commutative topological algebra. Mem. Amer. Math. Soc. **7**, 39 pp (1951)
105. Kakutani, S.: Ein Beweis des Satzes von M. Eidelheit über konvexe Mengen. Proc. Imp. Acad. Tokyo **13**(4), 93–94 (1937)
106. Khovanskii, A.G.: Fewnomials, *Translations of Mathematical Monographs*, vol. 88. American Mathematical Society, Providence, RI (1991). Translated from the Russian by S. Zdravkovska
107. Klep, I., Scheiderer, C., Volčič, J.: Globally trace-positive noncommutative polynomials and the unbounded tracial moment problem. Math. Ann. **387**(3-4), 1403–1433 (2023)
108. de Klerk, E.: Aspects of semidefinite programming. Interior point algorithms and selected applications, *Applied Optimization*, vol. 65. Kluwer Academic Publishers, Dordrecht (2002)
109. Knebusch, M., Scheiderer, C.: Real algebra. A first course. Universitext. Springer, Cham (2022). Translated and with contributions by Thomas Unger
110. Knutson, A., Tao, T.: The honeycomb model of $\mathrm{GL}_n(\mathbf{C})$ tensor products. I. Proof of the saturation conjecture. J. Amer. Math. Soc. **12**(4), 1055–1090 (1999)
111. Köthe, G.: Topological vector spaces. I, *Grundlehren der mathematischen Wissenschaften in Einzeldarstellungen*, vol. 159. Springer-Verlag New York, Inc., New York (1969). Translated from the German by D. J. H. Garling
112. Kriel, T.L., Schweighofer, M.: On the exactness of Lasserre relaxations for compact convex basic closed semialgebraic sets. SIAM J. Optim. **28**(2), 1796–1816 (2018)
113. Krivine, J.L.: Anneaux préordonnés. J. Analyse Math. **12**, 307–326 (1964)
114. Krull, W.: Allgemeine Bewertungstheorie. J. Reine Angew. Math. **167**, 160–196 (1932)
115. Kuhlmann, S., Marshall, M.: Positivity, sums of squares and the multi-dimensional moment problem. Trans. Amer. Math. Soc. **354**(11), 4285–4301 (2002)
116. Kuhlmann, S., Marshall, M., Schwartz, N.: Positivity, sums of squares and the multi-dimensional moment problem. II. Adv. Geom. **5**(4), 583–606 (2005)
117. Kunz, E.: Introduction to commutative algebra and algebraic geometry. Birkhäuser Boston, Inc., Boston, MA (1985). Translated from the German by Michael Ackerman. With a preface by David Mumford
118. Lam, T.Y.: Orderings, valuations and quadratic forms, *CBMS Regional Conference Series in Mathematics*, vol. 52. Published for the Conference Board of the Mathematical Sciences, Washington, DC; by the American Mathematical Society, Providence, RI (1983)

119. Lam, T.Y.: Introduction to quadratic forms over fields, *Graduate Studies in Mathematics*, vol. 67. American Mathematical Society, Providence, RI (2005)
120. Lang, S.: The theory of real places. Ann. of Math. (2) **57**, 378–391 (1953)
121. Lasserre, J.B.: Global optimization with polynomials and the problem of moments. SIAM J. Optim. **11**(3), 796–817 (2000/01)
122. Lasserre, J.B.: Moments, positive polynomials and their applications, *Imperial College Press Optimization Series*, vol. 1. Imperial College Press, London (2010)
123. Lasserre, J.B.: An introduction to polynomial and semi-algebraic optimization. Cambridge Texts in Applied Mathematics. Cambridge University Press, Cambridge (2015)
124. Laurent, M.: Sums of squares, moment matrices and optimization over polynomials. In: Emerging applications of algebraic geometry, *IMA Vol. Math. Appl.*, vol. 149, pp. 157–270. Springer, New York (2009)
125. Lax, A., Lax, P.D.: On sums of squares. Linear Algebra Appl. **20**(1), 71–75 (1978)
126. Lax, P.D.: Differential equations, difference equations and matrix theory. Comm. Pure Appl. Math. **11**, 175–194 (1958)
127. Liu, Q.: Algebraic geometry and arithmetic curves, *Oxford Graduate Texts in Mathematics*, vol. 6. Oxford University Press, Oxford (2002). Translated from the French by R. Erné
128. Lombardi, H., Perrucci, D., Roy, M.F.: An elementary recursive bound for effective Positivstellensatz and Hilbert's 17th problem. Mem. Amer. Math. Soc. **263**(1277), v+125 pp (2020)
129. Lombardi, H., Roy, M.F.: Elementary constructive theory of ordered fields. In: Effective methods in algebraic geometry (Castiglioncello, 1990), *Progr. Math.*, vol. 94, pp. 249–262. Birkhäuser Boston, Boston, MA (1991)
130. L'vovsky, S.: On inflection points, monomial curves, and hypersurfaces containing projective curves. Math. Ann. **306**(4), 719–735 (1996)
131. Mahé, L.: Une démonstration élémentaire du théorème de Bröcker-Scheiderer. C. R. Acad. Sci. Paris Sér. I Math. **309**(9), 613–616 (1989)
132. Mangolte, F.: Real algebraic varieties. Springer Monographs in Mathematics. Springer, Cham (2020). Translated from the 2017 French original by C. Maclean
133. Marker, D.: Model theory. An introduction, *Graduate Texts in Mathematics*, vol. 217. Springer-Verlag, New York (2002)
134. Marshall, M.: Minimal generation of basic sets in the real spectrum of a commutative ring. In: Recent advances in real algebraic geometry and quadratic forms (Berkeley, CA, 1990/1991; San Francisco, CA, 1991), *Contemp. Math.*, vol. 155, pp. 207–219. Amer. Math. Soc., Providence, RI (1994)
135. Marshall, M.: Spaces of orderings and abstract real spectra, *Lecture Notes in Mathematics*, vol. 1636. Springer-Verlag, Berlin (1996)
136. Marshall, M.: A real holomorphy ring without the Schmüdgen property. Canad. Math. Bull. **42**(3), 354–358 (1999)
137. Marshall, M.: Representations of non-negative polynomials having finitely many zeros. Ann. Fac. Sci. Toulouse Math. (6) **15**(3), 599–609 (2006)
138. Marshall, M.: Positive polynomials and sums of squares, *Mathematical Surveys and Monographs*, vol. 146. American Mathematical Society, Providence, RI (2008)
139. Marshall, M.: Polynomials non-negative on a strip. Proc. Amer. Math. Soc. **138**(5), 1559–1567 (2010)
140. Matsumura, H.: Commutative ring theory, *Cambridge Studies in Advanced Mathematics*, vol. 8, second edn. Cambridge University Press, Cambridge (1989). Translated from the Japanese by M. Reid
141. Motzkin, T.S.: The arithmetic-geometric inequality. In: Inequalities (Proc. Sympos. Wright-Patterson Air Force Base, Ohio, 1965), pp. 205–224. Academic Press, New York (1967)
142. Nemirovski, A.: Advances in convex optimization: conic programming. In: International Congress of Mathematicians. Vol. I, pp. 413–444. Eur. Math. Soc., Zürich (2007)
143. Netzer, T.: On semidefinite representations of non-closed sets. Linear Algebra Appl. **432**(12), 3072–3078 (2010)

144. Netzer, T., Plaumann, D.: Geometry of linear matrix inequalities. A course in convexity and real algebraic geometry with a view towards optimization. Compact Textbooks in Mathematics. Birkhäuser/Springer, Cham (2023)
145. Netzer, T., Plaumann, D., Schweighofer, M.: Exposed faces of semidefinitely representable sets. SIAM J. Optim. **20**(4), 1944–1955 (2010)
146. Netzer, T., Sinn, R.: A note on the convex hull of finitely many projections of spectrahedra (2009). URL `https://arxiv.org/pdf/0908.3386`
147. Nie, J., Schweighofer, M.: On the complexity of Putinar's Positivstellensatz. J. Complexity **23**(1), 135–150 (2007)
148. Parrilo, P.A.: Structured semidefinite programs and semialgebraic geometry methods in robustness and optimization. Ph.D. thesis, California Institute of Technology (2000)
149. Pfister, A.: Zur Darstellung definiter Funktionen als Summe von Quadraten. Invent. Math. **4**, 229–237 (1967)
150. Pfister, A.: On Hilbert's theorem about ternary quartics. In: Algebraic and arithmetic theory of quadratic forms, *Contemp. Math.*, vol. 344, pp. 295–301. Amer. Math. Soc., Providence, RI (2004)
151. Pfister, A., Scheiderer, C.: An elementary proof of Hilbert's theorem on ternary quartics. J. Algebra **371**, 1–25 (2012)
152. Pincus, D.: Zermelo-Fraenkel consistency results by Fraenkel-Mostowski methods. J. Symbolic Logic **37**, 721–743 (1972)
153. Plaumann, D.: Sums of squares on reducible real curves. Math. Z. **265**(4), 777–797 (2010)
154. Pólik, I., Terlaky, T.: A survey of the S-lemma. SIAM Rev. **49**(3), 371–418 (2007)
155. Pólya, G.: Über positive Darstellung von Polynomen. Vierteljahrsschrift Zürich 73, 141-145 (1928)
156. Powers, V., Reznick, B.: A new bound for Pólya's theorem with applications to polynomials positive on polyhedra. J. Pure Appl. Algebra **164**(1-2), 221–229 (2001). Effective methods in algebraic geometry (Bath, 2000)
157. Powers, V., Reznick, B., Scheiderer, C., Sottile, F.: A new approach to Hilbert's theorem on ternary quartics. C. R. Math. Acad. Sci. Paris **339**(9), 617–620 (2004)
158. Prestel, A.: Quadratische Semi-Ordnungen und quadratische Formen. Math. Z. **133**, 319–342 (1973)
159. Prestel, A., Delzell, C.N.: Positive polynomials. From Hilbert's 17th problem to real algebra. Springer Monographs in Mathematics. Springer-Verlag, Berlin (2001)
160. Prestel, A., Delzell, C.N.: Mathematical logic and model theory. A brief introduction. Universitext. Springer, London (2011). Expanded translation of the 1986 German original
161. Putinar, M.: Positive polynomials on compact semi-algebraic sets. Indiana Univ. Math. J. **42**(3), 969–984 (1993)
162. Quillen, D.G.: On the representation of hermitian forms as sums of squares. Invent. Math. **5**, 237–242 (1968)
163. Ramana, M., Goldman, A.J.: Some geometric results in semidefinite programming. J. Global Optim. **7**(1), 33–50 (1995)
164. Reznick, B.: Extremal PSD forms with few terms. Duke Math. J. **45**(2), 363–374 (1978)
165. Reznick, B.: Uniform denominators in Hilbert's seventeenth problem. Math. Z. **220**(1), 75–97 (1995)
166. Reznick, B.: Some concrete aspects of Hilbert's 17th Problem. In: Real algebraic geometry and ordered structures (Baton Rouge, LA, 1996), *Contemp. Math.*, vol. 253, pp. 251–272. Amer. Math. Soc., Providence, RI (2000)
167. Reznick, B.: On Hilbert's construction of positive polynomials (2007). URL `https://arxiv.org/pdf/0707.2156.pdf`
168. Riesz, F.: Über ein Problem des Herrn Carathéodory. J. Reine Angew. Math. **146**, 83–87 (1916)
169. Risler, J.J.: Une caractérisation des idéaux des variétés algébriques réelles. C. R. Acad. Sci. Paris Sér. A-B **271**, A1171–A1173 (1970)

170. Robinson, R.M.: Some definite polynomials which are not sums of squares of real polynomials. In: Selected questions of algebra and logic (a collection dedicated to the memory of A. I. Mal'cev) (Russian), pp. 264–282. Izdat. "Nauka" Sibirsk. Otdel., Novosibirsk (1973)
171. Rosenblum, M., Rovnyak, J.: The factorization problem for nonnegative operator valued functions. Bull. Amer. Math. Soc. **77**, 287–318 (1971)
172. Rudin, W.: Principles of mathematical analysis, third edn. McGraw-Hill Book Co., New York (1976)
173. Rudin, W.: Functional analysis, second edn. International Series in Pure and Applied Mathematics. McGraw-Hill, Inc., New York (1991)
174. Rudin, W.: Sums of squares of polynomials. Amer. Math. Monthly **107**(9), 813–821 (2000)
175. Sander, T.: Existence and uniqueness of the real closure of an ordered field without Zorn's lemma. J. Pure Appl. Algebra **73**(2), 165–180 (1991)
176. Scharlau, W.: Quadratic and Hermitian forms, *Grundlehren der mathematischen Wissenschaften*, vol. 270. Springer-Verlag, Berlin (1985)
177. Scheiderer, C.: Stability index of real varieties. Invent. Math. **97**(3), 467–483 (1989)
178. Scheiderer, C.: Sums of squares of regular functions on real algebraic varieties. Trans. Amer. Math. Soc. **352**(3), 1039–1069 (2000)
179. Scheiderer, C.: On sums of squares in local rings. J. Reine Angew. Math. **540**, 205–227 (2001)
180. Scheiderer, C.: Sums of squares on real algebraic curves. Math. Z. **245**(4), 725–760 (2003)
181. Scheiderer, C.: Distinguished representations of non-negative polynomials. J. Algebra **289**(2), 558–573 (2005)
182. Scheiderer, C.: Non-existence of degree bounds for weighted sums of squares representations. J. Complexity **21**(6), 823–844 (2005)
183. Scheiderer, C.: Sums of squares on real algebraic surfaces. Manuscripta Math. **119**(4), 395–410 (2006)
184. Scheiderer, C.: A Positivstellensatz for projective real varieties. Manuscripta Math. **138**(1-2), 73–88 (2012)
185. Scheiderer, C.: Sum of squares length of real forms. Math. Z. **286**(1-2), 559–570 (2017)
186. Scheiderer, C.: Semidefinite representation for convex hulls of real algebraic curves. SIAM J. Appl. Algebra Geom. **2**(1), 1–25 (2018)
187. Scheiderer, C.: Spectrahedral shadows. SIAM J. Appl. Algebra Geom. **2**(1), 26–44 (2018)
188. Scheiderer, C.: Second-order cone representation for convex sets in the plane. SIAM J. Appl. Algebra Geom. **5**(1), 114–139 (2021)
189. Scheiderer, C., Wenzel, S.: Polynomials nonnegative on the cylinder. In: Ordered algebraic structures and related topics, *Contemp. Math.*, vol. 697, pp. 291–300. Amer. Math. Soc., Providence, RI (2017)
190. Schmüdgen, K.: The K-moment problem for compact semi-algebraic sets. Math. Ann. **289**(2), 203–206 (1991)
191. Schmüdgen, K.: The moment problem, *Graduate Texts in Mathematics*, vol. 277. Springer, Cham (2017)
192. Schmüdgen, K., Schötz, M.: Positivstellensätze for semirings. Math. Ann. **389**(1), 947–985 (2024)
193. Schweighofer, M.: An algorithmic approach to Schmüdgen's Positivstellensatz. J. Pure Appl. Algebra **166**(3), 307–319 (2002)
194. Schweighofer, M.: On the complexity of Schmüdgen's positivstellensatz. J. Complexity **20**(4), 529–543 (2004)
195. Segal, I.E.: Irreducible representations of operator algebras. Bull. Amer. Math. Soc. **53**, 73–88 (1947)
196. Seidenberg, A.: A new decision method for elementary algebra. Ann. of Math. (2) **60**, 365–374 (1954)
197. Shafarevich, I.R.: Basic algebraic geometry 1. Varieties in projective space, third edn. Springer, Heidelberg (2013). Translated from the Russian by M. Reid
198. Shafarevich, I.R.: Basic algebraic geometry 2. Schemes and complex manifolds, third edn. Springer, Heidelberg (2013). Translated from the Russian by M. Reid

199. Slot, L.: Sum-of-squares hierarchies for polynomial optimization and the Christoffel-Darboux kernel. SIAM J. Optim. **32**(4), 2612–2635 (2022)
200. Sottile, F.: Real solutions to equations from geometry, *University Lecture Series*, vol. 57. American Mathematical Society, Providence, RI (2011)
201. Springer, T.A.: Sur les formes quadratiques d'indice zéro. C. R. Acad. Sci. Paris **234**, 1517–1519 (1952)
202. Stengle, G.: A nullstellensatz and a positivstellensatz in semialgebraic geometry. Math. Ann. **207**, 87–97 (1974)
203. Stengle, G.: Complexity estimates for the Schmüdgen Positivstellensatz. J. Complexity **12**(2), 167–174 (1996)
204. Stone, M.H.: A general theory of spectra. I. Proc. Nat. Acad. Sci. U.S.A. **26**, 280–283 (1940)
205. Sturmfels, B.: Gröbner bases and convex polytopes, *University Lecture Series*, vol. 8. American Mathematical Society, Providence, RI (1996)
206. Swan, R.G.: Hilbert's theorem on positive ternary quartics. In: Quadratic forms and their applications (Dublin, 1999), *Contemp. Math.*, vol. 272, pp. 287–292. Amer. Math. Soc., Providence, RI (2000)
207. Tarski, A.: A decision method for elementary algebra and geometry. RAND Corporation, Santa Monica, Calif. (1948)
208. Theobald, T.: Real algebraic geometry and optimization, *Graduate Studies in Mathematics*, vol. 241. American Mathematical Society, Providence, RI (2024)
209. Tyaglov, M.: On the number of real critical points of logarithmic derivatives and the Hawaii conjecture. J. Anal. Math. **114**, 1–62 (2011)
210. Vakil, R.: The rising sea. Foundations of algebraic geometry. URL `https://math.stanford.edu/~vakil/216blog/`
211. Vinnikov, V.: LMI representations of convex semialgebraic sets and determinantal representations of algebraic hypersurfaces: past, present, and future. In: Mathematical methods in systems, optimization, and control, *Oper. Theory Adv. Appl.*, vol. 222, pp. 325–349. Birkhäuser/Springer Basel AG, Basel (2012)
212. Webster, R.: Convexity. Oxford Science Publications. The Clarendon Press, Oxford University Press, New York (1994)
213. Wolkowicz, H., Saigal, R., Vandenberghe, L. (eds.): Handbook of semidefinite programming. Theory, algorithms, and applications, *International Series in Operations Research & Management Science*, vol. 27. Kluwer Academic Publishers, Boston, MA (2000)
214. Ziegler, G.M.: Lectures on polytopes, *Graduate Texts in Mathematics*, vol. 152. Springer-Verlag, New York (1995)

Name Index

C. Scheiderer, *A Course in Real Algebraic Geometry*, Graduate Texts in Mathematics 303,
https://doi.org/10.1007/978-3-031-69213-0

T

V

W

Notation Index

For general notational conventions see the end of the introduction. The following list of more specific symbols and notation is sorted by first occurrence.

C. Scheiderer, *A Course in Real Algebraic Geometry*, Graduate Texts in Mathematics 303,
https://doi.org/10.1007/978-3-031-69213-0

Subject Index

C. Scheiderer, *A Course in Real Algebraic Geometry*, Graduate Texts in Mathematics 303,
https://doi.org/10.1007/978-3-031-69213-0

D

E

F

G

H

I

K

L

M

N

T

GPSR Compliance

The European Union's (EU) General Product Safety Regulation (GPSR) is a set of rules that requires consumer products to be safe and our obligations to ensure this.

If you have any concerns about our products, you can contact us on ProductSafety@springernature.com

In case Publisher is established outside the EU, the EU authorized representative is:

Springer Nature Customer Service Center GmbH
Europaplatz 3
69115 Heidelberg, Germany

Batch number: 10370734

Printed by Printforce, the Netherlands